Open the door to the fascinating world of

physics

Physics, the most fundamental of all natural sciences, will reveal to you the basic principles of the Universe. And while physics can seem challenging, its true beauty lies in the sheer simplicity of fundamental physical theories—theories and concepts that can alter and expand your view of the world around you. Other courses that follow will use the same principles, so it is important that you understand and are able to apply the various concepts and theories discussed in the text. **Physics for Scientists and Engineers, Sixth Edition** is your guide to this fascinating science.

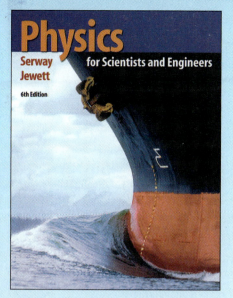

Physics for Scientists and Engineers
Serway
Jewett
6th Edition

Your quick start for studying smart

Achieve success in your physics course by making the most of what **Physics for Scientists and Engineers, Sixth Edition** has to offer you. From a host of in-text features to a range of Web resources, you'll have everything you need to understand the natural forces and principles of physics:

▶ **Dynamic built-in study aids.** Throughout every chapter the authors have built in a wide range of examples, exercises, and illustrations that will help you understand and appreciate the laws of physics. *See pages 2 and 3 for more information.*

▶ **A powerful Web-based learning system**. The text is fully integrated with **PhysicsNow**, an interactive learning system that tailors itself to your needs in the course. It's like having a personal tutor available whenever you need it! *See pages 4–7 to explore* **PhysicsNow**.

Your *Quick Start for Studying Smart* begins with this special tour through the book. On the following pages you'll discover how **Physics for Scientists and Engineers, Sixth Edition** and PhysicsNow not only enhance your experience in this course, but help you to succeed!

Quick Start for Studying Smart!

2

GENERAL PROBLEM-SOLVING STRATEGY

Conceptualize

- The first thing to do when approaching a problem is to *think about* and *understand* the situation. Study carefully any diagrams, graphs, tables, or photographs that accompany the problem. Imagine a movie, running in your mind, of what happens in the problem.
- If a diagram is not provided, you should almost always make a quick drawing of the situation. Indicate any known values, perhaps in a table or directly on your sketch.
- Now focus on what algebraic or numerical information is given in the problem. Carefully read the problem statement, looking for key phrases such as "starts from at rest" ($v_i = 0$), "stops" ($v_f = 0$), or "freely falls" ($a_y = -g = -9.80$ m/s^2).
- Now focus on the expected result of solving the problem. Exactly what is the question asking? Will the final result be numerical or algebraic? Do you know what units to expect?
- Don't forget to incorporate information from your own experiences and common sense. What should a reasonable answer look like? You wouldn't expect to calculate the speed of an automobile to be

Analyze

- Now you must analyze the problem and strive for a mathematical solution. Because you have already categorized the problem, it should not be too difficult to select relevant equations that apply to the type of situation in the problem. For example, if the problem involves a particle moving under constant acceleration, Equations 2.9 to 2.13 are relevant.
- Use algebra (and calculus, if necessary) to solve symbolically for the unknown variable in terms of what is given. Substitute in the appropriate numbers, calculate the result, and round it to the proper number of significant figures.

Finalize

- This is the most important part. Examine your numerical answer. Does it have the correct units? Does it meet your expectations from your conceptualization of the problem? What about the algebraic form of the result — before you substituted numerical values? Does it make sense? Examine the variables in the problem to see whether the answer would change in a physically meaningful way if they were drastically increased or decreased or even became zero. Looking at limiting cases to see whether they yield expected values is a very useful way to make sure that you are obtaining reasonable results.

  Think about how this problem compares with others you have done. How was it similar? In what critical ways did it differ? Why was this problem assigned? You should have learned something by doing it. Can you figure out what? If it is a new category of problem, be sure you understand it so that you can use it as a model for solving future problems in the same category.

  When solving complex problems, you may need to identify a series of sub-problems and apply the problem-solv-

PROBLEM-SOLVING HINTS

Applying Newton's Laws

The following procedure is recommended when dealing with problems involving Newton's laws:

- Draw a simple, neat diagram of the system to help *conceptualize* the problem.
- *Categorize* the problem: if any acceleration component is zero, the particle is in equilibrium in this direction and $\Sigma F = 0$. If not, the particle is undergoing an acceleration, the problem is one of nonequilibrium in this direction, and $\Sigma F = ma$.
- *Analyze* the problem by isolating the object whose motion is being analyzed. Draw a free-body diagram for this object. For systems containing more than one object, draw *separate* free-body diagrams for each object. *Do not* include in the free-body diagram forces exerted by the object on its surroundings.
- Establish convenient coordinate axes for each object and find the components of the forces along these axes. Apply Newton's second law, $\Sigma F = ma$, in component form. Check your dimensions to make sure that all terms have units of force.
- Solve the comp... have as many i... complete soluti...
- *Finalize* by maki... Also check the ... variables. By do...

Example 4.3   The Long Jump

A long-jumper (Fig. 4.12) leaves the ground at an angle of 20.0° above the horizontal and at a speed of 11.0 m/s.

(A) How far does he jump in the horizontal direction? (Assume his motion is equivalent to that of a particle.)

Solution We *conceptualize* the motion of the long-jumper as equivalent to that of a simple projectile such as the ball in Example 4.2, and *categorize* this problem as a projectile motion problem. Because the initial speed and launch angle are given, and because the final height is the same as the initial height, we further categorize this problem as satisfying the conditions for which Equations 4.13 and 4.14 can be used. This is the most direct way to *analyze* this problem... describi... the gen...

Figure 4.12 (Example 4.3) Mike Powell, current holder of the world long jump record of 8.95 m.

This is the time at which the long-jumper is at the *top* of the jump. Because of the symmetry of the vertical motion,

provides a graphical representation of the flight of the long-jumper. As before, we set our origin of coordinates at the takeoff point and label the peak as Ⓐ and the landing point as Ⓑ. The horizontal motion is described by Equation 4.11:

$$x_f = x_B = (v_i \cos \theta_i)\, t_B = (11.0\,\text{m/s})(\cos 20.0°)\, t_B$$

The value of x_B can be found if the time of landing t_B is known. We can find t_B by remembering that $a_y = -g$ and by using the y part of Equation 4.8a. We also note that at the top of the jump the vertical component of velocity v_{yA} is zero:

$$v_{yf} = v_{yA} = v_i \sin \theta_i - g t_A$$

another 0.384 s passes before the jumper returns to the ground. Therefore, the time at which the jumper lands is $t_B = 2t_A = 0.768$ s. Substituting this value into the above expression for x_f gives

$$x_f = x_B = (11.0\,\text{m/s})(\cos 20.0°)(0.768\,\text{s}) = \boxed{7.94\ \text{m}}$$

This is a reasonable distance for a world-class athlete.

(B) What is the maximum height reached?

Solution We find the maximum height reached by using Equation 4.12:

$$y_{max} = y_A = (v_i \sin \theta_i)\, t_A - \tfrac{1}{2} g t_A^2$$
$$= (11.0\,\text{m/s})(\sin 20.0°)(0.384\,\text{s})$$
$$- \tfrac{1}{2}(9.80\,\text{m/s}^2)(0.384\,\text{s})^2 = \boxed{0.722\ \text{m}}$$

To *finalize* this problem, find the answers to parts (a) and (b) using Equations 4.13 and 4.14. The results should agree. Treating the long-jumper as a particle is an oversimplification. Nevertheless, the values obtained are consistent with experience in sports. We learn that we can model a complicated system such as a long-jumper as a particle and still obtain results that are reasonable.

where k is a dimensionless constant of proportionality. Knowing the dimensions of a, r, and v, we see that the dimensional equation must be

$$\frac{L}{T^2} = L^n \left(\frac{L}{T}\right)^m = \frac{L^{n+m}}{T^m}$$

$n = -1$, and we can write the acceleration expression as

$$a = kr^{-1}v^2 = k\frac{v^2}{r}$$

When we discuss uniform circular motion later, we shall see that $k = 1$ if a consistent set of units is used. The constant k would not equal 1 if, for example, v were in km/h and you wanted a in m/s^2.

▲ **PITFALL PREVENTION**

1.5 Always Include Units

When performing calculations, include the units for every quantity and carry the units through the entire calculation. Avoid the temptation to drop the units early and then attach the expected units once you have an answer. By including the units in every step, you can detect errors if the units for the answer turn out to be incorrect.

1.5 Conversion of Units

Sometimes it is necessary to convert units from one measurement system to another, or to convert within a system, for example, from kilometers to meters. Equalities between SI and U.S. customary units of length are as follows:

1 mile = 1 609 m = 1.609 km	1 ft = 0.304 8 m = 30.48 cm
1 m = 39.37 in. = 3.281 ft	1 in. = 0.025 4 m = 2.54 cm (exactly)

A more complete list of conversion factors can be found in Appendix A.

Units can be treated as algebraic quantities that can cancel each other. For example, suppose we wish to convert 15.0 in. to centimeters. Because 1 in. is defined as exactly 2.54 cm, we find that

---

Example 4.5 That's Quite an Arm!

A stone is thrown from the top of a building upward at an angle of 30.0° to the horizontal with an initial speed of 20.0 m/s, as shown in Figure 4.14. If the height of the building is 45.0 m,

(A) how long before the stone hits the ground?

Solution We *conceptualize* the problem by studying Figure 4.14, in which we have indicated the various parameters. By now, it should be natural to *categorize* this as a projectile motion problem.

To *analyze* the problem, let us once again separate motion into two components. The initial x and y components of the stone's velocity are

$$v_{xi} = v_i \cos\theta_i = (20.0 \text{ m/s})\cos 30.0° = 17.3 \text{ m/s}$$

$$v_{yi} = v_i \sin\theta_i = (20.0 \text{ m/s})\sin 30.0° = 10.0 \text{ m/s}$$

To find t, we can use $y_f = y_i + v_{yi}t + \frac{1}{2}a_yt^2$(Eq. 4.9a) with $y_i = 0$, $y_f = -45.0$ m, $a_y = -g$, and $v_{yi} = 10.0$ m/s (there is a negative sign on the numerical value of y_f because we have chosen the top of the building as the origin):

$$-45.0 \text{ m} = (10.0 \text{ m/s})t - \frac{1}{2}(9.80 \text{ m/s}^2)t^2$$

Solving the quadratic equation for t gives, for the positive root, $t = \boxed{4.22 \text{ s}}$. To *finalize* this part, think: Does the negative root have any physical meaning?

(B) What is the speed of the stone just before it strikes the ground?

Solution We can use Equation 4.8a, $v_{yf} = v_{yi} + a_yt$, with $t = 4.22$ s to obtain the y component of the velocity just before the stone strikes the ground:

$$v_{yf} = 10.0 \text{ m/s} - (9.80 \text{ m/s}^2)(4.22 \text{ s}) = -31.4 \text{ m/s}$$

Because $v_{xf} = v_{xi} = 17.3$ m/s, the required speed is

$$v_f = \sqrt{v_{xf}^2 + v_{yf}^2} = \sqrt{(17.3)^2 + (-31.4)^2} \text{ m/s} = \boxed{35.9 \text{ m/s}}$$

What If? What if a horizontal wind is blowing in the same direction as the ball is thrown and it causes the ball to have a horizontal acceleration component $a_x = 0.500$ m/s^2. Which part of this example, (a) or (b), will have a different answer?

Answer Recall that the motions in the x and y directions are independent. Thus, the horizontal wind cannot affect the vertical motion. The vertical motion determines the time of the projectile in the air, so the answer to (a) does not change. The wind will cause the horizontal velocity component to increase with time, so that the final speed will change in part (b).

We can find the new final horizontal velocity component by using Equation 4.8a:

$$v_{xf} = v_{xi} + a_xt = 17.3 \text{ m/s} + (0.500 \text{ m/s}^2)(4.22 \text{ s})$$
$$= 19.4 \text{ m/s}$$

and the new final speed:

$$v_f = \sqrt{v_{xf}^2 + v_{yf}^2} = \sqrt{(19.4)^2 + (-31.4)^2} \text{ m/s} = 36.9 \text{ m/s}$$

Figure 4.14 (Example 4.5) A stone is thrown from the top of a building.

Explore projectile motion of the ball in this example link at http://www.pse6.com.

Quick Quiz 5.2 An object experiences no acceleration. Which of the following *cannot* be true for the object? (a) A single force acts on the object. (b) No forces act on the object. (c) Forces act on the object, but the forces cancel.

Quick Quiz 5.3 An object experiences a net force and exhibits an acceleration in response. Which of the following statements is *always* true? (a) The object moves in the direction of the force. (b) The acceleration is in the same direction as the velocity. (c) The acceleration is in the same direction as the force. (d) The velocity of the object increases.

Quick Quiz 5.4 You push an object, initially at rest, across a frictionless floor with a constant force for a time interval Δt, resulting in a final speed of v for the object. You repeat the experiment, but with a force that is twice as large. What time interval is now required to reach the same final speed v? (a) $4\Delta t$ (b) $2\Delta t$ (c) Δt (d) $\Delta t/2$ (e) $\Delta t/4$.

Answers to Quick Quizzes

5.1 (d). Choice (a) is true. Newton's first law tells us that motion requires no force: an object in motion continues to move at constant velocity in the absence of external forces. Choice (b) is also true. A stationary object can have several forces acting on it, but if the vector sum of all these external forces is zero, there is no net force and the object remains stationary.

5.2 (a). If a single force acts, this force constitutes the net force and there is an acceleration according to Newton's second law.

5.3 (c). Newton's second law relates only the force and the acceleration. Direction of motion is part of an object's *velocity*, and force determines the direction of acceleration, not that of velocity.

5.4 (d). With twice the force, the object will experience twice the acceleration. Because the force is constant, the acceler-

---

PITFALL PREVENTIONS

An easy way to make sure you avoid common mistakes—while studying *and* while taking exams! These helpful notes, found in the margins, allow you to correct misconceptions before applying them to new material.

WHAT IF?

What if you could get even more out of the text's worked examples? You can with this feature. The authors change some of the data or assumptions in about one-third of the *Worked Examples* and then explore the consequences. This allows you to apply concepts and problem-solving skills to new situations as well as test the final result to see if it's realistic.

QUICK QUIZZES

Test your understanding of the concepts before investigating the worked examples. Found throughout each chapter, these quizzes give you an opportunity to test your conceptual understanding while also making the material more interactive. Comprehensive answers are found at the end of each chapter. A supplementary set of *Quick Quizzes* is available on the text's Web site, accessible through **www.pse6.com**.

Quick Start for Studying Smart!

"You do not know anything until you have practiced."

R. P. Feynman, Nobel Laureate in Physics

Take a practice test for this chapter by clicking on the Practice Test link at http://www.pse6.com.

SUMMARY

Scalar quantities are those that have only magnitude and no associated direction. **Vector quantities** have both magnitude and direction and obey the laws of vector addition. The magnitude of a vector is *always* a positive number.

When two or more vectors are added together, all of them must have the same units and all of them must be the same type of quantity. We can add two vectors $\mathbf{A}$ and $\mathbf{B}$ graphically. In this method (Fig. 3.6), the resultant vector $\mathbf{R} = \mathbf{A} + \mathbf{B}$ runs from the tail of $\mathbf{A}$ to the tip of $\mathbf{B}$.

A second method of adding vectors involves **components** of the vectors. The x component A_x of the vector $\mathbf{A}$ is equal to the projection of $\mathbf{A}$ along the x axis of a coordinate system, as shown in Figure 3.13, where $A_x = A\cos\theta$. The y component A_y of $\mathbf{A}$ is the projection of $\mathbf{A}$ along the y axis, where $A_y = A\sin\theta$. Be sure you can determine which trigonometric functions you should use in all situations, especially when θ is defined as something other than the counterclockwise angle from the positive x axis.

If a vector $\mathbf{A}$ has an x component A_x and a y component A_y, the vector can be expressed in unit–vector form as $\mathbf{A} = A_x\mathbf{i} + A_y\mathbf{j}$. In this notation, $\mathbf{i}$ is a unit vector pointing in the positive x direction, and $\mathbf{j}$ is a unit vector pointing in the positive y direction. Because $\mathbf{i}$ and $\mathbf{j}$ are unit vectors, $|\mathbf{i}| = |\mathbf{j}| = 1$.

We can find the resultant of two or more vectors by resolving all vectors into their x and y components, adding their resultant x and y components, and then using the Pythagorean theorem to find the magnitude of the resultant vector. We can find the angle that the resultant vector makes with respect to the x axis by using a suitable trigonometric function.

QUESTIONS

1. Two vectors have unequal magnitudes. Can their sum be zero? Explain.

2. Can the magnitude of a particle's displacement be greater

4. Which of the following are vectors and which are not: force, temperature, the volume of water in a can, the ratings of a TV show, the height of a building, the velocity of the Universe?

plane. For what orientations of $\mathbf{A}$ ents be negative? For what orientas have opposite signs?

Example 4.5   That's Quite an Arm!

A stone is thrown from the top of a building upward at an angle of 30.0° to the horizontal with an initial speed of 20.0 m/s, as shown in Figure 4.14. If the height of the building is 45.0 m,

(A) how long before the stone hits the ground?

Solution We *conceptualize* the problem by studying Figure 4.14, in which we have indicated the various parameters. By now, it should be natural to *categorize* this as a projectile motion problem.

To *analyze* the problem, let us once again separate motion into two components. The initial x and y components of the stone's velocity are

$$v_{xi} = v_i\cos\theta_i = (20.0 \text{ m/s})\cos 30.0° = 17.3 \text{ m/s}$$

$$v_{yi} = v_i\sin\theta_i = (20.0 \text{ m/s})\sin 30.0° = 10.0 \text{ m/s}$$

To find t, we can use $y_f = y_i + v_{yi}t + \frac{1}{2}a_yt^2$ (Eq. 4.9a) with $y_i = 0$, $y_f = -45.0$ m, $a_y = -g$, and $v_{yi} = 10.0$ m/s (there is a negative sign on the numerical value of y_f because we have chosen the top of the building as the origin):

$$-45.0 \text{ m} = (10.0 \text{ m/s})t - \tfrac{1}{2}(9.80 \text{ m/s}^2)t^2$$

Solving the quadratic equation for t gives, for the positive root, $t = $ 4.22 s . To *finalize* this part, think: Does the negative root have any physical meaning?

(B) What is the speed of the stone just before it strikes the ground?

Solution We can use Equation 4.8a, $v_{yf} = v_{yi} + a_yt$, with $t = 4.22$ s to obtain the y component of the velocity just before the stone strikes the ground:

$$v_{yf} = 10.0 \text{ m/s} - (9.80 \text{ m/s}^2)(4.22 \text{ s}) = -31.4 \text{ m/s}$$

Because $v_{xf} = v_{xi} = 17.3$ m/s, the required speed is

$$v_f = \sqrt{v_{xf}^2 + v_{yf}^2} = \sqrt{(17.3)^2 + (-31.4)^2} \text{ m/s} = \boxed{35.9 \text{ m/s}}$$

To *finalize* this part, is it reasonable that the y component of the final velocity is negative? Is it reasonable that the final speed is larger than the initial speed of 20.0 m/s?

What If? What if a horizontal wind is blowing in the same direction as the ball is thrown and it causes the ball to have a horizontal acceleration component $a_x = 0.500$ m/s². Which part of this example, (a) or (b), will have a different answer?

Answer Recall that the motions in the x and y directions are independent. Thus, the horizontal wind cannot affect the vertical motion. The vertical motion determines the time of the projectile in the air, so the answer to (a) does not change. The wind will cause the horizontal velocity component to increase with time, so that the final speed will change in part (b).

We can find the new final horizontal velocity component by using Equation 4.8a:

$$v_{xf} = v_{xi} + a_xt = 17.3 \text{ m/s} + (0.500 \text{ m/s}^2)(4.22 \text{ s})$$
$$= 19.4 \text{ m/s}$$

and the new final speed:

$$v_f = \sqrt{v_{xf}^2 + v_{yf}^2} = \sqrt{(19.4)^2 + (-31.4)^2} \text{ m/s} = 36.9 \text{ m/s}$$

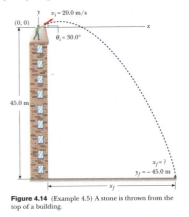

Figure 4.14 (Example 4.5) A stone is thrown from the top of a building.

Investigate this situation at the Interactive Worked Example link at http://www.pse6.com.

GO ONLINE AT www.pse6.com

Log on to **PhysicsNow** at **www.pse6.com** by using the free pincode packaged with this text.* You'll immediately notice the system's easy-to-use, browser-based format. Getting to where you need to go is as easy as a click of the mouse. The **PhysicsNow** system is made up of three interrelated parts:

▶ **How Much Do I Know?**

▶ **What Do I Need to Learn?**

▶ **What Have I Learned?**

These three interrelated elements work together, but are distinct enough to allow you the freedom to explore only those assets that meet your personal needs. You can use **PhysicsNow** like a traditional Web site, accessing all assets of a particular chapter and exploring on your own. The best way to maximize the system and *your* time is to start by taking the *Pre-Test*.

* Free PIN codes are only available with new copies of
Physics for Scientists and Engineers, Sixth Edition.

HOW MUCH DO I KNOW?

The Pre-Test is the first step in creating your *Personalized Learning Plan.* Each *Pre-Test* is based on the end-of-chapter homework problems and includes approximately 15 questions.

Once you've completed the *Pre-Test* you'll be presented with a detailed *Learning Plan* that outlines the elements you need to review to master the chapter's most essential concepts.

At each stage, the text is referenced to reinforce its value as a learning tool.

Turn the page to view problems from a sample *Personalized Learning Plan.*

WHAT DO I NEED TO LEARN?

Once you've completed the *Pre-Test* you're ready to work the problems in your *Personalized Learning Plan*—problems that will help you master concepts essential to your success in this course.

More than 200 *Active Figures* are taken from the text and animated to help you visualize physics in action. Each figure is paired with a question to help you focus on physics at work, and a brief quiz ensures that you understand the concept played out in the animations.

You'll continue to master the concepts though *Coached Problems*. These engaging problems reinforce the lessons in the text by taking a step-by-step approach to problem-solving methodology. Each *Coached Problem* gives you the option of working a question and receiving feedback, or seeing a solution worked for you. You'll find approximately five *Coached Problems* per chapter.

You'll strengthen your problem-solving and visualization skills by working through the *Interactive Examples*. Each step in the examples uses the authors' problem-solving methodology that is introduced in the text (see page 2 of this Visual Preface). You'll find *Interactive Examples* for each chapter of the text.

Quick Start for Studying Smart!

WHAT HAVE I LEARNED?

After working through the problems highlighted in your personal *Learning Plan* you'll move on to a *Chapter Quiz*. These multiple-choice quizzes present you with questions that are similar to those you might find in an exam. You can even e-mail your quiz results to your instructor.

Once you've completed the quiz you'll receive your results in the form of a percentage. If you need to improve your score, **PhysicsNow** will take you back through the system, beginning with *What Do I Know?*, and work with you as you continue to build your knowledge and skills and master concepts.

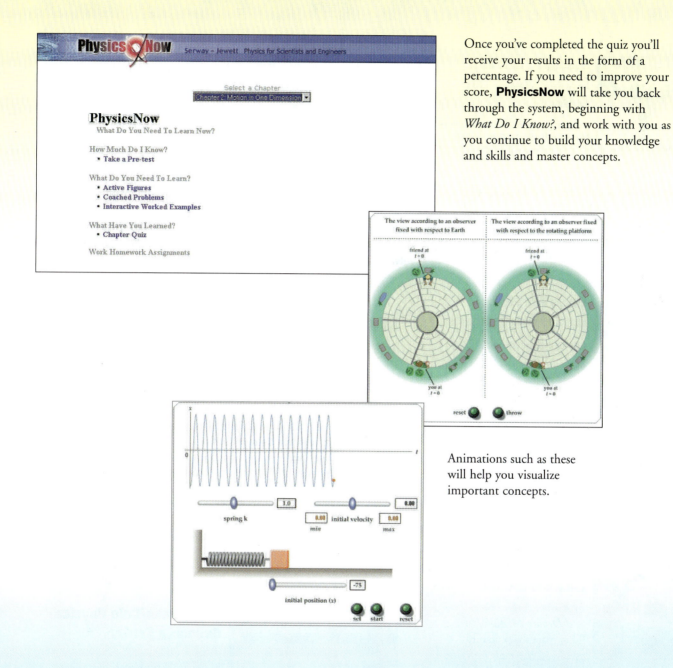

Animations such as these will help you visualize important concepts.

Chart your own course for success . . .

Log on to **www.pse6.com** to take advantage of **PhysicsNow!**

Make the most of the course and your time with these exclusive study tools

Enrich your experience outside of the classroom with a host of resources designed to help you excel in the course. To purchase any of these supplements, contact your campus bookstore or visit our online BookStore at www.brookscole.com.

Student Solutions Manual with Study Guide

Volume I ISBN: 0-534-40855-9
Volume II ISBN: 0-534-40856-7

by John R. Gordon, Ralph McGrew, and Raymond Serway This two-volume manual features detailed solutions to 20% of the end-of-chapter problems from the text. The manual also features a list of important equations, concepts, and answers to selected end-of-chapter questions.

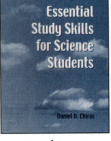

Essential Study Skills for Science Students

ISBN: 0-534-37595-2

by Daniel D. Chiras Written specifically for science students, this book discusses how to develop good study habits, sharpen memory, learn more quickly, get the most out of lectures, prepare for tests, produce excellent term papers, and improve critical-thinking skills.

The Physics Toolbox: A Survival Guide for Introductory Physics

ISBN: 0-03-034652-5

by Kirsten A. Hubbard and Debora M. Katz This "paperback mentor" gives you the material critical for success in physics, including an introduction to the nature of physics and science, a look at what to expect and how to succeed, a verbal overview of the concepts you'll encounter, and an extensive review of the math you'll need to solve the problems.

WebTUTOR Advantage

WebTutor™ Advantage on WebCT and Blackboard

WebCT ISBN: 0-534-40859-1
Blackboard ISBN: 0-534-40950-4

WebTutor Advantage offers real-time access to a full array of study tools, including chapter outlines, summaries, learning objectives, glossary flashcards (with audio), practice quizzes, **InfoTrac® College Edition** exercises, and Web links. **WebTutor Advantage** also provides robust communication tools, such as a course calendar, asynchronous discussion, real-time chat, a whiteboard, and an integrated e-mail system. Also new to **WebTutor Advantage** is access to *NewsEdge,* an online news service that brings the latest news to the **WebTutor Advantage** site daily.

Contact your instructor for more information.

Additional Web resources . . .

available FREE to you with each new copy of this text:

InfoTrac® College Edition

When you purchase a new copy of **Physics for Scientists and Engineers, Sixth Edition,** you automatically receive a FREE four-month subscription to **InfoTrac College Edition!** Newly improved, this extensive online library opens the door to the full text (not just abstracts) of countless articles from thousands of publications including *American Scientist, Physical Review, Science, Science Weekly,* and more! Use the passcode included with the new copy of this text and log on to **www.infotrac-college.com** to explore the wealth of resources available to you—24 hours a day and seven days a week!

Available only to college and university students. Journals subject to change.

The Brooks/Cole Physics Resource Center

http://physics.brookscole.com

Here you'll find even more opportunities to hone your skills and expand your knowledge. **The Brooks/Cole Physics Resource Center** is filled with helpful content that will engage you while you master the material. You'll find additional online quizzes, Web links, animations, and *NewEdge*—an online news service that brings the latest news to this site daily.

*We dedicate this book to the courageous astronauts
who died on the space shuttle* Columbia *on February 1, 2003.
The women and men of the international team lost their lives
not in a contest between countries or a struggle for
necessities but in advancing one of humankind's noblest
creations—science.*

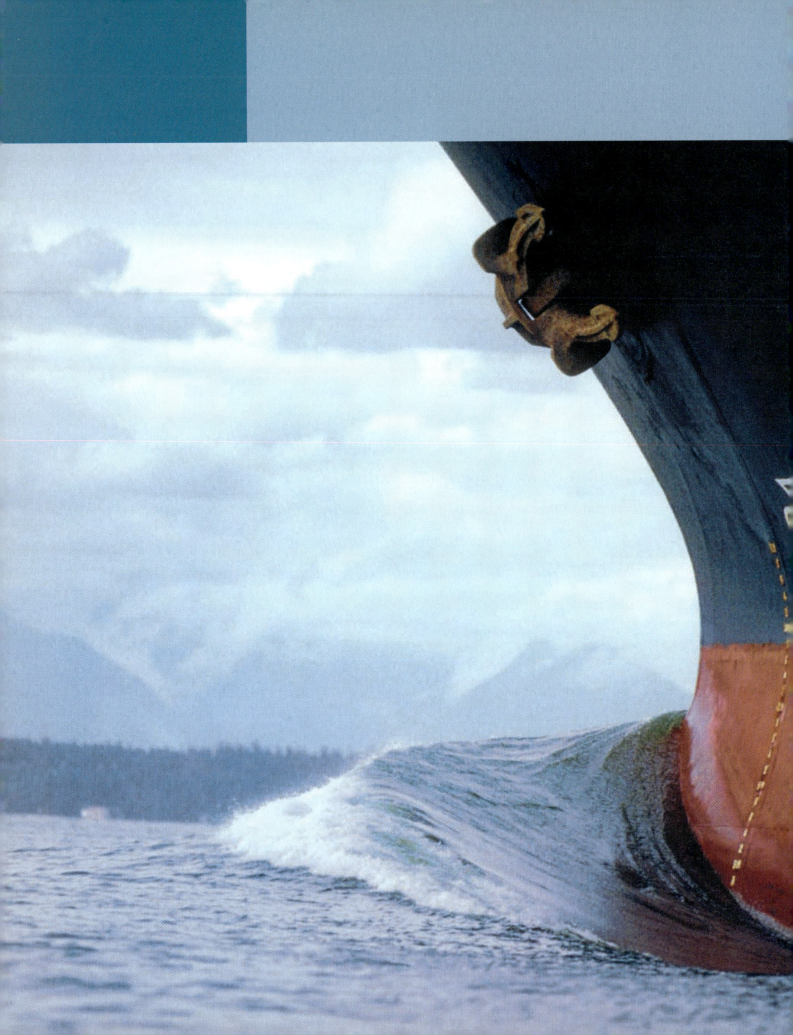

6th Edition

PHYSICS

for Scientists and Engineers

with Modern Physics

Chapters 39 – 46

5

Raymond A. Serway

John W. Jewett, Jr.

California State Polytechnic University–Pomona

THOMSON ™

BROOKS/COLE

Australia • Canada • Mexico • Singapore • Spain
United Kingdom • United States

THOMSON
BROOKS/COLE
™

Editor-in-Chief: Michelle Julet
Publisher: David Harris
Physics Editor: Chris Hall
Development Editor: Susan Dust Pashos
Assistant Editor: Rebecca Heider, Alyssa White
Editorial Assistant: Seth Dobrin, Jessica Howard
Technology Project Manager: Sam Subity
Marketing Manager: Kelley McAllister
Marketing Assistant: Sandra Perin
Advertising Project Manager: Stacey Purviance
Project Manager, Editorial Production: Teri Hyde
Print/Media Buyer: Barbara Britton
Permissions Editor: Joohee Lee

Production Service: Sparkpoint Communications, a division of J. B. Woolsey Associates
Text Designer: Lisa Devenish
Photo Researcher: Terri Wright
Copy Editor: Andrew Potter
Illustrator: Rolin Graphics
Cover Designer: Lisa Devenish
Cover Image: Water Displaced by Oil Tanker, © Stuart Westmorland/CORBIS
Compositor: Progressive Information Technologies
Cover Printer: Quebecor World, Versailles
Printer: Quebecor World, Versailles

Printed in the United States of America
1 2 3 4 5 6 7 07 06 05 04 03

For more information about our products, contact us at:
Thomson Learning Academic Resource Center
1-800-423-0563

For permission to use material from this text, contact us by:
Phone: 1-800-730-2214
Fax: 1-800-730-2215
Web: http://www.thomsonrights.com

Library of Congress Control Number: 2003100126

PHYSICS FOR SCIENTISTS AND ENGINEERS WITH MODERN PHYSICS, Sixth Edition
0-534-40854-0, Volume 5 (Chapters 39–46)

Brooks/Cole—Thomson Learning
10 Davis Drive
Belmont, CA 94002
USA

Asia
Thomson Learning
5 Shenton Way #01-01
UIC Building
Singapore 068808

Australia
Nelson Thomson Learning
102 Dodds Street
South Melbourne, Victoria 3205
Australia

Canada
Nelson Thomson Learning
1120 Birchmount Road
Toronto, Ontario M1K 5G4
Canada

Europe/Middle East/Africa
Thomson Learning
High Holborn House
50/51 Bedford Row
London WC1R 4LR
United Kingdom

Latin America
Thomson Learning
Seneca, 53
Colonia Polanco
11560 Mexico D.F.
Mexico

Spain
Paraninfo Thomson Learning
Calle/Magallanes, 25
28015 Madrid, Spain

Contents Overview

Steve Niedorf/Getty Images

Table of Contents

The Telegraph Colour Library/Getty Images

Don Bonsey/Getty Images

PART 4   Electricity and Magnetism   705

© 1973 Kim Vandiver & Harold E. Edgerton/Courtesy of Palm Press, Inc.

Austin McCrae

About the Authors

Raymond A. Serway received his doctorate at Illinois Institute of Technology and is Professor Emeritus at James Madison University. Dr. Serway began his teaching career at Clarkson University, where he conducted research and taught from 1967 to 1980. His second academic appointment was at James Madison University as Professor of Physics and Head of the Physics Department from 1980 to 1986. He remained at James Madison University until his retirement in 1997. He was the recipient of the Madison Scholar Award at James Madison University in 1990, the Distinguished Teaching Award at Clarkson University in 1977, and the Alumni Achievement Award from Utica College in 1985. As Guest Scientist at the IBM Research Laboratory in Zurich, Switzerland, he worked with K. Alex Müller, 1987 Nobel Prize recipient. Dr. Serway also held research appointments at Rome Air Development Center from 1961 to 1963, at IIT Research Institute from 1963 to 1967, and as a visiting scientist at Argonne National Laboratory, where he collaborated with his mentor and friend, Sam Marshall. In addition to earlier editions of this textbook, Dr. Serway is the co-author of the high-school textbook *Physics* with Jerry Faughn, published by Holt, Rinehart, & Winston and co-author of the third edition of *Principles of Physics* with John Jewett, the sixth edition of *College Physics* with Jerry Faughn, and the second edition of *Modern Physics* with Clem Moses and Curt Moyer. In addition, Dr. Serway has published more than 40 research papers in the field of condensed matter physics and has given more than 70 presentations at professional meetings. Dr. Serway and his wife Elizabeth enjoy traveling, golfing, gardening, and spending quality time with their four children and five grandchildren.

John W. Jewett, Jr. earned his doctorate at Ohio State University, specializing in optical and magnetic properties of condensed matter. Dr. Jewett began his academic career at Richard Stockton College of New Jersey, where he taught from 1974 to 1984. He is currently Professor of Physics at California State Polytechnic University, Pomona. Throughout his teaching career, Dr. Jewett has been active in promoting science education. In addition to receiving four National Science Foundation grants, he helped found and direct the Southern California Area Modern Physics Institute (SCAMPI). He also directed Science IMPACT (Institute for Modern Pedagogy and Creative Teaching), which works with teachers and schools to develop effective science curricula. Dr. Jewett's honors include the Stockton Merit Award at Richard Stockton College, the Outstanding Professor Award at California State Polytechnic University for 1991–1992, and the Excellence in Undergraduate Physics Teaching Award from the American Association of Physics Teachers (AAPT) in 1998. He has given over 80 presentations at professional meetings, including presentations at international conferences in China and Japan. In addition to his work on this textbook, he is co-author of the third edition of *Principles of Physics* with Ray Serway and author of *The World of Physics . . . Mysteries, Magic, and Myth.* Dr. Jewett enjoys playing piano, traveling, and collecting antiques that can be used as demonstration apparatus in physics lectures, as well as spending time with his wife Lisa and their children and grandchildren.

Preface

In writing this sixth edition of *Physics for Scientists and Engineers with Modern Physics,* we continue our ongoing efforts to improve the clarity of presentation and we again include new pedagogical features that help support the learning and teaching processes. Drawing on positive feedback from users of the fifth edition and reviewers' suggestions, we have refined the text in order to better meet the needs of students and teachers. We have for the first time integrated a powerful collection of media resources into many of the illustrations, examples, and end-of-chapter problems in the text. These resources compose the Web-based learning system *PhysicsNow* and are flagged by the media icon 🌐. Further details are described below.

This textbook is intended for a course in introductory physics for students majoring in science or engineering. The entire contents of the text in its extended version could be covered in a three-semester course, but it is possible to use the material in shorter sequences with the omission of selected chapters and sections. The mathematical background of the student taking this course should ideally include one semester of calculus. If that is not possible, the student should be enrolled in a concurrent course in introductory calculus.

Objectives

This introductory physics textbook has two main objectives: to provide the student with a clear and logical presentation of the basic concepts and principles of physics, and to strengthen an understanding of the concepts and principles through a broad range of interesting applications to the real world. To meet these objectives, we have placed emphasis on sound physical arguments and problem-solving methodology. At the same time, we have attempted to motivate the student through practical examples that demonstrate the role of physics in other disciplines, including engineering, chemistry, and medicine.

Changes in the Sixth Edition

A large number of changes and improvements have been made in preparing the sixth edition of this text. Some of the new features are based on our experiences and on current trends in science education. Other changes have been incorporated in response to comments and suggestions offered by users of the fifth edition and by reviewers of the manuscript. The following represent the major changes in the sixth edition:

Active Figures   Many diagrams from the text have been animated to form **Active Figures,** part of the *PhysicsNow* integrated Web-based learning system. By visualizing phenomena and processes that cannot be fully represented on a static page, students greatly increase their conceptual understanding. **Active Figures** are identified with the media icon 🌐. An addition to the figure caption in **blue type** describes briefly the nature and contents of the animation.

Interactive Worked Examples   Approximately 90 of the worked examples in the text have been identified as interactive, labeled with the media icon 🌐. As part of the *PhysicsNow* Web-based learning system, students can engage in an extension of the problem solved in the example. This often includes elements of both visualization and calculation, and may also involve prediction and intuition building. Often the interactivity is inspired by the **"What If?"** question we posed in the example text.

What If?   Approximately one-third of the worked examples in the text contain this new feature. At the completion of the example solution, a **What If?** question offers a variation on the situation posed in the text of the example. For instance, this feature might explore the effects of changing the conditions of the situation, determine what happens when a quantity is taken to a particular limiting value, or question whether additional information can be determined about the problem situation. The answer to the question generally includes both a conceptual response and a mathematical response. This feature encourages students to think about the results of the example and assists in conceptual understanding of the principles. It also prepares students to encounter novel problems featured on exams. Some of the end-of-chapter problems also carry the **"What If?"** feature.

Quick Quizzes   The number of Quick Quiz questions in each chapter has been increased. Quick Quizzes provide students with opportunities to test their understanding of the physical concepts presented. The questions require students to make decisions on the basis of sound reasoning, and some of them have been written to help students overcome common misconceptions. Quick Quizzes have been cast in an objective format, including multiple choice, true–false, and ranking. Answers to all Quick Quiz questions are found at the end of each chapter. Additional Quick Quizzes that can be used in classroom teaching are available on the instructor's companion Web site. Many instructors choose to use such questions in a "peer instruction" teaching style, but they can be used in standard quiz format as well.

Pitfall Preventions   These new features are placed in the margins of the text and address common student misconceptions and situations in which students often follow unproductive paths. Over 200 Pitfall Preventions are provided to help students avoid common mistakes and misunderstandings.

General Problem-Solving Strategy   A general strategy to be followed by the student is outlined at the end of Chapter 2 and provides students with a structured process for solving problems. In Chapters 3 through 5, the strategy is employed explicitly in every example so that students learn how it is applied. In the remaining chapters, the strategy appears explicitly in one example per chapter so that students are encouraged throughout the course to follow the procedure.

Line-by-Line Revision   The entire text has been carefully edited to improve clarity of presentation and precision of language. We hope that the result is a book that is both accurate and enjoyable to read.

Problems   A substantial revision of the end-of-chapter problems was made in an effort to improve their variety and interest, while maintaining their clarity and quality. Approximately 17% of the problems (about 550) are new. All problems have been carefully edited. Solutions to approximately 20% of the end-of-chapter problems are included in the *Student Solutions Manual and Study Guide*. These problems are identified by boxes around their numbers. A smaller subset of solutions, identified by the media icon 🌐, are available on the World Wide Web (**http://www.pse6.com**) as coached solutions with hints. Targeted feedback is provided for students whose instructors adopt *Physics for Scientists and Engineers with Modern Physics*, sixth edition. See the next section for a complete description of other features of the problem set.

Content Changes   The content and organization of the textbook is essentially the same as that of the fifth edition. An exception is that Chapter 13 (Oscillatory Motion) in the fifth edition has been moved to the Chapter 15 position in the sixth edition, in order to form a cohesive four-chapter Part 2 on oscillations and waves. Many sections in various chapters have been streamlined, deleted, or combined with other sections to allow for a more balanced presentation. The chapters on Modern Physics, Chapters 39–46, have been extensively rewritten to provide more up-to-date material as well as modern applications. A more detailed list of content changes can be found on the instructor's companion Web site.

Content

The material in this book covers fundamental topics in classical physics and provides an introduction to modern physics. The book is divided into six parts. Part 1 (Chapters 1 to 14) deals with the fundamentals of Newtonian mechanics and the physics of fluids, Part 2 (Chapters 15 to 18) covers oscillations, mechanical waves, and sound, Part 3 (Chapters 19 to 22) addresses heat and thermodynamics, Part 4 (Chapters 23 to 34) treats electricity and magnetism, Part 5 (Chapters 35 to 38) covers light and optics, and Part 6 (Chapters 39 to 46) deals with relativity and modern physics. Each part opener includes an overview of the subject matter covered in that part, as well as some historical perspectives.

Text Features

Most instructors would agree that the textbook selected for a course should be the student's primary guide for understanding and learning the subject matter. Furthermore, the textbook should be easily accessible and should be styled and written to facilitate instruction and learning. With these points in mind, we have included many pedagogical features in the textbook that are intended to enhance its usefulness to both students and instructors. These features are as follows:

Style   To facilitate rapid comprehension, we have attempted to write the book in a style that is clear, logical, and engaging. We have chosen a writing style that is somewhat informal and relaxed so that students will find the text appealing and enjoyable to read. New terms are carefully defined, and we have avoided the use of jargon.

Previews   All chapters begin with a brief preview that includes a discussion of the chapter's objectives and content.

Important Statements and Equations   Most important statements and definitions are set in **boldface** type or are highlighted with a background screen for added emphasis and ease of review. Similarly, important equations are highlighted with a background screen to facilitate location.

Bruce Ayers/Getty Images

Problem-Solving Hints   In several chapters, we have included general strategies for solving the types of problems featured both in the examples and in the end-of-chapter problems. This feature helps students to identify necessary steps in problem solving and to eliminate any uncertainty they might have. Problem-solving strategies are highlighted with a light red background screen for emphasis and ease of location.

Marginal Notes   Comments and notes appearing in blue type in the margin can be used to locate important statements, equations, and concepts in the text.

Pedagogical Use of Color   Readers should consult the **pedagogical color chart** (second page inside the front cover) for a listing of the color-coded symbols used in the text diagrams, Web-based **Active Figures,** and diagrams within **Interactive Worked Examples.** This system is followed consistently whenever possible, with slight variations made necessary by the complexity of physical situations depicted in Part 4.

Mathematical Level   We have introduced calculus gradually, keeping in mind that students often take introductory courses in calculus and physics concurrently. Most steps are shown when basic equations are developed, and reference is often made to mathematical appendices at the end of the textbook. Vector products are introduced later in the text, where they are needed in physical applications. The dot product is introduced in Chapter 7, which addresses energy and energy transfer; the cross product is introduced in Chapter 11, which deals with angular momentum.

Worked Examples   A large number of worked examples of varying difficulty are presented to promote students' understanding of concepts. In many cases, the examples serve as models for solving the end-of-chapter problems. Because of the increased emphasis on understanding physical concepts, many examples are conceptual in nature and are labeled as such. The examples are set off in boxes, and the answers to examples with numerical solutions are highlighted with a background screen. We have already mentioned that a number of examples are designated as interactive and are part of the *PhysicsNow* Web-based learning system.

Questions   Questions of a conceptual nature requiring verbal or written responses are provided at the end of each chapter. Over 1 000 questions are included in this edition. Some questions provide the student with a means of self-testing the concepts presented in the chapter. Others could serve as a basis for initiating classroom discussions. Answers to selected questions are included in the *Student Solutions Manual and Study Guide,* and answers to all questions are found in the *Instructor's Solutions Manual.*

Significant Figures   Significant figures in both worked examples and end-of-chapter problems have been handled with care. Most numerical examples are worked out to either two or three significant figures, depending on the precision of the data provided. End-of-chapter problems regularly state data and answers to three-digit precision.

Problems   An extensive set of problems is included at the end of each chapter; in all, over 3 000 problems are given throughout the text. Answers to odd-numbered problems are provided at the end of the book in a section whose pages have colored edges for ease of location. For the convenience of both the student and the instructor, about two thirds of the problems are keyed to specific sections of the chapter. The remaining problems, labeled "Additional Problems," are not keyed to specific sections.

Usually, the problems within a given section are presented so that the straightforward problems (those with black problem numbers) appear first. For ease of identification, the numbers of intermediate-level problems are printed in blue, and those of challenging problems are printed in magenta.

- **Review Problems**   Many chapters include review problems requiring the student to combine concepts covered in the chapter with those discussed in previous chapters. These problems reflect the cohesive nature of the principles in the text and verify that physics is not a scattered set of ideas. When facing real-world issues such as global warming or nuclear weapons, it may be necessary to call on ideas in physics from several parts of a textbook such as this one.

- **Paired Problems**   To allow focused practice in solving problems stated in symbolic terms, some end-of-chapter numerical problems are paired with the same problems in symbolic form. Paired problems are identified by a common light red background screen.

- **Computer- and Calculator-Based Problems**   Many chapters include one or more problems whose solution requires the use of a computer or graphing calculator. Computer modeling of physical phenomena enables students to obtain graphical representations of variables and to perform numerical analyses.

- **Coached Problems with Hints**   These have been described above as part of the *PhysicsNow* Web-based learning system. These problems are identified by the media icon 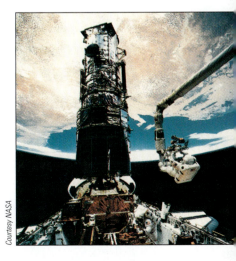 and targeted feedback is provided to students of instructors adopting the sixth edition.

Units   The international system of units (SI) is used throughout the text. The U.S. customary system of units is used only to a limited extent in the chapters on mechanics, heat, and thermodynamics.

Summaries   Each chapter contains a summary that reviews the important concepts and equations discussed in that chapter. A marginal note in **blue type** next to each chapter summary directs students to a practice test (Post-Test) for the chapter.

Appendices and Endpapers   Several appendices are provided at the end of the textbook. Most of the appendix material represents a review of mathematical concepts and techniques used in the text, including scientific notation, algebra, geometry, trigonometry, differential calculus, and integral calculus. Reference to these appendices is made

Courtesy NASA

throughout the text. Most mathematical review sections in the appendices include worked examples and exercises with answers. In addition to the mathematical reviews, the appendices contain tables of physical data, conversion factors, atomic masses, and the SI units of physical quantities, as well as a periodic table of the elements. Other useful information, including fundamental constants and physical data, planetary data, a list of standard prefixes, mathematical symbols, the Greek alphabet, and standard abbreviations of units of measure, appears on the endpapers.

Student Ancillaries

Student Solutions Manual and Study Guide by John R. Gordon, Ralph McGrew, and Raymond Serway. This two-volume manual features detailed solutions to 20% of the end-of-chapter problems from the text. The manual also features a list of important equations, concepts, and notes from key sections of the text, in addition to answers to selected end-of-chapter questions. Volume 1 contains Chapters 1 through 22 and Volume 2 contains Chapters 23 through 46.

WebTutor™ on WebCT and Blackboard   WebTutor offers students real-time access to a full array of study tools, including chapter outlines, summaries, learning objectives, glossary flashcards (with audio), practice quizzes, **InfoTrac® College Edition** exercises, and Web links.

InfoTrac® College Edition   Adopters and their students automatically receive a four-month subscription to **InfoTrac® College Edition** with every new copy of this book. Newly improved, this extensive online library opens the door to the full text (not just abstracts) of countless articles from thousands of publications including *American Scientist, Physical Review, Science, Science Weekly,* and more! Available only to college and university students. Journals subject to change.

The Brooks/Cole Physics Resource Center   You will find additional online quizzes, Web links and animations at **http://physics.brookscole.com.**

Ancillaries for Instructors

The first four ancillaries below are available to qualified adopters. Please consult your local sales representative for details.

Instructor's Solutions Manual by Ralph McGrew and James A. Currie. This two-volume manual contains complete worked solutions to all of the end-of-chapter problems in the textbook as well as answers to even-numbered problems. The solutions to problems new to the sixth edition are marked for easy identification by the instructor. New to this edition are complete answers to the conceptual questions in the main text. Volume 1 contains Chapters 1 through 22 and Volume 2 contains Chapters 23 through 46.

Printed Test Bank by Edward Adelson. This two-volume test bank contains approximately 2 300 multiple-choice questions. These questions are also available in electronic format with complete answers and solutions in the Brooks/Cole Assessment test program. Volume 1 contains Chapters 1 through 22 and Volume 2 contains Chapters 23 through 46.

Multimedia Manager   This easy-to-use multimedia lecture tool allows you to quickly assemble art and database files with notes to create fluid lectures. The CD-ROM set (Volume 1, Chapters 1–22; Volume 2, Chapters 23–46) includes a database of animations, video clips, and digital art from the text as well as electronic files of the *Instructor's Solutions Manual and Test Bank.* The simple interface makes it easy for you to incorporate graphics, digital video, animations, and audio clips into your lectures.

Transparency Acetates   Each volume contains approximately 100 acetates featuring art from the text. Volume 1 contains Chapters 1 through 22 and Volume 2 contains Chapters 23 through 46.

Brooks/Cole Assessment With a balance of efficiency, high performance, simplicity and versatility, **Brooks/Cole Assessment (BCA)** gives you the power to transform the learning and teaching experience. **BCA** is fully integrated testing, tutorial, and course management software accessible by instructors and students anytime, anywhere. Delivered for FREE in a browser-based format without the need for any proprietary software or plug-ins, **BCA** uses correct scientific notation to provide the drill of basic skills that students need, enabling the instructor to focus more time in higher-level learning activities (i.e., concepts and applications). Students can have unlimited practice in questions and problems, building their own confidence and skills. Results flow automatically to a grade book for tracking so that instructors will be better able to assess student understanding of the material, even prior to class or an actual test.

George Sample

WebTutor™ on WebCT and Blackboard With **WebTutor's** text-specific, preformatted content and total flexibility, instructors can easily create and manage their own personal Web site. **WebTutor's** course management tool gives instructors the ability to provide virtual office hours, post syllabi, set up threaded discussions, track student progress with the quizzing material, and much more. **WebTutor** also provides robust communication tools, such as a course calendar, asynchronous discussion, real-time chat, a whiteboard, and an integrated e-mail system.

Additional Options for Online Homework For detailed information and demonstrations, contact your Thomson • Brooks/Cole representative or visit the following:

- WebAssign: A Web-based Homework System
  http://www.webassign.net or contact WebAssign at *webassign@ncsu.edu*
- Homework Service
  http://hw.ph.utexas.edu/hw.html or contact *moore@physics.utexas.edu*
- CAPA: A Computer-Assisted Personalized Approach
  http://capa4.lite.msu.edu/homepage/

Instructor's Companion Web Site Consult the instructor's site at *http://www.pse6.com* for additional Quick Quiz questions, a detailed list of content changes since the fifth edition, a problem correlation guide, images from the text, and sample PowerPoint lectures. Instructors adopting the sixth edition of *Physics for Scientists and Engineers with Modern Physics* may download these materials after securing the appropriate password from their local Thomson • Brooks/Cole sales representative.

Teaching Options

The topics in this textbook are presented in the following sequence: classical mechanics, oscillations and mechanical waves, and heat and thermodynamics followed by electricity and magnetism, electromagnetic waves, optics, relativity, and modern physics. This presentation represents a traditional sequence, with the subject of mechanical waves being presented before electricity and magnetism. Some instructors may prefer to cover this material after completing electricity and magnetism (i.e., after Chapter 34). The chapter on relativity is placed near the end of the text because this topic often is treated as an introduction to the era of "modern physics." If time permits, instructors may choose to cover Chapter 39 after completing Chapter 13, as it concludes the material on Newtonian mechanics.

For those instructors teaching a two-semester sequence, some sections and chapters could be deleted without any loss of continuity. The following sections can be considered optional for this purpose:

2.7	Kinematic Equations Derived from Calculus	**6.4**	Motion in the Presence of Resistive Forces
4.6	Relative Velocity and Relative Acceleration	**6.5**	Numerical Modeling in Particle Dynamics
6.3	Motion in Accelerated Frames	**7.9**	Energy and the Automobile

Again we stress that Chapters 40–46 on modern physics have been thoroughly revised and updated. For those instructors covering modern physics and finding themselves pressed for time, the following sections could be deleted without loss of continuity:

Acknowledgments

The sixth edition of this textbook was prepared with the guidance and assistance of many professors who reviewed selections of the manuscript, the pre-revision text, or both. We wish to acknowledge the following scholars and express our sincere appreciation for their suggestions, criticisms, and encouragement:

Edward Adelson, *Ohio State University*

Michael R. Cohen, *Shippensburg University*

Jerry D. Cook, *Eastern Kentucky University*

J. William Dawicke, *Milwaukee School of Engineering*

N. John DiNardo, *Drexel University*

Andrew Duffy, *Boston University*

Robert J. Endorf, *University of Cincinnati*

F. Paul Esposito, *University of Cincinnati*

Joe L. Ferguson, *Mississippi State University*

Perry Ganas, *California State University, Los Angeles*

John C. Hardy, *Texas A&M University*

Michael Hayes, *University of Pretoria (South Africa)*

John T. Ho, *The State University of New York, Buffalo*

Joseph W. Howard, *Salisbury University*

Robert Hunt, *Johnson County Community College*

Walter S. Jaronski, *Radford University*

Sangyong Jeon, *McGill University, Quebec*

Stan Jones, *University of Alabama*

L. R. Jordan, *Palm Beach Community College*

Teruki Kamon, *Texas A & M University*

Louis E. Keiner, *Coastal Carolina University*

Topham Picturepoint/The Image Works

Mario Klarič, *Midlands Technical College*

Laird Kramer, *Florida International University*

Edwin H. Lo, *American University*

James G. McLean, *The State University of New York, Geneseo*

Richard E. Miers, *Indiana University–Purdue University, Fort Wayne*

Oscar Romulo Ochoa, *The College of New Jersey*

Paul S. Ormsby, *Moraine Valley Community College*

Didarul I. Qadir, *Central Michigan University*

Judith D. Redling, *New Jersey Institute of Technology*

Richard W. Robinett, *Pennsylvania State University*

Om P. Rustgi, *SUNY College at Buffalo*

Mesgun Sebhatu, *Winthrop University*

Natalia Semushkina, *Shippensburg University*

Daniel Stump, *Michigan State University*

Uwe C. Täuber, *Virginia Polytechnic Institute*

Perry A. Tompkins, *Samford University*

Doug Welch, *McMaster University, Ontario*

Augden Windelborn, *Northern Illinois University*

Jerzy M. Wrobel, *University of Missouri, Kansas City*

Jianshi Wu, *Fayetteville State University*

Michael Zincani, *University of Dallas*

Frank Oberle/Getty Images

This title was carefully checked for accuracy by Michael Kotlarchyk *(Rochester Institute of Technology)*, Chris Vuille *(Embry-Riddle Aeronautical University)*, Laurencin Dunbar *(St. Louis Community College)*, William Dawicke *(Milwaukee School of Engineering)*, Ioan Kosztin *(University of Missouri)*, Tom Barrett *(Ohio State University)*, Z. M. Stadnik *(University of Ottawa)*, Ronald E. Jodoin *(Rochester Institute of Technology)*, Brian A. Raue *(Florida International University)*, Peter Moeck *(Portland State University)*, and Grant Hart *(Brigham Young University)*. We thank them for their diligent efforts under schedule pressure!

We are grateful to Ralph McGrew for organizing the end-of-chapter problems, writing many new problems, and his excellent suggestions for improving the content of the textbook. Problems new to this edition were written by Edward Adelson, Ronald Bieniek, Michael Browne, Andrew Duffy, Robert Forsythe, Perry Ganas, Michael Hones, John Jewett, Boris Korsunsky, Edwin Lo, Ralph McGrew, Raymond Serway, and Jerzy Wrobel, with the help of Bennett Simpson and JoAnne Maniago. Students Alexander Coto, Karl Payne, and Eric Peterman made corrections to problems taken from previous editions, as did teachers David Aspnes, Robert Beichner, Joseph Biegen, Tom Devlin, Vasili Haralambous, Frank Hayes, Erika Hermon, Ken Menningen, Henry Nebel, and Charles Teague. We are grateful to authors John R. Gordon and Ralph McGrew and compositor Michael Rudmin for preparing the *Student Solutions Manual and Study Guide*. Authors Ralph McGrew and James Currie and compositor Mary Toscano have prepared an excellent *Instructor's Solutions Manual*, and we thank them. Edward Adelson has carefully edited and improved the Test Bank for the sixth edition. Kurt Vandervoort prepared extra Quick Quiz questions for the instructor's companion Web site.

Special thanks and recognition go to the professional staff at the Brooks/Cole Publishing Company—in particular Susan Pashos, Rebecca Heider and Alyssa White (who managed the ancillary program and so much more), Jessica Howard, Seth Dobrin, Peter McGahey, Teri Hyde, Michelle Julet, David Harris, and Chris Hall—for their fine work during the development and production of this textbook. We are most appreciative of Sam Subity's masterful management of the *PhysicsNow* media program. Kelley McAllister is our energetic Marketing Manager, and Stacey Purviance coordinates our marketing communications. We recognize the skilled production service provided by the staff at Sparkpoint Communications, the excellent artwork produced by Rolin Graphics, and the dedicated photo research efforts of Terri Wright.

Finally, we are deeply indebted to our wives and children for their love, support, and long-term sacrifices.

Raymond A. Serway

Leesburg, Virginia

John W. Jewett, Jr.

Pomona, California

To the Student

It is appropriate to offer some words of advice that should be of benefit to you, the student. Before doing so, we assume that you have read the Preface, which describes the various features of the text that will help you through the course.

How to Study

Very often instructors are asked, "How should I study physics and prepare for examinations?" There is no simple answer to this question, but we would like to offer some suggestions that are based on our own experiences in learning and teaching over the years.

First and foremost, maintain a positive attitude toward the subject matter, keeping in mind that physics is the most fundamental of all natural sciences. Other science courses that follow will use the same physical principles, so it is important that you understand and are able to apply the various concepts and theories discussed in the text.

Concepts and Principles

It is essential that you understand the basic concepts and principles before attempting to solve assigned problems. You can best accomplish this goal by carefully reading the textbook before you attend your lecture on the covered material. When reading the text, you should jot down those points that are not clear to you. We've purposely left wide margins in the text to give you space for making notes. Also be sure to make a diligent attempt at answering the questions in the Quick Quizzes as you come to them in your reading. We have worked hard to prepare questions that help you judge for yourself how well you understand the material. Study carefully the **What If?** features that appear with many of the worked examples. These will help you to extend your understanding beyond the simple act of arriving at a numerical result. The Pitfall Preventions will also help guide you away from common misunderstandings about physics. During class, take careful notes and ask questions about those ideas that are unclear to you. Keep in mind that few people are able to absorb the full meaning of scientific material after only one reading. Several readings of the text and your notes may be necessary. Your lectures and laboratory work supplement reading of the textbook and should clarify some of the more difficult material. You should minimize your memorization of material. Successful memorization of passages from the text, equations, and derivations does not necessarily indicate that you understand the material. Your understanding of the material will be enhanced through a combination of efficient study habits, discussions with other students and with instructors, and your ability to solve the problems presented in the textbook. Ask questions whenever you feel clarification of a concept is necessary.

Study Schedule

It is important that you set up a regular study schedule, preferably a daily one. Make sure that you read the syllabus for the course and adhere to the schedule set by your instructor. The lectures will make much more sense if you read the corresponding text material before attending them. As a general rule, you should devote about two hours of study time for every hour you are in class. If you are having trouble with the course, seek the advice of the instructor or other students who have taken the course. You may find it necessary to seek further instruction from experienced students. Very often, instructors offer review sessions in addition to regular class periods. It is important that you avoid the practice of delaying study until a day or two before an exam. More often than not, this approach has disastrous results. Rather than undertake an all-night study session, briefly review the basic concepts and equations, and get a good night's rest. If you feel you need additional help in understanding the concepts, in preparing for exams, or in problem solving, we suggest that you acquire a copy of the *Student Solutions Manual and Study Guide* that accompanies this textbook; this manual should be available at your college bookstore.

Use the Features

You should make full use of the various features of the text discussed in the Preface. For example, marginal notes are useful for locating and describing important equations and concepts, and **boldfaced** type indicates important statements and definitions. Many useful tables are contained in the Appendices, but most are incorporated in the text where they are most often referenced. Appendix B is a convenient review of mathematical techniques.

George Semple

Answers to odd-numbered problems are given at the end of the textbook, answers to Quick Quizzes are located at the end of each chapter, and answers to selected end-of-chapter questions are provided in the *Student Solutions Manual and Study Guide.* Problem-Solving Strategies and Hints are included in selected chapters throughout the text and give you additional information about how you should solve problems. The Table of Contents provides an overview of the entire text, while the Index enables you to locate specific material quickly. Footnotes sometimes are used to supplement the text or to cite other references on the subject discussed.

After reading a chapter, you should be able to define any new quantities introduced in that chapter and to discuss the principles and assumptions that were used to arrive at certain key relations. The chapter summaries and the review sections of the *Student Solutions Manual and Study Guide* should help you in this regard. In some cases, it may be necessary for you to refer to the index of the text to locate certain topics. You should be able to associate with each physical quantity the correct symbol used to represent that quantity and the unit in which the quantity is specified. Furthermore, you should be able to express each important equation in a concise and accurate prose statement.

Problem Solving

R. P. Feynman, Nobel laureate in physics, once said, "You do not know anything until you have practiced." In keeping with this statement, we strongly advise that you develop the skills necessary to solve a wide range of problems. Your ability to solve problems will be one of the main tests of your knowledge of physics, and therefore you should try to solve as many problems as possible. It is essential that you understand basic concepts and principles before attempting to solve problems. It is good practice to try to find alternate solutions to the same problem. For example, you can solve problems in mechanics using Newton's laws, but very often an alternative method that draws on energy considerations is more direct. You should not deceive yourself into thinking that you understand a problem merely because you have seen it solved in class. You must be able to solve the problem and similar problems on your own.

The approach to solving problems should be carefully planned. A systematic plan is especially important when a problem involves several concepts. First, read the problem several times until you are confident you understand what is being asked. Look for any key words that will help you interpret the problem and perhaps allow you to make certain assumptions. Your ability to interpret a question properly is an integral part of problem solving. Second, you should acquire the habit of writing down the information given in a problem and those quantities that need to be found; for example, you might construct a table listing both the quantities given and the quantities to be found. This procedure is sometimes used in the worked examples of the textbook. Finally, after you have decided on the method you feel is appropriate for a given problem, proceed with your solution. Specific problem-solving strategies (Hints) of this type are included in the text and are highlighted with a light red screen. We have also developed a General Problem-Solving Strategy to help guide you through complex problems. If you follow the steps of this procedure *(Conceptualize, Categorize, Analyze, Finalize)*, you will not only find it easier to come up with a solution, but you will also gain more from your efforts. This Strategy is located at the end of Chapter 2 (page 47) and is used in all worked examples in Chapters 3 through 5 so that you can learn how to apply it. In the remaining chapters, the Strategy is used in one example per chapter as a reminder of its usefulness.

Often, students fail to recognize the limitations of certain equations or physical laws in a particular situation. It is very important that you understand and remember the assumptions that underlie a particular theory or formalism. For example, certain equations in kinematics apply only to a particle moving with constant acceleration. These equations are not valid for describing motion whose acceleration is not constant, such as the motion of an object connected to a spring or the motion of an object through a fluid.

Experiments

Physics is a science based on experimental observations. In view of this fact, we recommend that you try to supplement the text by performing various types of "hands-on" experiments, either at home or in the laboratory. These can be used to test ideas and models discussed in class or in the textbook. For example, the common Slinky™ toy is excellent for studying traveling waves; a ball swinging on the end of a long string can be used to investigate pendulum motion; various masses attached to the end of a vertical spring or rubber band can be used to determine their elastic nature; an old pair of Polaroid sunglasses and some discarded lenses and a magnifying glass are the components of various experiments in optics; and an approximate measure of the free-fall acceleration can be determined simply by measuring with a stopwatch the time it takes for a ball to drop from a known height. The list of such experiments is endless. When physical models are not available, be imaginative and try to develop models of your own.

New Media

We strongly encourage you to use the *PhysicsNow* Web-based learning system that accompanies this textbook. It is far easier to understand physics if you see it in action, and these new materials will enable you to become a part of that action. *PhysicsNow* media described in the Preface are accessed at the URL *http://www.pse6.com,* and feature a three-step learning process consisting of a Pre-Test, a personalized learning plan, and a Post-Test.

In addition to other elements, *PhysicsNow* includes the following Active Figures and Interactive Worked Examples:

© Phil Degginger/Stone/Getty

Chapter 2
Active Figures 2.1, 2.3, 2.9, 2.10, 2.11, and 2.13
Examples 2.8 and 2.12

Chapter 3
Active Figures 3.2, 3.3, 3.6, and 3.16
Example 3.5

Chapter 4
Active Figures 4.5, 4.7, and 4.11
Examples 4.4, 4.5, and 4.18

Chapter 5
Active Figure 5.16
Examples 5.9, 5.10, 5.12, and 5.14

Chapter 6
Active Figures 6.2, 6.8, 6.12, and 6.15
Examples 6.4, 6.5, and 6.7

Chapter 7
Active Figure 7.10
Examples 7.9 and 7.11

Chapter 8
Active Figures 8.3, 8.4, and 8.16
Examples 8.2 and 8.4

Chapter 9
Active Figures 9.8, 9.9, 9.13, 9.16, and 9.17
Examples 9.1, 9.5, and 9.8

Chapter 10
Active Figures 10.4, 10.14, and 10.30
Examples 10.12, 10.13, and 10.14

Chapter 11
Active Figures 11.1, 11.3, and 11.4
Examples 11.6 and 11.10

Chapter 12
Active Figures 12.14, 12.16, 12.17, and 12.18
Examples 12.3 and 12.4

An Invitation to Physics

It is our sincere hope that you too will find physics an exciting and enjoyable experience and that you will profit from this experience, regardless of your chosen profession. Welcome to the exciting world of physics!

The scientist does not study nature because it is useful; he studies it because he delights in it, and he delights in it because it is beautiful. If nature were not beautiful, it would not be worth knowing, and if nature were not worth knowing, life would not be worth living.

—Henri Poincaré

PART 6

Modern Physics

At the end of the nineteenth century, many scientists believed that they had learned most of what there was to know about physics. Newton's laws of motion and his theory of universal gravitation, Maxwell's theoretical work in unifying electricity and magnetism, the laws of thermodynamics and kinetic theory, and the principles of optics were highly successful in explaining a variety of phenomena.

As the nineteenth century turned to the twentieth, however, a major revolution shook the world of physics. In 1900 Planck provided the basic ideas that led to the formulation of the quantum theory, and in 1905 Einstein formulated his brilliant special theory of relativity. The excitement of the times is captured in Einstein's own words: "It was a marvelous time to be alive." Both ideas were to have a profound effect on our understanding of nature. Within a few decades, these two theories inspired new developments and theories in the fields of atomic physics, nuclear physics, and condensed-matter physics.

In Chapter 39 we introduce the special theory of relativity. The theory provides us with a new and deeper view of physical laws. Although the concepts underlying this theory often violate our common sense, the theory correctly predicts the results of experiments involving speeds near the speed of light. In the extended version of this textbook, *Physics for Scientists and Engineers with Modern Physics,* we cover the basic concepts of quantum mechanics and their application to atomic and molecular physics, and we introduce solid-state physics, nuclear physics, particle physics, and cosmology.

You should keep in mind that, although the physics that was developed during the twentieth century has led to a multitude of important technological achievements, the story is still incomplete. Discoveries will continue to evolve during our lifetimes, and many of these discoveries will deepen or refine our understanding of nature and the world around us. It is still a "marvelous time to be alive." ■

◄ *A portion of the accelerator tunnel at Fermilab, near Chicago, Illinois. The tunnel is circular and 1.9 km in diameter. Using electric and magnetic fields, protons and antiprotons are accelerated to speeds close to that of light and then allowed to collide head-on, in order to investigate the production of new particles. (Fermilab Photo)*

Chapter 39

Relativity

▲ Standing on the shoulders of a giant. *David Serway, son of one of the authors, watches over his children, Nathan and Kaitlyn, as they frolic in the arms of Albert Einstein at the Einstein memorial in Washington, D.C. It is well known that Einstein, the principal architect of relativity, was very fond of children. (Emily Serway)*

Our everyday experiences and observations have to do with objects that move at speeds much less than the speed of light. Newtonian mechanics was formulated by observing and describing the motion of such objects, and this formalism is very successful in describing a wide range of phenomena that occur at low speeds. However, it fails to describe properly the motion of objects whose speeds approach that of light.

Experimentally, the predictions of Newtonian theory can be tested at high speeds by accelerating electrons or other charged particles through a large electric potential difference. For example, it is possible to accelerate an electron to a speed of $0.99c$ (where c is the speed of light) by using a potential difference of several million volts. According to Newtonian mechanics, if the potential difference is increased by a factor of 4, the electron's kinetic energy is four times greater and its speed should double to $1.98c$. However, experiments show that the speed of the electron—as well as the speed of any other object in the Universe—always remains less than the speed of light, regardless of the size of the accelerating voltage. Because it places no upper limit on speed, Newtonian mechanics is contrary to modern experimental results and is clearly a limited theory.

In 1905, at the age of only 26, Einstein published his special theory of relativity. Regarding the theory, Einstein wrote:

> The relativity theory arose from necessity, from serious and deep contradictions in the old theory from which there seemed no escape. The strength of the new theory lies in the consistency and simplicity with which it solves all these difficulties . . . .[1]

Although Einstein made many other important contributions to science, the special theory of relativity alone represents one of the greatest intellectual achievements of all time. With this theory, experimental observations can be correctly predicted over the range of speeds from $v = 0$ to speeds approaching the speed of light. At low speeds, Einstein's theory reduces to Newtonian mechanics as a limiting situation. It is important to recognize that Einstein was working on electromagnetism when he developed the special theory of relativity. He was convinced that Maxwell's equations were correct, and in order to reconcile them with one of his postulates, he was forced into the revolutionary notion of assuming that space and time are not absolute.

This chapter gives an introduction to the special theory of relativity, with emphasis on some of its consequences. The special theory covers phenomena such as the slowing down of moving clocks and the contraction of moving lengths. We also discuss the relativistic forms of momentum and energy.

In addition to its well-known and essential role in theoretical physics, the special theory of relativity has practical applications, including the design of nuclear power plants and modern global positioning system (GPS) units. These devices do not work if designed in accordance with nonrelativistic principles.

[1]   A. Einstein and L. Infeld, *The Evolution of Physics*, New York, Simon and Schuster, 1961.

39.1 The Principle of Galilean Relativity

To describe a physical event, we must establish a frame of reference. You should recall from Chapter 5 that an inertial frame of reference is one in which an object is observed to have no acceleration when no forces act on it. Furthermore, any system moving with constant velocity with respect to an inertial frame must also be in an inertial frame.

There is no absolute inertial reference frame. This means that the results of an experiment performed in a vehicle moving with uniform velocity will be identical to the results of the same experiment performed in a stationary vehicle. The formal statement of this result is called the **principle of Galilean relativity:**

Principle of Galilean relativity

> The laws of mechanics must be the same in all inertial frames of reference.

Let us consider an observation that illustrates the equivalence of the laws of mechanics in different inertial frames. A pickup truck moves with a constant velocity, as shown in Figure 39.1a. If a passenger in the truck throws a ball straight up, and if air effects are neglected, the passenger observes that the ball moves in a vertical path. The motion of the ball appears to be precisely the same as if the ball were thrown by a person at rest on the Earth. The law of universal gravitation and the equations of motion under constant acceleration are obeyed whether the truck is at rest or in uniform motion.

Both observers agree on the laws of physics—they each throw a ball straight up and it rises and falls back into their hand. What about the *path* of the ball thrown by the observer in the truck? Do the observers agree on the path? The observer on the ground sees the path of the ball as a parabola, as illustrated in Figure 39.1b, while, as mentioned earlier, the observer in the truck sees the ball move in a vertical path. Furthermore, according to the observer on the ground, the ball has a horizontal component of velocity equal to the velocity of the truck. Although the two observers disagree on certain aspects of the situation, they agree on the validity of Newton's laws and on such classical principles as conservation of energy and conservation of linear momentum. This agreement implies that no mechanical experiment can detect any difference between the two inertial frames. The only thing that can be detected is the relative motion of one frame with respect to the other.

> **Quick Quiz 39.1** Which observer in Figure 39.1 sees the ball's *correct* path? (a) the observer in the truck (b) the observer on the ground (c) both observers.

(a)

(b)

Figure 39.1 (a) The observer in the truck sees the ball move in a vertical path when thrown upward. (b) The Earth observer sees the path of the ball as a parabola.

Suppose that some physical phenomenon, which we call an *event,* occurs and is observed by an observer at rest in an inertial reference frame. The event's location and time of occurrence can be specified by the four coordinates (x, y, z, t). We would like to be able to transform these coordinates from those of an observer in one inertial frame to those of another observer in a frame moving with uniform relative velocity compared to the first frame. When we say an observer is "in a frame," we mean that the observer is at rest with respect to the origin of that frame.

Consider two inertial frames S and S′ (Fig. 39.2). The frame S′ moves with a constant velocity **v** along the common x and x' axes, where **v** is measured relative to S. We assume that the origins of S and S′ coincide at $t = 0$ and that an event occurs at point P in space at some instant of time. An observer in S describes the event with space–time coordinates (x, y, z, t), whereas an observer in S′ uses the coordinates (x', y', z', t') to describe the same event. As we see from the geometry in Figure 39.2, the relationships among these various coordinates can be written

$$x' = x - vt \qquad y' = y \qquad z' = z \qquad t' = t \qquad (39.1)$$

These equations are the **Galilean space–time transformation equations.** Note that time is assumed to be the same in both inertial frames. That is, within the framework of classical mechanics, all clocks run at the same rate, regardless of their velocity, so that the time at which an event occurs for an observer in S is the same as the time for the same event in S′. Consequently, the time interval between two successive events should be the same for both observers. Although this assumption may seem obvious, it turns out to be incorrect in situations where v is comparable to the speed of light.

Now suppose that a particle moves through a displacement of magnitude dx along the x axis in a time interval dt as measured by an observer in S. It follows from Equations 39.1 that the corresponding displacement dx' measured by an observer in S′ is $dx' = dx - v\,dt$, where frame S′ is moving with speed v in the x direction relative to frame S. Because $dt = dt'$, we find that

$$\frac{dx'}{dt'} = \frac{dx}{dt} - v$$

or

$$u'_x = u_x - v \qquad (39.2)$$

where u_x and u'_x are the x components of the velocity of the particle measured by observers in S and S′, respectively. (We use the symbol **u** for particle velocity rather than **v**, which is used for the relative velocity of two reference frames.) This is the **Galilean velocity transformation equation.** It is consistent with our intuitive notion of time and space as well as with our discussions in Section 4.6. As we shall soon see, however, it leads to serious contradictions when applied to electromagnetic waves.

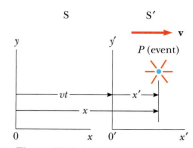

Figure 39.2 An event occurs at a point P. The event is seen by two observers in inertial frames S and S′, where S′ moves with a velocity **v** relative to S.

Galilean transformation equations

⚠ **PITFALL PREVENTION**

39.1 The Relationship Between the S and S′ Frames

Many of the mathematical representations in this chapter are true *only* for the specified relationship between the S and S′ frames. The x and x' axes coincide, except that their origins are different. The y and y' axes (and the z and z' axes), are parallel, but do not coincide due to the displacement of the origin of S′ with respect to that of S. We choose the time $t = 0$ to be the instant at which the origins of the two coordinate systems coincide. If the S′ frame is moving in the positive x direction relative to S, v is positive; otherwise it is negative.

Quick Quiz 39.2 A baseball pitcher with a 90-mi/h fastball throws a ball while standing on a railroad flatcar moving at 110 mi/h. The ball is thrown in the same direction as that of the velocity of the train. Applying the Galilean velocity transformation equation, the speed of the ball relative to the Earth is (a) 90 mi/h (b) 110 mi/h (c) 20 mi/h (d) 200 mi/h (e) impossible to determine.

The Speed of Light

It is quite natural to ask whether the principle of Galilean relativity also applies to electricity, magnetism, and optics. Experiments indicate that the answer is no. Recall from Chapter 34 that Maxwell showed that the speed of light in free space is $c = 3.00 \times 10^8$ m/s. Physicists of the late 1800s thought that light waves moved through a medium called the *ether* and that the speed of light was c only in a special, absolute frame

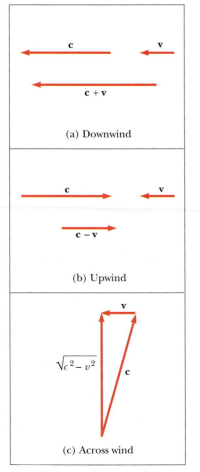

(a) Downwind

(b) Upwind

(c) Across wind

Figure 39.3 If the velocity of the ether wind relative to the Earth is **v** and the velocity of light relative to the ether is **c**, then the speed of light relative to the Earth is (a) $c + v$ in the downwind direction, (b) $c - v$ in the upwind direction, and (c) $(c^2 - v^2)^{1/2}$ in the direction perpendicular to the wind.

at rest with respect to the ether. The Galilean velocity transformation equation was expected to hold for observations of light made by an observer in any frame moving at speed v relative to the absolute ether frame. That is, if light travels along the x axis and an observer moves with velocity **v** along the x axis, the observer will measure the light to have speed $c \pm v$, depending on the directions of travel of the observer and the light.

Because the existence of a preferred, absolute ether frame would show that light was similar to other classical waves and that Newtonian ideas of an absolute frame were true, considerable importance was attached to establishing the existence of the ether frame. Prior to the late 1800s, experiments involving light traveling in media moving at the highest laboratory speeds attainable at that time were not capable of detecting differences as small as that between c and $c \pm v$. Starting in about 1880, scientists decided to use the Earth as the moving frame in an attempt to improve their chances of detecting these small changes in the speed of light.

As observers fixed on the Earth, we can take the view that we are stationary and that the absolute ether frame containing the medium for light propagation moves past us with speed v. Determining the speed of light under these circumstances is just like determining the speed of an aircraft traveling in a moving air current, or wind; consequently, we speak of an "ether wind" blowing through our apparatus fixed to the Earth.

A direct method for detecting an ether wind would use an apparatus fixed to the Earth to measure the ether wind's influence on the speed of light. If v is the speed of the ether relative to the Earth, then light should have its maximum speed $c + v$ when propagating downwind, as in Figure 39.3a. Likewise, the speed of light should have its minimum value $c - v$ when the light is propagating upwind, as in Figure 39.3b, and an intermediate value $(c^2 - v^2)^{1/2}$ in the direction perpendicular to the ether wind, as in Figure 39.3c. If the Sun is assumed to be at rest in the ether, then the velocity of the ether wind would be equal to the orbital velocity of the Earth around the Sun, which has a magnitude of approximately 3×10^4 m/s. Because $c = 3 \times 10^8$ m/s, it is necessary to detect a change in speed of about 1 part in 10^4 for measurements in the upwind or downwind directions. However, while such a change is experimentally measurable, all attempts to detect such changes and establish the existence of the ether wind (and hence the absolute frame) proved futile! We explore the classic experimental search for the ether in Section 39.2.

The principle of Galilean relativity refers only to the laws of mechanics. If it is assumed that the laws of electricity and magnetism are the same in all inertial frames, a paradox concerning the speed of light immediately arises. We can understand this by recognizing that Maxwell's equations seem to imply that the speed of light always has the fixed value 3.00×10^8 m/s in all inertial frames, a result in direct contradiction to what is expected based on the Galilean velocity transformation equation. According to Galilean relativity, the speed of light should *not* be the same in all inertial frames.

To resolve this contradiction in theories, we must conclude that either (1) the laws of electricity and magnetism are not the same in all inertial frames or (2) the Galilean velocity transformation equation is incorrect. If we assume the first alternative, then a preferred reference frame in which the speed of light has the value c must exist and the measured speed must be greater or less than this value in any other reference frame, in accordance with the Galilean velocity transformation equation. If we assume the second alternative, then we are forced to abandon the notions of absolute time and absolute length that form the basis of the Galilean space–time transformation equations.

39.2   The Michelson–Morley Experiment

The most famous experiment designed to detect small changes in the speed of light was first performed in 1881 by Albert A. Michelson (see Section 37.7) and later repeated under various conditions by Michelson and Edward W. Morley (1838–1923). We state at the outset that the outcome of the experiment contradicted the ether hypothesis.

The experiment was designed to determine the velocity of the Earth relative to that of the hypothetical ether. The experimental tool used was the Michelson interferometer, which was discussed in Section 37.7 and is shown again in Figure 39.4. Arm 2 is aligned along the direction of the Earth's motion through space. The Earth moving through the ether at speed v is equivalent to the ether flowing past the Earth in the opposite direction with speed v. This ether wind blowing in the direction opposite the direction of Earth's motion should cause the speed of light measured in the Earth frame to be $c - v$ as the light approaches mirror M_2 and $c + v$ after reflection, where c is the speed of light in the ether frame.

The two light beams reflect from M_1 and M_2 and recombine, and an interference pattern is formed, as discussed in Section 37.7. The interference pattern is observed while the interferometer is rotated through an angle of 90°. This rotation interchanges the speed of the ether wind between the arms of the interferometer. The rotation should cause the fringe pattern to shift slightly but measurably. Measurements failed, however, to show any change in the interference pattern! The Michelson–Morley experiment was repeated at different times of the year when the ether wind was expected to change direction and magnitude, but the results were always the same: **no fringe shift of the magnitude required was ever observed.**[2]

The negative results of the Michelson–Morley experiment not only contradicted the ether hypothesis but also showed that it was impossible to measure the absolute velocity of the Earth with respect to the ether frame. However, Einstein offered a postulate for his special theory of relativity that places quite a different interpretation on these null results. In later years, when more was known about the nature of light, the idea of an ether that permeates all of space was abandoned. **Light is now understood to be an electromagnetic wave, which requires no medium for its propagation.** As a result, the idea of an ether in which these waves travel became unnecessary.

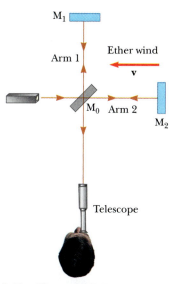

Active Figure 39.4 According to the ether wind theory, the speed of light should be $c - v$ as the beam approaches mirror M_2 and $c + v$ after reflection.

At the Active Figures link at http://www.pse6.com, you can adjust the speed of the ether wind to see the effect on the light beams if there were an ether.

Details of the Michelson–Morley Experiment

To understand the outcome of the Michelson–Morley experiment, let us assume that the two arms of the interferometer in Figure 39.4 are of equal length L. We shall analyze the situation as if there were an ether wind, because that is what Michelson and Morley expected to find. As noted above, the speed of the light beam along arm 2 should be $c - v$ as the beam approaches M_2 and $c + v$ after the beam is reflected. Thus, the time interval for travel to the right is $L/(c - v)$, and the time interval for travel to the left is $L/(c + v)$. The total time interval for the round trip along arm 2 is

$$\Delta t_{\text{arm 2}} = \frac{L}{c + v} + \frac{L}{c - v} = \frac{2Lc}{c^2 - v^2} = \frac{2L}{c}\left(1 - \frac{v^2}{c^2}\right)^{-1}$$

Now consider the light beam traveling along arm 1, perpendicular to the ether wind. Because the speed of the beam relative to the Earth is $(c^2 - v^2)^{1/2}$ in this case (see Fig. 39.3), the time interval for travel for each half of the trip is $L/(c^2 - v^2)^{1/2}$, and the total time interval for the round trip is

$$\Delta t_{\text{arm 1}} = \frac{2L}{(c^2 - v^2)^{1/2}} = \frac{2L}{c}\left(1 - \frac{v^2}{c^2}\right)^{-1/2}$$

Thus, the time difference Δt between the horizontal round trip (arm 2) and the vertical round trip (arm 1) is

$$\Delta t = \Delta t_{\text{arm 2}} - \Delta t_{\text{arm 1}} = \frac{2L}{c}\left[\left(1 - \frac{v^2}{c^2}\right)^{-1} - \left(1 - \frac{v^2}{c^2}\right)^{-1/2}\right]$$

---

[2] From an Earth observer's point of view, changes in the Earth's speed and direction of motion in the course of a year are viewed as ether wind shifts. Even if the speed of the Earth with respect to the ether were zero at some time, six months later the speed of the Earth would be 60 km/s with respect to the ether, and as a result a fringe shift should be noticed. No shift has ever been observed, however.

Because $v^2/c^2 \ll 1$, we can simplify this expression by using the following binomial expansion after dropping all terms higher than second order:

$$(1 - x)^n \approx 1 - nx \qquad (\text{for } x \ll 1)$$

In our case, $x = v^2/c^2$, and we find that

$$\Delta t = \Delta t_{\text{arm 2}} - \Delta t_{\text{arm 1}} \approx \frac{Lv^2}{c^3} \tag{39.3}$$

This time difference between the two instants at which the reflected beams arrive at the viewing telescope gives rise to a phase difference between the beams, producing an interference pattern when they combine at the position of the telescope. A shift in the interference pattern should be detected when the interferometer is rotated through 90° in a horizontal plane, so that the two beams exchange roles. This rotation results in a time difference twice that given by Equation 39.3. Thus, the path difference that corresponds to this time difference is

$$\Delta d = c \, (2 \, \Delta t) = \frac{2Lv^2}{c^2}$$

Because a change in path length of one wavelength corresponds to a shift of one fringe, the corresponding fringe shift is equal to this path difference divided by the wavelength of the light:

$$\text{Shift} = \frac{2Lv^2}{\lambda c^2} \tag{39.4}$$

In the experiments by Michelson and Morley, each light beam was reflected by mirrors many times to give an effective path length L of approximately 11 m. Using this value and taking v to be equal to 3.0×10^4 m/s, the speed of the Earth around the Sun, we obtain a path difference of

$$\Delta d = \frac{2(11 \text{ m})(3.0 \times 10^4 \text{ m/s})^2}{(3.0 \times 10^8 \text{ m/s})^2} = 2.2 \times 10^{-7} \text{ m}$$

This extra travel distance should produce a noticeable shift in the fringe pattern. Specifically, using 500-nm light, we expect a fringe shift for rotation through 90° of

$$\text{Shift} = \frac{\Delta d}{\lambda} = \frac{2.2 \times 10^{-7} \text{ m}}{5.0 \times 10^{-7} \text{ m}} \approx 0.44$$

The instrument used by Michelson and Morley could detect shifts as small as 0.01 fringe. However, **it detected no shift whatsoever in the fringe pattern.** Since then, the experiment has been repeated many times by different scientists under a wide variety of conditions, and no fringe shift has ever been detected. Thus, it was concluded that the motion of the Earth with respect to the postulated ether cannot be detected.

Many efforts were made to explain the null results of the Michelson–Morley experiment and to save the ether frame concept and the Galilean velocity transformation equation for light. All proposals resulting from these efforts have been shown to be wrong. No experiment in the history of physics received such valiant efforts to explain the absence of an expected result as did the Michelson–Morley experiment. The stage was set for Einstein, who solved the problem in 1905 with his special theory of relativity.

39.3   Einstein's Principle of Relativity

In the previous section we noted the impossibility of measuring the speed of the ether with respect to the Earth and the failure of the Galilean velocity transformation equation in the case of light. Einstein proposed a theory that boldly removed these difficulties and at the same time completely altered our notion of space and time.[3] He based his special theory of relativity on two postulates:

1. **The principle of relativity:** The laws of physics must be the same in all inertial reference frames.
2. **The constancy of the speed of light:** The speed of light in vacuum has the same value, $c = 3.00 \times 10^8$ m/s, in all inertial frames, regardless of the velocity of the observer or the velocity of the source emitting the light.

The first postulate asserts that *all* the laws of physics—those dealing with mechanics, electricity and magnetism, optics, thermodynamics, and so on—are the same in all reference frames moving with constant velocity relative to one another. This postulate is a sweeping generalization of the principle of Galilean relativity, which refers only to the laws of mechanics. From an experimental point of view, Einstein's principle of relativity means that any kind of experiment (measuring the speed of light, for example) performed in a laboratory at rest must give the same result when performed in a laboratory moving at a constant velocity with respect to the first one. Hence, no preferred inertial reference frame exists, and it is impossible to detect absolute motion.

Note that postulate 2 is required by postulate 1: if the speed of light were not the same in all inertial frames, measurements of different speeds would make it possible to distinguish between inertial frames; as a result, a preferred, absolute frame could be identified, in contradiction to postulate 1.

Although the Michelson–Morley experiment was performed before Einstein published his work on relativity, it is not clear whether or not Einstein was aware of the details of the experiment. Nonetheless, the null result of the experiment can be readily understood within the framework of Einstein's theory. According to his principle of relativity, the premises of the Michelson–Morley experiment were incorrect. In the process of trying to explain the expected results, we stated that when light traveled against the ether wind its speed was $c - v$, in accordance with the Galilean velocity transformation equation. However, if the state of motion of the observer or of the source has no influence on the value found for the speed of light, one always measures the value to be c. Likewise, the light makes the return trip after reflection from the mirror at speed c, not at speed $c + v$. Thus, the motion of the Earth does not influence the fringe pattern observed in the Michelson–Morley experiment, and a null result should be expected.

If we accept Einstein's theory of relativity, we must conclude that relative motion is unimportant when measuring the speed of light. At the same time, we shall see that we must alter our common-sense notion of space and time and be prepared for some surprising consequences. It may help as you read the pages ahead to keep in mind that our common-sense ideas are based on a lifetime of everyday experiences and not on observations of objects moving at hundreds of thousands of kilometers per second. Thus, these results will seem strange, but that is only because we have no experience with them.

39.4 Consequences of the Special Theory of Relativity

Before we discuss the consequences of Einstein's special theory of relativity, we must first understand how an observer located in an inertial reference frame describes an event. As mentioned earlier, an event is an occurrence describable by three space

Albert Einstein
German-American Physicist
(1879–1955)

Einstein, one of the greatest physicists of all times, was born in Ulm, Germany. In 1905, at the age of 26, he published four scientific papers that revolutionized physics. Two of these papers were concerned with what is now considered his most important contribution: the special theory of relativity.

In 1916, Einstein published his work on the general theory of relativity. The most dramatic prediction of this theory is the degree to which light is deflected by a gravitational field. Measurements made by astronomers on bright stars in the vicinity of the eclipsed Sun in 1919 confirmed Einstein's prediction, and as a result Einstein became a world celebrity.

Einstein was deeply disturbed by the development of quantum mechanics in the 1920s despite his own role as a scientific revolutionary. In particular, he could never accept the probabilistic view of events in nature that is a central feature of quantum theory. The last few decades of his life were devoted to an unsuccessful search for a unified theory that would combine gravitation and electromagnetism. *(AIP Niels Bohr Library)*

---

[3] A. Einstein, "On the Electrodynamics of Moving Bodies," *Ann. Physik* 17:891, 1905. For an English translation of this article and other publications by Einstein, see the book by H. Lorentz, A. Einstein, H. Minkowski, and H. Weyl, *The Principle of Relativity*, Dover, 1958.

coordinates and one time coordinate. Observers in different inertial frames will describe the same event with coordinates that have different values.

As we examine some of the consequences of relativity in the remainder of this section, we restrict our discussion to the concepts of simultaneity, time intervals, and lengths, all three of which are quite different in relativistic mechanics from what they are in Newtonian mechanics. For example, in relativistic mechanics the distance between two points and the time interval between two events depend on the frame of reference in which they are measured. That is, **in relativistic mechanics there is no such thing as an absolute length or absolute time interval.** Furthermore, **events at different locations that are observed to occur simultaneously in one frame are not necessarily observed to be simultaneous in another frame moving uniformly with respect to the first.**

Simultaneity and the Relativity of Time

A basic premise of Newtonian mechanics is that a universal time scale exists that is the same for all observers. In fact, Newton wrote that "Absolute, true, and mathematical time, of itself, and from its own nature, flows equably without relation to anything external." Thus, Newton and his followers simply took simultaneity for granted. In his special theory of relativity, Einstein abandoned this assumption.

Einstein devised the following thought experiment to illustrate this point. A boxcar moves with uniform velocity, and two lightning bolts strike its ends, as illustrated in Figure 39.5a, leaving marks on the boxcar and on the ground. The marks on the boxcar are labeled A' and B', and those on the ground are labeled A and B. An observer O' moving with the boxcar is midway between A' and B', and a ground observer O is midway between A and B. The events recorded by the observers are the striking of the boxcar by the two lightning bolts.

The light signals emitted from A and B at the instant at which the two bolts strike reach observer O at the same time, as indicated in Figure 39.5b. This observer realizes that the signals have traveled at the same speed over equal distances, and so rightly concludes that the events at A and B occurred simultaneously. Now consider the same events as viewed by observer O'. By the time the signals have reached observer O, observer O' has moved as indicated in Figure 39.5b. Thus, the signal from B' has already swept past O', but the signal from A' has not yet reached O'. In other words, O' sees the signal from B' before seeing the signal from A'. According to Einstein, *the two observers must find that light travels at the same speed*. Therefore, observer O' concludes that the lightning strikes the front of the boxcar before it strikes the back.

This thought experiment clearly demonstrates that the two events that appear to be simultaneous to observer O do not appear to be simultaneous to observer O'.

▲ PITFALL PREVENTION

39.2 Who's Right?

You might wonder which observer in Fig. 39.5 is correct concerning the two lightning strikes. *Both are correct*, because the principle of relativity states that *there is no preferred inertial frame of reference.* Although the two observers reach different conclusions, both are correct in their own reference frame because the concept of simultaneity is not absolute. This, in fact, is the central point of relativity—any uniformly moving frame of reference can be used to describe events and do physics.

(a)   (b)

Figure 39.5 (a) Two lightning bolts strike the ends of a moving boxcar. (b) The events appear to be simultaneous to the stationary observer O, standing midway between A and B. The events do not appear to be simultaneous to observer O', who claims that the front of the car is struck before the rear. Note that in (b) the leftward-traveling light signal has already passed O' but the rightward-traveling signal has not yet reached O'.

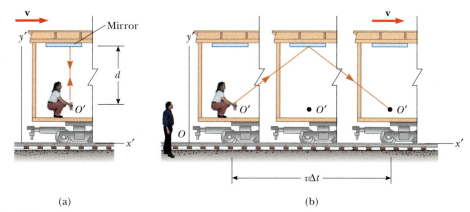

(a)   (b)   (c)

Active Figure 39.6 (a) A mirror is fixed to a moving vehicle, and a light pulse is sent out by observer O' at rest in the vehicle. (b) Relative to a stationary observer O standing alongside the vehicle, the mirror and O' move with a speed v. Note that what observer O measures for the distance the pulse travels is greater than $2d$. (c) The right triangle for calculating the relationship between Δt and Δt_p.

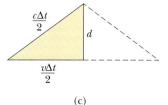

At the Active Figures link at http://www.pse6.com, you can observe the bouncing of the light pulse for various speeds of the train.

In other words,

> two events that are simultaneous in one reference frame are in general not simultaneous in a second frame moving relative to the first. That is, simultaneity is not an absolute concept but rather one that depends on the state of motion of the observer.

Einstein's thought experiment demonstrates that two observers can disagree on the simultaneity of two events. **This disagreement, however, depends on the transit time of light to the observers and, therefore, does *not* demonstrate the deeper meaning of relativity.** In relativistic analyses of high-speed situations, relativity shows that simultaneity is relative even when the transit time is subtracted out. In fact, all of the relativistic effects that we will discuss from here on will assume that we are ignoring differences caused by the transit time of light to the observers.

Time Dilation

We can illustrate the fact that observers in different inertial frames can measure different time intervals between a pair of events by considering a vehicle moving to the right with a speed v, such as the boxcar shown in Figure 39.6a. A mirror is fixed to the ceiling of the vehicle, and observer O' at rest in the frame attached to the vehicle holds a flashlight a distance d below the mirror. At some instant, the flashlight emits a pulse of light directed toward the mirror (event 1), and at some later time after reflecting from the mirror, the pulse arrives back at the flashlight (event 2). Observer O' carries a clock and uses it to measure the time interval Δt_p between these two events. (The subscript p stands for *proper*, as we shall see in a moment.) Because the light pulse has a speed c, the time interval required for the pulse to travel from O' to the mirror and back is

$$\Delta t_p = \frac{\text{distance traveled}}{\text{speed}} = \frac{2d}{c} \tag{39.5}$$

Now consider the same pair of events as viewed by observer O in a second frame, as shown in Figure 39.6b. According to this observer, the mirror and flashlight are moving to the right with a speed v, and as a result the sequence of events appears entirely different. By the time the light from the flashlight reaches the mirror, the mirror has moved to the right a distance $v\,\Delta t/2$, where Δt is the time interval required for the light to travel from O' to the mirror and back to O' as measured by O. In other words, O concludes that, because of the motion of the vehicle, if the light is to hit the mirror, it must leave the flashlight at an angle with respect to the vertical direction. Comparing Figure 39.6a and b, we see that the light must travel farther in (b) than in (a). (Note that neither observer "knows" that he or she is moving. Each is at rest in his or her own inertial frame.)

According to the second postulate of the special theory of relativity, both observers must measure c for the speed of light. Because the light travels farther according to O, it follows that the time interval Δt measured by O is longer than the time interval Δt_p measured by O'. To obtain a relationship between these two time intervals, it is convenient to use the right triangle shown in Figure 39.6c. The Pythagorean theorem gives

$$\left(\frac{c\,\Delta t}{2}\right)^2 = \left(\frac{v\,\Delta t}{2}\right)^2 + d^2$$

Solving for Δt gives

$$\Delta t = \frac{2d}{\sqrt{c^2 - v^2}} = \frac{2d}{c\sqrt{1 - \dfrac{v^2}{c^2}}} \tag{39.6}$$

Because $\Delta t_p = 2d/c$, we can express this result as

Time dilation

$$\Delta t = \frac{\Delta t_p}{\sqrt{1 - \dfrac{v^2}{c^2}}} = \gamma\,\Delta t_p \tag{39.7}$$

where

$$\gamma = \frac{1}{\sqrt{1 - \dfrac{v^2}{c^2}}} \tag{39.8}$$

Because γ is always greater than unity, this result says that **the time interval Δt measured by an observer moving with respect to a clock is longer than the time interval Δt_p measured by an observer at rest with respect to the clock.** This effect is known as **time dilation.**

We can see that time dilation is not observed in our everyday lives by considering the factor γ. This factor deviates significantly from a value of 1 only for very high speeds, as shown in Figure 39.7 and Table 39.1. For example, for a speed of $0.1c$, the value of γ is 1.005. Thus, there is a time dilation of only 0.5% at one-tenth the speed of light. Speeds that we encounter on an everyday basis are far slower than this, so we do not see time dilation in normal situations.

The time interval Δt_p in Equations 39.5 and 39.7 is called the **proper time interval.** (In German, Einstein used the term *Eigenzeit*, which means "own-time.") In

Table 39.1

Approximate Values for γ at Various Speeds	
v/c	γ
0.001 0	1.000 000 5
0.010	1.000 05
0.10	1.005
0.20	1.021
0.30	1.048
0.40	1.091
0.50	1.155
0.60	1.250
0.70	1.400
0.80	1.667
0.90	2.294
0.92	2.552
0.94	2.931
0.96	3.571
0.98	5.025
0.99	7.089
0.995	10.01
0.999	22.37

Figure 39.7 Graph of γ versus v. As the speed approaches that of light, γ increases rapidly.

general, **the proper time interval is the time interval between two events measured by an observer who sees the events occur at the same point in space.**

If a clock is moving with respect to you, the time interval between ticks of the moving clock is observed to be longer than the time interval between ticks of an identical clock in your reference frame. Thus, it is often said that a moving clock is measured to run more slowly than a clock in your reference frame by a factor γ. This is true for mechanical clocks as well as for the light clock just described. We can generalize this result by stating that all physical processes, including chemical and biological ones, are measured to slow down when those processes occur in a frame moving with respect to the observer. For example, the heartbeat of an astronaut moving through space would keep time with a clock inside the spacecraft. Both the astronaut's clock and heartbeat would be measured to slow down according to an observer on Earth comparing time intervals with his own clock (although the astronaut would have no sensation of life slowing down in the spacecraft).

PITFALL PREVENTION

39.3 The Proper Time Interval

It is *very* important in relativistic calculations to correctly identify the observer who measures the proper time interval. The proper time interval between two events is always the time interval measured by an observer for whom the two events take place at the same position.

> **Quick Quiz 39.3** Suppose the observer O' on the train in Figure 39.6 aims her flashlight at the far wall of the boxcar and turns it on and off, sending a pulse of light toward the far wall. Both O' and O measure the time interval between when the pulse leaves the flashlight and it hits the far wall. Which observer measures the proper time interval between these two events? (a) O' (b) O (c) both observers (d) neither observer.

> **Quick Quiz 39.4** A crew watches a movie that is two hours long in a spacecraft that is moving at high speed through space. Will an Earthbound observer, who is watching the movie through a powerful telescope, measure the duration of the movie to be (a) longer than, (b) shorter than, or (c) equal to two hours?

Strange as it may seem, time dilation is a verifiable phenomenon. An experiment reported by Hafele and Keating provided direct evidence of time dilation.[4] Time intervals measured with four cesium atomic clocks in jet flight were compared with time intervals measured by Earth-based reference atomic clocks. In order to compare these results with theory, many factors had to be considered, including periods of speeding up and slowing down relative to the Earth, variations in direction of travel, and the fact that the gravitational field experienced by the flying clocks was weaker than that experienced by the Earth-based clock. The results were in good agreement with the predictions of the special theory of relativity and can be explained in terms of the relative motion between the Earth and the jet aircraft. In their paper, Hafele and Keating stated that "Relative to the atomic time scale of the U.S. Naval Observatory, the flying clocks lost 59 ± 10 ns during the eastward trip and gained 273 ± 7 ns during the westward trip. . . . These results provide an unambiguous empirical resolution of the famous clock paradox with macroscopic clocks."

Another interesting example of time dilation involves the observation of *muons,* unstable elementary particles that have a charge equal to that of the electron and a mass 207 times that of the electron. (We will study the muon and other particles in Chapter 46.) Muons can be produced by the collision of cosmic radiation with atoms high in the atmosphere. Slow-moving muons in the laboratory have a lifetime which is measured to be the proper time interval $\Delta t_p = 2.2$ μs. If we assume that the speed of atmospheric muons is close to the speed of light, we find that these particles can travel a distance of approximately $(3.0 \times 10^8 \text{ m/s})(2.2 \times 10^{-6} \text{ s}) \approx 6.6 \times 10^2$ m before they decay (Fig. 39.8a). Hence, they are unlikely to reach the surface of the Earth from

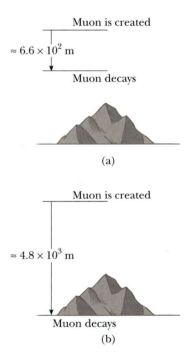

Figure 39.8 (a) Without relativistic considerations, muons created in the atmosphere and traveling downward with a speed of $0.99c$ travel only about 6.6×10^2 m before decaying with an average lifetime of 2.2 μs. Thus, very few muons reach the surface of the Earth. (b) With relativistic considerations, the muon's lifetime is dilated according to an observer on Earth. As a result, according to this observer, the muon can travel about 4.8×10^3 m before decaying. This results in many of them arriving at the surface.

---

[4] J. C. Hafele and R. E. Keating, "Around the World Atomic Clocks: Relativistic Time Gains Observed," *Science,* 177:168, 1972.

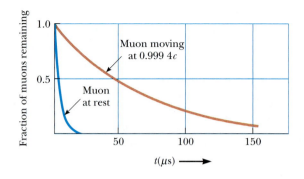

Figure 39.9 Decay curves for muons at rest and for muons traveling at a speed of 0.999 4*c*.

high in the atmosphere where they are produced. However, experiments show that a large number of muons *do* reach the surface. The phenomenon of time dilation explains this effect. As measured by an observer on Earth, the muons have a dilated lifetime equal to $\gamma \Delta t_p$. For example, for $v = 0.99c$, $\gamma \approx 7.1$ and $\gamma \Delta t_p \approx 16$ μs. Hence, the average distance traveled by the muons in this time as measured by an observer on Earth is approximately $(0.99)(3.0 \times 10^8 \text{ m/s})(16 \times 10^{-6} \text{ s}) \approx 4.8 \times 10^3$ m, as indicated in Figure 39.8b.

In 1976, at the laboratory of the European Council for Nuclear Research (CERN) in Geneva, muons injected into a large storage ring reached speeds of approximately 0.999 4*c*. Electrons produced by the decaying muons were detected by counters around the ring, enabling scientists to measure the decay rate and hence the muon lifetime. The lifetime of the moving muons was measured to be approximately 30 times as long as that of the stationary muon (Fig. 39.9), in agreement with the prediction of relativity to within two parts in a thousand.

Example 39.1   What Is the Period of the Pendulum?

The period of a pendulum is measured to be 3.00 s in the reference frame of the pendulum. What is the period when measured by an observer moving at a speed of 0.950*c* relative to the pendulum?

Solution To conceptualize this problem, let us change frames of reference. Instead of the observer moving at 0.950*c*, we can take the equivalent point of view that the observer is at rest and the pendulum is moving at 0.950*c* past the stationary observer. Hence, the pendulum is an example of a clock moving at high speed with respect to an observer and we can categorize this problem as one involving time dilation.

To analyze the problem, note that the proper time interval, measured in the rest frame of the pendulum, is $\Delta t_p = 3.00$ s. Because a clock moving with respect to an observer is measured to run more slowly than a stationary clock by a factor γ, Equation 39.7 gives

$$\Delta t = \gamma \Delta t_p = \frac{1}{\sqrt{1 - \frac{(0.950c)^2}{c^2}}} \Delta t_p = \frac{1}{\sqrt{1 - 0.902}} \Delta t_p$$

$$= (3.20)(3.00 \text{ s}) = \boxed{9.60 \text{ s}}$$

To finalize this problem, we see that indeed a moving pendulum is measured to take longer to complete a period

than a pendulum at rest does. The period increases by a factor of $\gamma = 3.20$. We see that this is consistent with Table 39.1, where this value lies between those for γ for $v/c = 0.94$ and $v/c = 0.96$.

What If? What if we increase the speed of the observer by 5.00%? Does the dilated time interval increase by 5.00%?

Answer Based on the highly nonlinear behavior of γ as a function of v in Figure 39.7, we would guess that the increase in Δt would be different from 5.00%. Increasing v by 5.00% gives us

$$v_{\text{new}} = (1.050\,0)(0.950c) = 0.997\,5c$$

(Because γ varies so rapidly with v when v is this large, we will keep one additional significant figure until the final answer.) If we perform the time dilation calculation again, we find that

$$\Delta t = \gamma \Delta t_p = \frac{1}{\sqrt{1 - \frac{(0.997\,5c)^2}{c^2}}} \Delta t_p = \frac{1}{\sqrt{1 - 0.995\,0}} \Delta t_p$$

$$= (14.15)(3.00 \text{ s}) = 42.5 \text{ s}$$

Thus, the 5.00% increase in speed has caused over a 300% increase in the dilated time!

Example 39.2 How Long Was Your Trip?

Suppose you are driving your car on a business trip and are traveling at 30 m/s. Your boss, who is waiting at your destination, expects the trip to take 5.0 h. When you arrive late, your excuse is that your car clock registered the passage of 5.0 h but that you were driving fast and so your clock ran more slowly than your boss's clock. If your car clock actually did indicate a 5.0-h trip, how much time passed on your boss's clock, which was at rest on the Earth?

Solution We begin by calculating γ from Equation 39.8:

$$\gamma = \frac{1}{\sqrt{1 - \dfrac{v^2}{c^2}}} = \frac{1}{\sqrt{1 - \dfrac{(3 \times 10^1 \text{ m/s})^2}{(3 \times 10^8 \text{ m/s})^2}}}$$

$$= \frac{1}{\sqrt{1 - 10^{-14}}}$$

If you try to determine this value on your calculator, you will probably obtain $\gamma = 1$. However, if we perform a binomial expansion, we can more precisely determine the value as

$$\gamma = (1 - 10^{-14})^{-1/2} \approx 1 + \tfrac{1}{2}(10^{-14}) = 1 + 5.0 \times 10^{-15}$$

This result indicates that at typical automobile speeds, γ is not much different from 1.

Applying Equation 39.7, we find Δt, the time interval measured by your boss, to be

$$\Delta t = \gamma \, \Delta t_p = (1 + 5.0 \times 10^{-15})(5.0 \text{ h})$$

$$= 5.0 \text{ h} + 2.5 \times 10^{-14} \text{ h} = \boxed{5.0 \text{ h} + 0.09 \text{ ns}}$$

Your boss's clock would be only 0.09 ns ahead of your car clock. You might want to think of another excuse!

The Twin Paradox

An intriguing consequence of time dilation is the so-called *twin paradox* (Fig. 39.10). Consider an experiment involving a set of twins named Speedo and Goslo. When they are 20 yr old, Speedo, the more adventuresome of the two, sets out on an epic journey to Planet X, located 20 ly from the Earth. (Note that 1 lightyear (ly) is the distance light travels through free space in 1 year.) Furthermore, Speedo's spacecraft is capable of reaching a speed of 0.95c relative to the inertial frame of his twin brother back home. After reaching Planet X, Speedo becomes homesick and immediately returns to the Earth at the same speed 0.95c. Upon his return, Speedo is shocked to discover that Goslo has aged 42 yr and is now 62 yr old. Speedo, on the other hand, has aged only 13 yr.

At this point, it is fair to raise the following question—which twin is the traveler and which is really younger as a result of this experiment? From Goslo's frame of reference, he was at rest while his brother traveled at a high speed away from him and then came back. According to Speedo, however, he himself remained stationary while Goslo and the Earth raced away from him and then headed back. This leads to an apparent

| Speedo | Goslo | | Speedo | Goslo |
| | (a) | | | (b) |

Figure 39.10 (a) As one twin leaves his brother on the Earth, both are the same age. (b) When Speedo returns from his journey to Planet X, he is younger than his twin Goslo.

contradiction due to the apparent symmetry of the observations. Which twin has developed signs of excess aging?

The situation in our current problem is actually not symmetrical. To resolve this apparent paradox, recall that the special theory of relativity describes observations made in inertial frames of reference moving relative to each other. Speedo, the space traveler, must experience a series of accelerations during his journey because he must fire his rocket engines to slow down and start moving back toward Earth. As a result, his speed is not always uniform, and consequently he is not in an inertial frame. Therefore, there is no paradox—only Goslo, who is always in a single inertial frame, can make correct predictions based on special relativity. During each passing year noted by Goslo, slightly less than 4 months elapses for Speedo.

Only Goslo, who is in a single inertial frame, can apply the simple time-dilation formula to Speedo's trip. Thus, Goslo finds that instead of aging 42 yr, Speedo ages only $(1 - v^2/c^2)^{1/2}(42 \text{ yr}) = 13 \text{ yr}$. Thus, according to Goslo, Speedo spends 6.5 yr traveling to Planet X and 6.5 yr returning, for a total travel time of 13 yr, in agreement with our earlier statement.

Quick Quiz 39.5 Suppose astronauts are paid according to the amount of time they spend traveling in space. After a long voyage traveling at a speed approaching c, would a crew rather be paid according to (a) an Earth-based clock, (b) their spacecraft's clock, or (c) either clock?

Length Contraction

▲ **PITFALL PREVENTION**

39.4   The Proper Length

As with the proper time interval, it is *very* important in relativistic calculations to correctly identify the observer who measures the proper length. The proper length between two points in space is always the length measured by an observer at rest with respect to the points. Often the proper time interval and the proper length are *not* measured by the same observer.

The measured distance between two points also depends on the frame of reference. **The proper length L_p of an object is the length measured by someone at rest relative to the object.** The length of an object measured by someone in a reference frame that is moving with respect to the object is always less than the proper length. This effect is known as **length contraction.**

Consider a spacecraft traveling with a speed v from one star to another. There are two observers: one on the Earth and the other in the spacecraft. The observer at rest on the Earth (and also assumed to be at rest with respect to the two stars) measures the distance between the stars to be the proper length L_p. According to this observer, the time interval required for the spacecraft to complete the voyage is $\Delta t = L_p/v$. The passages of the two stars by the spacecraft occur at the same position for the space traveler. Thus, the space traveler measures the proper time interval Δt_p. Because of time dilation, the proper time interval is related to the Earth-measured time interval by $\Delta t_p = \Delta t/\gamma$. Because the space traveler reaches the second star in the time Δt_p, he or she concludes that the distance L between the stars is

$$L = v \Delta t_p = v \frac{\Delta t}{\gamma}$$

Because the proper length is $L_p = v \Delta t$, we see that

Length contraction

$$L = \frac{L_p}{\gamma} = L_p \sqrt{1 - \frac{v^2}{c^2}} \tag{39.9}$$

where $\sqrt{1 - v^2/c^2}$ is a factor less than unity. **If an object has a proper length L_p when it is measured by an observer at rest with respect to the object, then when it moves with speed v in a direction parallel to its length, its length L is measured to be shorter according to $L = L_p \sqrt{1 - v^2/c^2} = L_p/\gamma$.**

For example, suppose that a meter stick moves past a stationary Earth observer with speed v, as in Figure 39.11. The length of the stick as measured by an observer in a frame attached to the stick is the proper length L_p shown in Figure 39.11a. The length of the stick L measured by the Earth observer is shorter than L_p by the factor $(1 - v^2/c^2)^{1/2}$. Note that **length contraction takes place only along the direction of motion.**

The proper length and the proper time interval are defined differently. The proper length is measured by an observer for whom the end points of the length remain fixed in space. The proper time interval is measured by someone for whom the two events take place at the same position in space. As an example of this point, let us return to the decaying muons moving at speeds close to the speed of light. An observer in the muon's reference frame would measure the proper lifetime, while an Earth-based observer would measure the proper length (the distance from creation to decay in Figure 39.8). In the muon's reference frame, there is no time dilation but the distance of travel to the surface is observed to be shorter when measured in this frame. Likewise, in the Earth observer's reference frame, there is time dilation, but the distance of travel is measured to be the proper length. Thus, when calculations on the muon are performed in both frames, the outcome of the experiment in one frame is the same as the outcome in the other frame—more muons reach the surface than would be predicted without relativistic effects.

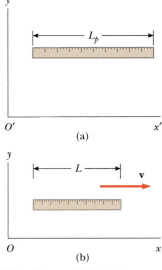

Active Figure 39.11 (a) A meter stick measured by an observer in a frame attached to the stick (that is, both have the same velocity) has its proper length L_p. (b) The stick measured by an observer in a frame in which the stick has a velocity **v** relative to the frame is measured to be shorter than its proper length L_p by a factor $(1 - v^2/c^2)^{1/2}$.

At the Active Figures link at http://www.pse6.com, you can view the meter stick from the points of view of two observers to compare the measured length of the stick.

> **Quick Quiz 39.6** You are packing for a trip to another star. During the journey, you will be traveling at $0.99c$. You are trying to decide whether you should buy smaller sizes of your clothing, because you will be thinner on your trip, due to length contraction. Also, you are considering saving money by reserving a smaller cabin to sleep in, because you will be shorter when you lie down. Should you (a) buy smaller sizes of clothing, (b) reserve a smaller cabin, (c) do neither of these, or (d) do both of these?

> **Quick Quiz 39.7** You are observing a spacecraft moving away from you. You measure it to be shorter than when it was at rest on the ground next to you. You also see a clock through the spacecraft window, and you observe that the passage of time on the clock is measured to be slower than that of the watch on your wrist. Compared to when the spacecraft was on the ground, what do you measure if the spacecraft turns around and comes *toward* you at the same speed? (a) The spacecraft is measured to be longer and the clock runs faster. (b) The spacecraft is measured to be longer and the clock runs slower. (c) The spacecraft is measured to be shorter and the clock runs faster. (d) The spacecraft is measured to be shorter and the clock runs slower.

Space–Time Graphs

It is sometimes helpful to make a *space–time graph*, in which ct is the ordinate and position x is the abscissa. The twin paradox is displayed in such a graph in Figure 39.12

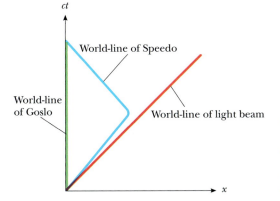

Figure 39.12 The twin paradox on a space–time graph. The twin who stays on the Earth has a world-line along the ct axis. The path of the traveling twin through space–time is represented by a world-line that changes direction.

from the point of view of Goslo. A path through space–time is called a **world-line.** At the origin, the world-lines of Speedo and Goslo coincide because the twins are in the same location at the same time. After Speedo leaves on his trip, his world-line diverges from that of his brother. Goslo's world-line is vertical because he remains fixed in location. At their reunion, the two world-lines again come together. Note that it would be impossible for Speedo to have a world-line that crossed the path of a light beam that left the Earth when he did. To do so would require him to have a speed greater than c (not possible, as shown in Sections 39.6 and 39.7).

World-lines for light beams are diagonal lines on space–time graphs, typically drawn at 45° to the right or left of vertical (assuming that the x and ct axes have the same scales), depending on whether the light beam is traveling in the direction of increasing or decreasing x. These two world-lines mean that all possible future events for Goslo and Speedo lie within two 45° lines extending from the origin. Either twin's presence at an event outside this "light cone" would require that twin to move at a speed greater than c, which we have said is not possible. Also, the only past events that Goslo and Speedo could have experienced occurred within two similar 45° world-lines that approach the origin from below the x axis.

Example 39.3   The Contraction of a Spacecraft

A spacecraft is measured to be 120.0 m long and 20.0 m in diameter while at rest relative to an observer. If this spacecraft now flies by the observer with a speed of $0.99c$, what length and diameter does the observer measure?

Solution From Equation 39.9, the length measured by the observer is

$$L = L_p \sqrt{1 - \frac{v^2}{c^2}} = (120.0\text{ m}) \sqrt{1 - \frac{(0.99c)^2}{c^2}} = \boxed{17\text{ m}}$$

The diameter measured by the observer is still 20.0 m because the diameter is a dimension perpendicular to the motion and length contraction occurs only along the direction of motion.

Example 39.4   The Pole-in-the-Barn Paradox   `Interactive`

The twin paradox, discussed earlier, is a classic "paradox" in relativity. Another classic "paradox" is this: Suppose a runner moving at $0.75c$ carries a horizontal pole 15 m long toward a barn that is 10 m long. The barn has front and rear doors. An observer on the ground can instantly and simultaneously open and close the two doors by remote control. When the runner and the pole are inside the barn, the ground observer closes and then opens both doors so that the runner and pole are momentarily captured inside the barn and then proceed to exit the barn from the back door. Do both the runner and the ground observer agree that the runner makes it safely through the barn?

Solution From our everyday experience, we would be surprised to see a 15-m pole fit inside a 10-m barn. But the pole is in motion with respect to the ground observer, who measures the pole to be contracted to a length L_{pole}, where

$$L_{\text{pole}} = L_p \sqrt{1 - \frac{v^2}{c^2}} = (15\text{ m}) \sqrt{1 - (0.75)^2} = 9.9\text{ m}$$

Thus, the ground observer measures the pole to be slightly shorter than the barn and there is no problem with momentarily capturing the pole inside it. The "paradox" arises when we consider the runner's point of view. The runner

sees the barn contracted to

$$L_{\text{barn}} = L_p \sqrt{1 - \frac{v^2}{c^2}} = (10\text{ m}) \sqrt{1 - (0.75)^2} = \boxed{6.6\text{ m}}$$

Because the pole is in the rest frame of the runner, the runner measures it to have its proper length of 15 m. How can a 15-m pole fit inside a 6.6-m barn? While this is the classic question that is often asked, this is not the question we have asked, because it is not the important question. We asked *if the runner can make it safely through the barn.*

The resolution of the "paradox" lies in the relativity of simultaneity. The closing of the two doors is measured to be simultaneous by the ground observer. Because the doors are at different positions, however, they do not close simultaneously as measured by the runner. The rear door closes and then opens first, allowing the leading edge of the pole to exit. The front door of the barn does not close until the trailing edge of the pole passes by.

We can analyze this using a space-time graph. Figure 39.13a is a space–time graph from the ground observer's point of view. We choose $x = 0$ as the position of the front door of the barn and $t = 0$ as the instant at which the leading end of the pole is located at the front door of the barn. The world-lines for the two ends of the barn are separated by 10 m and are vertical because the barn is not moving relative to this observer. For the pole, we follow two tilted world-lines, one

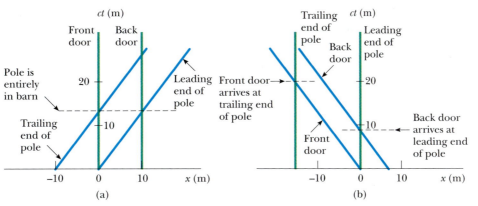

Figure 39.13 (Example 39.4) Space–time graphs for the pole-in-the-barn paradox.
(a) From the ground observer's point of view, the world-lines for the front and back
doors of the barn are vertical lines. The world-lines for the ends of the pole are
tilted and are 9.9 m apart horizontally. The front door of the barn is at $x = 0$, and
the leading end of the pole enters the front door at $t = 0$. The entire pole is inside
the barn at the time indicated by the dashed line. (b) From the runner's point of
view, the world-lines for the ends of the pole are vertical. The barn is moving in the
negative direction, so the world-lines for the front and back doors are tilted to the
left. The leading end of the pole exits the back door before the trailing end arrives
at the front door.

for each end of the moving pole. These world-lines are 9.9 m
apart horizontally, which is the contracted length seen by the
ground observer. As seen in Figure 39.13a, at one instant, the
pole is entirely within the barn.

Figure 39.13b shows the space–time graph according to
the runner. Here, the world-lines for the pole are separated by
15 m and are vertical because the pole is at rest in the runner's
frame of reference. The barn is hurtling *toward* the runner, so
the world-lines for the front and rear doors of the barn are
tilted in the opposite direction compared to Figure 39.13a.
The world-lines for the barn are separated by 6.6 m, the
contracted length as seen by the runner. Notice that the front
of the pole leaves the rear door of the barn long before the
back of the pole enters the barn. Thus, the opening of the
rear door occurs before the closing of the front door.

From the ground observer's point of view, the time at
which the trailing end of the pole enters the barn is found
from

$$\Delta t = t - 0 = t = \frac{\Delta x}{v} = \frac{9.9 \text{ m}}{0.75c} = \frac{13.2 \text{ m}}{c}$$

Thus, the pole should be completely inside the barn at a
time corresponding to $ct = 13.2$ m. This is consistent with
the point on the ct axis in Figure 39.13a where the pole is
inside the barn.

From the runner's point of view, the time at which the
leading end of the pole leaves the barn is found from

$$\Delta t = t - 0 = t = \frac{\Delta x}{v} = \frac{6.6 \text{ m}}{0.75c} = \frac{8.8 \text{ m}}{c}$$

leading to $ct = 8.8$ m. This is consistent with the point on
the ct axis in Figure 39.13b where the back door of the barn
arrives at the leading end of the pole. Finally, the time at
which the trailing end of the pole enters the front door of
the barn is found from

$$\Delta t = t - 0 = t = \frac{\Delta x}{v} = \frac{15 \text{ m}}{0.75c} = \frac{20 \text{ m}}{c}$$

This gives $ct = 20$ m, which agrees with the instant shown in
Figure 39.13b.

Investigate the pole-in-the-barn paradox at the Interactive Worked Example link at **http://www.pse6.com.**

Example 39.5   A Voyage to Sirius

An astronaut takes a trip to Sirius, which is located a
distance of 8 lightyears from the Earth. The astronaut
measures the time of the one-way journey to be 6 yr. If the
spaceship moves at a constant speed of $0.8c$, how can the 8-ly
distance be reconciled with the 6-yr trip time measured by
the astronaut?

Solution The distance of 8 ly represents the proper length
from the Earth to Sirius measured by an observer seeing
both objects nearly at rest. The astronaut sees Sirius
approaching her at $0.8c$ but also sees the distance

contracted to

$$\frac{8 \text{ ly}}{\gamma} = (8 \text{ ly}) \sqrt{1 - \frac{v^2}{c^2}} = (8 \text{ ly}) \sqrt{1 - \frac{(0.8c)^2}{c^2}} = 5 \text{ ly}$$

Thus, the travel time measured on her clock is

$$\Delta t = \frac{d}{v} = \frac{5 \text{ ly}}{0.8c} = 6 \text{ yr}$$

Note that we have used the value for the speed of light as
$c = 1$ ly/yr.

What If? What if this trip is observed with a very powerful telescope by a technician in Mission Control on Earth? At what time will this technician *see* that the astronaut has arrived at Sirius?

Answer The time interval that the technician will measure for the astronaut to arrive is

$$\Delta t = \frac{d}{v} = \frac{8 \text{ ly}}{0.8c} = 10 \text{ yr}$$

In order for the technician to *see* the arrival, the light from the scene of the arrival must travel back to Earth and enter

the telescope. This will require a time interval of

$$\Delta t = \frac{d}{v} = \frac{8 \text{ ly}}{c} = 8 \text{ yr}$$

Thus, the technician sees the arrival after 10 yr + 8 yr = 18 yr. Notice that if the astronaut immediately turns around and comes back home, she arrives, according to the technician, 20 years after leaving, only 2 years after he *saw* her arrive! In addition, she would have aged by only 12 years.

The Relativistic Doppler Effect

Another important consequence of time dilation is the shift in frequency found for light emitted by atoms in motion as opposed to light emitted by atoms at rest. This phenomenon, known as the Doppler effect, was introduced in Chapter 17 as it pertains to sound waves. In the case of sound, the motion of the source with respect to the medium of propagation can be distinguished from the motion of the observer with respect to the medium. Light waves must be analyzed differently, however, because they require no medium of propagation, and no method exists for distinguishing the motion of a light source from the motion of the observer.

If a light source and an observer approach each other with a relative speed v, the frequency f_{obs} measured by the observer is

$$f_{obs} = \frac{\sqrt{1 + v/c}}{\sqrt{1 - v/c}} f_{source} \tag{39.10}$$

where f_{source} is the frequency of the source measured in its rest frame. Note that this relativistic Doppler shift equation, unlike the Doppler shift equation for sound, depends only on the relative speed v of the source and observer and holds for relative speeds as great as c. As you might expect, the equation predicts that $f_{obs} > f_{source}$ when the source and observer approach each other. We obtain the expression for the case in which the source and observer recede from each other by substituting negative values for v in Equation 39.10.

The most spectacular and dramatic use of the relativistic Doppler effect is the measurement of shifts in the frequency of light emitted by a moving astronomical object such as a galaxy. Light emitted by atoms and normally found in the extreme violet region of the spectrum is shifted toward the red end of the spectrum for atoms in other galaxies—indicating that these galaxies are *receding* from us. The American astronomer Edwin Hubble (1889–1953) performed extensive measurements of this *red shift* to confirm that most galaxies are moving away from us, indicating that the Universe is expanding.

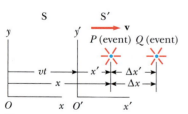

Figure 39.14 Events occur at points *P* and *Q* and are observed by an observer at rest in the S frame and another in the S′ frame, which is moving to the right with a speed *v*.

39.5   The Lorentz Transformation Equations

Suppose an event that occurs at some point P is reported by two observers, one at rest in a frame S and another in a frame S′ that is moving to the right with speed v as in Figure 39.14. The observer in S reports the event with space–time coordinates (x, y, z, t), while the observer in S′ reports the same event using the coordinates (x', y', z', t'). If two events occur at P and Q, Equation 39.1 predicts that $\Delta x = \Delta x'$, that is, the distance between the two points in space at which the events occur does not depend on motion of the observer. Because this is contradictory to the notion of length contraction, the Galilean transformation is not valid when v approaches the speed of light. In this section, we state the correct transformation equations that apply for all speeds in the range $0 \leq v < c$.

The equations that are valid for all speeds and enable us to transform coordinates from S to S′ are the **Lorentz transformation equations:**

$$x' = \gamma(x - vt) \qquad y' = y \qquad z' = z \qquad t' = \gamma\left(t - \frac{v}{c^2}x\right) \qquad \text{(39.11)}$$

Lorentz transformation for S → S′

These transformation equations were developed by Hendrik A. Lorentz (1853–1928) in 1890 in connection with electromagnetism. However, it was Einstein who recognized their physical significance and took the bold step of interpreting them within the framework of the special theory of relativity.

Note the difference between the Galilean and Lorentz time equations. In the Galilean case, $t = t'$, but in the Lorentz case the value for t' assigned to an event by an observer O' in the S′ frame in Figure 39.14 depends both on the time t and on the coordinate x as measured by an observer O in the S frame. This is consistent with the notion that an event is characterized by four space–time coordinates (x, y, z, t). In other words, in relativity, space and time are *not* separate concepts but rather are closely interwoven with each other.

If we wish to transform coordinates in the S′ frame to coordinates in the S frame, we simply replace v by $-v$ and interchange the primed and unprimed coordinates in Equations 39.11:

$$x = \gamma(x' + vt') \qquad y = y' \qquad z = z' \qquad t = \gamma\left(t' + \frac{v}{c^2}x'\right) \qquad \text{(39.12)}$$

Inverse Lorentz transformation for S′ → S

When $v \ll c$, the Lorentz transformation equations should reduce to the Galilean equations. To verify this, note that as v approaches zero, $v/c \ll 1$; thus, $\gamma \rightarrow 1$, and Equations 39.11 reduce to the Galilean space–time transformation equations:

$$x' = x - vt \qquad y' = y \qquad z' = z \qquad t' = t$$

In many situations, we would like to know the difference in coordinates between two events or the time interval between two events as seen by observers O and O'. We can accomplish this by writing the Lorentz equations in a form suitable for describing pairs of events. From Equations 39.11 and 39.12, we can express the differences between the four variables x, x', t, and t' in the form

$$\left.\begin{array}{l} \Delta x' = \gamma(\Delta x - v\,\Delta t) \\[2mm] \Delta t' = \gamma\left(\Delta t - \dfrac{v}{c^2}\,\Delta x\right) \end{array}\right\} \text{S} \rightarrow \text{S}' \qquad \text{(39.13)}$$

$$\left.\begin{array}{l} \Delta x = \gamma(\Delta x' + v\,\Delta t') \\[2mm] \Delta t = \gamma\left(\Delta t' + \dfrac{v}{c^2}\,\Delta x'\right) \end{array}\right\} \text{S}' \rightarrow \text{S} \qquad \text{(39.14)}$$

where $\Delta x' = x'_2 - x'_1$ and $\Delta t' = t'_2 - t'_1$ are the differences measured by observer O' and $\Delta x = x_2 - x_1$ and $\Delta t = t_2 - t_1$ are the differences measured by observer O. (We have not included the expressions for relating the y and z coordinates because they are unaffected by motion along the x direction.[5])

---

[5] Although relative motion of the two frames along the x axis does not change the y and z coordinates of an object, it does change the y and z velocity components of an object moving in either frame, as noted in Section 39.6.

Example 39.6   Simultaneity and Time Dilation Revisited

Use the Lorentz transformation equations in difference form to show that

(A) simultaneity is not an absolute concept and that

(B) a moving clock is measured to run more slowly than a clock that is at rest with respect to an observer.

Solution (A) Suppose that two events are simultaneous and separated in space such that $\Delta t' = 0$ and $\Delta x' \neq 0$ according to an observer O' who is moving with speed v relative to O. From the expression for Δt given in Equation 39.14, we see that in this case the time interval Δt measured by observer O is $\Delta t = \gamma v \, \Delta x'/c^2$. That is, the

time interval for the same two events as measured by O is nonzero, and so the events do not appear to be simultaneous to O.

(B) Suppose that observer O' carries a clock that he uses to measure a time interval $\Delta t'$. He finds that two events occur at the same place in his reference frame ($\Delta x' = 0$) but at different times ($\Delta t' \neq 0$). Observer O' is moving with speed v relative to O, who measures the time interval between the events to be Δt. In this situation, the expression for Δt given in Equation 39.14 becomes $\Delta t = \gamma \, \Delta t'$. This is the equation for time dilation found earlier (Eq. 39.7), where $\Delta t' = \Delta t_p$ is the proper time measured by the clock carried by observer O'. Thus, O measures the moving clock to run slow.

39.6 The Lorentz Velocity Transformation Equations

Suppose two observers in relative motion with respect to each other are both observing the motion of an object. Previously, we defined an event as occurring at an instant of time. Now, we wish to interpret the "event" as the motion of the object. We know that the Galilean velocity transformation (Eq. 39.2) is valid for low speeds. How do the observers' measurements of the velocity of the object relate to each other if the speed of the object is close to that of light? Once again S′ is our frame moving at a speed v relative to S. Suppose that an object has a velocity component u'_x measured in the S′ frame, where

$$u'_x = \frac{dx'}{dt'} \tag{39.15}$$

Using Equation 39.11, we have

$$dx' = \gamma(dx - v\,dt)$$

$$dt' = \gamma\left(dt - \frac{v}{c^2}\,dx\right)$$

Substituting these values into Equation 39.15 gives

$$u'_x = \frac{dx'}{dt'} = \frac{dx - v\,dt}{dt - \dfrac{v}{c^2}\,dx} = \frac{\dfrac{dx}{dt} - v}{1 - \dfrac{v}{c^2}\dfrac{dx}{dt}}$$

But dx/dt is just the velocity component u_x of the object measured by an observer in S, and so this expression becomes

Lorentz velocity transformation for S → S′

$$u'_x = \frac{u_x - v}{1 - \dfrac{u_x v}{c^2}} \tag{39.16}$$

If the object has velocity components along the y and z axes, the components as measured by an observer in S′ are

$$u'_y = \frac{u_y}{\gamma\left(1 - \dfrac{u_x v}{c^2}\right)} \quad \text{and} \quad u'_z = \frac{u_z}{\gamma\left(1 - \dfrac{u_x v}{c^2}\right)} \tag{39.17}$$

Note that u'_y and u'_z do not contain the parameter v in the numerator because the relative velocity is along the x axis.

When v is much smaller than c (the nonrelativistic case), the denominator of Equation 39.16 approaches unity, and so $u'_x \approx u_x - v$, which is the Galilean velocity transformation equation. In another extreme, when $u_x = c$, Equation 39.16 becomes

$$u'_x = \frac{c - v}{1 - \dfrac{cv}{c^2}} = \frac{c\left(1 - \dfrac{v}{c}\right)}{1 - \dfrac{v}{c}} = c$$

From this result, we see that a speed measured as c by an observer in S is also measured as c by an observer in S'—independent of the relative motion of S and S'. Note that this conclusion is consistent with Einstein's second postulate—that the speed of light must be c relative to all inertial reference frames. Furthermore, we find that the speed of an object can never be measured as larger than c. That is, the speed of light is the ultimate speed. We return to this point later.

To obtain u_x in terms of u'_x, we replace v by $-v$ in Equation 39.16 and interchange the roles of u_x and u'_x:

$$u_x = \frac{u'_x + v}{1 + \dfrac{u'_x v}{c^2}} \tag{39.18}$$

▲ **PITFALL PREVENTION**

39.5 What Can the Observers Agree On?

We have seen several measurements that the two observers O and O' do *not* agree on: (1) the time interval between events that take place in the same position in one of the frames, (2) the distance between two points that remain fixed in one of their frames, (3) the velocity components of a moving particle, and (4) whether two events occurring at different locations in both frames are simultaneous or not. Note that the two observers *can* agree on (1) their relative speed of motion v with respect to each other, (2) the speed c of any ray of light, and (3) the simultaneity of two events which take place at the same position *and* time in some frame.

Quick Quiz 39.8 You are driving on a freeway at a relativistic speed. Straight ahead of you, a technician standing on the ground turns on a searchlight and a beam of light moves exactly vertically upward, as seen by the technician. As you observe the beam of light, you measure the magnitude of the vertical component of its velocity as (a) equal to c (b) greater than c (c) less than c.

Quick Quiz 39.9 Consider the situation in Quick Quiz 39.8 again. If the technician aims the searchlight directly at you instead of upward, you measure the magnitude of the horizontal component of its velocity as (a) equal to c (b) greater than c (c) less than c.

Example 39.7   Relative Velocity of Two Spacecraft

Two spacecraft A and B are moving in opposite directions, as shown in Figure 39.15. An observer on the Earth measures the speed of craft A to be $0.750c$ and the speed of craft B to be $0.850c$. Find the velocity of craft B as observed by the crew on craft A.

Figure 39.15 (Example 39.7) Two spacecraft A and B move in opposite directions. The speed of B relative to A is *less* than c and is obtained from the relativistic velocity transformation equation.

Solution To conceptualize this problem, we carefully identify the observers and the event. The two observers are on the Earth and on spacecraft A. The event is the motion of spacecraft B. Because the problem asks to find an observed velocity, we categorize this problem as one requiring the Lorentz velocity transformation. To analyze the problem, we note that the Earth observer makes two measurements, one of each spacecraft. We identify this observer as being at rest in the S frame. Because the velocity of spacecraft B is what we wish to measure, we identify the speed u_x as $-0.850c$. The velocity of spacecraft A is also the velocity of the observer at rest in the S' frame, which is attached to the spacecraft, relative to the observer at rest in S. Thus, $v = 0.750c$. Now we can obtain the velocity u'_x of craft B relative to craft A by using Equation 39.16:

$$u'_x = \frac{u_x - v}{1 - \dfrac{u_x v}{c^2}} = \frac{-0.850c - 0.750c}{1 - \dfrac{(-0.850c)(0.750c)}{c^2}}$$

$$= -0.977c$$

To finalize this problem, note that the negative sign indicates that craft B is moving in the negative *x* direction as observed by the crew on craft A. Is this consistent with your expectation from Figure 39.15? Note that the speed is less than *c*. That is, an object whose speed is less than *c* in one frame of reference must have a speed less than *c* in any other frame. (If the Galilean velocity transformation equation were used in this example, we would find that $u'_x = u_x - v = -0.850c - 0.750c = -1.60c$, which is impossible. The Galilean transformation equation does not work in relativistic situations.)

What If? What if the two spacecraft pass each other? Now what is their relative speed?

Answer The calculation using Equation 39.16 involves only the velocities of the two spacecraft and does not depend on their locations. After they pass each other, they have the same velocities, so the velocity of craft B as observed by the crew on craft A is the same, $-0.977c$. The only difference after they pass is that B is receding from A whereas it was approaching A before it passed.

Example 39.8   The Speeding Motorcycle

Imagine a motorcycle moving with a speed $0.80c$ past a stationary observer, as shown in Figure 39.16. If the rider tosses a ball in the forward direction with a speed of $0.70c$ relative to himself, what is the speed of the ball relative to the stationary observer?

Solution The speed of the motorcycle relative to the stationary observer is $v = 0.80c$. The speed of the ball in the frame of reference of the motorcyclist is $u'_x = 0.70c$. Therefore, the speed u_x of the ball relative to the stationary observer is

0.80c
0.70c

$$u_x = \frac{u'_x + v}{1 + \dfrac{u'_x v}{c^2}} = \frac{0.70c + 0.80c}{1 + \dfrac{(0.70c)(0.80c)}{c^2}} = 0.96c$$

Figure 39.16 (Example 39.8) A motorcyclist moves past a stationary observer with a speed of $0.80c$ and throws a ball in the direction of motion with a speed of $0.70c$ relative to himself.

Example 39.9   Relativistic Leaders of the Pack

Interactive

Two motorcycle pack leaders named David and Emily are racing at relativistic speeds along perpendicular paths, as shown in Figure 39.17. How fast does Emily recede as seen by David over his right shoulder?

Solution Figure 39.17 represents the situation as seen by a police officer at rest in frame S, who observes the following:

David: $u_x = 0.75c$ $u_y = 0$

Emily: $u_x = 0$ $u_y = -0.90c$

To calculate Emily's speed of recession as seen by David, we take S′ to move along with David and then calculate u'_x and

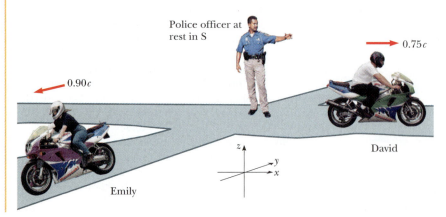

Police officer at rest in S

0.75c

0.90c

David

Emily

Figure 39.17 (Example 39.9) David moves to the east with a speed $0.75c$ relative to the police officer, and Emily travels south at a speed $0.90c$ relative to the officer.

u'_y for Emily using Equations 39.16 and 39.17:

$$u'_x = \frac{u_x - v}{1 - \dfrac{u_x v}{c^2}} = \frac{0 - 0.75c}{1 - \dfrac{(0)(0.75c)}{c^2}} = -0.75c$$

$$u'_y = \frac{u_y}{\gamma\left(1 - \dfrac{u_x v}{c^2}\right)} = \frac{\sqrt{1 - \dfrac{(0.75c)^2}{c^2}}\,(-0.90c)}{\left(1 - \dfrac{(0)(0.75c)}{c^2}\right)}$$

$$= -0.60c$$

Thus, the speed of Emily as observed by David is

$$u' = \sqrt{(u'_x)^2 + (u'_y)^2} = \sqrt{(-0.75c)^2 + (-0.60c)^2}$$

$$= \boxed{0.96c}$$

Note that this speed is less than c, as required by the special theory of relativity.

Investigate this situation with various speeds of David and Emily at the Interactive Worked Example link at http://www.pse6.com.

39.7 Relativistic Linear Momentum and the Relativistic Form of Newton's Laws

We have seen that in order to describe properly the motion of particles within the framework of the special theory of relativity, we must replace the Galilean transformation equations by the Lorentz transformation equations. Because the laws of physics must remain unchanged under the Lorentz transformation, we must generalize Newton's laws and the definitions of linear momentum and energy to conform to the Lorentz transformation equations and the principle of relativity. These generalized definitions should reduce to the classical (nonrelativistic) definitions for $v \ll c$.

First, recall that the law of conservation of linear momentum states that when two particles (or objects that can be modeled as particles) collide, the total momentum of the isolated system of the two particles remains constant. Suppose that we observe this collision in a reference frame S and confirm that the momentum of the system is conserved. Now imagine that the momenta of the particles are measured by an observer in a second reference frame S′ moving with velocity **v** relative to the first frame. Using the Lorentz velocity transformation equation and the classical definition of linear momentum, $\mathbf{p} = m\mathbf{u}$ (where **u** is the velocity of a particle), we find that linear momentum is *not* measured to be conserved by the observer in S′. However, because the laws of physics are the same in all inertial frames, linear momentum of the system must be conserved in all frames. We have a contradiction. In view of this contradiction and assuming that the Lorentz velocity transformation equation is correct, we must modify the definition of linear momentum to satisfy the following conditions:

- The linear momentum of an isolated system must be conserved in all collisions.

- The relativistic value calculated for the linear momentum **p** of a particle must approach the classical value $m\mathbf{u}$ as **u** approaches zero.

For any particle, the correct relativistic equation for linear momentum that satisfies these conditions is

$$\mathbf{p} \equiv \frac{m\mathbf{u}}{\sqrt{1 - \dfrac{u^2}{c^2}}} = \gamma m\mathbf{u} \qquad (39.19)$$

where **u** is the velocity of the particle and m is the mass of the particle. When u is much less than c, $\gamma = (1 - u^2/c^2)^{-1/2}$ approaches unity and **p** approaches $m\mathbf{u}$. Therefore,

▲ **PITFALL PREVENTION**

39.6 Watch Out for "Relativistic Mass"

Some older treatments of relativity maintained the conservation of momentum principle at high speeds by using a model in which the mass of a particle increases with speed. You might still encounter this notion of "relativistic mass" in your outside reading, especially in older books. Be aware that this notion is no longer widely accepted and mass is considered as *invariant*, independent of speed. The mass of an object in all frames is considered to be the mass as measured by an observer at rest with respect to the object.

Definition of relativistic linear momentum

SPEED LIMIT
3×10^8 m/s

The speed of light is the speed limit of the Universe. It is the maximum possible speed for energy transfer and for information transfer. Any object with mass must move at a lower speed.

the relativistic equation for **p** does indeed reduce to the classical expression when u is much smaller than c.

The relativistic force **F** acting on a particle whose linear momentum is **p** is defined as

$$\mathbf{F} \equiv \frac{d\mathbf{p}}{dt} \tag{39.20}$$

where **p** is given by Equation 39.19. This expression, which is the relativistic form of Newton's second law, is reasonable because it preserves classical mechanics in the limit of low velocities and is consistent with conservation of linear momentum for an isolated system (**F** = 0) both relativistically and classically.

It is left as an end-of-chapter problem (Problem 69) to show that under relativistic conditions, the acceleration **a** of a particle decreases under the action of a constant force, in which case $a \propto (1 - u^2/c^2)^{3/2}$. From this proportionality, we see that as the particle's speed approaches c, the acceleration caused by any finite force approaches zero. Hence, it is impossible to accelerate a particle from rest to a speed $u \geq c$. This argument shows that the speed of light is the ultimate speed, as noted at the end of the preceding section.

Example 39.10   Linear Momentum of an Electron

An electron, which has a mass of 9.11×10^{-31} kg, moves with a speed of $0.750c$. Find its relativistic momentum and compare this value with the momentum calculated from the classical expression.

Solution Using Equation 39.19 with $u = 0.750c$, we have

$$p = \frac{m_e u}{\sqrt{1 - \dfrac{u^2}{c^2}}}$$

$$p = \frac{(9.11 \times 10^{-31} \text{ kg})(0.750)(3.00 \times 10^8 \text{ m/s})}{\sqrt{1 - \dfrac{(0.750c)^2}{c^2}}}$$

$$p = \quad 3.10 \times 10^{-22} \text{ kg·m/s}$$

The classical expression (used incorrectly here) gives

$$p_{\text{classical}} = m_e u = 2.05 \times 10^{-22} \text{ kg·m/s}$$

Hence, the correct relativistic result is 50% greater than the classical result!

39.8   Relativistic Energy

We have seen that the definition of linear momentum requires generalization to make it compatible with Einstein's postulates. This implies that most likely the definition of kinetic energy must also be modified.

To derive the relativistic form of the work–kinetic energy theorem, let us imagine a particle moving in one dimension along the x axis. A force in the x direction causes the momentum of the particle to change according to Equation 39.20. The work done by the force F on the particle is

$$W = \int_{x_1}^{x_2} F \, dx = \int_{x_1}^{x_2} \frac{dp}{dt} \, dx \tag{39.21}$$

In order to perform this integration and find the work done on the particle and the relativistic kinetic energy as a function of u, we first evaluate dp/dt:

$$\frac{dp}{dt} = \frac{d}{dt} \frac{mu}{\sqrt{1 - \dfrac{u^2}{c^2}}} = \frac{m(du/dt)}{\left(1 - \dfrac{u^2}{c^2}\right)^{3/2}}$$

Substituting this expression for dp/dt and $dx = u\,dt$ into Equation 39.21 gives

$$W = \int_0^t \frac{m(du/dt)\,u\,dt}{\left(1 - \dfrac{u^2}{c^2}\right)^{3/2}} = m \int_0^u \frac{u}{\left(1 - \dfrac{u^2}{c^2}\right)^{3/2}}\,du$$

where we use the limits 0 and u in the integral because the integration variable has been changed from t to u. We assume that the particle is accelerated from rest to some final speed u. Evaluating the integral, we find that

$$W = \frac{mc^2}{\sqrt{1 - \dfrac{u^2}{c^2}}} - mc^2 \qquad (39.22)$$

Recall from Chapter 7 that the work done by a force acting on a system consisting of a single particle equals the change in kinetic energy of the particle. Because we assumed that the initial speed of the particle is zero, we know that its initial kinetic energy is zero. We therefore conclude that the work W in Equation 39.22 is equivalent to the relativistic kinetic energy K:

$$K = \frac{mc^2}{\sqrt{1 - \dfrac{u^2}{c^2}}} - mc^2 = \gamma mc^2 - mc^2 = (\gamma - 1)\,mc^2 \qquad (39.23)$$

Relativistic kinetic energy

This equation is routinely confirmed by experiments using high-energy particle accelerators.

At low speeds, where $u/c \ll 1$, Equation 39.23 should reduce to the classical expression $K = \frac{1}{2}mu^2$. We can check this by using the binomial expansion $(1 - \beta^2)^{-1/2} \approx 1 + \frac{1}{2}\beta^2 + \cdots$ for $\beta \ll 1$, where the higher-order powers of β are neglected in the expansion. (In treatments of relativity, β is a common symbol used to represent u/c or v/c.) In our case, $\beta = u/c$, so that

$$\gamma = \frac{1}{\sqrt{1 - \dfrac{u^2}{c^2}}} = \left(1 - \frac{u^2}{c^2}\right)^{-1/2} \approx 1 + \frac{1}{2}\frac{u^2}{c^2}$$

Substituting this into Equation 39.23 gives

$$K \approx \left[\left(1 + \frac{1}{2}\frac{u^2}{c^2}\right) - 1\right]mc^2 = \frac{1}{2}mu^2 \qquad \text{(for } u/c \ll 1\text{)}$$

which is the classical expression for kinetic energy. A graph comparing the relativistic and nonrelativistic expressions is given in Figure 39.18. In the relativistic case, the particle speed never exceeds c, regardless of the kinetic energy. The two curves are in good agreement when $u \ll c$.

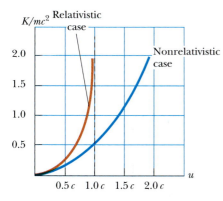

Figure 39.18 A graph comparing relativistic and nonrelativistic kinetic energy of a moving particle. The energies are plotted as a function of particle speed u. In the relativistic case, u is always less than c.

The constant term mc^2 in Equation 39.23, which is independent of the speed of the particle, is called the **rest energy** E_R of the particle:

Rest energy

$$E_R = mc^2 \qquad (39.24)$$

The term γmc^2, which does depend on the particle speed, is therefore the sum of the kinetic and rest energies. We define γmc^2 to be the **total energy** E:

Total energy = kinetic energy + rest energy

$$E = K + mc^2 \qquad (39.25)$$

or

Total energy of a relativistic particle

$$E = \frac{mc^2}{\sqrt{1 - \dfrac{u^2}{c^2}}} = \gamma mc^2 \qquad (39.26)$$

The relationship $E = K + mc^2$ shows that **mass is a form of energy,** where c^2 in the rest energy term is just a constant conversion factor. This expression also shows that a small mass corresponds to an enormous amount of energy, a concept fundamental to nuclear and elementary-particle physics.

In many situations, the linear momentum or energy of a particle is measured rather than its speed. It is therefore useful to have an expression relating the total energy E to the relativistic linear momentum p. This is accomplished by using the expressions $E = \gamma mc^2$ and $p = \gamma mu$. By squaring these equations and subtracting, we can eliminate u (Problem 43). The result, after some algebra, is[6]

Energy–momentum relationship for a relativistic particle

$$E^2 = p^2c^2 + (mc^2)^2 \qquad (39.27)$$

When the particle is at rest, $p = 0$ and so $E = E_R = mc^2$.

In Section 35.1, we introduced the concept of a particle of light, called a **photon.** For particles that have zero mass, such as photons, we set $m = 0$ in Equation 39.27 and find that

$$E = pc \qquad (39.28)$$

This equation is an exact expression relating total energy and linear momentum for photons, which always travel at the speed of light (in vacuum).

Finally, note that because the mass m of a particle is independent of its motion, m must have the same value in all reference frames. For this reason, m is often called the **invariant mass.** On the other hand, because the total energy and linear momentum of a particle both depend on velocity, these quantities depend on the reference frame in which they are measured.

When we are dealing with subatomic particles, it is convenient to express their energy in electron volts (Section 25.1) because the particles are usually given this energy by acceleration through a potential difference. The conversion factor, as you recall from Equation 25.5, is

$$1 \text{ eV} = 1.60 \times 10^{-19} \text{ J}$$

For example, the mass of an electron is 9.11×10^{-31} kg. Hence, the rest energy of the electron is

$$m_e c^2 = (9.11 \times 10^{-31} \text{ kg})(3.00 \times 10^8 \text{ m/s})^2 = 8.20 \times 10^{-14} \text{ J}$$
$$= (8.20 \times 10^{-14} \text{ J})(1 \text{ eV}/1.60 \times 10^{-19} \text{ J}) = 0.511 \text{ MeV}$$

---

[6]   One way to remember this relationship is to draw a right triangle having a hypotenuse of length E and legs of lengths pc and mc^2.

Quick Quiz 39.10 The following *pairs* of energies represent the rest energy and total energy of three different particles: particle 1: *E*, 2*E*; particle 2: *E*, 3*E*; particle 3: 2*E*, 4*E*. Rank the particles, from greatest to least, according to their (a) mass; (b) kinetic energy; (c) speed.

Example 39.11   The Energy of a Speedy Electron

An electron in a television picture tube typically moves with a speed $u = 0.250c$. Find its total energy and kinetic energy in electron volts.

Solution Using the fact that the rest energy of the electron is 0.511 MeV together with Equation 39.26, we have

$$E = \frac{m_e c^2}{\sqrt{1 - \dfrac{u^2}{c^2}}} = \frac{0.511 \text{ MeV}}{\sqrt{1 - \dfrac{(0.250c)^2}{c^2}}}$$

$$= 1.03(0.511 \text{ MeV}) = \boxed{0.528 \text{ MeV}}$$

This is 3% greater than the rest energy.

We obtain the kinetic energy by subtracting the rest energy from the total energy:

$$K = E - m_e c^2 = 0.528 \text{ MeV} - 0.511 \text{ MeV}$$

$$= \boxed{0.017 \text{ MeV}}$$

Example 39.12   The Energy of a Speedy Proton

(A) Find the rest energy of a proton in electron volts.

Solution Using Equation 39.24,

$$E_R = m_p c^2 = (1.67 \times 10^{-27} \text{ kg})(3.00 \times 10^8 \text{ m/s})^2$$

$$= (1.50 \times 10^{-10} \text{ J}) \left(\frac{1.00 \text{ eV}}{1.60 \times 10^{-19} \text{ J}} \right)$$

$$= \boxed{938 \text{ MeV}}$$

(B) If the total energy of a proton is three times its rest energy, what is the speed of the proton?

Solution Equation 39.26 gives

$$E = 3m_p c^2 = \frac{m_p c^2}{\sqrt{1 - \dfrac{u^2}{c^2}}}$$

$$3 = \frac{1}{\sqrt{1 - \dfrac{u^2}{c^2}}}$$

Solving for *u* gives

$$\left(1 - \frac{u^2}{c^2} \right) = \frac{1}{9}$$

$$\frac{u^2}{c^2} = \frac{8}{9}$$

$$u = \frac{\sqrt{8}}{3} c = 0.943c = \boxed{2.83 \times 10^8 \text{ m/s}}$$

(C) Determine the kinetic energy of the proton in electron volts.

Solution From Equation 39.25,

$$K = E - m_p c^2 = 3m_p c^2 - m_p c^2 = 2m_p c^2$$

Because $m_p c^2 = 938$ MeV, we see that $K = \boxed{1\,880 \text{ MeV}}$.

(D) What is the proton's momentum?

Solution We can use Equation 39.27 to calculate the momentum with $E = 3m_p c^2$:

$$E^2 = p^2 c^2 + (m_p c^2)^2 = (3m_p c^2)^2$$

$$p^2 c^2 = 9(m_p c^2)^2 - (m_p c^2)^2 = 8(m_p c^2)^2$$

$$p = \sqrt{8} \frac{m_p c^2}{c} = \sqrt{8} \frac{(938 \text{ MeV})}{c} = \boxed{2\,650 \text{ MeV}/c}$$

The unit of momentum is written MeV/*c*, which is a common unit in particle physics.

What If? In classical physics, if the momentum of a particle doubles, the kinetic energy increases by a factor of 4. What happens to the kinetic energy of the speedy proton in this example if its momentum doubles?

Answer Based on what we have seen so far in relativity, it is likely that you would predict that its kinetic energy does not increase by a factor of 4. If the momentum doubles, the new momentum is

$$p_{\text{new}} = 2 \left(\sqrt{8} \frac{m_p c^2}{c} \right) = 4\sqrt{2} \frac{m_p c^2}{c}$$

Using Equation 39.27, we find the square of the new total energy:

$$E_{\text{new}}^2 = p_{\text{new}}^2 c^2 + (m_p c^2)^2$$

$$E_{new}^2 = \left(4\sqrt{2} \, \frac{m_p c^2}{c}\right)^2 c^2 + (m_p c^2)^2 = 33(m_p c^2)^2$$

$$E_{new} = \sqrt{33}(m_p c^2) = 5.7 m_p c^2$$

Now, using Equation 39.25, we find the new kinetic energy:

$$K_{new} = E_{new} - m_p c^2 = 5.7 m_p c^2 - m_p c^2 = 4.7 m_p c^2$$

Notice that this is only 2.35 times as large as the kinetic energy we found in part (C), not four times as large. In general, the factor by which the kinetic energy increases if the momentum doubles will depend on the initial momentum, but will approach 4 as the momentum approaches zero. In this latter situation, classical physics correctly describes the situation.

39.9 Mass and Energy

Equation 39.26, $E = \gamma m c^2$, which represents the total energy of a particle, suggests that even when a particle is at rest ($\gamma = 1$) it still possesses enormous energy through its mass. The clearest experimental proof of the equivalence of mass and energy occurs in nuclear and elementary particle interactions in which the conversion of mass into kinetic energy takes place. Because of this, in relativistic situations, we cannot use the principle of conservation of energy as it was outlined in Chapters 7 and 8. We must include rest energy as another form of energy storage.

This concept is important in atomic and nuclear processes, in which the change in mass is a relatively large fraction of the initial mass. For example, in a conventional nuclear reactor, the uranium nucleus undergoes *fission*, a reaction that results in several lighter fragments having considerable kinetic energy. In the case of ^{235}U, which is used as fuel in nuclear power plants, the fragments are two lighter nuclei and a few neutrons. The total mass of the fragments is less than that of the ^{235}U by an amount Δm. The corresponding energy $\Delta m c^2$ associated with this mass difference is exactly equal to the total kinetic energy of the fragments. The kinetic energy is absorbed as the fragments move through water, raising the internal energy of the water. This internal energy is used to produce steam for the generation of electrical power.

Next, consider a basic *fusion* reaction in which two deuterium atoms combine to form one helium atom. The decrease in mass that results from the creation of one helium atom from two deuterium atoms is $\Delta m = 4.25 \times 10^{-29}$ kg. Hence, the corresponding energy that results from one fusion reaction is $\Delta m c^2 = 3.83 \times 10^{-12}$ J = 23.9 MeV. To appreciate the magnitude of this result, if only 1 g of deuterium is converted to helium, the energy released is on the order of 10^{12} J! At the year 2003 cost of electrical energy, this would be worth about $30 000. We shall present more details of these nuclear processes in Chapter 45 of the extended version of this textbook.

Example 39.13   Mass Change in a Radioactive Decay

The ^{216}Po nucleus is unstable and exhibits radioactivity (Chapter 44). It decays to ^{212}Pb by emitting an alpha particle, which is a helium nucleus, ^4He. Find

(A) the mass change in this decay and

(B) the energy that this represents.

Solution Using values in Table A.3, we see that the initial and final masses are

$$m_i = m(^{216}Po) = 216.001\,905 \text{ u}$$

$$m_f = m(^{212}Pb) + m(^4He) = 211.991\,888 \text{ u} + 4.002\,603 \text{ u}$$

$$= 215.994\,491 \text{ u}$$

Thus, the mass change is

$$\Delta m = 216.001\,905 \text{ u} - 215.994\,491 \text{ u} = 0.007\,414 \text{ u}$$

$$= \boxed{1.23 \times 10^{-29} \text{ kg}}$$

(B) The energy associated with this mass change is

$$E = \Delta m c^2 = (1.23 \times 10^{-29} \text{ kg})(3.00 \times 10^8 \text{ m/s})^2$$

$$= 1.11 \times 10^{-12} \text{ J} = \boxed{6.92 \text{ MeV}}$$

This energy appears as the kinetic energy of the alpha particle and the ^{212}Pb nucleus after the decay.

39.10   The General Theory of Relativity

Up to this point, we have sidestepped a curious puzzle. Mass has two seemingly different properties: a *gravitational attraction* for other masses and an *inertial* property that represents a resistance to acceleration. To designate these two attributes, we use the subscripts g and i and write

$$\text{Gravitational property} \qquad F_g = m_g g$$

$$\text{Inertial property} \qquad \sum F = m_i a$$

The value for the gravitational constant G was chosen to make the magnitudes of m_g and m_i numerically equal. Regardless of how G is chosen, however, the strict proportionality of m_g and m_i has been established experimentally to an extremely high degree: a few parts in 10^{12}. Thus, it appears that gravitational mass and inertial mass may indeed be exactly proportional.

But why? They seem to involve two entirely different concepts: a force of mutual gravitational attraction between two masses, and the resistance of a single mass to being accelerated. This question, which puzzled Newton and many other physicists over the years, was answered by Einstein in 1916 when he published his theory of gravitation, known as the *general theory of relativity*. Because it is a mathematically complex theory, we offer merely a hint of its elegance and insight.

In Einstein's view, the dual behavior of mass was evidence for a very intimate and basic connection between the two behaviors. He pointed out that no mechanical experiment (such as dropping an object) could distinguish between the two situations illustrated in Figures 39.19a and 39.19b. In Figure 39.19a, a person is standing in an elevator on the surface of a planet, and feels pressed into the floor, due to the gravitational force. In Figure 39.19b, the person is in an elevator in empty space accelerating upward with $a = g$. The person feels pressed into the floor with the same force as in Figure 39.19a. In each case, an object released by the observer undergoes a downward acceleration of magnitude g relative to the floor. In Figure 39.19a, the person is in an inertial frame in a gravitational field. In Figure 39.19b, the person is in a noninertial frame accelerating in gravity-free space. Einstein's claim is that these two situations are completely equivalent.

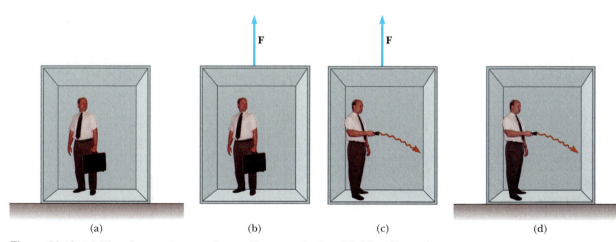

| (a) | (b) | (c) | (d) |

Figure 39.19 (a) The observer is at rest in a uniform gravitational field **g**, directed downward. (b) The observer is in a region where gravity is negligible, but the frame is accelerated by an external force **F** that produces an acceleration **g** directed upward. According to Einstein, the frames of reference in parts (a) and (b) are equivalent in every way. No local experiment can distinguish any difference between the two frames. (c) In the accelerating frame, a ray of light would appear to bend downward due to the acceleration of the elevator. (d) If parts (a) and (b) are truly equivalent, as Einstein proposed, then part (c) suggests that a ray of light would bend downward in a gravitational field.

Einstein carried this idea further and proposed that *no* experiment, mechanical or otherwise, could distinguish between the two cases. This extension to include all phenomena (not just mechanical ones) has interesting consequences. For example, suppose that a light pulse is sent horizontally across an elevator that is accelerating upward in empty space, as in Figure 39.19c. From the point of view of an observer in an inertial frame outside of the elevator, the light travels in a straight line while the floor of the elevator accelerates upward. According to the observer on the elevator, however, the trajectory of the light pulse bends downward as the floor of the elevator (and the observer) accelerates upward. Therefore, based on the equality of parts (a) and (b) of the figure for all phenomena, Einstein proposed that **a beam of light should also be bent downward by a gravitational field,** as in Figure 39.19d. Experiments have verified the effect, although the bending is small. A laser aimed at the horizon falls less than 1 cm after traveling 6 000 km. (No such bending is predicted in Newton's theory of gravitation.)

The two postulates of Einstein's **general theory of relativity** are

Postulates of the general theory of relativity

- All the laws of nature have the same form for observers in any frame of reference, whether accelerated or not.

- In the vicinity of any point, a gravitational field is equivalent to an accelerated frame of reference in the absence of gravitational effects. (This is the **principle of equivalence.**)

One interesting effect predicted by the general theory is that time is altered by gravity. A clock in the presence of gravity runs slower than one located where gravity is negligible. Consequently, the frequencies of radiation emitted by atoms in the presence of a strong gravitational field are *red-shifted* to lower frequencies when compared with the same emissions in the presence of a weak field. This gravitational red shift has been detected in spectral lines emitted by atoms in massive stars. It has also been verified on the Earth by comparing the frequencies of gamma rays emitted from nuclei separated vertically by about 20 m.

The second postulate suggests that a gravitational field may be "transformed away" at any point if we choose an appropriate accelerated frame of reference—a freely falling one. Einstein developed an ingenious method of describing the acceleration necessary to make the gravitational field "disappear." He specified a concept, the *curvature of space–time,* that describes the gravitational effect at every point. In fact, the curvature of space–time completely replaces Newton's gravitational theory. According to Einstein, there is no such thing as a gravitational force. Rather, the presence of a mass causes a curvature of space–time in the vicinity of the mass, and this curvature dictates the space–time path that all freely moving objects must follow. In 1979, John Wheeler summarized Einstein's general theory of relativity in a single sentence: "Space tells matter how to move and matter tells space how to curve."

As an example of the effects of curved space–time, imagine two travelers moving on parallel paths a few meters apart on the surface of the Earth and maintaining an exact northward heading along two longitude lines. As they observe each other near the equator, they will claim that their paths are exactly parallel. As they approach the North Pole, however, they notice that they are moving closer together, and they will actually meet at the North Pole. Thus, they will claim that they moved along parallel paths, but moved toward each other, *as if there were an attractive force between them.* They will make this conclusion based on their everyday experience of moving on flat surfaces. From our mental representation, however, we realize that they are walking on a curved surface, and it is the geometry of the curved surface that causes them to converge, rather than an attractive force. In a similar way, general relativity replaces the notion of forces with the movement of objects through curved space–time.

One prediction of the general theory of relativity is that a light ray passing near the Sun should be deflected in the curved space–time created by the Sun's mass. This prediction was confirmed when astronomers detected the bending of starlight near the

Courtesy of NASA

Einstein's cross. The four bright spots are images of the same galaxy that have been bent around a massive object located between the galaxy and the Earth. The massive object acts like a lens, causing the rays of light that were diverging from the distant galaxy to converge on the Earth. (If the intervening massive object had a uniform mass distribution, we would see a bright ring instead of four spots.)

Figure 39.20 Deflection of starlight passing near the Sun. Because of this effect, the Sun or some other remote object can act as a *gravitational lens*. In his general theory of relativity, Einstein calculated that starlight just grazing the Sun's surface should be deflected by an angle of 1.75 s of arc.

Sun during a total solar eclipse that occurred shortly after World War I (Fig. 39.20). When this discovery was announced, Einstein became an international celebrity.

If the concentration of mass becomes very great, as is believed to occur when a large star exhausts its nuclear fuel and collapses to a very small volume, a **black hole** may form. Here, the curvature of space–time is so extreme that, within a certain distance from the center of the black hole, all matter and light become trapped, as discussed in Section 13.7.

SUMMARY

The two basic postulates of the special theory of relativity are

- The laws of physics must be the same in all inertial reference frames.

- The speed of light in vacuum has the same value, $c = 3.00 \times 10^8$ m/s, in all inertial frames, regardless of the velocity of the observer or the velocity of the source emitting the light.

Three consequences of the special theory of relativity are

- Events that are measured to be simultaneous for one observer are not necessarily measured to be simultaneous for another observer who is in motion relative to the first.

- Clocks in motion relative to an observer are measured to run slower by a factor $\gamma = (1 - v^2/c^2)^{-1/2}$. This phenomenon is known as **time dilation.**

- The length of objects in motion are measured to be contracted in the direction of motion by a factor $1/\gamma = (1 - v^2/c^2)^{1/2}$. This phenomenon is known as **length contraction.**

To satisfy the postulates of special relativity, the Galilean transformation equations must be replaced by the **Lorentz transformation equations:**

$$x' = \gamma(x - vt) \qquad y' = y \qquad z' = z \qquad t' = \gamma\left(t - \frac{v}{c^2}x\right) \qquad (39.11)$$

where $\gamma = (1 - v^2/c^2)^{-1/2}$ and the S′ frame moves in the x direction relative to the S frame.

The relativistic form of the **velocity transformation equation** is

$$u_x' = \frac{u_x - v}{1 - \dfrac{u_x v}{c^2}} \qquad (39.16)$$

where u_x is the speed of an object as measured in the S frame and u_x' is its speed measured in the S′ frame.

Take a practice test for this chapter by clicking on the Practice Test link at http://www.pse6.com.

The relativistic expression for the **linear momentum** of a particle moving with a velocity **u** is

$$\mathbf{p} \equiv \frac{m\mathbf{u}}{\sqrt{1 - \dfrac{u^2}{c^2}}} = \gamma m \mathbf{u} \tag{39.19}$$

The relativistic expression for the **kinetic energy** of a particle is

$$K = \frac{mc^2}{\sqrt{1 - \dfrac{u^2}{c^2}}} - mc^2 = (\gamma - 1)\, mc^2 \tag{39.23}$$

The constant term mc^2 in Equation 39.23 is called the **rest energy** E_R of the particle:

$$E_R = mc^2 \tag{39.24}$$

The total energy E of a particle is given by

$$E = \frac{mc^2}{\sqrt{1 - \dfrac{u^2}{c^2}}} = \gamma mc^2 \tag{39.26}$$

The relativistic linear momentum of a particle is related to its total energy through the equation

$$E^2 = p^2 c^2 + (mc^2)^2 \tag{39.27}$$

QUESTIONS

1. What two speed measurements do two observers in relative motion always agree on?

2. A spacecraft with the shape of a sphere moves past an observer on Earth with a speed $0.5c$. What shape does the observer measure for the spacecraft as it moves past?

3. The speed of light in water is 230 Mm/s. Suppose an electron is moving through water at 250 Mm/s. Does this violate the principle of relativity?

4. Two identical clocks are synchronized. One is then put in orbit directed eastward around the Earth while the other remains on the Earth. Which clock runs slower? When the moving clock returns to the Earth, are the two still synchronized?

5. Explain why it is necessary, when defining the length of a rod, to specify that the positions of the ends of the rod are to be measured simultaneously.

6. A train is approaching you at very high speed as you stand next to the tracks. Just as an observer on the train passes you, you both begin to play the same Beethoven symphony on portable compact disc players. (a) According to you, whose CD player finishes the symphony first? (b) **What If?** According to the observer on the train, whose CD player finishes the symphony first? (c) Whose CD player really finishes the symphony first?

7. List some ways our day-to-day lives would change if the speed of light were only 50 m/s.

8. Does saying that a moving clock runs slower than a stationary one imply that something is physically unusual about the moving clock?

9. How is acceleration indicated on a space–time graph?

10. A particle is moving at a speed less than $c/2$. If the speed of the particle is doubled, what happens to its momentum?

11. Give a physical argument that shows that it is impossible to accelerate an object of mass m to the speed of light, even with a continuous force acting on it.

12. The upper limit of the speed of an electron is the speed of light c. Does that mean that the momentum of the electron has an upper limit?

13. Because mass is a measure of energy, can we conclude that the mass of a compressed spring is greater than the mass of the same spring when it is not compressed?

14. It is said that Einstein, in his teenage years, asked the question, "What would I see in a mirror if I carried it in my hands and ran at the speed of light?" How would you answer this question?

15. Some distant astronomical objects, called quasars, are receding from us at half the speed of light (or greater). What is the speed of the light we receive from these quasars?

16. Photons of light have zero mass. How is it possible that they have momentum?

17. "Newtonian mechanics correctly describes objects moving at ordinary speeds and relativistic mechanics correctly describes objects moving very fast." "Relativistic mechanics must make a smooth transition as it reduces to Newtonian mechanics in a case where the speed of an object becomes small compared to the speed of light." Argue for or against each of these two statements.

Figure Q39.18

18. Two cards have straight edges. Suppose that the top edge of one card crosses the bottom edge of another card at a small angle, as in Figure Q39.18a. A person slides the cards together at a moderately high speed. In what direction does the intersection point of the edges move? Show that it can move at a speed greater than the speed of light.

 A small flashlight is suspended in a horizontal plane and set into rapid rotation. Show that the spot of light it produces on a distant screen can move across the screen at a speed greater than the speed of light. (If you use a laser pointer, as in Figure Q39.18b, make sure the direct laser light cannot enter a person's eyes.) Argue that these experiments do not invalidate the principle that no material, no energy, and no information can move faster than light moves in a vacuum.

19. Describe how the results of Example 39.7 would change if, instead of fast space vehicles, two ordinary cars were approaching each other at highway speeds.

20. Two objects are identical except that one is hotter than the other. Compare how they respond to identical forces.

21. With regard to reference frames, how does general relativity differ from special relativity?

22. Two identical clocks are in the same house, one upstairs in a bedroom, and the other downstairs in the kitchen. Which clock runs more slowly? Explain.

23. *A thought experiment.* Imagine ants living on a merry-go-round turning at relativistic speed, which is their two-dimensional world. From measurements on small circles they are thoroughly familiar with the number π. When they measure the circumference of their world, and divide it by the diameter, they expect to calculate the number $\pi = 3.141\,59.\ldots$. We see the merry-go-round turning at relativistic speed. From our point of view, the ants' measuring rods on the circumference are experiencing length contraction in the tangential direction; hence the ants will need some extra rods to fill that entire distance. The rods measuring the diameter, however, do not contract, because their motion is perpendicular to their lengths. As a result, the computed ratio does not agree with the number π. If you were an ant, you would say that the rest of the universe is spinning in circles, and your disk is stationary. What possible explanation can you then give for the discrepancy, in light of the general theory of relativity?

PROBLEMS

1, 2, 3 = straightforward, intermediate, challenging   ☐ = full solution available in the *Student Solutions Manual and Study Guide*

🌐 = coached solution with hints available at http://www.pse6.com   💻 = computer useful in solving problem

▢ = paired numerical and symbolic problems

Section 39.1   The Principle of Galilean Relativity

1. A 2 000-kg car moving at 20.0 m/s collides and locks together with a 1 500-kg car at rest at a stop sign. Show that momentum is conserved in a reference frame moving at 10.0 m/s in the direction of the moving car.

2. A ball is thrown at 20.0 m/s inside a boxcar moving along the tracks at 40.0 m/s. What is the speed of the ball relative to the ground if the ball is thrown (a) forward (b) backward (c) out the side door?

3. In a laboratory frame of reference, an observer notes that Newton's second law is valid. Show that it is also valid for an observer moving at a constant speed, small compared with the speed of light, relative to the laboratory frame.

4. Show that Newton's second law is *not* valid in a reference frame moving past the laboratory frame of Problem 3 with a constant acceleration.

Section 39.2 The Michelson–Morley Experiment
Section 39.3 Einstein's Principle of Relativity
Section 39.4 Consequences of the Special Theory of Relativity

> Problem 43 in Chapter 4 can be assigned with this section.

5. How fast must a meter stick be moving if its length is measured to shrink to 0.500 m?

6. At what speed does a clock move if it is measured to run at a rate that is half the rate of a clock at rest with respect to an observer?

7. An astronaut is traveling in a space vehicle that has a speed of $0.500c$ relative to the Earth. The astronaut measures her pulse rate at 75.0 beats per minute. Signals generated by the astronaut's pulse are radioed to Earth when the vehicle is moving in a direction perpendicular to the line that connects the vehicle with an observer on the Earth. (a) What pulse rate does the Earth observer measure? (b) **What If?** What would be the pulse rate if the speed of the space vehicle were increased to $0.990c$?

8. An astronomer on Earth observes a meteoroid in the southern sky approaching the Earth at a speed of $0.800c$. At the time of its discovery the meteoroid is 20.0 ly from the Earth. Calculate (a) the time interval required for the meteoroid to reach the Earth as measured by the Earth-bound astronomer, (b) this time interval as measured by a tourist on the meteoroid, and (c) the distance to the Earth as measured by the tourist.

9. An atomic clock moves at 1 000 km/h for 1.00 h as measured by an identical clock on the Earth. How many nanoseconds slow will the moving clock be compared with the Earth clock, at the end of the 1.00-h interval?

10. A muon formed high in the Earth's atmosphere travels at speed $v = 0.990c$ for a distance of 4.60 km before it decays into an electron, a neutrino, and an antineutrino $(\mu^- \rightarrow e^- + \nu + \overline{\nu})$. (a) How long does the muon live, as measured in its reference frame? (b) How far does the Earth travel, as measured in the frame of the muon?

11. A spacecraft with a proper length of 300 m takes 0.750 μs to pass an Earth observer. Determine the speed of the spacecraft as measured by the Earth observer.

12. (a) An object of proper length L_p takes a time interval Δt to pass an Earth observer. Determine the speed of the object as measured by the Earth observer. (b) A column of tanks, 300 m long, takes 75.0 s to pass a child waiting at a street corner on her way to school. Determine the speed of the armored vehicles. (c) Show that the answer to part (a) includes the answer to Problem 11 as a special case, and includes the answer to part (b) as another special case.

13. **Review problem.** In 1963 Mercury astronaut Gordon Cooper orbited the Earth 22 times. The press stated that for each orbit he aged 2 millionths of a second less than he would have if he had remained on the Earth. (a) Assuming that he was 160 km above the Earth in a circular orbit, determine the time difference between someone on the Earth and the orbiting astronaut for the 22 orbits. You will need to use the approximation $\sqrt{1 - x} \approx 1 - x/2$, for small x. (b) Did the press report accurate information? Explain.

14. For what value of v does $\gamma = 1.010\,0$? Observe that for speeds lower than this value, time dilation and length contraction are effects amounting to less than 1%.

15. A friend passes by you in a spacecraft traveling at a high speed. He tells you that his craft is 20.0 m long and that the identically constructed craft you are sitting in is 19.0 m long. According to your observations, (a) how long is your spacecraft, (b) how long is your friend's craft, and (c) what is the speed of your friend's craft?

16. The identical twins Speedo and Goslo join a migration from the Earth to Planet X. It is 20.0 ly away in a reference frame in which both planets are at rest. The twins, of the same age, depart at the same time on different spacecraft. Speedo's craft travels steadily at $0.950c$, and Goslo's at $0.750c$. Calculate the age difference between the twins after Goslo's spacecraft lands on Planet X. Which twin is the older?

17. An interstellar space probe is launched from the Earth. After a brief period of acceleration it moves with a constant velocity, with a magnitude of 70.0% of the speed of light. Its nuclear-powered batteries supply the energy to keep its data transmitter active continuously. The batteries have a lifetime of 15.0 yr as measured in a rest frame. (a) How long do the batteries on the space probe last as measured by Mission Control on the Earth? (b) How far is the probe from the Earth when its batteries fail, as measured by Mission Control? (c) How far is the probe from the Earth when its batteries fail, as measured by its built-in trip odometer? (d) For what total time interval after launch are data received from the probe by Mission Control? Note that radio waves travel at the speed of light and fill the space between the probe and the Earth at the time of battery failure.

18. **Review problem.** An alien civilization occupies a brown dwarf, nearly stationary relative to the Sun, several lightyears away. The extraterrestrials have come to love original broadcasts of *I Love Lucy*, on our television channel 2, at carrier frequency 57.0 MHz. Their line of sight to us is in the plane of the Earth's orbit. Find the difference between the highest and lowest frequencies they receive due to the Earth's orbital motion around the Sun.

19. Police radar detects the speed of a car (Fig. P39.19) as follows. Microwaves of a precisely known frequency are broadcast toward the car. The moving car reflects the microwaves with a Doppler shift. The reflected waves are received and combined with an attenuated version of the transmitted wave. Beats occur between the two microwave signals. The beat frequency is measured. (a) For an electromagnetic wave reflected back to its source from a mirror approaching at speed v, show that the reflected

wave has frequency

$$f = f_{source} \frac{c + v}{c - v}$$

where f_{source} is the source frequency. (b) When v is much less than c, the beat frequency is much smaller than the transmitted frequency. In this case use the approximation $f + f_{source} \approx 2 f_{source}$ and show that the beat frequency can be written as $f_{beat} = 2v/\lambda$. (c) What beat frequency is measured for a car speed of 30.0 m/s if the microwaves have frequency 10.0 GHz? (d) If the beat frequency measurement is accurate to ± 5 Hz, how accurate is the velocity measurement?

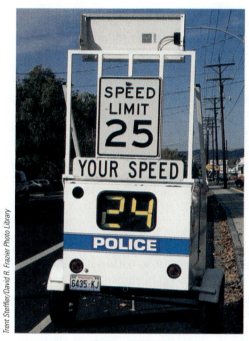

Trent Steffler/David R. Frazier Photo Library

Figure P39.19

20. *The red shift.* A light source recedes from an observer with a speed v_{source} that is small compared with c. (a) Show that the fractional shift in the measured wavelength is given by the approximate expression

$$\frac{\Delta\lambda}{\lambda} \approx \frac{v_{source}}{c}$$

This phenomenon is known as the red shift, because the visible light is shifted toward the red. (b) Spectroscopic measurements of light at $\lambda = 397$ nm coming from a galaxy in Ursa Major reveal a red shift of 20.0 nm. What is the recessional speed of the galaxy?

21. A physicist drives through a stop light. When he is pulled over, he tells the police officer that the Doppler shift made the red light of wavelength 650 nm appear green to him, with a wavelength of 520 nm. The police officer writes out a traffic citation for speeding. How fast was the physicist traveling, according to his own testimony?

Section 39.5   The Lorentz Transformation Equations

22. Suzanne observes two light pulses to be emitted from the same location, but separated in time by 3.00 μs. Mark sees

the emission of the same two pulses separated in time by 9.00 μs. (a) How fast is Mark moving relative to Suzanne? (b) According to Mark, what is the separation in space of the two pulses?

23. A moving rod is observed to have a length of 2.00 m and to be oriented at an angle of 30.0° with respect to the direction of motion, as shown in Figure P39.23. The rod has a speed of 0.995c. (a) What is the proper length of the rod? (b) What is the orientation angle in the proper frame?

2.00 m

30.0°

Direction of motion

Figure P39.23

24. An observer in reference frame S sees two events as simultaneous. Event A occurs at the point (50.0 m, 0, 0) at the instant 9:00:00 Universal time, 15 January 2004. Event B occurs at the point (150 m, 0, 0) at the same moment. A second observer, moving past with a velocity of $0.800c\hat{\mathbf{i}}$, also observes the two events. In her reference frame S', which event occurred first and what time interval elapsed between the events?

25. A red light flashes at position $x_R = 3.00$ m and time $t_R = 1.00 \times 10^{-9}$ s, and a blue light flashes at $x_B = 5.00$ m and $t_B = 9.00 \times 10^{-9}$ s, all measured in the S reference frame. Reference frame S' has its origin at the same point as S at $t = t' = 0$; frame S' moves uniformly to the right. Both flashes are observed to occur at the same place in S'. (a) Find the relative speed between S and S'. (b) Find the location of the two flashes in frame S'. (c) At what time does the red flash occur in the S' frame?

Section 39.6   The Lorentz Velocity Transformation Equations

26. A Klingon spacecraft moves away from the Earth at a speed of 0.800c (Fig. P39.26). The starship *Enterprise* pursues at a speed of 0.900c relative to the Earth. Observers on the Earth see the *Enterprise* overtaking the Klingon craft at a relative speed of 0.100c. With what speed is the *Enterprise* overtaking the Klingon craft as seen by the crew of the *Enterprise*?

S

S'

$v = 0.800c$

$u = 0.900c$

x

x'

Figure P39.26

27. Two jets of material from the center of a radio galaxy are ejected in opposite directions. Both jets move at 0.750c

relative to the galaxy. Determine the speed of one jet relative to the other.

28. A spacecraft is launched from the surface of the Earth with a velocity of $0.600c$ at an angle of $50.0°$ above the horizontal positive x axis. Another spacecraft is moving past, with a velocity of $0.700c$ in the negative x direction. Determine the magnitude and direction of the velocity of the first spacecraft as measured by the pilot of the second spacecraft.

Section 39.7 Relativistic Linear Momentum and the Relativistic Form of Newton's Laws

29. Calculate the momentum of an electron moving with a speed of (a) $0.010\ 0c$, (b) $0.500c$, and (c) $0.900c$.

30. The nonrelativistic expression for the momentum of a particle, $p = mu$, agrees with experiment if $u \ll c$. For what speed does the use of this equation give an error in the momentum of (a) 1.00% and (b) 10.0%?

31. A golf ball travels with a speed of 90.0 m/s. By what fraction does its relativistic momentum magnitude p differ from its classical value mu? That is, find the ratio $(p - mu)/mu$.

32. Show that the speed of an object having momentum of magnitude p and mass m is

$$u = \frac{c}{\sqrt{1 + (mc/p)^2}}$$

33. [www] An unstable particle at rest breaks into two fragments of unequal mass. The mass of the first fragment is 2.50×10^{-28} kg, and that of the other is 1.67×10^{-27} kg. If the lighter fragment has a speed of $0.893c$ after the breakup, what is the speed of the heavier fragment?

Section 39.8 Relativistic Energy

34. Determine the energy required to accelerate an electron from (a) $0.500c$ to $0.900c$ and (b) $0.900c$ to $0.990c$.

35. A proton in a high-energy accelerator moves with a speed of $c/2$. Use the work–kinetic energy theorem to find the work required to increase its speed to (a) $0.750c$ and (b) $0.995c$.

36. Show that, for any object moving at less than one-tenth the speed of light, the relativistic kinetic energy agrees with the result of the classical equation $K = \frac{1}{2}mu^2$ to within less than 1%. Thus for most purposes, the classical equation is good enough to describe these objects, whose motion we call *nonrelativistic*.

37. Find the momentum of a proton in MeV/c units assuming its total energy is twice its rest energy.

38. Find the kinetic energy of a 78.0-kg spacecraft launched out of the solar system with speed 106 km/s by using (a) the classical equation $K = \frac{1}{2}mu^2$. (b) **What If?** Calculate its kinetic energy using the relativistic equation.

39. [www] A proton moves at $0.950c$. Calculate its (a) rest energy, (b) total energy, and (c) kinetic energy.

40. A cube of steel has a volume of $1.00\ \text{cm}^3$ and a mass of 8.00 g when at rest on the Earth. If this cube is now given a speed $u = 0.900c$, what is its density as measured by a stationary observer? Note that relativistic density is defined as E_R/c^2V.

41. An unstable particle with a mass of 3.34×10^{-27} kg is initially at rest. The particle decays into two fragments that fly off along the x axis with velocity components $0.987c$ and $-0.868c$. Find the masses of the fragments. (*Suggestion:* Conserve both energy and momentum.)

42. An object having mass 900 kg and traveling at speed $0.850c$ collides with a stationary object having mass 1 400 kg. The two objects stick together. Find (a) the speed and (b) the mass of the composite object.

43. Show that the energy–momentum relationship $E^2 = p^2c^2 + (mc^2)^2$ follows from the expressions $E = \gamma mc^2$ and $p = \gamma mu$.

44. In a typical color television picture tube, the electrons are accelerated through a potential difference of 25 000 V. (a) What speed do the electrons have when they strike the screen? (b) What is their kinetic energy in joules?

45. Consider electrons accelerated to an energy of 20.0 GeV in the 3.00-km-long Stanford Linear Accelerator. (a) What is the γ factor for the electrons? (b) What is their speed? (c) How long does the accelerator appear to them?

46. Compact high-power lasers can produce a 2.00-J light pulse of duration 100 fs, focused to a spot 1 μm in diameter. (See Mourou and Umstader, "Extreme Light," *Scientific American*, May 2002, page 81.) The electric field in the light accelerates electrons in the target material to near the speed of light. (a) What is the average power of the laser during the pulse? (b) How many electrons can be accelerated to $0.999\ 9c$ if 0.010 0% of the pulse energy is converted into energy of electron motion?

47. A pion at rest ($m_\pi = 273m_e$) decays to a muon ($m_\mu = 207m_e$) and an antineutrino ($m_{\bar{\nu}} \approx 0$). The reaction is written $\pi^- \rightarrow \mu^- + \bar{\nu}$. Find the kinetic energy of the muon and the energy of the antineutrino in electron volts. (*Suggestion:* Conserve both energy and momentum.)

48. According to observer A, two objects of equal mass and moving along the x axis collide head on and stick to each other. Before the collision, this observer measures that object 1 moves to the right with a speed of $3c/4$, while object 2 moves to the left with the same speed. According to observer B, however, object 1 is initially at rest. (a) Determine the speed of object 2 as seen by observer B. (b) Compare the total initial energy of the system in the two frames of reference.

Section 39.9 Mass and Energy

49. Make an order-of-magnitude estimate of the ratio of mass increase to the original mass of a flag, as you run it up a flagpole. In your solution explain what quantities you take as data and the values you estimate or measure for them.

50. When 1.00 g of hydrogen combines with 8.00 g of oxygen, 9.00 g of water is formed. During this chemical reaction, 2.86×10^5 J of energy is released. How much mass do the constituents of this reaction lose? Is the loss of mass likely to be detectable?

51. In a nuclear power plant the fuel rods last 3 yr before they are replaced. If a plant with rated thermal power 1.00 GW operates at 80.0% capacity for 3.00 yr, what is the loss of mass of the fuel?

52. **Review problem.** The total volume of water in the oceans is approximately 1.40×10^9 km^3. The density of sea water is 1 030 kg/m^3, and the specific heat of the water is 4 186 J/(kg·°C). Find the increase in mass of the oceans produced by an increase in temperature of 10.0°C.

53. The power output of the Sun is 3.77×10^{26} W. How much mass is converted to energy in the Sun each second?

54. A gamma ray (a high-energy photon) can produce an electron (e$^-$) and a positron (e$^+$) when it enters the electric field of a heavy nucleus: $\gamma \rightarrow e^+ + e^-$. What minimum gamma-ray energy is required to accomplish this task? (*Note:* The masses of the electron and the positron are equal.)

Section 39.10 The General Theory of Relativity

55. An Earth satellite used in the global positioning system moves in a circular orbit with period 11 h 58 min. (a) Determine the radius of its orbit. (b) Determine its speed. (c) The satellite contains an oscillator producing the principal nonmilitary GPS signal. Its frequency is 1 575.42 MHz in the reference frame of the satellite. When it is received on the Earth's surface, what is the fractional change in this frequency due to time dilation, as described by special relativity? (d) The gravitational blue shift of the frequency according to general relativity is a separate effect. The magnitude of that fractional change is given by

$$\frac{\Delta f}{f} = \frac{\Delta U_g}{mc^2}$$

where ΔU_g is the change in gravitational potential energy of an object–Earth system when the object of mass m is moved between the two points at which the signal is observed. Calculate this fractional change in frequency. (e) What is the overall fractional change in frequency? Superposed on both of these relativistic effects is a Doppler shift that is generally much larger. It can be a red shift or a blue shift, depending on the motion of a particular satellite relative to a GPS receiver (Fig. P39.55).

Figure P39.55 This global positioning system (GPS) receiver incorporates relativistically corrected time calculations in its analysis of signals it receives from orbiting satellites. This allows the unit to determine its position on the Earth's surface to within a few meters. If these corrections were not made, the location error would be about 1 km.

Additional Problems

56. An astronaut wishes to visit the Andromeda galaxy, making a one-way trip that will take 30.0 yr in the spacecraft's frame of reference. Assume that the galaxy is 2.00×10^6 ly away and that the astronaut's speed is constant. (a) How fast must he travel relative to the Earth? (b) What will be the kinetic energy of his 1 000-metric-ton spacecraft? (c) What is the cost of this energy if it is purchased at a typical consumer price for electric energy: $0.130/kWh?

57. The cosmic rays of highest energy are protons that have kinetic energy on the order of 10^{13} MeV. (a) How long would it take a proton of this energy to travel across the Milky Way galaxy, having a diameter $\sim 10^5$ ly, as measured in the proton's frame? (b) From the point of view of the proton, how many kilometers across is the galaxy?

58. An electron has a speed of $0.750c$. (a) Find the speed of a proton that has the same kinetic energy as the electron. (b) **What If?** Find the speed of a proton that has the same momentum as the electron.

59. Ted and Mary are playing a game of catch in frame S′, which is moving at $0.600c$ with respect to frame S, while Jim, at rest in frame S, watches the action (Fig. P39.59). Ted throws the ball to Mary at $0.800c$ (according to Ted) and their separation (measured in S′) is 1.80×10^{12} m. (a) According to Mary, how fast is the ball moving? (b) According to Mary, how long does it take the ball to reach her? (c) According to Jim, how far apart are Ted and Mary, and how fast is the ball moving? (d) According to Jim, how long does it take the ball to reach Mary?

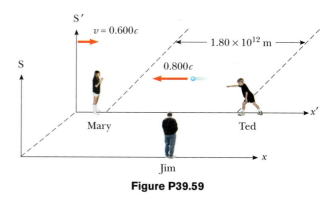

Figure P39.59

60. A rechargeable AA battery with a mass of 25.0 g can supply a power of 1.20 W for 50.0 min. (a) What is the difference in mass between a charged and an uncharged battery? (b) What fraction of the total mass is this mass difference?

61. The net nuclear fusion reaction inside the Sun can be written as $4\,^1\text{H} \rightarrow\,^4\text{He} + \Delta E$. The rest energy of each hydrogen atom is 938.78 MeV and the rest energy of the helium-4 atom is 3 728.4 MeV. Calculate the percentage of the starting mass that is transformed to other forms of energy.

62. An object disintegrates into two fragments. One of the fragments has mass 1.00 MeV/c^2 and momentum 1.75 MeV/c in the positive x direction. The other fragment has mass 1.50 MeV/c^2 and momentum 2.00 MeV/c in the positive y direction. Find (a) the mass and (b) the speed of the original object.

63. An alien spaceship traveling at $0.600c$ toward the Earth launches a landing craft with an advance guard of purchasing agents and physics teachers. The lander travels in the same direction with a speed of $0.800c$ relative to the mother ship. As observed on the Earth, the spaceship is 0.200 ly from the Earth when the lander is launched. (a) What speed do the Earth observers measure for the approaching lander? (b) What is the distance to the Earth at the time of lander launch, as observed by the aliens? (c) How long does it take the lander to reach the Earth as observed by the aliens on the mother ship? (d) If the lander has a mass of 4.00×10^5 kg, what is its kinetic energy as observed in the Earth reference frame?

64. A physics professor on the Earth gives an exam to her students, who are in a spacecraft traveling at speed v relative to the Earth. The moment the craft passes the professor, she signals the start of the exam. She wishes her students to have a time interval T_0 (spacecraft time) to complete the exam. Show that she should wait a time interval (Earth time) of

$$T = T_0 \sqrt{\frac{1 - v/c}{1 + v/c}}$$

before sending a light signal telling them to stop. (*Suggestion:* Remember that it takes some time for the second light signal to travel from the professor to the students.)

65. Spacecraft I, containing students taking a physics exam, approaches the Earth with a speed of $0.600c$ (relative to the Earth), while spacecraft II, containing professors proctoring the exam, moves at $0.280c$ (relative to the Earth) directly toward the students. If the professors stop the exam after 50.0 min have passed on their clock, how long does the exam last as measured by (a) the students (b) an observer on the Earth?

66. Energy reaches the upper atmosphere of the Earth from the Sun at the rate of 1.79×10^{17} W. If all of this energy were absorbed by the Earth and not re-emitted, how much would the mass of the Earth increase in 1.00 yr?

67. A supertrain (proper length 100 m) travels at a speed of $0.950c$ as it passes through a tunnel (proper length 50.0 m). As seen by a trackside observer, is the train ever completely within the tunnel? If so, with how much space to spare?

68. Imagine that the entire Sun collapses to a sphere of radius R_g such that the work required to remove a small mass m from the surface would be equal to its rest energy mc^2. This radius is called the *gravitational radius* for the Sun. Find R_g. (It is believed that the ultimate fate of very massive stars is to collapse beyond their gravitational radii into black holes.)

69. A particle with electric charge q moves along a straight line in a uniform electric field $\mathbf{E}$ with a speed of u. The electric force exerted on the charge is $q\mathbf{E}$. The motion and the electric field are both in the x direction. (a) Show that the acceleration of the particle in the x direction is given by

$$a = \frac{du}{dt} = \frac{qE}{m}\left(1 - \frac{u^2}{c^2}\right)^{3/2}$$

(b) Discuss the significance of the dependence of the acceleration on the speed. (c) **What If?** If the particle starts from rest at $x = 0$ at $t = 0$, how would you proceed to find the speed of the particle and its position at time t?

70. An observer in a coasting spacecraft moves toward a mirror at speed v relative to the reference frame labeled by S in Figure P39.70. The mirror is stationary with respect to S. A light pulse emitted by the spacecraft travels toward the mirror and is reflected back to the craft. The front of the craft is a distance d from the mirror (as measured by observers in S) at the moment the light pulse leaves the craft. What is the total travel time of the pulse as measured by observers in (a) the S frame and (b) the front of the spacecraft?

$v = 0.800c$

Figure P39.70

71. The creation and study of new elementary particles is an important part of contemporary physics. Especially interesting is the discovery of a very massive particle. To create a particle of mass M requires an energy Mc^2. With enough energy, an exotic particle can be created by allowing a fast moving particle of ordinary matter, such as a proton, to collide with a similar target particle. Let us consider a perfectly inelastic collision between two protons: an incident proton with mass m_p, kinetic energy K, and momentum magnitude p joins with an originally stationary target proton to form a single product particle of mass M. You might think that the creation of a new product particle, nine times more massive than in a previous experiment, would require just nine times more energy for the incident proton. Unfortunately not all of the kinetic energy of the incoming proton is available to create the product particle, since conservation of momentum requires that after the collision the system as a whole still must have some kinetic energy. Only a fraction of the energy of the incident particle is thus available to create a new particle. You will determine how the energy available for particle creation depends on the energy of the moving proton. Show that the energy available to create a product particle is given by

$$Mc^2 = 2m_p c^2 \sqrt{1 + \frac{K}{2m_p c^2}}$$

From this result, when the kinetic energy K of the incident proton is large compared to its rest energy $m_p c^2$, we see that M approaches $(2m_p K)^{1/2}/c$. Thus if the energy of the incoming proton is increased by a factor of nine, the mass you can create increases only by a factor of three. This disappointing result is the main reason that most modern accelerators, such as those at CERN (in Europe), at Fermilab (near Chicago), at SLAC (at Stanford), and at DESY (in Germany), use *colliding beams*. Here the total momentum of a pair of interacting particles can be zero. The center of mass can be at rest after the collision, so in principle all of the initial kinetic energy can be used for particle creation, according to

$$Mc^2 = 2mc^2 + K = 2mc^2 \left(1 + \frac{K}{2mc^2}\right)$$

where K is the total kinetic energy of two identical colliding particles. Here if $K \gg mc^2$, we have M directly proportional to K, as we would desire. These machines are difficult to build and to operate, but they open new vistas in physics.

72. A particle of mass m moving along the x axis with a velocity component $+u$ collides head-on and sticks to a particle of mass $m/3$ moving along the x axis with the velocity component $-u$. What is the mass M of the resulting particle?

73. A rod of length L_0 moving with a speed v along the horizontal direction makes an angle θ_0 with respect to the x' axis. (a) Show that the length of the rod as measured by a stationary observer is $L = L_0[1 - (v^2/c^2)\cos^2\theta_0]^{1/2}$. (b) Show that the angle that the rod makes with the x axis is given by $\tan\theta = \gamma\tan\theta_0$. These results show that the rod is both contracted and rotated. (Take the lower end of the rod to be at the origin of the primed coordinate system.)

74. Suppose our Sun is about to explode. In an effort to escape, we depart in a spacecraft at $v = 0.800c$ and head toward the star Tau Ceti, 12.0 ly away. When we reach the midpoint of our journey from the Earth, we see our Sun explode and, unfortunately, at the same instant we see Tau Ceti explode as well. (a) In the spacecraft's frame of reference, should we conclude that the two explosions occurred simultaneously? If not, which occurred first? (b) **What If?** In a frame of reference in which the Sun and Tau Ceti are at rest, did they explode simultaneously? If not, which exploded first?

75. A ^{57}Fe nucleus at rest emits a 14.0-keV photon. Use conservation of energy and momentum to deduce the kinetic energy of the recoiling nucleus in electron volts. (Use $Mc^2 = 8.60 \times 10^{-9}$ J for the final state of the ^{57}Fe nucleus.)

76. 💻 Prepare a graph of the relativistic kinetic energy and the classical kinetic energy, both as a function of speed, for an object with a mass of your choice. At what speed does the classical kinetic energy underestimate the experimental value by 1%? by 5%? by 50%?

Answers to Quick Quizzes

39.1 (c). While the observers' measurements differ, both are correct.

39.2 (d). The Galilean velocity transformation gives us $u_x = u'_x + v = 110$ mi/h $+ 90$ mi/h $= 200$ mi/h.

39.3 (d). The two events (the pulse leaving the flashlight and the pulse hitting the far wall) take place at different locations for both observers, so neither measures the proper time interval.

39.4 (a). The two events are the beginning and the end of the movie, both of which take place at rest with respect to the spacecraft crew. Thus, the crew measures the proper

time interval of 2 h. Any observer in motion with respect to the spacecraft, which includes the observer on Earth, will measure a longer time interval due to time dilation.

39.5 (a). If their on-duty time is based on clocks that remain on the Earth, they will have larger paychecks. A shorter time interval will have passed for the astronauts in their frame of reference than for their employer back on the Earth.

39.6 (c). Both your body and your sleeping cabin are at rest in your reference frame; thus, they will have their proper length according to you. There will be no change in measured lengths of objects, including yourself, within your spacecraft.

39.7 (d). Time dilation and length contraction depend only on the relative speed of one observer relative to another, not on whether the observers are receding or approaching each other.

39.8 (c). Because of your motion toward the source of the light, the light beam has a horizontal component of velocity as measured by you. The magnitude of the vector sum of the horizontal and vertical component vectors must be equal to c, so the magnitude of the vertical component must be smaller than c.

39.9 (a). In this case, there is only a horizontal component of the velocity of the light, and you must measure a speed of c.

39.10 (a) $m_3 > m_2 = m_1$; the rest energy of particle 3 is $2E$, while it is E for particles 1 and 2. (b) $K_3 = K_2 > K_1$; the kinetic energy is the difference between the total energy and the rest energy. The kinetic energy is $4E - 2E = 2E$ for particle 3, $3E - E = 2E$ for particle 2, and $2E - E = E$ for particle 1. (c) $u_2 > u_3 = u_1$; from Equation 39.26, $E = \gamma E_R$. Solving this for the square of the particle speed u, we find $u^2 = c^2(1 - (E_R/E)^2)$. Thus, the particle with the smallest ratio of rest energy to total energy will have the largest speed. Particles 1 and 3 have the same ratio as each other, and the ratio of particle 2 is smaller.

Chapter 40

Introduction to Quantum Physics

▲ This lightbulb filament glows with an orange color. Why? Classical physics is unable to explain the experimentally observed wavelength distribution of electromagnetic radiation from a hot object. A theory proposed in 1900 and describing the radiation from such objects represents the dawn of quantum physics. (Steve Cole/Getty Images)

In the preceding chapter, we discussed the fact that Newtonian mechanics must be replaced by Einstein's special theory of relativity when dealing with particle speeds comparable to the speed of light. As the twentieth century progressed, many experimental and theoretical problems were resolved by the special theory of relativity. There were many other problems, however, for which neither relativity nor classical physics could provide a theoretical answer. Attempts to apply the laws of classical physics to explain the behavior of matter on the atomic scale were consistently unsuccessful. For example, the emission of discrete wavelengths of light from atoms in a high-temperature gas could not be explained within the framework of classical physics. While models based on classical physics predicted light frequencies with the correct order of magnitude, they predicted that the frequencies would be continuous rather than discrete.

As physicists sought new ways to solve these puzzles, another revolution took place in physics between 1900 and 1930. A new theory called *quantum mechanics* was highly successful in explaining the behavior of particles of microscopic size. Like the special theory of relativity, the quantum theory requires a modification of our ideas concerning the physical world.

The first explanation of a phenomenon using quantum theory was introduced by Max Planck, but most of the subsequent mathematical developments and interpretations were made by a number of other distinguished physicists, including Einstein, Bohr, de Broglie, Schrödinger, Heisenberg, Born, and Dirac. Despite the great success of the quantum theory, Einstein frequently played the role of its critic, especially with regard to the manner in which the theory was interpreted.

Because an extensive study of quantum theory is beyond the scope of this book, this chapter is simply an introduction to its underlying principles.

40.1 Blackbody Radiation and Planck's Hypothesis

An object at any temperature emits **thermal radiation** from its surface, as discussed in Section 20.7. The characteristics of this radiation depend on the temperature and properties of the surface. Careful study shows that the thermal radiation it emits consists of a continuous distribution of wavelengths from all portions of the electromagnetic spectrum. If the object is at room temperature, the wavelengths of thermal radiation are mainly in the infrared region, and hence the radiation is not detected by the eye. As the surface temperature of the object increases, the object eventually begins to glow visibly red. At sufficiently high temperatures, the glowing object appears white, as in the hot tungsten filament of a lightbulb.

From a classical viewpoint, thermal radiation originates from accelerated charged particles in the atoms near the surface of the object; those charged particles emit radiation much as small antennas do. The thermally agitated accelerating particles can have a distribution of energies, which accounts for the continuous spectrum of radiation emitted by the object. By the end of the nineteenth century, however, it became apparent that the classical theory of thermal radiation was inadequate. The basic problem was in understanding the observed distribution of wavelengths in the radiation emitted by a black

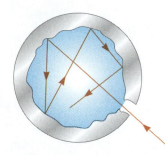

Figure 40.1 The opening to the cavity inside a hollow object is a good approximation of a black body. Light entering the small opening strikes the interior walls, where some is absorbed and some is reflected at a random angle. The cavity walls re-radiate at wavelengths corresponding to their temperature, producing standing waves in the cavity. Some of the energy from these standing waves can leave through the opening.

Stefan's law

body. As defined in Section 20.7, a **black body** is an ideal system that absorbs all radiation incident on it. The electromagnetic radiation emitted by the black body is called **blackbody radiation.**

A good approximation of a black body is a small hole leading to the inside of a hollow object, as shown in Figure 40.1. Any radiation incident on the hole from outside the cavity enters the hole and is absorbed or reflected a number of times from the walls of the cavity; hence the hole acts as a perfect absorber. The nature of the radiation leaving the cavity through the hole depends only on the temperature of the cavity walls and not on the material of which the walls are made. The spaces between lumps of hot charcoal (Fig. 40.2) emit light that is very much like blackbody radiation.

The radiation emitted by oscillators in the cavity walls experiences boundary conditions—it reflects from the walls of the cavity. Consequently, standing electromagnetic waves are established within the three-dimensional interior of the cavity. Many standing-wave modes are possible, and the distribution of the energy in the cavity among these modes determines the wavelength distribution of the radiation leaving the cavity through the hole.

The wavelength distribution of radiation from cavities was studied extensively in the late nineteenth century. Figure 40.3 shows how the intensity of blackbody radiation is found to vary with temperature and wavelength. The following two consistent experimental findings were seen as especially significant:

1. **The total power of the emitted radiation increases with temperature.**

We discussed this briefly in Chapter 20, where we introduced **Stefan's law,**

$$\mathcal{P} = \sigma A e T^4 \tag{40.1}$$

where $\mathcal{P}$ is the power in watts radiated from the surface of an object, σ is the Stefan–Boltzmann constant, equal to 5.670×10^{-8} W/m^2·K^4, A is the surface area of the object in square meters, e is the emissivity of the surface, and T is the surface temperature in kelvins. For a black body, the emissivity is $e = 1$ exactly. Recalling that $I \equiv \mathcal{P}/A$ is the definition of intensity and that $e = 1$ for a black body, we can write Stefan's law in the form $I = \sigma T^4$ at the surface of the object.

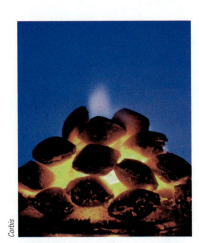

Figure 40.2 The glow emanating from the spaces between these hot charcoal briquettes is, to a close approximation, blackbody radiation. The color of the light depends only upon the temperature of the briquettes.

Active Figure 40.3 Intensity of blackbody radiation versus wavelength at three temperatures. Note that the amount of radiation emitted (the area under a curve) increases with increasing temperature. The visible range of wavelengths is between 0.4 μm and 0.7 μm. Thus, the 4 000-K curve has a peak that is near the visible range and represents an object that would glow with a yellowish-white appearance. At about 6 000 K, the peak is in the center of the visible wavelengths and the object appears white.

At the Active Figures link at http:// www.pse6.com, **you can adjust the temperature of the black body and study the radiation emitted from it.**

2. The peak of the wavelength distribution shifts to shorter wavelengths as the temperature increases.

This behavior was found to be described by the following relationship, called **Wien's displacement law:**

$$\lambda_{max} T = 2.898 \times 10^{-3} \text{ m} \cdot \text{K} \qquad (40.2)$$

Wien's displacement law

where λ_{max} is the wavelength at which the curve peaks and T is the absolute temperature of the surface of the object emitting the radiation. The wavelength at the peak of the curve is inversely proportional to the absolute temperature; that is, as the temperature increases, the peak is "displaced" to shorter wavelengths (Fig. 40.3).

This second finding (Wien's displacement law) is consistent with the behavior of the object mentioned at the beginning of this section. At room temperature, it does not appear to glow (peak in the infrared, $\lambda > 0.7 \ \mu$m). At higher temperature it glows red (peak in the near infrared with some radiation at the red end of the visible spectrum, $\lambda \approx 0.7 \ \mu$m), and at still higher temperature it glows white (peak in the visible, $0.4 \ \mu$m $< \lambda < 0.7 \ \mu$m).

> **Quick Quiz 40.1** Figure 40.4 shows two stars in the constellation Orion. Betelgeuse appears to glow red, while Rigel looks blue in color. Which star has a higher surface temperature? (a) Betelgeuse (b) Rigel (c) They both have the same surface temperature. (d) Impossible to determine.

Figure 40.4 (Quick Quiz 40.1) Which star is hotter?

A successful theory for blackbody radiation must predict the shape of the curves in Figure 40.3, the temperature dependence expressed in Stefan's law, and the shift of the peak with temperature described by Wien's displacement law. Early attempts to use classical ideas to explain the shapes of the curves in Figure 40.3 failed.

We consider one of these early attempts here. To describe the distribution of energy from a black body, it is useful to define $I(\lambda, T) \, d\lambda$ to be the intensity, or power per unit area, emitted in the wavelength interval $d\lambda$. The result of a calculation based on a classical theory of blackbody radiation known as the **Rayleigh–Jeans law** is

$$I(\lambda, T) = \frac{2\pi c k_B T}{\lambda^4} \qquad (40.3)$$

Rayleigh–Jeans law

where k_B is Boltzmann's constant. The black body is modeled as the hole leading into a cavity supporting many modes of oscillation of the electromagnetic field caused by accelerated charges in the cavity walls, resulting in the emission of electromagnetic waves at all wavelengths. In the classical theory used to derive Equation 40.3, the average energy for each wavelength of the standing wave modes is assumed to be proportional to $k_B T$, based on the theorem of equipartition of energy that we discussed in Section 21.1.

An experimental plot of the blackbody radiation spectrum is shown in Figure 40.5, together with the theoretical prediction of the Rayleigh–Jeans law. At long wavelengths, the Rayleigh–Jeans law is in reasonable agreement with experimental data, but at short wavelengths major disagreement is apparent.

Note that as λ approaches zero, the function $I(\lambda, T)$ given by Equation 40.3 approaches infinity. Hence, according to classical theory, not only should short wavelengths predominate in a blackbody spectrum, but also the energy emitted by any black body should become infinite in the limit of zero wavelength. In contrast to this prediction, the experimental data plotted in Figure 40.5 show that as λ approaches zero, $I(\lambda, T)$ also approaches zero. This mismatch of theory and experiment was so disconcerting that scientists called it the *ultraviolet catastrophe*. (This "catastrophe"—infinite energy—occurs as the wavelength approaches zero; the word "ultraviolet" was applied because ultraviolet wavelengths are short.)

Figure 40.5 Comparison of experimental results and the curve predicted by the Rayleigh–Jeans law for the distribution of blackbody radiation.

Max Planck

German Physicist (1858–1947)

Planck introduced the concept of "quantum of action" (Planck's constant, h) in an attempt to explain the spectral distribution of blackbody radiation, which laid the foundations for quantum theory. In 1918 he was awarded the Nobel Prize for this discovery of the quantized nature of energy. (©Bettmann/CORBIS)

In 1900, Max Planck developed a theory of blackbody radiation that leads to an equation for $I(\lambda, T)$ that is in complete agreement with experimental results at all wavelengths. Planck assumed that the cavity radiation came from atomic oscillators in the cavity walls in Figure 40.1. Planck made two bold and controversial assumptions concerning the nature of the oscillators in the cavity walls:

• The energy of an oscillator can have only certain *discrete* values E_n:

$$E_n = nhf \tag{40.4}$$

where n is a positive integer called a **quantum number,**[1] f is the frequency of oscillation, and h is a parameter that Planck introduced and is now called **Planck's constant.** Because the energy of each oscillator can have only discrete values given by Equation 40.4, we say the energy is **quantized.** Each discrete energy value corresponds to a different **quantum state,** represented by the quantum number n. When the oscillator is in the $n = 1$ quantum state, its energy is hf; when it is in the $n = 2$ quantum state, its energy is $2hf$; and so on.

• The oscillators emit or absorb energy when making a transition from one quantum state to another. The entire energy difference between the initial and final states in the transition is emitted or absorbed as a single quantum of radiation. If the transition is from one state to a lower adjacent state—say, from the $n = 3$ state to the $n = 2$ state—Equation 40.4 shows that the amount of energy emitted by the oscillator is

$$E = hf \tag{40.5}$$

An oscillator emits or absorbs energy only when it changes quantum states. If it remains in one quantum state, no energy is absorbed or emitted. Figure 40.6 is an **energy-level diagram** showing the quantized energy levels and allowed transitions proposed by Planck. This is an important semigraphical representation that is used often in quantum physics.[2] The vertical axis is linear in energy, and the allowed energy levels are represented as horizontal lines. The quantized system can have only the energies represented by the horizontal lines.

The key point in Planck's theory is the radical assumption of quantized energy states. This development was a clear deviation from classical physics and marked the birth of the quantum theory.

In the Rayleigh–Jeans model, the average energy associated with a particular wavelength of standing waves in the cavity is the same for all wavelengths and equal to $k_B T$. Planck used the same classical ideas as in the Rayleigh–Jeans model to arrive at the energy density as a product of constants and the average energy for a given wavelength,

▲ **PITFALL PREVENTION**

40.2 n Is Again an Integer

In the preceding chapters on optics, we used the symbol n for the index of refraction, which was not an integer. Here we are again using n as we did in Chapter 18 to indicate the standing-wave mode on a string or in an air column. In quantum physics, n is often used as an integer quantum number to identify a particular quantum state of a system.

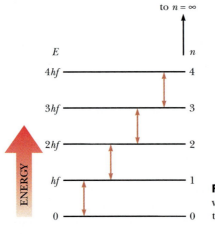

Figure 40.6 Allowed energy levels for an oscillator with frequency f. Allowed transitions are indicated by the double-headed arrows.

[1]   A quantum number is generally an integer (although half–integer quantum numbers can occur) that describes an allowed state of a system, like the values of n describing the normal modes of oscillation of a string fixed at both ends, as discussed in Section 18.3.

[2]   We first saw an energy-level diagram in Section 21.4.

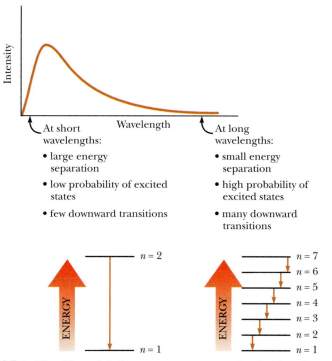

Active Figure 40.7 In Planck's model, the average energy associated with a given wavelength is the product of the energy of a transition and a factor related to the probability of the transition occurring. As the energy levels move farther apart at shorter wavelengths (higher energy), the probability of excitation decreases, as does the probability of a transition from the excited state.

Investigate the energy levels and observe the emission of radiation of different wavelengths at the Active Figures link at http://www.pse6.com.

but the average energy is not given by the equipartition theorem. The average energy of a wave is the average energy difference between levels of the oscillator, *weighted according to the probability of the wave being emitted.* This weighting is based on the occupation of higher-energy states as described by the Boltzmann distribution law, which we presented in Section 21.5. According to this law, the probability of a state being occupied is proportional to the factor $e^{-E/k_B T}$ where E is the energy of the state.

At low frequencies, the energy levels are close together, as on the right in Figure 40.7, and many of the energy states are excited because the Boltzmann factor $e^{-E/k_B T}$ is relatively large for these states. Thus, there are many contributions to the outgoing radiation, although each contribution has very low energy. Now, consider high-frequency radiation, that is, radiation with short wavelength. In order to obtain this radiation, the allowed energies are very far apart, as on the left in Figure 40.7. The probability of thermal agitation exciting these high energy levels is small because of the small value of the Boltzmann factor for large values of E. At high frequencies, the low probability of excitation results in very little contribution to the total energy, even though each quantum is of large energy. This "turns the curve over" and brings it down to zero again at short wavelengths.

Using this approach, Planck was able to generate a theoretical expression for the wavelength distribution that agreed remarkably well with the experimental curves in Figure 40.3:

$$I(\lambda, T) = \frac{2\pi h c^2}{\lambda^5 (e^{hc/\lambda k_B T} - 1)}$$

(40.6)

Planck's wavelength distribution function

This function includes the parameter h, which Planck adjusted so that his curve matched the experimental data at all wavelengths. The value of this parameter is found to be independent of the material of which the black body is made and independent of the temperature—it is a fundamental constant of nature. The value of h, Planck's constant, which we saw briefly first in Chapter 11 and again in Chapter 35, is

$$h = 6.626 \times 10^{-34} \text{ J} \cdot \text{s}$$

(40.7)

Planck's constant

At long wavelengths, Equation 40.6 reduces to the Rayleigh–Jeans expression, Equation 40.3 (see Problem 12), and at short wavelengths it predicts an exponential decrease in $I(\lambda, T)$ with decreasing wavelength, in agreement with experimental results.

When Planck presented his theory, most scientists (including Planck!) did not consider the quantum concept to be realistic. They believed it was a mathematical trick that happened to predict the correct results. Hence, Planck and others continued to search for a more "rational" explanation of blackbody radiation. However, subsequent developments showed that a theory based on the quantum concept (rather than on classical concepts) had to be used to explain not only blackbody radiation but also a number of other phenomena at the atomic level.

In 1905, Einstein rederived Planck's results by assuming that the cavity oscillations of the electromagnetic field were themselves quantized. In other words, he proposed that quantization is a fundamental property of light and other electromagnetic radiation. This led to the concept of photons as will be discussed in Section 40.2. Critical to the success of the quantum or photon theory was the relation between energy and frequency, which classical theory completely failed to predict.

You may have had your body temperature measured at the doctor's office by an *ear thermometer*, which can read your temperature in a matter of seconds (Fig. 40.8). This type of thermometer measures the amount of infrared radiation emitted by the eardrum in a fraction of a second. It then converts the amount of radiation into a temperature reading. This thermometer is very sensitive because temperature is raised to the fourth power in Stefan's law. Suppose that you have a fever of 1°C above normal. Because absolute temperatures are found by adding 273 to Celsius temperatures, the ratio of your fever temperature to normal body temperature of 37°C is

$$\frac{T_{\text{fever}}}{T_{\text{normal}}} = \frac{38°C + 273°C}{37°C + 273°C} = 1.0032$$

This is only a 0.32% increase in temperature. The increase in radiated power, however, is proportional to the fourth power of temperature, so

$$\frac{\mathcal{P}_{\text{fever}}}{\mathcal{P}_{\text{normal}}} = \left(\frac{38°C + 273°C}{37°C + 273°C}\right)^4 = 1.013$$

This is a 1.3% increase in radiated power, which is easily measured by modern infrared radiation sensors.

Photodisc/Getty Images

Figure 40.8 An ear thermometer measures a patient's temperature by detecting the intensity of infrared radiation leaving the eardrum.

Example 40.1   Thermal Radiation from Different Objects

Find the peak wavelength of the blackbody radiation emitted by each of the following:

(A) The human body when the skin temperature is 35°C

(B) The tungsten filament of a lightbulb, which operates at 2 000 K

(C) The Sun, which has a surface temperature of about 5 800 K

Solution (A) From Wien's displacement law (Eq. 40.2), we have $\lambda_{\max}T = 2.898 \times 10^{-3}$ m · K. Solving for $\lambda_{\max}$ and noting that 35°C corresponds to an absolute temperature of 308 K, we have

$$\lambda_{\max} = \frac{2.898 \times 10^{-3}\,\text{m} \cdot \text{K}}{308\,\text{K}} = \boxed{9.4\ \mu\text{m}}$$

This radiation is in the infrared region of the spectrum and is invisible to the human eye. Some animals (pit vipers, for

instance) are able to detect radiation of this wavelength and therefore can locate warm-blooded prey even in the dark.

(B) Following the same procedure as in part (A), we find

$$\lambda_{\max} = \frac{2.898 \times 10^{-3}\,\text{m} \cdot \text{K}}{2\,000\,\text{K}} = \boxed{1.4\ \mu\text{m}}$$

This also is in the infrared, meaning that most of the energy emitted by a lightbulb is not visible to us.

(C) Again following the same procedure, we have

$$\lambda_{\max} = \frac{2.898 \times 10^{-3}\,\text{m} \cdot \text{K}}{5\,800\,\text{K}} = \boxed{0.50\ \mu\text{m}}$$

This is near the center of the visible spectrum, about the color of a yellow-green tennis ball. Because it is the most prevalent color in sunlight, our eyes have evolved to be most sensitive to light of approximately this wavelength.

Example 40.2   The Quantized Oscillator

A 2.0-kg block is attached to a massless spring that has a force constant of $k = 25$ N/m. The spring is stretched 0.40 m from its equilibrium position and released.

(A) Find the total energy of the system and the frequency of oscillation according to classical calculations.

Solution Because of our study of oscillating blocks in Chapter 15, this problem is easy to conceptualize. The phrase "according to classical calculations" tells us that we should categorize this part of the problem as a classical analysis of the oscillator. To analyze the problem, we know that the total energy of a simple harmonic oscillator having an amplitude A is $\frac{1}{2}kA^2$ (Eq. 15.21). Therefore,

$$E = \tfrac{1}{2}kA^2 = \tfrac{1}{2}(25 \text{ N/m})(0.40 \text{ m})^2 = \boxed{2.0 \text{ J}}$$

The frequency of oscillation is (Eq. 15.14)

$$f = \frac{1}{2\pi}\sqrt{\frac{k}{m}} = \frac{1}{2\pi}\sqrt{\frac{25 \text{ N/m}}{2.0 \text{ kg}}} = \boxed{0.56 \text{ Hz}}$$

(B) Assuming that the energy is quantized, find the quantum number n for the system oscillating with this amplitude.

Solution This part of the problem is categorized as a quantum analysis of the oscillator. To analyze the problem, we note that the energy of the oscillator is quantized according to Equation 40.4. Thus,

$$E_n = nhf = n(6.626 \times 10^{-34} \text{ J}\cdot\text{s})(0.56 \text{ Hz}) = 2.0 \text{ J}$$

$$n = \boxed{5.4 \times 10^{33}}$$

To finalize this problem, note that 5.4×10^{33} is a very large quantum number, which is typical for macroscopic systems. Changes from one quantum state to another appear as smooth transitions rather than discontinuous changes as is the case for a microscopic system. Changes between quantum states for the oscillator are explored in the following **What If?**

What If? Suppose the oscillator makes a transition from the $n = 5.4 \times 10^{33}$ state to the state corresponding to $n = 5.4 \times 10^{33} - 1$. By how much does the energy of the oscillator change in this one-quantum change?

Answer From Equation 40.5, the energy difference between states differing in n by 1 is

$$E = hf = (6.626 \times 10^{-34} \text{ J}\cdot\text{s})(0.56 \text{ Hz}) = 3.7 \times 10^{-34} \text{ J}$$

This energy change due to a one-quantum change is fractionally equal to 3.7×10^{-34} J/2.0 J, or about one part in 10^{34}! It is such a small fraction of the total energy of the oscillator that we cannot detect it. Thus, even though the energy of a macroscopic spring–block system is quantized and does indeed decrease by small quantum jumps, our senses perceive the decrease as continuous. Quantum effects become important and measurable only on the submicroscopic level of atoms and molecules.

40.2   The Photoelectric Effect

Blackbody radiation was historically the first phenomenon to be explained with a quantum model. In the latter part of the nineteenth century, at the same time that data were taken on thermal radiation, experiments showed that light incident on certain metallic surfaces causes electrons to be emitted from those surfaces. This phenomenon, which we first met in Section 35.1, is known as the **photoelectric effect,** and the emitted electrons are called **photoelectrons.**[3]

Figure 40.9 is a diagram of an apparatus for studying the photoelectric effect. An evacuated glass or quartz tube contains a metallic plate E connected to the negative terminal of a battery and another metallic plate C that is connected to the positive terminal of the battery. When the tube is kept in the dark, the ammeter reads zero, indicating no current in the circuit. However, when plate E is illuminated by light having an appropriate wavelength, a current is detected by the ammeter, indicating a flow of charges across the gap between plates E and C. This current arises from photoelectrons emitted from the negative plate (the emitter) and collected at the positive plate (the collector).

Figure 40.10 is a plot of photoelectric current versus potential difference ΔV applied between plates E and C for two light intensities. At large values of ΔV, the current

---

[3] Photoelectrons are not different from other electrons. They are given this name solely because of their ejection from a metal by light in the photoelectric effect.

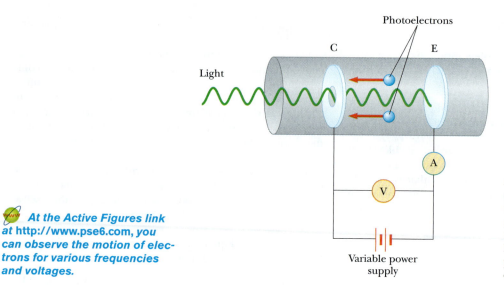

Active Figure 40.9 A circuit diagram for studying the photoelectric effect. When light strikes the plate E (the emitter), photoelectrons are ejected from the plate. Electrons moving from plate E to plate C (the collector) constitute a current in the circuit.

At the Active Figures link at http://www.pse6.com, you can observe the motion of electrons for various frequencies and voltages.

Active Figure 40.10 Photoelectric current versus applied potential difference for two light intensities. The current increases with intensity but reaches a saturation level for large values of ΔV. At voltages equal to or more negative than $-\Delta V_s$, where ΔV_s is the stopping potential, the current is zero.

At the Active Figures link at http://www.pse6.com, you can sweep through the voltage range and observe the current curve for different intensities of radiation.

reaches a maximum value; all of the electrons emitted from E are collected at C, and the current cannot increase further. In addition, the maximum current increases as the intensity of the incident light increases, as you might expect—more electrons are ejected by the higher-intensity light. Finally, when ΔV is negative—that is, when the battery in the circuit is reversed to make plate E positive and plate C negative—the current drops because many of the photoelectrons emitted from E are repelled by the now negative plate C. In this situation, only those photoelectrons having a kinetic energy greater than $e|\Delta V|$ reach plate C, where e is the magnitude of the charge on the electron. When ΔV is equal to or more negative than $-\Delta V_s$, where ΔV_s is the **stopping potential,** no photoelectrons reach C and the current is zero.

Let us consider the combination of the electric field between the plates and an electron ejected from plate E to be an isolated system. Suppose this electron stops just as it reaches plate C. Because the system of the electron and the electric field between the plates is isolated, the total mechanical energy of the system must be conserved:

$$K_1 + U_1 = K_2 + U_2$$

where configuration 1 refers to the instant that the electron leaves the metal with kinetic energy K_1 and configuration 2 is when the electron stops just before touching plate C. If we define the electric potential energy of the system in configuration 1 to be zero, we have

$$K_1 + 0 = 0 + (-e)(-\Delta V)$$

Now suppose we increase the potential difference ΔV in the negative direction just until no current flows. In this case the electron that stops just before reaching plate C will have the maximum possible kinetic energy upon leaving the metal surface, and ΔV will be the stopping potential ΔV_s. The previous equation can then be written as

$$K_{max} = e\,\Delta V_s \qquad (40.8)$$

This equation allows us to measure K_{max} experimentally by determining the voltage ΔV_s at which the current drops to zero.

Below we list several features of the photoelectric effect. For each feature, we compare the predictions made by a classical approach, using the wave model for light, with the experimental results.

1. Dependence of photoelectron kinetic energy on light intensity

   Classical prediction: Electrons should absorb energy continuously from the electromagnetic waves. As the light intensity incident on a metal is increased, energy should be transferred into the metal at a higher rate and the electrons should be ejected with more kinetic energy.

Experimental result: The maximum kinetic energy of photoelectrons is *independent* of light intensity. This is shown in Figure 40.10 by the fact that both curves fall to zero at the *same* negative voltage. According to Equation 40.8, the maximum kinetic energy is proportional to the stopping potential.

2. Time interval between incidence of light and ejection of photoelectrons

   Classical prediction: At low light intensities, a measurable time interval should pass between the instant the light is turned on and the time an electron is ejected from the metal. This time interval is required for the electron to absorb the incident radiation before it acquires enough energy to escape from the metal.

   Experimental result: Electrons are emitted from the surface of the metal almost *instantaneously* (less than 10^{-9} s after the surface is illuminated), even at very low light intensities.

3. Dependence of ejection of electrons on light frequency

   Classical prediction: Electrons should be ejected from the metal at any incident light frequency, as long as the light intensity is high enough, because energy is transferred to the metal regardless of the incident light frequency.

   Experimental result: No electrons are emitted if the incident light frequency falls below some **cutoff frequency** f_c, whose value is characteristic of the material being illuminated. No electrons are ejected below this cutoff frequency *regardless* of the light intensity.

4. Dependence of photoelectron kinetic energy on light frequency

   Classical prediction: There should be *no* relationship between the frequency of the light and the electron kinetic energy. The kinetic energy should be related to the intensity of the light.

   Experimental result: The maximum kinetic energy of the photoelectrons increases with increasing light frequency.

Notice that experimental results contradict *all four* classical predictions. A successful explanation of the photoelectric effect was given by Einstein in 1905, the same year he published his special theory of relativity. As part of a general paper on electromagnetic radiation, for which he received a Nobel prize in 1921, Einstein extended Planck's concept of quantization to electromagnetic waves, as mentioned in Section 40.1. He assumed that light (or any other electromagnetic wave) of frequency f can be considered a stream of quanta, regardless of the source of the radiation. Today we call these quanta **photons.** Each photon has an energy E, given by Equation 40.5, $E = hf$. Each photon moves in a vacuum at the speed of light c, where $c = 3.00 \times 10^8$ m/s.

Quick Quiz 40.2 While standing outdoors one evening, you are exposed to the following four types of electromagnetic radiation: yellow light from a sodium street lamp, radio waves from an AM radio station, radio waves from an FM radio station, and microwaves from an antenna of a communications system. Rank these types of waves in terms of increasing photon energy, lowest first.

In Einstein's model of the photoelectric effect, a photon of the incident light gives *all* its energy hf to a *single* electron in the metal. Thus, the absorption of energy by the electrons is not a continuous process as envisioned in the wave model, but rather a discontinuous process in which energy is delivered to the electrons in discrete bundles. The energy transfer is accomplished via a one photon–one electron event.[4]

Electrons ejected from the surface of the metal and not making collisions with other metal atoms before escaping possess the maximum kinetic energy K_{max}. Accord-

---

[4] In principle, two photons could combine to provide an electron with their combined energy. This is highly improbable, however, without the high intensity of radiation available from very strong lasers.

ing to Einstein, the maximum kinetic energy for these liberated electrons is

Photoelectric effect equation

$$K_{max} = hf - \phi \qquad (40.9)$$

where ϕ is called the **work function** of the metal. **The work function represents the minimum energy with which an electron is bound in the metal** and is on the order of a few electron volts. Table 40.1 lists work functions for various metals.

We can understand Equation 40.9 by rearranging it as follows:

$$K_{max} + \phi = hf$$

In this form, we see that Einstein's equation is Equation 7.17 applied to the system of the electron and the metal. K_{max} is the change in kinetic energy of the electron, assuming it begins at rest, ϕ is the change in potential energy of the system, assuming we define the potential energy when the electron is within the metal as zero, and hf is the transfer of energy into the system by electromagnetic radiation.

Using the photon model of light, one can explain the observed features of the photoelectric effect that could not be understood using classical concepts:

1. Dependence of photoelectron kinetic energy on light intensity

    Equation 40.9 shows that K_{max} is independent of the light intensity. The maximum kinetic energy of any one electron, which equals $hf - \phi$, depends only on the light frequency and the work function. If the light intensity is doubled, the number of photons arriving per unit time is doubled, which doubles the rate at which photoelectrons are emitted. The maximum kinetic energy of any one photoelectron, however, is unchanged.

2. Time interval between incidence of light and ejection of photoelectrons

    Near-instantaneous emission of electrons is consistent with the photon model of light. The incident energy appears in small packets, and there is a one-to-one interaction between photons and electrons. Even if the incident light has very low intensity, there may be very few photons arriving per unit time interval, but each one can have sufficient energy to eject an electron immediately.

3. Dependence of ejection of electrons on light frequency

    Failure to observe the photoelectric effect below a certain cutoff frequency follows from the fact that the photon must have energy greater than the work function ϕ in order to eject an electron. If the energy of an incoming photon does not satisfy this requirement, an electron cannot be ejected from the surface, regardless of light intensity.

4. Dependence of photoelectron kinetic energy on light frequency

    That K_{max} increases with increasing frequency is easily understood with Equation 40.9.

Einstein's model predicts a linear relationship (Eq. 40.9) between the maximum electron kinetic energy K_{max} and the light frequency f. Experimental observation of a linear relationship between K_{max} and f would be a final confirmation of Einstein's theory. Indeed, such a linear relationship is observed, as sketched in Figure 40.11, and the slope of the lines in such a plot is Planck's constant h. The intercept on the horizontal axis gives the cutoff frequency below which no photoelectrons are emitted, regardless of light intensity. The cutoff frequency is related to the work function through the relationship $f_c = \phi/h$. The cutoff frequency corresponds to a **cutoff wavelength** λ_c, where

Cutoff wavelength

$$\lambda_c = \frac{c}{f_c} = \frac{c}{\phi/h} = \frac{hc}{\phi} \qquad (40.10)$$

and c is the speed of light. Wavelengths greater than λ_c incident on a material having a work function ϕ do not result in the emission of photoelectrons.

The combination hc occurs often when relating the energy of a photon to its wavelength. A common shortcut to use in solving problems is to express this

Table 40.1

Work Functions of Selected Metals[a]	
Metal	ϕ **(eV)**
Na	2.46
Al	4.08
Cu	4.70
Zn	4.31
Ag	4.73
Pt	6.35
Pb	4.14
Fe	4.50

[a] Values are typical for metals listed. Actual values may vary depending on whether the metal is a single crystal or polycrystalline. Values may also depend on the face from which electrons are ejected from crystalline metals. Furthermore, different experimental procedures may produce differing values.

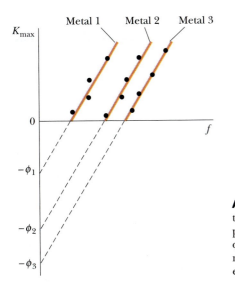

Active Figure 40.11 A plot of K_{max} of photoelectrons versus frequency of incident light in a typical photoelectric effect experiment. Photons with frequency less than the cutoff frequency for a given metal do not have sufficient energy to eject an electron from the metal.

At the Active Figures link at http://www.pse6.com, you can sweep through the frequency range and observe the curve for different target metals.

combination in useful units according to the following approximation:

$$hc = 1\,240 \text{ eV} \cdot \text{nm}$$

One of the first practical uses of the photoelectric effect was as the detector in a light meter of a camera. Light reflected from the object to be photographed strikes a photoelectric surface in the meter, causing it to emit photoelectrons that then pass through a sensitive ammeter. The magnitude of the current in the ammeter depends on the light intensity.

The phototube, another early application of the photoelectric effect, acts much like a switch in an electric circuit. It produces a current in the circuit when light of sufficiently high frequency falls on a metal plate in the phototube, but produces no current in the dark. Phototubes were used in burglar alarms and in the detection of the soundtrack on motion picture film. Modern semiconductor devices have now replaced older devices based on the photoelectric effect.

The photoelectric effect is used today in the operation of photomultiplier tubes. Figure 40.12 shows the structure of such a device. A photon striking the photocathode ejects an electron by means of the photoelectric effect. This electron is accelerated across the potential difference between the photocathode and the first *dynode*, shown as being at +200 V relative to the photocathode in Figure 40.12. This high-energy

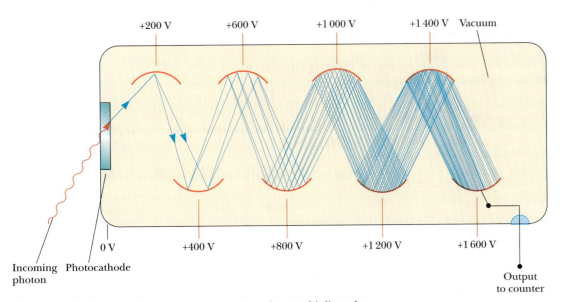

Figure 40.12 The multiplication of electrons in a photomultiplier tube.

electron strikes the dynode and ejects several more electrons. This process is repeated through a series of dynodes at ever higher potentials until an electrical pulse is produced as millions of electrons strike the last dynode. This is why the tube is called a *multiplier*—one photon at the input has resulted in millions of electrons at the output.

The photomultiplier tube is used in nuclear detectors to detect photons produced by the interaction of energetic charged particles or gamma rays with certain materials. It is also used in astronomy in a technique called *photoelectric photometry*. In this technique, the light collected by a telescope from a single star is allowed to fall on a photomultiplier tube for a time interval. The tube measures the total light energy during the time interval, which can then be converted to a luminosity of the star.

The photomultiplier tube is being replaced in many astronomical observations with a *charge-coupled device* (CCD), which is the same device that is used in a digital camera. In this device, an array of pixels is formed on the silicon surface of an integrated circuit (Section 43.7). When the surface is exposed to light from an astronomical scene through a telescope or a terrestrial scene through a digital camera, electrons generated by the photoelectric effect are caught in "traps" beneath the surface. The number of electrons is related to the intensity of the light striking the surface. A signal processor measures the number of electrons associated with each pixel and converts this information into a digital code that a computer can use to reconstruct and display the scene.

The *electron bombardment CCD camera* allows higher sensitivity than a conventional CCD. In this device, electrons ejected from a photocathode by the photoelectric effect are accelerated through a high voltage before striking a CCD array. The higher energy of the electrons results in a very sensitive detector of low-intensity radiation.

The explanation of the photoelectric effect with a quantum model, combined with Planck's quantum model for blackbody radiation, laid a strong foundation for further investigation into quantum physics. In the next section, we present a third experimental result that provides further strong evidence of the quantum nature of light.

Quick Quiz 40.3 Consider one of the curves in Figure 40.10. Suppose the intensity of the incident light is held fixed but its frequency is increased. The stopping potential in Figure 40.10 (a) remains fixed (b) moves to the right (c) moves to the left.

Quick Quiz 40.4 Suppose classical physicists had come up with the idea of predicting the appearance of a plot of K_{max} versus f, as in Figure 40.11. Draw a graph of what their expected plot would look like, based on the wave model for light.

Example 40.3   The Photoelectric Effect for Sodium

Interactive

A sodium surface is illuminated with light having a wavelength of 300 nm. The work function for sodium metal is 2.46 eV. Find

(A) the maximum kinetic energy of the ejected photoelectrons and

(B) the cutoff wavelength for sodium.

Solution (A) The energy of each photon in the illuminating light beam is

$$E = hf = \frac{hc}{\lambda} = \frac{1\ 240\ \text{eV} \cdot \text{nm}}{300\ \text{nm}} = 4.13\ \text{eV}$$

Using Equation 40.9 gives

$$K_{max} = hf - \phi = 4.13\ \text{eV} - 2.46\ \text{eV} = \boxed{1.67\ \text{eV}}$$

(B) We can calculate the cutoff wavelength from Equation 40.10:

$$\lambda_c = \frac{hc}{\phi} = \frac{1\ 240\ \text{eV} \cdot \text{nm}}{2.46\ \text{eV}} = \boxed{504\ \text{nm}}$$

This wavelength is in the yellow-green region of the visible spectrum.

Investigate the photoelectric effect for different materials and different wavelengths of light at the Interactive Worked Example link at http://www.pse6.com.

40.3 The Compton Effect

In 1919, Einstein concluded that a photon of energy E travels in a single direction and carries a momentum equal to $E/c = hf/c$. In 1923, Arthur Holly Compton (1892–1962) and Peter Debye (1884–1966) independently carried Einstein's idea of photon momentum further.

Prior to 1922, Compton and his co-workers had accumulated evidence showing that the classical wave theory of light failed to explain the scattering of x-rays from electrons. According to classical theory, electromagnetic waves of frequency f_0 incident on electrons should have two effects, as shown in Figure 40.13a: (1) radiation pressure (see Section 34.4) should cause the electrons to accelerate in the direction of propagation of the waves, and (2) the oscillating electric field of the incident radiation should set the electrons into oscillation at the apparent frequency f', where f' is the frequency in the frame of the moving electrons. This apparent frequency f' is different from the frequency f_0 of the incident radiation because of the Doppler effect (see Section 17.4). Each electron first absorbs radiation as a moving particle and then re-radiates as a moving particle, thereby exhibiting two Doppler shifts in the frequency of radiation.

Because different electrons will move at different speeds after the interaction, depending on the amount of energy absorbed from the electromagnetic waves, the scattered wave frequency at a given angle to the incoming radiation should show a

Arthur Holly Compton

**American Physicist
(1892–1962)**

Compton was born in Wooster, Ohio, and attended Wooster College and Princeton University. He became the director of the laboratory at the University of Chicago, where experimental work concerned with sustained nuclear chain reactions was conducted. This work was of central importance to the construction of the first nuclear weapon. His discovery of the Compton effect led to his sharing of the 1927 Nobel Prize in physics with Charles Wilson. (*Courtesy of AIP Niels Bohr Library*)

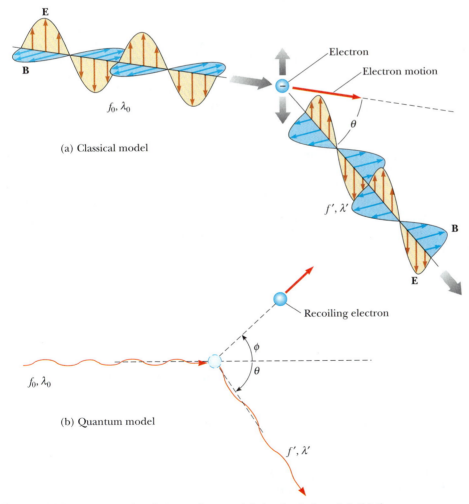

Figure 40.13 X-ray scattering from an electron: (a) the classical model; (b) the quantum model.

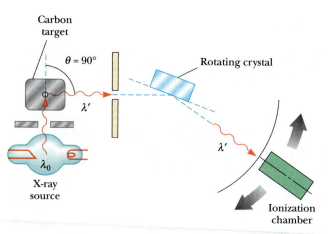

Figure 40.14 Schematic diagram of Compton's apparatus. The wavelength was measured with a rotating crystal spectrometer using graphite (carbon) as the target.

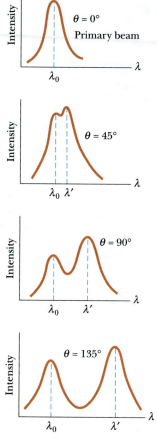

Figure 40.15 Scattered x-ray intensity versus wavelength for Compton scattering at $\theta = 0°$, $45°$, $90°$, and $135°$.

Compton shift equation

Compton wavelength

distribution of Doppler-shifted values. Contrary to this prediction, Compton's experiments showed that, at a given angle, only *one* frequency of radiation is observed. Compton and his co-workers realized that they could explain these experiments by treating photons not as waves but rather as point-like particles having energy hf and momentum hf/c, and by assuming that the energy and momentum of the isolated system of the colliding photon–electron pair are conserved. Compton adopted a particle model for something that was well known as a wave, and today this scattering phenomenon is known as the **Compton effect.** Figure 40.13b shows the quantum picture of the collision between an individual x-ray photon and an electron.

The second difference between the classical and quantum models is also shown in Figure 40.13b. In the classical model, the electron is pushed along the direction of propagation of the incident x-ray by radiation pressure. In the quantum model, the electron is scattered through an angle ϕ with respect to this direction, as if this were a billiard-ball type of collision. (The symbol ϕ used here is an angle and not to be confused with the work function, which was discussed in the preceding section.)

Figure 40.14 is a schematic diagram of the apparatus used by Compton. The x-rays, scattered from a graphite target, were analyzed with a rotating crystal spectrometer, and the intensity was measured with an ionization chamber that generated a current proportional to the intensity. The incident beam consisted of monochromatic x-rays of wavelength $\lambda_0 = 0.071$ nm. The experimental intensity-versus-wavelength plots observed by Compton for four scattering angles (corresponding to θ in Fig. 40.13) are shown in Figure 40.15. The graphs for the three nonzero angles show two peaks, one at λ_0 and one at $\lambda' > \lambda_0$. The shifted peak at λ' is caused by the scattering of x-rays from free electrons, and it was predicted by Compton to depend on scattering angle as

$$\lambda' - \lambda_0 = \frac{h}{m_e c}(1 - \cos\theta) \tag{40.11}$$

where m_e is the mass of the electron. This expression is known as the **Compton shift equation,** and the factor $h/m_e c$ is called the **Compton wavelength** of the electron. It has a currently accepted value of

$$\frac{h}{m_e c} = 0.002\ 43 \text{ nm}$$

The unshifted peak at λ_0 in Figure 40.15 is caused by x-rays scattered from electrons tightly bound to the target atoms. This unshifted peak also is predicted by Equation 40.11 if the electron mass is replaced with the mass of a carbon atom, which is about 23 000 times the mass of the electron. Thus, there is a wavelength shift for scattering from an electron bound to an atom, but it is so small that it was undetectable in Compton's experiment.

Compton's measurements were in excellent agreement with the predictions of Equation 40.11. It is fair to say that these results were the first that really convinced many physicists of the fundamental validity of quantum theory!

Quick Quiz 40.5 Note that for any given scattering angle θ, Equation 40.11 gives the same value for the Compton shift for any wavelength. Keeping this in mind, for which of the following types of radiation is the fractional shift in wavelength at a given scattering angle the largest? (a) radio waves (b) microwaves (c) visible light (d) x-rays.

Derivation of the Compton Shift Equation

We can derive the Compton shift equation by assuming that the photon behaves like a particle and collides elastically with a free electron initially at rest, as shown in Figure 40.13b. In this model, the photon is treated as a particle having energy $E = hf = hc/\lambda$ and zero rest energy. In the scattering process, the total energy and total linear momentum of the system must be conserved. Applying the principle of conservation of energy to this process gives

$$\frac{hc}{\lambda_0} = \frac{hc}{\lambda'} + K_e$$

where hc/λ_0 is the energy of the incident photon, hc/λ' is the energy of the scattered photon, and K_e is the kinetic energy of the recoiling electron. Because the electron may recoil at a speed comparable to that of light, we must use the relativistic expression $K_e = (\gamma - 1)m_e c^2$ (Eq. 39.23). Therefore,

$$\frac{hc}{\lambda_0} = \frac{hc}{\lambda'} + (\gamma - 1)m_e c^2 \qquad (40.12)$$

where $\gamma = 1/\sqrt{1 - (v^2/c^2)}$.

Next, we apply the law of conservation of momentum to this collision, noting that the x and y components of momentum are each conserved independently. Figure 40.16 shows the initial momentum of the incident photon and the vector sum of the final momenta of the recoiling electron and the scattered photon in the collision of Figure 40.13b. Equation 39.28 shows that the momentum of a photon has a magnitude $p = E/c$, and we know from Equation 40.5 that $E = hf$. Therefore, $p = hf/c$. Substituting λf for c (Eq. 16.12) in this expression gives us $p = h/\lambda$. Because the relativistic expression for the momentum of the recoiling electron is $p_e = \gamma m_e v$ (Eq. 39.19), we obtain the following expressions for the x and y components of linear momentum, where the angles are as described in Figure 40.16:

$$x \text{ component:} \qquad \frac{h}{\lambda_0} = \frac{h}{\lambda'} \cos\theta + \gamma m_e v \cos\phi \qquad (40.13)$$

$$y \text{ component:} \qquad 0 = \frac{h}{\lambda'} \sin\theta - \gamma m_e v \sin\phi \qquad (40.14)$$

By eliminating v and ϕ from Equations 40.12 to 40.14, we obtain a single expression that relates the remaining three variables (λ', λ_0, and θ). After some algebra (see Problem 58), we obtain the Compton shift equation:

$$\lambda' - \lambda_0 = \frac{h}{m_e c}(1 - \cos\theta)$$

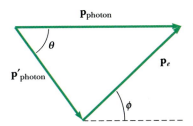

Figure 40.16 Momentum vectors for the Compton scattering event shown in Figure 40.13b.

Example 40.4 Compton Scattering at 45°

X-rays of wavelength $\lambda_0 = 0.200\,000$ nm are scattered from a block of material. The scattered x-rays are observed at an angle of 45.0° to the incident beam. Calculate their wavelength.

Solution The shift in wavelength of the scattered x-rays is given by Equation 40.11:

$$\lambda' - \lambda_0 = \frac{h}{m_e c}(1 - \cos\theta)$$

$$\lambda' = \lambda_0 + \frac{6.626 \times 10^{-34}\,\text{J}\cdot\text{s}}{(9.11 \times 10^{-31}\,\text{kg})(3.00 \times 10^8\,\text{m/s})}$$

$$\times\,(1 - \cos 45.0°)$$

$$= 0.200\,000 \times 10^{-9}\,\text{m} + 7.10 \times 10^{-13}\,\text{m}$$

$$= \boxed{0.200\,710\,\text{nm}}$$

What If? What if we move our detector so that scattered x-rays are detected at an angle larger than 45°? Does the wavelength of the scattered x-rays increase or decrease as the angle θ increases?

Answer In Equation 40.11, if the angle θ increases, the term $\cos\theta$ decreases. Consequently, the factor $(1 - \cos\theta)$ increases. Therefore, the magnitude of the Compton shift increases. Because the shift is defined as the scattered wavelength minus the incident wavelength, the scattered wavelength increases.

We could also apply an energy argument to achieve this same result. As the scattering angle increases, more energy is transferred from the incident photon to the electron. As a result, the energy of the scattered photon decreases with increasing scattering angle. Because $E = hf$, the frequency of the scattered photon decreases and, because $\lambda = c/f$, the wavelength increases.

Study Compton scattering for different angles at the Interactive Worked Example link at http://www.pse6.com.

40.4 Photons and Electromagnetic Waves

Phenomena such as the photoelectric effect and the Compton effect offer ironclad evidence that when light (or other forms of electromagnetic radiation) and matter interact, the light behaves as if it were composed of particles having energy hf and momentum h/λ. How can light be considered a photon (in other words, a particle) when we know it is a wave? On the one hand, we describe light in terms of photons having energy and momentum. On the other hand, we recognize that light and other electromagnetic waves exhibit interference and diffraction effects, which are consistent only with a wave interpretation.

Which model is correct? Is light a wave or a particle? The answer depends on the phenomenon being observed. Some experiments can be explained either better or solely with the photon model, whereas others are explained better or solely with the wave model. The end result is that **we must accept both models and admit that the true nature of light is not describable in terms of any single classical picture.** However, you should recognize that the same light beam that can eject photoelectrons from a metal (meaning that the beam consists of photons) can also be diffracted by a grating (meaning that the beam is a wave). In other words, **the particle model and the wave model of light complement each other.**

The success of the particle model of light in explaining the photoelectric effect and the Compton effect raises many other questions. If light is a particle, what is the meaning of the "frequency" and "wavelength" of the particle, and which of these two properties determines its energy and momentum? Is light *simultaneously* a wave and a particle? Although photons have no rest energy (a nonobservable quantity because a photon cannot be at rest!), is there a simple expression for the *effective mass* of a moving photon? If photons have effective mass, do they experience gravitational attraction? What is the spatial extent of a photon, and how does an electron absorb or scatter one photon? Some of these questions can be answered, but others demand a view of atomic processes that is too pictorial and literal. Many of these questions stem from classical analogies such as colliding billiard balls and water waves breaking on a shore. Quantum mechanics gives light a more fluid and flexible nature by treating the particle model and wave model of light as both necessary and complementary. Neither model can be used exclusively to describe all properties of light. A complete understanding of the observed behavior of light can be attained only if the two models are combined in a complementary manner. Before discussing this combination in more detail, we now turn our attention from electromagnetic waves to the behavior of entities that we have called particles.

40.5 The Wave Properties of Particles

Students introduced to the dual nature of light often find the concept difficult to accept. In the world around us, we are accustomed to regarding such things as baseballs solely as particles and other things such as sound waves solely as forms of wave motion. Every large-scale observation can be interpreted by considering either a wave explanation or a particle explanation, but in the world of photons and electrons, such distinctions are not as sharply drawn. Even more disconcerting is the fact that, under certain conditions, the things we unambiguously call "particles" exhibit wave characteristics!

In his 1923 doctoral dissertation, Louis de Broglie postulated that **because photons have both wave and particle characteristics, perhaps all forms of matter have both properties.** This was a highly revolutionary idea with no experimental confirmation at that time. According to de Broglie, electrons, just like light, have a dual particle–wave nature.

In Section 40.3, we found that the momentum of a photon can be expressed as

$$p = \frac{h}{\lambda}$$

From this equation we see that the photon wavelength can be specified by its momentum: $\lambda = h/p$. De Broglie suggested that material particles of momentum p have a characteristic wavelength that is given by the same expression, $\lambda = h/p$. Because the magnitude of the momentum of a particle of mass m and speed v is $p = mv$, the **de Broglie wavelength** of that particle is[5]

$$\lambda = \frac{h}{p} = \frac{h}{mv} \qquad (40.15)$$

de Broglie wavelength

Furthermore, in analogy with photons, de Broglie postulated that particles obey the Einstein relation $E = hf$, where E is the total energy of the particle. Then the frequency of a particle is

$$f = \frac{E}{h} \qquad (40.16)$$

The dual nature of matter is apparent in these last two equations because each contains both particle concepts (mv and E) and wave concepts (λ and f). The fact that these relationships are established experimentally for photons makes the de Broglie hypothesis somewhat easier to accept.

The Davisson–Germer Experiment

De Broglie's proposal in 1923 that matter exhibits both wave and particle properties was regarded as pure speculation. If particles such as electrons had wave properties, then under the correct conditions they should exhibit diffraction effects. Only three years later, C. J. Davisson (1881–1958) and L. H. Germer (1896–1971) succeeded in measuring the wavelength of electrons. Their important discovery provided the first experimental confirmation of the matter waves proposed by de Broglie.

Louis de Broglie
French Physicist (1892–1987)

De Broglie was born in Dieppe, France. At the Sorbonne in Paris, he studied history in preparation for what he hoped to be a career in the diplomatic service. The world of science is lucky that he changed his career path to become a theoretical physicist. De Broglie was awarded the Nobel Prize in 1929 for his prediction of the wave nature of electrons. (*AIP Niels Bohr Library*)

---

[5] The de Broglie wavelength for a particle moving at *any* speed v is $\lambda = h/\gamma mv$ where $\gamma = [1 - (v^2/c^2)]^{-1/2}$.

Interestingly, the intent of the initial Davisson–Germer experiment was not to confirm the de Broglie hypothesis. In fact, their discovery was made by accident (as is often the case). The experiment involved the scattering of low-energy electrons (about 54 eV) from a nickel target in a vacuum. During one experiment, the nickel surface was badly oxidized because of an accidental break in the vacuum system. After the target was heated in a flowing stream of hydrogen to remove the oxide coating, electrons scattered by it exhibited intensity maxima and minima at specific angles. The experimenters finally realized that the nickel had formed large crystalline regions upon heating and that the regularly spaced planes of atoms in these regions served as a diffraction grating for electron matter waves. (See the discussion of diffraction by crystals in Section 38.5.)

Shortly thereafter, Davisson and Germer performed more extensive diffraction measurements on electrons scattered from single-crystal targets. Their results showed conclusively the wave nature of electrons and confirmed the de Broglie relationship $p = h/\lambda$. In the same year, G. P. Thomson (1892–1975) of Scotland also observed electron diffraction patterns by passing electrons through very thin gold foils. Diffraction patterns have since been observed in the scattering of helium atoms, hydrogen atoms, and neutrons. Hence, the universal nature of matter waves has been established in various ways.

The problem of understanding the dual nature of matter and radiation is conceptually difficult because the two models seem to contradict each other. This problem as it applies to light was discussed earlier. The **principle of complementarity** states that the **wave and particle models of either matter or radiation complement each other.** Neither model can be used exclusively to describe matter or radiation adequately. Because humans tend to generate mental images based on their experiences from the everyday world (baseballs, water waves, and so forth), we use both descriptions in a complementary manner to explain any given set of data from the quantum world.

▲ **PITFALL PREVENTION**

40.3   What's Waving?

If particles have wave properties, what's waving? You are familiar with waves on strings, which are very concrete. Sound waves are more abstract, but you are likely comfortable with them. Electromagnetic waves are even more abstract, but at least they can be described in terms of physical variables and electric and magnetic fields. In contrast, waves associated with particles are completely abstract and cannot be associated with a physical variable. In Chapter 41, we will describe the wave associated with a particle in terms of probability.

Quick Quiz 40.6   An electron and a proton both moving at nonrelativistic speeds have the same de Broglie wavelength. Which of the following are also the same for the two particles? (a) speed (b) kinetic energy (c) momentum (d) frequency.

Quick Quiz 40.7   We have discussed two wavelengths associated with the electron—the Compton wavelength and the de Broglie wavelength. Which is an actual *physical* wavelength associated with the electron? (a) the Compton wavelength (b) the de Broglie wavelength (c) both wavelengths (d) neither wavelength.

Example 40.5   The Wavelength of an Electron

Calculate the de Broglie wavelength for an electron ($m_e = 9.11 \times 10^{-31}$ kg) moving at 1.00×10^7 m/s.

Solution   Equation 40.15 gives

$$\lambda = \frac{h}{m_e v} = \frac{6.63 \times 10^{-34} \text{ J} \cdot \text{s}}{(9.11 \times 10^{-31} \text{ kg})(1.00 \times 10^7 \text{ m/s})}$$

$$= \boxed{7.28 \times 10^{-11} \text{ m}}$$

Example 40.6   The Wavelength of a Rock

A rock of mass 50 g is thrown with a speed of 40 m/s. What is its de Broglie wavelength?

Solution   From Equation 40.15 we have

$$\lambda = \frac{h}{mv} = \frac{6.63 \times 10^{-34} \text{ J} \cdot \text{s}}{(50 \times 10^{-3} \text{ kg})(40 \text{ m/s})}$$

$$\lambda = \boxed{3.32 \times 10^{-34} \text{ m}}$$

This wavelength is much smaller than any aperture through which the rock could possibly pass. This means that we could not observe diffraction effects, and as a result the wave properties of large-scale objects cannot be observed.

Example 40.7   An Accelerated Charged Particle

A particle of charge q and mass m has been accelerated from rest to a nonrelativistic speed through a potential difference ΔV. Find an expression for its de Broglie wavelength.

Solution When a charged particle is accelerated from rest through a potential difference ΔV, its gain in kinetic energy $\frac{1}{2}mv^2$ must equal the loss in potential energy $q\,\Delta V$ of the charge–field system:

$$\tfrac{1}{2}mv^2 = q\,\Delta V$$

Because $p = mv$, we can express this equation in the form

$$\frac{p^2}{2m} = q\,\Delta V$$

$$p = \sqrt{2mq\,\Delta V}$$

Substituting this expression for p into Equation 40.15 gives

$$\lambda = \frac{h}{p} = \frac{h}{\sqrt{2mq\,\Delta V}}$$

The Electron Microscope

A practical device that relies on the wave characteristics of electrons is the **electron microscope.** A *transmission* electron microscope, used for viewing flat, thin samples, is shown in Figure 40.17. In many respects it is similar to an optical microscope, but the electron microscope has a much greater resolving power because it can accelerate electrons to very high kinetic energies, giving them very short wavelengths. No microscope can resolve details that are significantly smaller than the wavelength of the waves used to illuminate the object. Typically, the wavelengths of electrons are about 100 times shorter than those

Electron gun

Cathode

Anode

Electromagnetic lens

Electromagnetic condenser lens

Screen

Visual transmission

Vacuum

Core

Coil

Electron beam

Specimen goes here

Specimen chamber door

Projector lens

Photo chamber

© David Parker/Photo Researchers, Inc.

(a)   (b)

Figure 40.17 (a) Diagram of a transmission electron microscope for viewing a thinly sectioned sample. The "lenses" that control the electron beam are magnetic deflection coils. (b) An electron microscope.

Figure 40.18 A color-enhanced electron microscope photograph shows significant detail of a storage mite, *Lepidoglyphus destructor.* The mite is so small, with a maximum length of 0.75 mm, that ordinary microscopes do not reveal minute anatomical details.

of the visible light used in optical microscopes. As a result, an electron microscope with ideal lenses would be able to distinguish details about 100 times smaller than those distinguished by an optical microscope. (Electromagnetic radiation of the same wavelength as the electrons in an electron microscope is in the x-ray region of the spectrum.)

The electron beam in an electron microscope is controlled by electrostatic or magnetic deflection, which acts on the electrons to focus the beam and form an image. Rather than examining the image through an eyepiece as in an optical microscope, the viewer looks at an image formed on a monitor or other type of display screen. Figure 40.18 shows the amazing detail available with an electron microscope.

40.6 The Quantum Particle

The discussions presented in previous sections may be quite disturbing, because we considered the particle and wave models to be distinct in the past. The notion that both light and material particles have both particle and wave properties does not fit with this distinction. We have experimental evidence, however, that this is just what we must accept. The recognition of this dual nature leads to a new model, the **quantum particle.** In this model, entities have both particle and wave characteristics, and we must choose one appropriate behavior—particle or wave—in order to understand a particular phenomenon.

In this section, we shall explore this model in a way that might make you more comfortable with this idea. We shall do this by demonstrating that we can construct from waves an entity that exhibits properties of a particle.

Let us first recall some characteristics of ideal particles and ideal waves. An ideal particle has zero size. Thus, an essential feature of a particle is that it is *localized* in space. An ideal wave has a single frequency and is infinitely long, as suggested by Figure 40.19a.

(a)

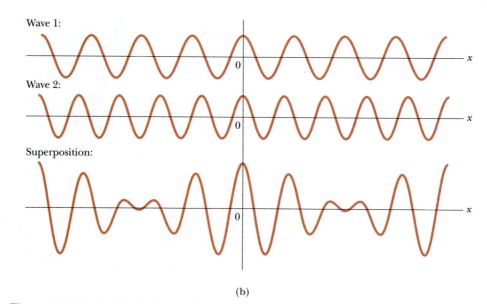

Wave 1:

Wave 2:

Superposition:

(b)

Figure 40.19 (a) An idealized wave of an exact single frequency is the same throughout space and time. (b) If two ideal waves with slightly different frequencies are combined, beats result (Section 18.7). The regions of space at which there is constructive interference are different from those at which there is destructive interference.

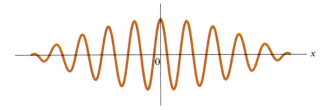

Active Figure 40.20 If a large number of waves are combined, the result is a wave packet, which represents a particle.

Choose the number of waves to add together and observe the resulting wave packet at the Active Figures link at **http://www.pse6.com.**

Thus, an ideal wave is *unlocalized* in space. Let us show that we can build a localized entity from infinitely long waves. Imagine drawing one wave along the x axis, with one of its crests located at $x = 0$, as at the top of Figure 40.19b. Now draw a second wave, of the same amplitude but a different frequency, with one of its crests also at $x = 0$. The result of the superposition of these two waves is a *beat*, as the waves are alternately in phase and out of phase. (Beats were discussed in Section 18.7.) The lowest curve in Figure 40.19b shows the results of superposing these two waves.

Notice that we have already introduced some localization by superposing the two waves. A single wave has the same amplitude everywhere in space—no point in space is any different from any other point. By adding a second wave, however, there is something different about the in-phase points compared to the out-of-phase points.

Now imagine that more and more waves are added to our original two, each new wave having a new frequency. Each new wave is added so that one of its crests is at $x = 0$. The result is that all of the waves add constructively at $x = 0$. When we consider a large number of waves, the probability of a positive value of a wave function at any point $x \neq 0$ is equal to the probability of a negative value, and there is destructive interference *everywhere* except near $x = 0$, where we superposed all of the crests. The result of this is shown in Figure 40.20. The small region of constructive interference is called a **wave packet.** This is a localized region of space that is different from all other regions. We can identify the wave packet as a particle—it has the localized nature of what we have come to recognize as a particle! The location of the wave packet corresponds to the position of the particle.

The localized nature of this entity is the *only* characteristic of a particle that was generated with this process. We have not addressed how the wave packet might achieve such particle characteristics as mass, electric charge, spin, etc. Thus, you may not be completely convinced that we have built a particle. As further evidence that the wave packet can represent the particle, let us show that the wave packet has another characteristic of a particle.

We return to our combination of two waves in order to make the mathematical representation simple. Consider two waves with equal amplitudes but different frequencies f_1 and f_2. We can represent the waves mathematically as

$$y_1 = A\cos(k_1 x - \omega_1 t) \qquad \text{and} \qquad y_2 = A\cos(k_2 x - \omega_2 t)$$

where, as in Chapter 16, $\omega = 2\pi f$ and $k = 2\pi/\lambda$. Using the superposition principle, we add the waves:

$$y = y_1 + y_2 = A\cos(k_1 x - \omega_1 t) + A\cos(k_2 x - \omega_2 t)$$

It is convenient to write this in a form that uses the trigonometric identity

$$\cos a + \cos b = 2\cos\left(\frac{a-b}{2}\right)\cos\left(\frac{a+b}{2}\right)$$

Letting $a = k_1 x - \omega_1 t$ and $b = k_2 x - \omega_2 t$, we find that

$$y = 2A\cos\left(\frac{(k_1 x - \omega_1 t) - (k_2 x - \omega_2 t)}{2}\right)\cos\left(\frac{(k_1 x - \omega_1 t) + (k_2 x - \omega_2 t)}{2}\right)$$

$$= \left[2A\cos\left(\frac{\Delta k}{2}x - \frac{\Delta\omega}{2}t\right)\right]\cos\left(\frac{k_1 + k_2}{2}x - \frac{\omega_1 + \omega_2}{2}t\right) \qquad (40.17)$$

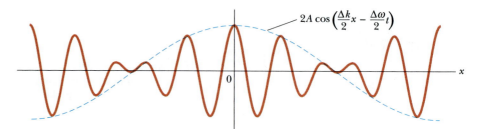

Active Figure 40.21 The beat pattern of Figure 40.19b, with an envelope function (blue curve) superimposed.

Observe the movement of the waves and the envelope at the Active Figures link at http://www.pse6.com.

where $\Delta k = k_1 - k_2$ and $\Delta\omega = \omega_1 - \omega_2$. The second cosine factor represents a wave with a wave number and frequency that are equal to the averages of the values for the individual waves.

The factor in square brackets represents the envelope of the wave, as shown by the blue curve in Figure 40.21. Notice that this factor also has the mathematical form of a wave. **This envelope of the combination can travel through space with a different speed than the individual waves.** As an extreme example of this possibility, imagine combining two identical waves moving in opposite directions. The two waves move with the same speed, but the envelope has a speed of *zero*, because we have built a standing wave, which we studied in Section 18.2.

For an individual wave, the speed is given by Equation 16.11,

Phase speed of a wave in a wave packet

$$v_{\text{phase}} = \frac{\omega}{k} \qquad (40.18)$$

This speed is called the **phase speed** because it is the rate of advance of a crest on a single wave, which is a point of fixed phase. This equation can be interpreted as follows: the phase speed of a wave is the ratio of the coefficient of the time variable t to the coefficient of the space variable x in the equation for the wave, $y = A\cos(kx - \omega t)$.

The factor in brackets in Equation 40.17 is of the form of a wave, so it moves with a speed given by this same ratio:

$$v_g = \frac{\text{coefficient of time variable } t}{\text{coefficient of space variable } x} = \frac{(\Delta\omega/2)}{(\Delta k/2)} = \frac{\Delta\omega}{\Delta k}$$

The subscript g on the speed indicates that this is commonly called the **group speed,** or the speed of the wave packet (the *group* of waves) that we have built. We have generated this expression for a simple addition of two waves. For a superposition of a very large number of waves to form a wave packet, this ratio becomes a derivative:

Group speed of a wave packet

$$v_g = \frac{d\omega}{dk} \qquad (40.19)$$

Let us multiply the numerator and the denominator by $\hbar$, where $\hbar = h/2\pi$:

$$v_g = \frac{\hbar\, d\omega}{\hbar\, dk} = \frac{d(\hbar\omega)}{d(\hbar k)} \qquad (40.20)$$

We look at the terms in the parentheses in the numerator and denominator of Equation 40.20 separately. For the numerator,

$$\hbar\omega = \frac{h}{2\pi}\,(2\pi f) = hf = E$$

For the denominator,

$$\hbar k = \frac{h}{2\pi}\left(\frac{2\pi}{\lambda}\right) = \frac{h}{\lambda} = p$$

Thus, Equation 40.20 can be written as

$$v_g = \frac{d(\hbar\omega)}{d(\hbar k)} = \frac{dE}{dp} \qquad (40.21)$$

Because we are exploring the possibility that the envelope of the combined waves represents the particle, consider a free particle moving with a speed u that is small compared to that of light. The energy of the particle is its kinetic energy:

$$E = \tfrac{1}{2}mu^2 = \frac{p^2}{2m}$$

Differentiating this equation with respect to p gives

$$v_g = \frac{dE}{dp} = \frac{d}{dp}\left(\frac{p^2}{2m}\right) = \frac{1}{2m}(2p) = u \qquad (40.22)$$

Thus, the group speed of the wave packet is identical to the speed of the particle that it is modeled to represent! This gives us further confidence that the wave packet is a reasonable way to build a particle.

Quick Quiz 40.8 As an analogy to wave packets, consider an "automobile packet" that occurs near the scene of an accident on a freeway. The phase speed is analogous to the speed of individual automobiles as they move through the backup caused by the accident. The group speed can be identified as the speed of the leading edge of the packet of cars. For the automobile packet, (a) the group speed is the same as the phase speed. (b) the group speed is less than the phase speed. (c) the group speed is greater than the phase speed.

Quick Quiz 40.9 As another analogy to wave packets, consider a "runner packet" that occurs at the start of a footrace of length L. As the runners begin the race, the packet of runners spreads in size as the faster runners outpace the slower runners. The phase speed is the speed of a single runner, while we can identify the group speed v_g as the speed with which the average position of the entire packet of runners moves. The time interval for the winning runner to run the race is (a) greater than L/v_g. (b) equal to L/v_g. (c) less than L/v_g.

40.7 The Double-Slit Experiment Revisited

Wave–particle duality is now a firmly accepted concept reinforced by experimental results including those of the Davisson–Germer experiment. However, as with the postulates of special relativity, this concept often leads to clashes with familiar thought patterns we hold from everyday experience.

One way to crystallize our ideas about the electron's wave–particle duality is to consider an experiment in which electrons are fired at a double slit. Consider a parallel beam of mono-energetic electrons that is incident on a double slit as in Figure 40.22. We shall assume that the slit widths are small compared to the electron wavelength, so that we need not worry about diffraction maxima and minima, as we discussed for light in Section 38.2. An electron detector is positioned far from the slits at a distance much greater than d, the separation distance of the slits. If the detector collects electrons for a long enough time, one finds a typical wave interference pattern for the counts per minute, or probability of arrival of electrons. Such an interference pattern would not be expected if the electrons behaved as classical particles. This is clear evidence that electrons are interfering—a distinct wave-like behavior.

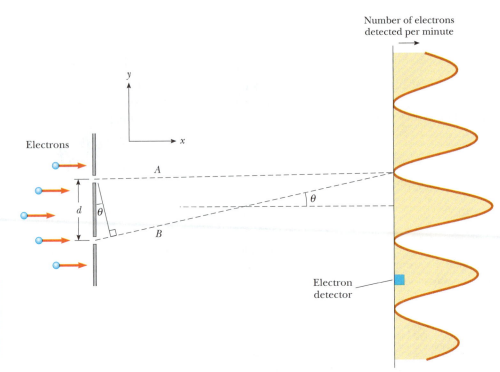

Number of electrons
detected per minute

Electrons

Electron
detector

Figure 40.22 Electron diffraction. The slit separation d is much greater than the individual slit widths and much less than the distance between the slit and the detector. The electron detector is movable along the y direction in the drawing and so can detect electrons diffracted at different values of θ. The detector acts like the "viewing screen" of Young's double-slit experiment with light discussed in Chapter 37.

(a) After 28 electrons

(b) After 1 000 electrons

(c) After 10 000 electrons

(d) Two-slit electron pattern

Active Figure 40.23 (a), (b), (c) Computer-simulated interference patterns for a beam of electrons incident on a double slit. (d) Photograph of a double-slit interference pattern produced by electrons.

At the Active Figures link at http://www.pse6.com, you can watch the interference pattern develop over time and see how it is destroyed by the action of keeping track of which slit an electron goes through.

If the experiment is carried out at lower electron beam intensities, the interference pattern is still observed if the time interval for the measurement is sufficiently long. This is illustrated by the computer-simulated patterns in Figure 40.23. Note that the interference pattern becomes clearer as the number of electrons reaching the screen increases.

If one imagines a single electron in the beam reaching one of the slits, standard wave theory can be used to find the angular separation θ between the central probability maximum and its neighboring minimum. The minimum occurs when the path length difference between A and B is half a wavelength, or when

$$d \sin \theta = \frac{\lambda}{2}$$

Because an electron's wavelength is given by $\lambda = h/p$, we see that for small θ,

$$\sin \theta \approx \theta = \frac{h}{2pd}$$

Thus, the dual nature of the electron is clearly shown in this experiment: **the electrons are detected as particles at a localized spot at some instant of time, but the probability of arrival at that spot is determined by finding the intensity of two interfering waves.**

Let us now look at this experiment from another point of view. If one slit is covered during the experiment, a symmetric curve peaked around the center of the open slit is observed—the central maximum of the single-slit diffraction pattern. Plots of the counts per minute (probability of arrival of electrons) with the lower or upper slit closed are shown as blue curves in the central part of Figure 40.24.

If another experiment is now performed with slit 2 blocked half of the time and then slit 1 blocked during the remaining time, the accumulated pattern of counts/min shown by the blue curve on the right side of Figure 40.24 is completely different from

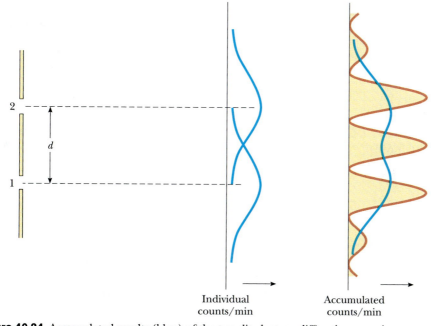

Figure 40.24 Accumulated results (blue) of the two-slit electron diffraction experiment with each slit closed half the time. The result with both slits open is shown in red.

the case with both slits open (red curve). There is no longer a maximum probability of arrival of an electron at $\theta = 0$. In fact, **the interference pattern has been lost and the accumulated result is simply the sum of the individual results.** When only one slit is open at a time, we know that the electron has the same localizability and indivisibility at the slits as we measure at the detector, because the electron clearly goes through slit 1 or slit 2. Thus, the total must be analyzed as the sum of those electrons that come through slit 1, and those that come through slit 2.

When both slits are open, it is tempting to assume that the electron goes through either slit 1 or slit 2, and that the counts/min are again given by the combination of the single-slit distributions. However, the experimental results indicated by the red interference pattern in Figure 40.24 contradict this assumption. Hence, our assumption that the electron is localized and goes through only one slit when both slits are open must be wrong (a painful conclusion!).

In order to interpret these results, we are forced to conclude that **an electron interacts with both slits simultaneously.** If we attempt to determine experimentally which slit the electron goes through, the act of measuring destroys the interference pattern. It is impossible to determine which slit the electron goes through. In effect, **we can say only that the electron passes through both slits!** The same arguments apply to photons.

If we restrict ourselves to a pure particle model, it is an uncomfortable notion that the electron can be present at both slits at once. From the quantum particle model, however, the particle can be considered to be built from waves that exist throughout space. Thus, the wave components of the electron are present at both slits at the same time, and this model leads to a more comfortable interpretation of this experiment.

40.8 The Uncertainty Principle

Whenever one measures the position or velocity of a particle at any instant, experimental uncertainties are built into the measurements. According to classical mechanics, there is no fundamental barrier to an ultimate refinement of the apparatus or experimental procedures. In other words, it is possible, in principle, to make such measurements with arbitrarily small uncertainty. Quantum theory predicts, however, that **it is fundamentally impossible to make simultaneous measurements of a particle's position and momentum with infinite accuracy.**

In 1927, Werner Heisenberg (1901–1976) introduced this notion, which is now known as the **Heisenberg uncertainty principle:**

> If a measurement of the position of a particle is made with uncertainty Δx and a simultaneous measurement of its x component of momentum is made with uncertainty Δp_x, the product of the two uncertainties can never be smaller than $\hbar/2$:

Heisenberg uncertainty principle

$$\Delta x \, \Delta p_x \geq \frac{\hbar}{2} \qquad (40.23)$$

That is, **it is physically impossible to measure simultaneously the exact position and exact momentum of a particle.** Heisenberg was careful to point out that the inescapable uncertainties Δx and Δp_x do not arise from imperfections in practical measuring instruments. Rather, **the uncertainties arise from the quantum structure of matter.**

To understand the uncertainty principle suppose the wavelength of a particle is known *exactly.* According to the de Broglie relation, $\lambda = h/p$, we would therefore know the momentum to be precisely $p = h/\lambda$.

In reality, a single-wavelength wave would exist throughout space. Any region along this wave is the same as any other region (Fig. 40.19a). If we were to ask, "Where is the particle that this wave represents?", there is no special location in space along the wave that could be identified with the particle—all points along the wave are the same. Therefore, we have *infinite* uncertainty in the position of the particle—we know nothing about its location. Perfect knowledge of the particle's momentum has cost us all information about its location.

In comparison, now consider a particle whose momentum is uncertain, so that it has a range of possible values of momentum. According to the de Broglie relation, this results in a range of wavelengths. Thus, the particle is not represented by a single wavelength, but a combination of wavelengths within this range. This combination forms a wave packet, as we discussed in Section 40.6 and illustrated in Figure 40.20. If we are asked to determine the location of the particle, we can only say that it is somewhere in the region defined by the wave packet—because there is a distinct difference between this region and the rest of space. Thus, by losing some information about the momentum of the particle, we have gained information about its position.

If we were to lose *all* information about the momentum, then we would be adding together waves of all possible wavelengths. This would result in a wave packet of zero length. Thus, if we know nothing about the momentum, we know exactly where the particle is.

The mathematical form of the uncertainty principle states that the product of the uncertainties in position and momentum is always larger than some minimum value. This value can be calculated from the types of arguments discussed above, which result in the value of $\hbar/2$ in Equation 40.23.

Another form of the uncertainty principle can be generated by reconsidering Figure 40.20. Imagine that the horizontal axis is time rather than spatial position x. We can then make the same arguments that we made about knowledge of wavelength and position in the time domain. The corresponding variables would be frequency and time. Because frequency is related to the energy of the particle by $E = hf$, the uncertainty principle in this form is

$$\Delta E \, \Delta t \geq \frac{\hbar}{2} \qquad (40.24)$$

Werner Heisenberg

German Theoretical Physicist (1901–1976)

Heisenberg obtained his Ph.D. in 1923 at the University of Munich. While other physicists tried to develop physical models of quantum phenomena, Heisenberg developed an abstract mathematical model called *matrix mechanics.* The more widely accepted physical models were shown to be equivalent to matrix mechanics. Heisenberg made many other significant contributions to physics, including his famous uncertainty principle for which he received a Nobel Prize in 1932, the prediction of two forms of molecular hydrogen, and theoretical models of the nucleus. (*Courtesy of the University of Hamburg*)

This form of the uncertainty principle suggests that energy conservation can appear to be violated by an amount ΔE as long as it is only for a short time interval Δt consistent with Equation 40.24. We shall use this notion to estimate the rest energies of particles in Chapter 46.

▲ **PITFALL PREVENTION**

40.4 The Uncertainty Principle

Some students incorrectly interpret the uncertainty principle as meaning that a measurement interferes with the system. For example, if an electron is observed in a hypothetical experiment using an optical microscope, the photon used to see the electron collides with it and makes it move, giving it an uncertainty in momentum. This is *not* the idea of the uncertainty principle. The uncertainty principle is independent of the measurement process and is grounded in the wave nature of matter.

Quick Quiz 40.10 The location of a particle is measured and specified as being exactly at $x = 0$, with *zero* uncertainty in the x direction. How does this affect the uncertainty of its velocity component in the y direction? (a) It does not affect it. (b) It makes it infinite. (c) It makes it zero.

Quick Quiz 40.11 A quantum argument for the phenomenon of diffraction of light claims that photons passing through a narrow slit have been localized to the width of the slit. Because we have gained information about their position, they must have a larger uncertainty in momentum along the plane of the screen in which the slit is cut. Thus, the photons gain momentum perpendicular to their original direction of propagation and spread out, forming on a screen a bright area that is wider than the slit. Suppose we are observing diffraction of light and suddenly Planck's constant drops to half its previous value. This quantum argument for diffraction would claim that (a) the bright area on the screen is unchanged (b) the bright area on the screen becomes wider (c) the bright area on the screen becomes narrower.

Example 40.8   Locating an Electron

The speed of an electron is measured to be 5.00×10^3 m/s to an accuracy of 0.003 00%. Find the minimum uncertainty in determining the position of this electron.

Solution Assuming that the electron is moving along the x axis, the x component of the momentum of the electron is

$$p_x = mv_x = (9.11 \times 10^{-31} \text{ kg})(5.00 \times 10^3 \text{ m/s})$$
$$= 4.56 \times 10^{-27} \text{ kg} \cdot \text{m/s}$$

The uncertainty in p_x is 0.003 00% of this value:

$$\Delta p_x = (0.000\,030\,0)(4.56 \times 10^{-27} \text{ kg} \cdot \text{m/s})$$
$$= 1.37 \times 10^{-31} \text{ kg} \cdot \text{m/s}$$

We can now calculate the minimum uncertainty in position by using this value of Δp_x and Equation 40.23:

$$\Delta x\, \Delta p_x \geq \frac{\hbar}{2}$$

$$\Delta x \geq \frac{\hbar}{2\,\Delta p_x} = \frac{1.05 \times 10^{-34} \text{ J} \cdot \text{s}}{2(1.37 \times 10^{-31} \text{ kg} \cdot \text{m/s})}$$

$$= \boxed{0.383 \text{ mm}}$$

Example 40.9   The Line Width of Atomic Emissions

Atoms have quantized energy levels similar to those of Planck's oscillators, although the energy levels of an atom are generally not evenly spaced. When an atom makes a transition between states, energy is emitted in the form of a photon. Although an excited atom can radiate at any time from $t = 0$ to $t = \infty$, the average time interval after excitation during which an atom radiates is called the **lifetime** τ. If $\tau = 1.0 \times 10^{-8}$ s, use the uncertainty principle to compute the line width Δf produced by this finite lifetime.

Solution We use $\Delta E\,\Delta t \geq \hbar/2$, where $\Delta E = h\,\Delta f$ and $\Delta t = 1.0 \times 10^{-8}$ s is the lifetime. Thus, the minimum value of

Δf is

$$\Delta f = \frac{\Delta E}{h} = \frac{\hbar}{2h\,\Delta t} = \frac{1}{4\pi\,\Delta t} = \frac{1}{4\pi(1.0 \times 10^{-8} \text{ s})}$$

$$= \boxed{8.0 \times 10^6 \text{ Hz}}$$

Note that ΔE is the uncertainty in the energy of the excited atom. It is also the uncertainty in the energy of the photon emitted by an atom in this state.

What If? What if this same lifetime were associated with a transition that emits a radio wave rather than a visible light wave from an atom? Is the fractional line width Δ*f*/*f* larger or smaller than for the visible light?

Answer Because we are assuming the same lifetime for both transitions, Δf is independent of the frequency of radiation. Radio waves have lower frequencies than light waves, so the ratio $\Delta f/f$ will be larger for the radio waves. Assuming a light-wave frequency f of 6.00×10^{14} Hz, the fractional line width is

$$\frac{\Delta f}{f} = \frac{8.0 \times 10^6 \text{ Hz}}{6.00 \times 10^{14} \text{ Hz}} = 1.3 \times 10^{-8}$$

This narrow fractional line width can be measured with a sensitive interferometer. Usually, however, temperature and pressure effects overshadow the natural line width and broaden the line through mechanisms associated with the Doppler effect and collisions.

Assuming a radio-wave frequency f of 94.7×10^6 Hz, the fractional line width is

$$\frac{\Delta f}{f} = \frac{8.0 \times 10^6 \text{ Hz}}{94.7 \times 10^6 \text{ Hz}} = 8.4 \times 10^{-2}$$

Thus, for the radio wave, this same absolute line width corresponds to a fractional line width of over 8%.

SUMMARY

Take a practice test for this chapter by clicking on the Practice Test link at http://www.pse6.com.

The characteristics of **blackbody radiation** cannot be explained using classical concepts. Planck introduced the quantum concept and Planck's constant when he assumed that atomic oscillators existing only in discrete energy states were responsible for this radiation. In Planck's model, radiation is emitted in single quantized packets whenever an oscillator makes a transition between discrete energy states. Einstein successfully extended Planck's quantum hypothesis to the standing waves of electromagnetic radiation in a cavity used in the blackbody radiation model.

The **photoelectric effect** is a process whereby electrons are ejected from a metal surface when light is incident on that surface. In Einstein's model, light is viewed as a stream of particles, or **photons,** each having energy $E = hf$, where f is the frequency and h is Planck's constant. The maximum kinetic energy of the ejected photoelectron is

$$K_{\max} = hf - \phi \tag{40.9}$$

where ϕ is the **work function** of the metal.

X-rays are scattered at various angles by electrons in a target. In such a scattering event, a shift in wavelength is observed for the scattered x-rays, and the phenomenon is known as the **Compton effect.** Classical physics does not predict the correct behavior in this effect. If the x-ray is treated as a photon, conservation of energy and linear momentum applied to the photon–electron collisions yields for the Compton shift:

$$\lambda' - \lambda_0 = \frac{h}{m_e c} (1 - \cos \theta) \tag{40.11}$$

where m_e is the mass of the electron, c is the speed of light, and θ is the scattering angle.

Light has a dual nature in that it has both wave and particle characteristics. Some experiments can be explained either better or solely by the particle model, whereas others can be explained either better or solely by the wave model.

Every object of mass m and momentum $p = mv$ has wave properties, with a **de Broglie wavelength** given by:

$$\lambda = \frac{h}{p} = \frac{h}{mv} \tag{40.15}$$

By combining a large number of waves, a single region of constructive interference, called a **wave packet,** can be created. The wave packet carries the characteristic of localization like a particle, but has wave properties because it is built from waves. For an individual wave in the wave packet, the **phase speed** is

$$v_{\text{phase}} = \frac{\omega}{k} \tag{40.18}$$

For the wave packet as a whole, the **group speed** is

$$v_g = \frac{d\omega}{dk} \tag{40.19}$$

For a wave packet representing a particle, the group speed can be shown to be the same as the speed of the particle.

The **Heisenberg uncertainty principle** states that if a measurement of position is made with uncertainty Δx and a simultaneous measurement of linear momentum is made with uncertainty Δp_x, the product of the two uncertainties is restricted to

$$\Delta x \, \Delta p_x \geq \frac{\hbar}{2} \tag{40.23}$$

Another form of the uncertainty principle relates measurements of energy and time:

$$\Delta E \, \Delta t \geq \frac{\hbar}{2} \tag{40.24}$$

QUESTIONS

1. What assumptions did Planck make in dealing with the problem of blackbody radiation? Discuss the consequences of these assumptions.

2. The classical model of blackbody radiation given by the Rayleigh–Jeans law has two major flaws. Identify them and explain how Planck's law deals with them.

3. All objects radiate energy. Why, then, are we not able to see all objects in a dark room?

4. The brightest star in the constellation Lyra is the bluish star Vega, and the brightest star in Boötes is the reddish star Arcturus. How do you account for the difference in color of the two stars?

5. If the photoelectric effect is observed for one metal, can you conclude that the effect will also be observed for another metal under the same conditions? Explain.

6. What does the slope of the line in Figure 40.11 represent? What does the y intercept represent? How would such graphs for different metals compare with one another?

7. Why does the existence of a cutoff frequency in the photoelectric effect favor a particle theory for light over a wave theory?

8. In the photoelectric effect, explain why the stopping potential depends on the frequency of light but not on the intensity.

9. A student claims that she is going to eject electrons from a piece of metal by placing a radio transmitter antenna adjacent to the metal and sending a strong AM radio signal into the antenna. The work function of a metal is typically a few electron volts. Will this work?

10. Suppose the photoelectric effect occurs in a gaseous target rather than a solid plate. Will photoelectrons be produced at all frequencies of the incident photon? Explain.

11. Which has more energy, a photon of ultraviolet radiation or a photon of yellow light?

12. Which is more likely to cause sunburn (because more energy is absorbed by individual molecules in skin cells): (a) infrared light, (b) visible light, or (c) ultraviolet light?

13. How does the Compton effect differ from the photoelectric effect?

14. What assumptions did Compton make in dealing with the scattering of a photon from an electron?

15. An x-ray photon is scattered by an electron. What happens to the frequency of the scattered photon relative to that of the incident photon?

16. Suppose a photograph were made of a person's face using only a few photons. Would the result be simply a very faint image of the face? Explain your answer.

17. Is light a wave or a particle? Support your answer by citing specific experimental evidence.

18. Is an electron a wave or a particle? Support your answer by citing some experimental results.

19. Why was the demonstration of electron diffraction by Davisson and Germer an important experiment?

20. An electron and a proton are accelerated from rest through the same potential difference. Which particle has the longer wavelength?

21. If matter has a wave nature, why is this wave-like characteristic not observable in our daily experiences?

22. Why is it impossible to measure simultaneously with infinite accuracy the position and speed of a particle?

23. In describing the passage of electrons through a slit and arriving at a screen, the physicist Richard Feynman said that "electrons arrive in lumps, like particles, but the probability of arrival of these lumps is determined as the intensity of the waves would be. It is in this sense that the electron behaves sometimes like a particle and sometimes like a wave." Elaborate on this point in your own words. (For a further discussion of this point, see R. Feynman, *The Character of Physical Law*, Cambridge, MA, MIT Press, 1980, Chapter 6.)

24. Why is an electron microscope more suitable than an optical microscope for "seeing" objects less than 1 μm in size?

25. Shown in the photographs on pages 1176 and 1177, *iridescence* is the phenomenon that gives shining colors to the feathers of peacocks, hummingbirds, resplendent quetzals, and even ducks and grackles. Without pigments, it colors Morpho butterflies, Urania moths, some beetles and flies, rainbow trout, and mother-of-pearl in abalone shells. Iridescent colors change as you turn an object. They can look different to your two eyes, so that the object appears to have metallic luster. Iridescent colors were first described in print not by an artist or biologist, but by the physicist Isaac Newton. They are produced by a wide variety of intricate structures in different species—Problem 56 in Chapter 38 describes those in a peacock feather. These structures were all unknown until the invention of the electron microscope. Explain why light microscopes cannot reveal them.

26. *Blacker than black, brighter than white.* (a) Take a large, closed, empty cardboard box. Cut a slot a few millimeters wide in one side. Use black pens, markers, and black material to make some stripes next to the slot, as shown in Figure Q40.26a. Inspect them with care and choose which is blackest—the figure may not show enough contrast to reveal which it is. Explain why it is blackest. (b) Locate an intri-

cately shaped compact fluorescent light fixture. Look at it through dark glasses and describe where it appears brightest. Explain why it is brightest there. Figure Q40.26b shows two such light fixtures held near each other. *Suggestion:* Gustav Kirchhoff, professor at Heidelberg and master of the obvious, gave the same answer to part (a) as you likely will. His answer to part (b) would begin like this: When electromagnetic radiation falls on its surface, an object reflects some fraction r of the energy and absorbs the rest. Whether the fraction reflected is 0.8 or 0.001, the fraction absorbed is $a = 1 - r$. Suppose the object and its surroundings are at the same temperature. The energy the object absorbs joins its fund of internal energy, but the second law of thermodynamics implies that the absorbed energy cannot raise the object's temperature. It does not produce a temperature increase because the object's energy budget has one more term: energy radiated . . . . You still have to make the observations and answer questions (a) and (b), but you can incorporate some of Kirchhoff's ideas into your answer if you wish.

(a)   (b)

Figure Q40.26

PROBLEMS

1, 2, 3 = straightforward, intermediate, challenging   ☐ = full solution available in the *Student Solutions Manual and Study Guide*

🌐 = coached solution with hints available at http://www.pse6.com   💻 = computer useful in solving problem

▭ = paired numerical and symbolic problems

Section 40.1   Blackbody Radiation and Planck's Hypothesis

1. The human eye is most sensitive to 560-nm light. What is the temperature of a black body that would radiate most intensely at this wavelength?

2. (a) Lightning produces a maximum air temperature on the order of 10^4 K, whereas (b) a nuclear explosion produces a temperature on the order of 10^7 K. Use Wien's displacement law to find the order of magnitude of the wavelength of the thermally produced photons radiated

with greatest intensity by each of these sources. Name the part of the electromagnetic spectrum where you would expect each to radiate most strongly.

3. A black body at 7 500 K consists of an opening of diameter 0.050 0 mm, looking into an oven. Find the number of photons per second escaping the hole and having wavelengths between 500 nm and 501 nm.

4. Consider a black body of surface area 20.0 cm² and temperature 5 000 K. (a) How much power does it radiate? (b) At what wavelength does it radiate most intensely? Find

the spectral power per wavelength at (c) this wavelength and at wavelengths of (d) 1.00 nm (an x- or γ ray), (e) 5.00 nm (ultraviolet light or an x-ray), (f) 400 nm (at the boundary between UV and visible light), (g) 700 nm (at the boundary between visible and infrared light), (h) 1.00 mm (infrared light or a microwave) and (i) 10.0 cm (a microwave or radio wave). (j) About how much power does the object radiate as visible light?

5. The radius of our Sun is 6.96×10^8 m, and its total power output is 3.77×10^{26} W. (a) Assuming that the Sun's surface emits as a black body, calculate its surface temperature. (b) Using the result of part (a), find λ_{max} for the Sun.

6. A sodium-vapor lamp has a power output of 10.0 W. Using 589.3 nm as the average wavelength of this source, calculate the number of photons emitted per second.

7. Calculate the energy, in electron volts, of a photon whose frequency is (a) 620 THz, (b) 3.10 GHz, (c) 46.0 MHz. (d) Determine the corresponding wavelengths for these photons and state the classification of each on the electromagnetic spectrum.

8. The average threshold of dark-adapted (scotopic) vision is 4.00×10^{-11} W/m^2 at a central wavelength of 500 nm. If light having this intensity and wavelength enters the eye and the pupil is open to its maximum diameter of 8.50 mm, how many photons per second enter the eye?

9. An FM radio transmitter has a power output of 150 kW and operates at a frequency of 99.7 MHz. How many photons per second does the transmitter emit?

10. A simple pendulum has a length of 1.00 m and a mass of 1.00 kg. The amplitude of oscillations of the pendulum is 3.00 cm. Estimate the quantum number for the pendulum.

11. **Review problem.** A star moving away from the Earth at $0.280c$ emits radiation that we measure to be most intense at the wavelength 500 nm. Determine the surface temperature of this star.

12. Show that at long wavelengths, Planck's radiation law (Equation 40.6) reduces to the Rayleigh–Jeans law (Equation 40.3).

Section 40.2   The Photoelectric Effect

13. Molybdenum has a work function of 4.20 eV. (a) Find the cutoff wavelength and cutoff frequency for the photoelectric effect. (b) What is the stopping potential if the incident light has a wavelength of 180 nm?

14. Electrons are ejected from a metallic surface with speeds ranging up to 4.60×10^5 m/s when light with a wavelength of 625 nm is used. (a) What is the work function of the surface? (b) What is the cutoff frequency for this surface?

15. Lithium, beryllium, and mercury have work functions of 2.30 eV, 3.90 eV, and 4.50 eV, respectively. Light with a wavelength of 400 nm is incident on each of these metals. Determine (a) which metals exhibit the photoelectric effect and (b) the maximum kinetic energy for the photoelectrons in each case.

16. A student studying the photoelectric effect from two different metals records the following information: (i) the stopping potential for photoelectrons released from metal 1 is 1.48 V larger than that for metal 2, and (ii) the threshold frequency for metal 1 is 40.0% smaller than that for metal 2. Determine the work function for each metal.

17. Two light sources are used in a photoelectric experiment to determine the work function for a particular metal surface. When green light from a mercury lamp (λ = 546.1 nm) is used, a stopping potential of 0.376 V reduces the photocurrent to zero. (a) Based on this measurement, what is the work function for this metal? (b) What stopping potential would be observed when using the yellow light from a helium discharge tube (λ = 587.5 nm)?

18. From the scattering of sunlight, Thomson calculated the classical radius of the electron as having a value of 2.82×10^{-15} m. Sunlight with an intensity of 500 W/m^2 falls on a disk with this radius. Calculate the time interval required to accumulate 1.00 eV of energy. Assume that light is a classical wave and that the light striking the disk is completely absorbed. How does your result compare with the observation that photoelectrons are emitted promptly (within 10^{-9} s)?

19. **Review problem.** An isolated copper sphere of radius 5.00 cm, initially uncharged, is illuminated by ultraviolet light of wavelength 200 nm. What charge will the photoelectric effect induce on the sphere? The work function for copper is 4.70 eV.

20. **Review problem.** A light source emitting radiation at 7.00×10^{14} Hz is incapable of ejecting photoelectrons from a certain metal. In an attempt to use this source to eject photoelectrons from the metal, the source is given a velocity toward the metal. (a) Explain how this procedure produces photoelectrons. (b) When the speed of the light source is equal to $0.280c$, photoelectrons just begin to be ejected from the metal. What is the work function of the metal? (c) When the speed of the light source is increased to $0.900c$, determine the maximum kinetic energy of the photoelectrons.

Section 40.3   The Compton Effect

21. Calculate the energy and momentum of a photon of wavelength 700 nm.

22. X-rays having an energy of 300 keV undergo Compton scattering from a target. The scattered rays are detected at 37.0° relative to the incident rays. Find (a) the Compton shift at this angle, (b) the energy of the scattered x-ray, and (c) the energy of the recoiling electron.

23. A 0.001 60-nm photon scatters from a free electron. For what (photon) scattering angle does the recoiling electron have kinetic energy equal to the energy of the scattered photon?

24. A 0.110-nm photon collides with a stationary electron. After the collision, the electron moves forward and the photon recoils backward. Find the momentum and the kinetic energy of the electron.

25. A 0.880-MeV photon is scattered by a free electron initially at rest such that the scattering angle of the scattered electron is equal to that of the scattered photon ($\theta = \phi$ in Fig. 40.13b). (a) Determine the angles θ and ϕ. (b) Determine the energy and momentum of the scattered photon. (c) Determine the kinetic energy and momentum of the scattered electron.

26. A photon having energy E_0 is scattered by a free electron initially at rest such that the scattering angle of the scattered electron is equal to that of the scattered photon ($\theta = \phi$ in Fig. 40.13b). (a) Determine the angles θ and ϕ. (b) Determine the energy and momentum of the scattered photon. (c) Determine the kinetic energy and momentum of the scattered electron.

27. In a Compton scattering experiment, an x-ray photon scatters through an angle of 17.4° from a free electron that is initially at rest. The electron recoils with a speed of 2 180 km/s. Calculate (a) the wavelength of the incident photon and (b) the angle through which the electron scatters.

28. A 0.700-MeV photon scatters off a free electron such that the scattering angle of the photon is twice the scattering angle of the electron (Fig. P40.28). Determine (a) the scattering angle for the electron and (b) the final speed of the electron.

Figure P40.28

29. A photon having wavelength λ scatters off a free electron at A (Fig. P40.29) producing a second photon having wavelength λ'. This photon then scatters off another free electron at B, producing a third photon having wavelength λ'' and moving in a direction directly opposite the original photon as shown in Figure P40.29. Determine the numerical value of $\Delta\lambda = \lambda'' - \lambda$.

Figure P40.29

30. Find the maximum fractional energy loss for a 0.511-MeV gamma ray that is Compton scattered from a free (a) electron (b) proton.

Section 40.4 Photons and Electromagnetic Waves

31. An electromagnetic wave is called *ionizing radiation* if its photon energy is larger than say 10.0 eV, so that a single photon has enough energy to break apart an atom. With reference to Figure 34.12, identify what regions of the electromagnetic spectrum fit this definition of ionizing radiation and what does not.

32. Review problem. A helium–neon laser produces a beam of diameter 1.75 mm, delivering 2.00×10^{18} photons/s. Each photon has a wavelength of 633 nm. (a) Calculate the amplitudes of the electric and magnetic fields inside the beam. (b) If the beam shines perpendicularly onto a perfectly reflecting surface, what force does it exert on the surface? (c) If the beam is absorbed by a block of ice at 0°C for 1.50 h, what mass of ice is melted?

Section 40.5 The Wave Properties of Particles

33. Calculate the de Broglie wavelength for a proton moving with a speed of 1.00×10^6 m/s.

34. Calculate the de Broglie wavelength for an electron that has kinetic energy (a) 50.0 eV and (b) 50.0 keV.

35. (a) An electron has a kinetic energy of 3.00 eV. Find its wavelength. (b) **What If?** A photon has energy 3.00 eV. Find its wavelength.

36. (a) Show that the wavelength of a nonrelativistic neutron is

$$\lambda = \frac{2.86 \times 10^{-11}}{\sqrt{K_n}} \text{ m}$$

where K_n is the kinetic energy of the neutron in electron volts. (b) What is the wavelength of a 1.00-keV neutron?

37. The nucleus of an atom is on the order of 10^{-14} m in diameter. For an electron to be confined to a nucleus, its de Broglie wavelength would have to be on this order of magnitude or smaller. (a) What would be the kinetic energy of an electron confined to this region? (b) Given that typical binding energies of electrons in atoms are measured to be on the order of a few eV, would you expect to find an electron in a nucleus? Explain.

38. In the Davisson–Germer experiment, 54.0-eV electrons were diffracted from a nickel lattice. If the first maximum in the diffraction pattern was observed at $\phi = 50.0°$ (Fig. P40.38),

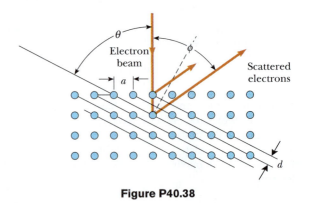

Figure P40.38

what was the lattice spacing a between the vertical rows of atoms in the figure? (It is not the same as the spacing between the horizontal rows of atoms.)

39. (a) Show that the frequency f and wavelength λ of a freely moving particle are related by the expression

$$\left(\frac{f}{c}\right)^2 = \frac{1}{\lambda^2} + \frac{1}{\lambda_C{}^2}$$

where $\lambda_C = h/mc$ is the Compton wavelength of the particle. (b) Is it ever possible for a particle having nonzero mass to have the same wavelength *and* frequency as a photon? Explain.

40. A photon has an energy equal to the kinetic energy of a particle moving with a speed of $0.900c$. (a) Calculate the ratio of the wavelength of the photon to the wavelength of the particle. (b) What would this ratio be for a particle having a speed of $0.001\,00c$? (c) **What If?** What value does the ratio of the two wavelengths approach at high particle speeds? (d) At low particle speeds?

41. The resolving power of a microscope depends on the wavelength used. If one wished to "see" an atom, a resolution of approximately 1.00×10^{-11} m would be required. (a) If electrons are used (in an electron microscope), what minimum kinetic energy is required for the electrons? (b) **What If?** If photons are used, what minimum photon energy is needed to obtain the required resolution?

42. After learning about de Broglie's hypothesis that particles of momentum p have wave characteristics with wavelength $\lambda = h/p$, an 80.0-kg student has grown concerned about being diffracted when passing through a 75.0-cm-wide doorway. Assume that significant diffraction occurs when the width of the diffraction aperture is less that 10.0 times the wavelength of the wave being diffracted. (a) Determine the maximum speed at which the student can pass through the doorway in order to be significantly diffracted. (b) With that speed, how long will it take the student to pass through the doorway if it is in a wall 15.0 cm thick? Compare your result to the currently accepted age of the Universe, which is 4×10^{17} s. (c) Should this student worry about being diffracted?

Section 40.6 The Quantum Particle

43. Consider a freely moving quantum particle with mass m and speed u. Its energy is $E = K = \frac{1}{2}mu^2$. Determine the phase speed of the quantum wave representing the particle and show that it is different from the speed at which the particle transports mass and energy.

44. For a free relativistic quantum particle moving with speed v, the total energy is $E = hf = \hbar\omega = \sqrt{p^2c^2 + m^2c^4}$ and the momentum is $p = h/\lambda = \hbar k = \gamma mv$. For the quantum wave representing the particle, the group speed is $v_g = d\omega/dk$. Prove that the group speed of the wave is the same as the speed of the particle.

Section 40.7 The Double-Slit Experiment Revisited

45. Neutrons traveling at 0.400 m/s are directed through a pair of slits having a 1.00-mm separation. An array of detectors is placed 10.0 m from the slits. (a) What is the de Broglie wavelength of the neutrons? (b) How far off axis is

the first zero-intensity point on the detector array? (c) When a neutron reaches a detector, can we say which slit the neutron passed through? Explain.

46. A modified oscilloscope is used to perform an electron interference experiment. Electrons are incident on a pair of narrow slits 0.060 0 μm apart. The bright bands in the interference pattern are separated by 0.400 mm on a screen 20.0 cm from the slits. Determine the potential difference through which the electrons were accelerated to give this pattern.

47. In a certain vacuum tube, electrons evaporate from a hot cathode at a slow, steady rate and accelerate from rest through a potential difference of 45.0 V. Then they travel 28.0 cm as they pass through an array of slits and fall on a screen to produce an interference pattern. If the beam current is below a certain value, only one electron at a time will be in flight in the tube. What is this value? In this situation, the interference pattern still appears, showing that each individual electron can interfere with itself.

Section 40.8 The Uncertainty Principle

48. Suppose Fuzzy, a quantum–mechanical duck, lives in a world in which $h = 2\pi$ J $\cdot$ s. Fuzzy has a mass of 2.00 kg and is initially known to be within a pond 1.00 m wide. (a) What is the minimum uncertainty in the component of his velocity parallel to the width of the pond? (b) Assuming that this uncertainty in speed prevails for 5.00 s, determine the uncertainty in his position after this time interval.

49. An electron $(m_e = 9.11 \times 10^{-31}$ kg$)$ and a bullet $(m = 0.020\,0$ kg$)$ each have a velocity of magnitude of 500 m/s, accurate to within 0.010 0%. Within what limits could we determine the position of the objects along the direction of the velocity?

50. An air rifle is used to shoot 1.00-g particles at 100 m/s through a hole of diameter 2.00 mm. How far from the rifle must an observer be in order to see the beam spread by 1.00 cm because of the uncertainty principle? Compare this answer with the diameter of the visible Universe $(2 \times 10^{26}$ m$)$.

51. Use the uncertainty principle to show that if an electron were confined inside an atomic nucleus of diameter 2×10^{-15} m, it would have to be moving relativistically, while a proton confined to the same nucleus can be moving nonrelativistically.

52. (a) Show that the kinetic energy of a nonrelativistic particle can be written in terms of its momentum as $K = p^2/2m$. (b) Use the results of (a) to find the minimum kinetic energy of a proton confined within a nucleus having a diameter of 1.00×10^{-15} m.

53. A woman on a ladder drops small pellets toward a point target on the floor. (a) Show that, according to the uncertainty principle, the average miss distance must be at least

$$\Delta x_f = \left(\frac{2\hbar}{m}\right)^{1/2} \left(\frac{2H}{g}\right)^{1/4}$$

where H is the initial height of each pellet above the floor and m is the mass of each pellet. Assume that the spread in impact points is given by $\Delta x_f = \Delta x_i + (\Delta v_x)t$. (b) If $H = 2.00$ m and $m = 0.500$ g, what is Δx_f?

Additional Problems

54. Figure P40.54 shows the stopping potential versus the incident photon frequency for the photoelectric effect for sodium. Use the graph to find (a) the work function, (b) the ratio h/e, and (c) the cutoff wavelength. The data are taken from R. A. Millikan, *Phys. Rev.* 7:362 (1916).

Figure P40.54

55. The following table shows data obtained in a photoelectric experiment. (a) Using these data, make a graph similar to Figure 40.11 that plots as a straight line. From the graph, determine (b) an experimental value for Planck's constant (in joule-seconds) and (c) the work function (in electron volts) for the surface. (Two significant figures for each answer are sufficient.)

Wavelength (nm)	Maximum Kinetic Energy of Photoelectrons (eV)
588	0.67
505	0.98
445	1.35
399	1.63

56. Review problem. Photons of wavelength λ are incident on a metal. The most energetic electrons ejected from the metal are bent into a circular arc of radius R by a magnetic field having a magnitude B. What is the work function of the metal?

57. A 200-MeV photon is scattered at 40.0° by a free proton initially at rest. (a) Find the energy (in MeV) of the scattered photon. (b) What kinetic energy (in MeV) does the proton acquire?

58. Derive the equation for the Compton shift (Eq. 40.11) from Equations 40.12, 40.13, and 40.14.

59. Show that a photon cannot transfer all of its energy to a free electron. (*Suggestion:* Note that system energy and momentum must be conserved.)

60. Show that the speed of a particle having de Broglie wavelength λ and Compton wavelength $\lambda_C = h/(mc)$ is

$$v = \frac{c}{\sqrt{1 + (\lambda/\lambda_C)^2}}$$

61. The total power per unit area radiated by a black body at a temperature T is the area under the $I(\lambda, T)$-versus-λ curve, as shown in Figure 40.3. (a) Show that this power per unit area is

$$\int_0^\infty I(\lambda, T)\, d\lambda = \sigma T^4$$

where $I(\lambda, T)$ is given by Planck's radiation law and σ is a constant independent of T. This result is known as Stefan's law. (See Section 20.7.) To carry out the integration, you should make the change of variable $x = hc/\lambda k_B T$ and use the fact that

$$\int_0^\infty \frac{x^3\, dx}{e^x - 1} = \frac{\pi^4}{15}$$

(b) Show that the Stefan–Boltzmann constant σ has the value

$$\sigma = \frac{2\pi^5 k_B^4}{15 c^2 h^3} = 5.67 \times 10^{-8} \text{ W/m}^2 \cdot \text{K}^4$$

62. Derive Wien's displacement law from Planck's law. Proceed as follows. In Figure 40.3 note that the wavelength at which a black body radiates with greatest intensity is the wavelength for which the graph of $I(\lambda, T)$ versus λ has a horizontal tangent. From Equation 40.6 evaluate the derivative $dI/d\lambda$. Set it equal to zero. Solve the resulting transcendental equation numerically to prove $hc/\lambda_{max} k_B T = 4.965 \ldots$, or $\lambda_{max} T = hc/4.965\, k_B$. Evaluate the constant as precisely as possible and compare it with Wien's experimental value.

63. The spectral distribution function $I(\lambda, T)$ for an ideal black body at absolute temperature T is shown in Figure P40.63. (a) Show that the percentage of the total power radiated per unit area in the range $0 \le \lambda \le \lambda_{max}$ is

$$\frac{A}{A + B} = 1 - \frac{15}{\pi^4} \int_0^{4.965} \frac{x^3}{e^x - 1}\, dx$$

independent of the value of T. (b) Using numerical integration, show that this ratio is approximately 1/4.

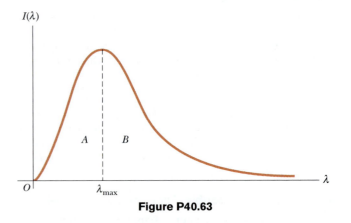

Figure P40.63

64. The neutron has a mass of 1.67×10^{-27} kg. Neutrons emitted in nuclear reactions can be slowed down via collisions with matter. They are referred to as thermal neutrons once they come into thermal equilibrium with their surroundings. The average kinetic energy $(3k_B T/2)$ of a thermal neutron is approximately 0.04 eV. Calculate the de Broglie wavelength of a neutron with a kinetic energy of 0.040 0 eV. How does it compare with the characteristic atomic spacing in a crystal? Would you expect thermal neutrons to exhibit diffraction effects when scattered by a crystal?

65. Show that the ratio of the Compton wavelength λ_C to the de Broglie wavelength $\lambda = h/p$ for a relativistic electron is

$$\frac{\lambda_C}{\lambda} = \left[\left(\frac{E}{m_e c^2} \right)^2 - 1 \right]^{1/2}$$

where E is the total energy of the electron and m_e is its mass.

66. Johnny Jumper's favorite trick is to step out of his 16th-story window and fall 50.0 m into a pool. A news reporter takes a picture of 75.0-kg Johnny just before he makes a splash, using an exposure time of 5.00 ms. Find (a) Johnny's de Broglie wavelength at this moment, (b) the uncertainty of his kinetic energy measurement during such a period of time, and (c) the percent error caused by such an uncertainty.

67. A π^0 meson is an unstable particle produced in high-energy particle collisions. Its rest energy is about 135 MeV, and it exists for an average lifetime of only 8.70×10^{-17} s before decaying into two gamma rays. Using the uncertainty principle, estimate the fractional uncertainty $\Delta m/m$ in its mass determination.

68. A photon of initial energy E_0 undergoes Compton scattering at an angle θ from a free electron (mass m_e) initially at rest. Using relativistic equations for energy and momentum conservation, derive the following relationship for the final energy E' of the scattered photon:

$$E' = E_0 \left[1 + \left(\frac{E_0}{m_e c^2} \right) (1 - \cos \theta) \right]^{-1}.$$

69. **Review problem.** Consider an extension of Young's double-slit experiment performed with photons. Think of Figure 40.24 as a top view looking down on the apparatus. The viewing screen can be a large flat array of *charge-coupled detectors*. Each cell in the array registers individual photons with high efficiency, so we can see where individual photons strike the screen in real time. We cover slit 1 with a polarizer with its transmission axis horizontal, and slit 2 with a polarizer with vertical transmission axis. Any one photon is either absorbed by a polarizing filter or allowed to pass through. The photons that come through a polarizer have their electric field oscillating in the plane defined by their direction of motion and the filter axis. Now we place another large sheet of polarizing material just in front of the screen. For experimental trial 1, we make the transmission axis of this third polarizer horizontal. This choice in effect blocks slit 2. After many photons have been sent through the apparatus, their distribution on the viewing screen is shown by the lower blue curve in the middle of Figure 40.24. For trial 2, we turn the polarizer at the screen to make its transmission axis vertical. Then the screen receives photons only by way of slit 2, and their distribution is shown as the upper blue curve. For trial 3, we temporarily remove

the third sheet of polarizing material. Then the interference pattern shown by the red curve on the right in Figure 40.24 appears. (a) Is the light arriving at the screen to form the interference pattern polarized? Explain your answer. (b) Next, in trial 4 we replace the large square of polarizing material in front of the screen and set its transmission axis to 45°, halfway between horizontal and vertical. What appears on the screen? (c) Suppose we repeat all of trials 1 through 4 with very low light intensity, so that only one photon is present in the apparatus at a time. What are the results now? (d) We go back to high light intensity for convenience and in trial 5 make the large square of polarizer turn slowly and steadily about a rotation axis through its center and perpendicular to its area. What appears on the screen? (e) **What If?** At last, we go back to very low light intensity and replace the large square sheet of polarizing plastic with a flat layer of liquid crystal, to which we can apply an electric field in either a horizontal or a vertical direction. With the applied field we can very rapidly switch the liquid crystal to transmit only photons with horizontal electric field, to act as a polarizer with a vertical transmission axis, or to transmit all photons with high efficiency. We keep track of photons as they are emitted individually by the source. For each photon we wait until it has passed through the pair of slits. Then we quickly choose the setting of the liquid crystal and make that photon encounter a horizontal polarizer, a vertical polarizer, or no polarizer before it arrives at the detector array. We can alternate among the conditions we earlier set up in trials 1, 2, and 3. We keep track of our settings of the liquid crystal and sort out how photons behave under the different conditions, to end up with full sets of data for all three of those trials. What are the results?

70. A photon with wavelength λ_0 moves toward a free electron that is moving with speed u in the same direction as the photon (Fig. P40.70a). The photon scatters at an angle θ (Fig. P40.70b). Show that the wavelength of the scattered photon is

$$\lambda' = \lambda_0 \left(\frac{1 - (u/c) \cos \theta}{1 - (u/c)} \right) + \frac{h}{m_e c} \sqrt{\frac{1 + (u/c)}{1 - (u/c)}} (1 - \cos \theta)$$

(a)   (b)

Figure P40.70

Answers to Quick Quizzes

40.1 (b). A very hot star will have its peak in the blackbody intensity distribution curve at wavelengths shorter than the visible. As a result, more blue light is emitted than red light.

40.2 AM radio, FM radio, microwaves, sodium light. The order of photon energy will be the same as the order of frequency. See Figure 34.12 for a pictorial representation of electromagnetic radiation in order of frequency.

40.3 (c). When the frequency is increased, the photons each carry more energy, so a stopping potential larger in magnitude is required for the current to fall to zero.

40.4 Classical physics predicts that light of sufficient intensity causes emission of photoelectrons, independent of frequency and without a cutoff frequency. Also, the greater the intensity, the larger the maximum kinetic energy of the electrons, with some time delay in emission at low intensities. Thus the classical expectation (which did not match experiment) yields a graph that looks like this:

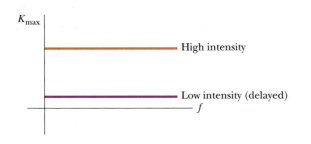

40.5 (d). The shift $\Delta\lambda$ is independent of λ. Thus, the largest fractional shift will correspond to the smallest wavelength.

40.6 (c). According to Equation 40.15, two particles with the same de Broglie wavelength will have the same momentum $p = mv$. If the electron and proton have the same momentum, they cannot have the same speed (a) because of the difference in their masses. For the same reason, because $K = p^2/2m$, they cannot have the same kinetic energy (b). Because the particles have different kinetic energies, Equation 40.16 tells us that the particles do not have the same frequency (d).

40.7 (b). The Compton wavelength (Section 40.3) is a combination of constants and has no relation to the motion of the electron. The de Broglie wavelength (Eq. 40.15) is associated with the motion of the electron through its momentum.

40.8 (b). The group speed is zero because the leading edge of the packet remains fixed at the location of the accident.

40.9 (c). The phase speed of the winning runner is larger than the average speed of all the runners, so the time interval for the winning runner is less than L/v_g.

40.10 (a). The uncertainty principle relates uncertainty in position and velocity along the same axis. The zero uncertainty in position along the x axis results in infinite uncertainty in its velocity component in the x direction, but it is unrelated to the y direction.

40.11 (c). According to the uncertainty principle, if Planck's constant is smaller, the uncertainty in momentum can be smaller and the momentum perpendicular to the original direction of propagation would be smaller. Note that classical wave theory does not include Planck's constant, so it would predict no effect.

ROSE IS ROSE reprinted by permission of United Feature Syndicate, Inc.

Chapter 41

Quantum Mechanics

▲ A quantum corral *shows two aspects of current technological advances in physics. The first aspect involves control over individual atoms. This corral is formed by positioning iron atoms in a stadium-shaped ring on a copper surface. The second aspect is the ability to image the individual atoms with a scanning tunneling microscope. The corral can be used to study the quantized states of electrons trapped in a small region. (Courtesy of IBM Research, Almaden Research Center. Unauthorized use prohibited.)*

In this chapter we introduce *quantum mechanics,* an extremely successful theory for explaining the behavior of microscopic particles. This theory, developed from 1925 to 1926 by Erwin Schrödinger, Werner Heisenberg, and others, enables us to understand a host of phenomena involving atoms, molecules, nuclei, and solids. The discussion in this chapter follows from the introductory quantum concepts that were developed in Chapter 40 and incorporates some of the features of waves under boundary conditions that were explored in Chapter 18. We shall also discuss practical applications, including the scanning tunneling microscope and nanoscale devices that may be used in future quantum computers. Finally, we shall return to the simple harmonic oscillator that was introduced in Chapter 15 and examine it from a quantum mechanical point of view. We shall find a strong connection between this discussion and Planck's model for blackbody radiation, which represented the beginning of our study of quantum physics in Chapter 40.

41.1   An Interpretation of Quantum Mechanics

The preceding chapter introduced some new and strange ideas. In particular, we concluded on the basis of experimental evidence that both matter and electromagnetic radiation are sometimes best described as particles and sometimes as waves, depending on the phenomenon being observed. We can improve our understanding of quantum physics by making another conceptual connection between particles and waves using the notion of probability, a term that was introduced in Chapter 40.

We begin by discussing electromagnetic radiation from the particle point of view. The probability per unit volume of finding a photon in a given region of space at an instant of time is proportional to the number N of photons per unit volume at that time:

$$\frac{\text{Probability}}{V} \propto \frac{N}{V}$$

The number of photons per unit volume is proportional to the intensity of the radiation:

$$\frac{N}{V} \propto I$$

Now, we form a connection between the particle model and the wave model by recalling that the intensity of electromagnetic radiation is proportional to the square of the electric field amplitude E for the electromagnetic wave (Equation 34.21):

$$I \propto E^2$$

Equating the beginning and the end of this string of proportionalities, we have

$$\frac{\text{Probability}}{V} \propto E^2 \qquad (41.1)$$

Thus, for electromagnetic radiation, the probability per unit volume of finding a particle associated with this radiation (the photon) is proportional to the square of the amplitude of the associated electromagnetic wave.

Recognizing the wave–particle duality of both electromagnetic radiation and matter, we should suspect a parallel proportionality for a material particle—the probability per unit volume of finding the particle is proportional to the square of the amplitude of a wave representing the particle. In the previous chapter, we learned that there is a de Broglie wave associated with every particle. The amplitude of the de Broglie wave associated with a particle is not a measurable quantity. (This is because the wave function representing a particle is generally a complex function, as we discuss below.) In contrast, the electric field is a measurable quantity for an electromagnetic wave. The matter analog to Equation 41.1 relates the square of the amplitude of the wave to the probability per unit volume of finding the particle. As a result, we simply call the amplitude of the wave associated with the particle the **probability amplitude,** or the **wave function,** and give it the symbol Ψ.

In general, the complete wave function Ψ for a system depends on the positions of all the particles in the system and on time, and therefore can be written $\Psi(\mathbf{r}_1, \mathbf{r}_2, \mathbf{r}_3, \ldots, \mathbf{r}_j, \ldots, t)$, where $\mathbf{r}_j$ is the position vector of the jth particle in the system. For many systems of interest, including all of those that we shall study in this text, the wave function Ψ is mathematically separable in space and time and can be written as a product of a space function ψ for one particle of the system and a complex time function[1]:

$$\Psi(\mathbf{r}_1, \mathbf{r}_2, \mathbf{r}_3, \ldots, \mathbf{r}_j, \ldots, t) = \psi(\mathbf{r}_j)\, e^{-i\omega t} \qquad (41.2)$$

Space- and time-dependent wave function Ψ

where $\omega(= 2\pi f)$ is the angular frequency of the wave function and $i = \sqrt{-1}$.

For any system in which the potential energy is time-independent and depends only on the positions of particles within the system, the important information about the system is contained within the space part of the wave function. The time part is simply the factor $e^{-i\omega t}$. Therefore, the understanding of ψ is the critical aspect of a given problem.

The wave function ψ is often complex-valued. The absolute square $|\psi|^2 = \psi^*\psi$, where ψ^* is the complex conjugate[2] of ψ, is always real and positive, and is proportional to the probability per unit volume of finding a particle at a given point at some instant. The wave function contains within it all the information that can be known about the particle.

This probabilistic interpretation of the wave function was first suggested by Max Born (1882–1970) in 1928. In 1926, Erwin Schrödinger (1887–1961) proposed a wave equation that describes the manner in which the wave function changes in space and time. The *Schrödinger wave equation*, which we shall examine in Section 41.4, represents a key element in the theory of quantum mechanics.

The concepts of quantum mechanics, strange as they sometimes may seem, developed from classical ideas. In fact, when the techniques of quantum mechanics are applied to macroscopic systems, the results are essentially identical to those of classical physics. This blending of the two approaches occurs when the de Broglie wavelength is small compared with the dimensions of the system. The situation is similar to the agreement between relativistic mechanics and classical mechanics when $v \ll c$.

---

[1] The standard form of a complex number is $a + ib$. The notation $e^{i\theta}$ is equivalent to the standard form as follows:

$$e^{i\theta} = \cos\theta + i\sin\theta$$

Thus, the notation $e^{-i\omega t}$ in Equation 41.2 is equivalent to $\cos(-\omega t) + i\sin(-\omega t) = \cos\omega t - i\sin\omega t$.

[2] For a complex number $z = a + ib$, the complex conjugate is found by changing i to $-i$: $z^* = a - ib$. The product of a complex number and its complex conjugate is always real and positive: $z^*z = (a - ib)(a + ib) = a^2 - (ib)^2 = a^2 - (i)^2 b^2 = a^2 + b^2$.

▲ **PITFALL PREVENTION**

41.1 The Wave Function Belongs to a System

The common language in quantum mechanics is to associate a wave function with a particle. The wave function, however, is determined by the particle *and* its interaction with its environment, so it more rightfully belongs to a system. In many cases, the particle is the only part of the system that experiences a change, which is why the common language has developed. You will see examples in the future in which it is more proper to think of the system wave function rather than the particle wave function.

In Section 40.5, we found that the de Broglie equation relates the momentum of a particle to its wavelength through the relation $p = h/\lambda$. If an ideal free particle has a precisely known momentum p_x, its wave function is an uninterrupted sinusoidal wave of wavelength $\lambda = h/p_x$, and the particle has equal probability of being at any point along the x axis (Fig. 40.19a). The wave function ψ for such a free particle moving along the x axis can be written as

Wave function for a free particle

$$\psi(x) = Ae^{ikx} \tag{41.3}$$

where $k = 2\pi/\lambda$ is the angular wave number (Eq. 16.8) of the wave representing the particle and A is a constant amplitude.[3]

Although we cannot measure ψ, we can measure the real quantity $|\psi|^2$, which can be interpreted as follows. If ψ represents a single particle, then $|\psi|^2$—called the **probability density**—is the relative probability per unit volume that the particle will be found at any given point in the volume. This interpretation can also be stated in the following manner. If dV is a small volume element surrounding some point, then the probability of finding the particle in that volume element is $|\psi|^2 \, dV$.

Probability density $|\psi|^2$

One-Dimensional Wave Functions and Expectation Values

In this section we deal only with one-dimensional systems, where the particle must be located along the x axis, so the probability $|\psi|^2 \, dV$ is modified to become $|\psi|^2 \, dx$. In this case, the probability $P(x) \, dx$ that the particle will be found in the infinitesimal interval dx around the point x is

$$P(x) \, dx = |\psi|^2 \, dx \tag{41.4}$$

Although it is not possible to specify the position of a particle with complete certainty, it is possible through $|\psi|^2$ to specify the probability of observing it in a region surrounding a given point x. **The probability of finding the particle in the arbitrary interval $a \leq x \leq b$ is**

$$P_{ab} = \int_a^b |\psi|^2 \, dx \tag{41.5}$$

Figure 41.1 The probability of a particle being in the interval $a \leq x \leq b$ is the area under the probability density curve from a to b.

The probability P_{ab} is the area under the curve of $|\psi|^2$ versus x between the points $x = a$ and $x = b$, as in Figure 41.1.

Experimentally, there is a finite probability of finding a particle in an interval near some point at some instant. The value of that probability must lie between the limits 0 and 1. For example, if the probability is 0.30, there is a 30% chance of finding the particle in the interval.

Because the particle must be somewhere along the x axis, the sum of the probabilities over all values of x must be 1:

Normalization condition on ψ

$$\int_{-\infty}^{\infty} |\psi|^2 \, dx = 1 \tag{41.6}$$

Any wave function satisfying Equation 41.6 is said to be **normalized.** Normalization is simply a statement that the particle exists at some point in space.

The wave function ψ satisfies a wave equation, just as the electric field associated with an electromagnetic wave satisfies a wave equation that follows from Maxwell's

---

[3] For the free particle, the full wave function, based on Equation 41.2, is

$$\Psi(x, t) = Ae^{ikx}e^{-i\omega t} = Ae^{i(kx - \omega t)} = A[\cos(kx - \omega t) + i\sin(kx - \omega t)]$$

The real part of this wave function has the same form as the waves that we added together to form wave packets in Section 40.6.

equations. The wave equation satisfied by ψ is the Schrödinger equation (Section 41.4), and ψ can be computed from it. Although ψ is not a measurable quantity, all the measurable quantities of a particle, such as its energy and momentum, can be derived from a knowledge of ψ. For example, once the wave function for a particle is known, it is possible to calculate the average position at which you would expect to find the particle after many measurements. This average position is called the **expectation value** of x and is defined by the equation

$$\langle x \rangle \equiv \int_{-\infty}^{\infty} \psi^* x \psi \, dx \qquad (41.7)$$

Expectation value for position *x*

(Brackets, $\langle \ldots \rangle$, are used to denote expectation values.) Furthermore, one can find the expectation value of any function $f(x)$ associated with the particle by using the following equation[4]:

$$\langle f(x) \rangle \equiv \int_{-\infty}^{\infty} \psi^* f(x) \psi \, dx \qquad (41.8)$$

Expectation value for a function *f(x)*

Quick Quiz 41.1 Consider the wave function for the free particle, Equation 41.3. At what value of x is the particle most likely to be found at a given time? (a) at $x = 0$ (b) at small nonzero values of x (c) at large values of x (d) anywhere along the x axis.

Example 41.1   A Wave Function for a Particle

Consider a particle whose wave function is given by the following expression:

$$\psi(x) = Ae^{-ax^2}$$

This wave function is graphed in Figure 41.2.

(A) What is the value of A if this wave function is normalized?

$\psi(x)$

$\psi(x) = Ae^{-ax^2}$

0   x

Figure 41.2 (Example 41.1) A symmetric wave function for a particle, given by $\psi(x) = Ae^{-ax^2}$.

Solution We use the normalization condition, Equation 41.6:

$$\int_{-\infty}^{\infty} |\psi|^2 \, dx = \int_{-\infty}^{\infty} (Ae^{-ax^2})^2 \, dx = A^2 \int_{-\infty}^{\infty} e^{-2ax^2} \, dx = 1$$

The integral can be expressed as the sum of two integrals:

$$A^2 \int_{-\infty}^{\infty} e^{-2ax^2} \, dx = A^2 \left(\int_{0}^{\infty} e^{-2ax^2} \, dx + \int_{-\infty}^{0} e^{-2ax^2} \, dx \right) = 1$$

Because of the x^2 factor in the exponential function, these two integrals have the same value whether x is positive or negative. Thus,

$$(1) \qquad 2A^2 \int_{0}^{\infty} e^{-2ax^2} \, dx = 1$$

We evaluate the integral with the help of Table B.6 in the Appendix:

$$\int_{0}^{\infty} e^{-2ax^2} \, dx = \frac{1}{2} \sqrt{\frac{\pi}{2a}}$$

Thus, Equation (1) becomes

$$2A^2 \left(\frac{1}{2} \sqrt{\frac{\pi}{2a}} \right) = 1$$

---

[4]   Expectation values are analogous to "weighted averages," in which each possible value of a function is multiplied by the probability of the occurrence of that value before summing over all possible values. We write the expectation value as $\int_{-\infty}^{\infty} \psi^* f(x) \psi \, dx$ rather than $\int_{-\infty}^{\infty} f(x) |\psi|^2 dx$ because in more advanced treatments of quantum mechanics, $f(x)$ may be represented by an operator (such as a derivative) rather than a simple multiplicative function. In these situations, the operator is applied only to ψ and not to ψ^*.

$$A = \left(\frac{2a}{\pi}\right)^{1/4}$$

(B) What is the expectation value of x for this particle?

Solution We evaluate the expectation value using Equation 41.7:

$$\langle x \rangle \equiv \int_{-\infty}^{\infty} \psi^* x \psi \, dx = \int_{-\infty}^{\infty} (Ae^{-ax^2}) \, x \, (Ae^{-ax^2}) \, dx$$

$$= A^2 \int_{-\infty}^{\infty} xe^{-2ax^2} dx$$

As in part (A), we express the integral as a sum of two integrals,

$$(2) \qquad \langle x \rangle = A^2 \left(\int_{0}^{\infty} xe^{-2ax^2} dx + \int_{-\infty}^{\infty} xe^{-2ax^2} dx \right)$$

If we change the integration variable from x to $-x$ in the second integral, it becomes

$$\int_{-\infty}^{0} xe^{-2ax^2} dx = \int_{\infty}^{0} -xe^{-2ax^2} (-dx) = \int_{\infty}^{0} xe^{-2ax^2} dx$$

Reversing the order of the limits introduces a negative sign:

$$\int_{\infty}^{0} xe^{-2ax^2} dx = -\int_{0}^{\infty} xe^{-2ax^2} dx$$

This is the negative of the first integral in Equation (2), so

$$\langle x \rangle = \boxed{0}$$

Given the symmetry of the wave function around $x = 0$ in Figure 41.2, it is not a surprising result that the average position of the particle is at $x = 0$.

In Section 41.8, we show that the wave function that we have studied in this example represents the lowest-energy state of the quantum harmonic oscillator. We shall explore more details of this system in that section.

We close this section by noting the important mathematical features of a physically reasonable wave function $\psi(x)$ for a system.

- $\psi(x)$ may be a complex function or a real function, depending on the system;
- $\psi(x)$ must be defined at all points in space and be single-valued;
- $\psi(x)$ must be normalized according to Equation 41.6;
- $\psi(x)$ must be continuous in space—there must be no discontinuous jumps in the value of the wave function at any point.[5]

41.2 A Particle in a Box

In this section, we shall apply some of the ideas we have developed to a simple physical problem—a particle confined to a one-dimensional region of space, called the *particle-in-a-box* problem (even though the "box" is one-dimensional!). From a classical viewpoint, if a particle is bouncing elastically back and forth along the x axis between two impenetrable walls separated by a distance L, as in Figure 41.3a, its motion is easy to describe. If the speed of the particle is v, the magnitude of its momentum mv remains constant, as does its kinetic energy. Classical physics places no restrictions on the values of a particle's momentum and energy. The quantum-mechanical approach to this problem is quite different and requires that we find the appropriate wave function consistent with the conditions of the situation.

Because the walls are impenetrable, there is zero probability of finding the particle outside the box, so the wave function $\psi(x)$ must be zero for $x < 0$ and for $x > L$. As we pointed out at the end of Section 41.1, a wave function must be continuous in space. Thus, if ψ is zero outside the walls, it must also be zero *at* the walls; that is, $\psi(0) = 0$ and $\psi(L) = 0$. Only those wave functions that satisfy these boundary conditions are allowed.

(a)

(b)

Figure 41.3 (a) A particle of mass m and velocity **v**, confined to bouncing between two impenetrable walls separated by a distance L. (b) The potential-energy function for the system.

---

[5] If the wave function were not continuous at a point, the derivative of the wave function at that point would be infinite. This leads to difficulties in the Schrödinger equation, for which the wave function is a solution as discussed in Section 41.4.

Figure 41.3b, a graphical representation of the particle-in-a-box problem, shows the potential energy of the particle–environment system as a function of the position of the particle. As long as the particle is inside the box, the potential energy of the system does not depend on the location of the particle and we can choose its value to be zero. Outside the box, we must ensure that the wave function is zero. We can do this by defining the potential energy of the system as infinitely large if the particle were outside the box. Therefore, the only way a particle could be outside the box is if the system has an infinite amount of energy, which is impossible.

The wave function for a particle in the box can be expressed as a real sinusoidal function[6]:

$$\psi(x) = A \sin\left(\frac{2\pi x}{\lambda}\right) \tag{41.9}$$

where λ is the de Broglie wavelength associated with the particle. This wave function must satisfy boundary conditions at the walls. The boundary condition at $x = 0$ is satisfied already because the sine function is zero when $x = 0$. For the boundary condition at $x = L$, we have

$$\psi(L) = 0 = A \sin\left(\frac{2\pi L}{\lambda}\right)$$

which can only be true if

$$\frac{2\pi L}{\lambda} = n\pi \quad \longrightarrow \quad \lambda = \frac{2L}{n} \tag{41.10}$$

where $n = 1, 2, 3, \ldots$. Thus, only certain wavelengths for the particle are allowed! Each of the allowed wavelengths corresponds to a quantum state for the system, and n is the quantum number. Expressing the wave function in terms of the quantum number n, we have

$$\psi(x) = A \sin\left(\frac{2\pi x}{\lambda}\right) = A \sin\left(\frac{2\pi x}{2L/n}\right) = A \sin\left(\frac{n\pi x}{L}\right) \tag{41.11}$$

Wave functions for a particle in a box

Figures 41.4a and b are graphical representations of ψ versus x and $|\psi|^2$ versus x for $n = 1, 2,$ and 3 for the particle in a box.[7] Note that although ψ can be positive or negative, $|\psi|^2$ is always positive. Because $|\psi|^2$ represents a probability density, a negative value for $|\psi|^2$ would be meaningless.

Further inspection of Figure 41.4b shows that $|\psi|^2$ is zero at the boundaries, satisfying our boundary conditions. In addition, $|\psi|^2$ is zero at other points, depending on the value of n. For $n = 2$, $|\psi|^2 = 0$ at $x = L/2$; for $n = 3$, $|\psi|^2 = 0$ at $x = L/3$ and $x = 2L/3$. The number of zero points increases by one each time the quantum number increases by one.

Because the wavelengths of the particle are restricted by the condition $\lambda = 2L/n$, the magnitude of the momentum of the particle is also restricted to specific values, which we can find from the expression for the de Broglie wavelength, Equation 40.15:

$$p = \frac{h}{\lambda} = \frac{h}{2L/n} = \frac{nh}{2L}$$

---

[6] We show this explicitly in Section 41.4.

[7] Note that $n = 0$ is not allowed because, according to Equation 41.11, the wave function would be $\psi = 0$. This is not a physically reasonable wave function; for example, it cannot be normalized because $\int_{-\infty}^{\infty} |\psi|^2 \, dx = 0$, but Equation 41.6 tells us that this integral must equal 1.

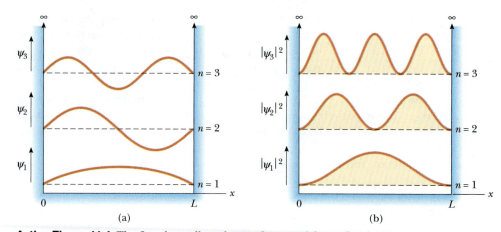

(a)   (b)

Active Figure 41.4 The first three allowed states for a particle confined to a one-dimensional box. The states are shown superimposed on the potential energy function of Figure 41.3b. (a) The wave functions ψ for $n = 1, 2$, and 3. (b) The probability densities $|\psi|^2$ for $n = 1, 2$, and 3. The wave functions and probability densities are plotted vertically from separate axes that are offset vertically for clarity. The positions of these axes on the potential-energy function suggest the relative energies of the states, but the positions are not shown to scale.

At the Active Figures link at http://www.pse6.com, you can measure the probability of a particle being between two points for the three quantum states in the figure.

We have chosen the potential energy of the system to be zero when the particle is inside the box. Therefore, the allowed values of the energy of the system, which is simply the kinetic energy of the particle, are

$$E_n = \tfrac{1}{2}mv^2 = \frac{p^2}{2m} = \frac{(nh/2L)^2}{2m}$$

Quantized energies for a particle in a box

$$E_n = \left(\frac{h^2}{8mL^2}\right) n^2 \qquad n = 1, 2, 3, \ldots \qquad (41.12)$$

As we see from this expression, **the energy of the particle is quantized.** The lowest allowed energy corresponds to the **ground state,** which is the lowest energy state for any system. For the particle in a box, the ground state corresponds to $n = 1$, for which $E_1 = h^2/8mL^2$. Because $E_n = n^2 E_1$, the **excited states** corresponding to $n = 2, 3, 4, \ldots$ have energies given by $4E_1, 9E_1, 16E_1, \ldots$.

Figure 41.5 is an energy-level diagram describing the energy values of the allowed states. Note that the state $n = 0$, for which E would be equal to zero, is *not* allowed. (See footnote 7 on page 1327.) This means that **according to quantum mechanics, the particle can never be at rest.** The smallest energy it can have, corresponding to $n = 1$, is called the **ground-state energy.** This result contradicts the classical viewpoint, in which $E = 0$ is an acceptable state, as are *all* positive values of E.

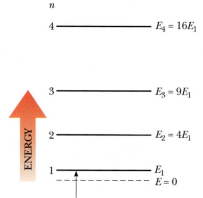

Ground-state energy > 0

Active Figure 41.5 Energy-level diagram for a particle confined to a one-dimensional box of length L. The lowest allowed energy is $E_1 = h^2/8mL^2$.

Adjust the length of the box and the mass of the particle to see the effect on the energy levels at the Active Figures link at http://www.pse6.com.

Quick Quiz 41.2 Redraw Figure 41.4b, the probability of finding a particle at a particular location in a box, on the basis of classical mechanics instead of quantum mechanics.

Quick Quiz 41.3 A particle is in a box of length L. Suddenly, the length of the box is increased to $2L$. What happens to the energy levels shown in Figure 41.5? (a) Nothing—they are unaffected. (b) They move farther apart. (c) They move closer together.

Example 41.2   A Bound Electron

An electron is confined between two impenetrable walls 0.200 nm apart. Determine the energy levels for the states $n = 1, 2,$ and 3.

Solution For the state $n = 1$, Equation 41.12 gives

$$E_1 = \frac{h^2}{8m_eL^2} = \frac{(6.63 \times 10^{-34} \, \text{J} \cdot \text{s})^2}{8(9.11 \times 10^{-31} \, \text{kg})(2.00 \times 10^{-10} \, \text{m})^2}$$

$$E_1 = 1.51 \times 10^{-18} \, \text{J} = \boxed{9.42 \, \text{eV}}$$

For $n = 2$ and $n = 3$, $E_2 = 4E_1 = \boxed{37.7 \, \text{eV}}$ and $E_3 = 9E_1 = \boxed{84.8 \, \text{eV}}$. Although this is a rather primitive model, it can be used to estimate the energy of an electron trapped in a vacant crystal site.

Investigate the energy levels of various particles trapped in a one-dimensional box at the Interactive Worked Example link at http://www.pse6.com

Example 41.3   Energy Quantization for a Macroscopic Object

A 0.500-kg baseball is confined between two rigid walls of a stadium that can be modeled as a box of length 100 m. Calculate the minimum speed of the baseball.

Solution The minimum speed corresponds to the state for which $n = 1$. Using Equation 41.12 with $n = 1$ gives the ground-state energy:

$$E_1 = \frac{h^2}{8mL^2} = \frac{(6.63 \times 10^{-34} \, \text{J} \cdot \text{s})^2}{8(0.500 \, \text{kg})(100 \, \text{m})^2} = 1.10 \times 10^{-71} \, \text{J}$$

Because $E = K = \frac{1}{2}mv^2$, we have

$$v = \left(\frac{2K}{m}\right)^{1/2} = \left[\frac{2(1.10 \times 10^{-71} \, \text{J})}{0.500 \, \text{kg}}\right]^{1/2}$$

$$= \boxed{6.63 \times 10^{-36} \, \text{m/s}}$$

This speed is so small that the object can be considered to be at rest, which is what one would expect for the minimum speed of a macroscopic object.

What If? What if a sharp line drive is hit so that the baseball is moving with a speed of 150 m/s? What is the quantum number of the state in which the baseball now resides?

Answer We expect the quantum number to be very large because the baseball is a macroscopic object. The kinetic energy of the baseball is

$$\tfrac{1}{2}mv^2 = \tfrac{1}{2}(0.500 \, \text{kg})(150 \, \text{m/s})^2 = 5.62 \times 10^3 \, \text{J}$$

From Equation 41.12, we find the quantum number n:

$$n = \sqrt{\frac{8mL^2E_n}{h^2}} = \sqrt{\frac{8(0.500 \, \text{kg})(100 \, \text{m})^2(5.62 \times 10^3 \, \text{J})}{(6.63 \times 10^{-34} \, \text{J} \cdot \text{s})^2}}$$

$$= 2.26 \times 10^{37}$$

This is a tremendously large quantum number. Thus, as the baseball pushes air out of the way, hits the ground, and rolls to a stop, it moves through more than 10^{37} quantum states. These states are so close together in energy that we do not see the transitions from one state to the next. We see what appears to be a smooth variation in the speed of the ball. The quantum nature of the universe is simply not evident in the motion of macroscopic objects.

Example 41.4   Model of an Atom

An atom can be viewed as several electrons moving around a positively charged nucleus, where the electrons are subject mainly to the electrical attraction of the nucleus. (This attraction is partially "screened" by the inner-core electrons and is therefore diminished.) Figure 41.6 represents the potential energy of an atom as a function of r, the distance between the electron and the nucleus.

(A) Use the simple model of a particle in a box to *estimate* the energy (in electron volts) required to raise an atom from the state $n = 1$ to the state $n = 2$, assuming that the atom has a radius of 0.100 nm and that the moving electron carries the energy that has been added to the atom.

Solution Using Equation 41.12 and taking the length L of the box to be 0.200 nm (the diameter of the atom) and $m = m_e = 9.11 \times 10^{-31}$ kg, we find that

Figure 41.6 (Example 41.4) Model of potential energy versus separation distance r for an atom. The allowed energy levels of the atom are shown superimposed on the potential energy function.

$$E_n = \left(\frac{h^2}{8m_e L^2}\right) n^2$$

$$= \frac{(6.63 \times 10^{-34}\,\text{J}\cdot\text{s})^2}{8(9.11 \times 10^{-31}\,\text{kg})(2.00 \times 10^{-10}\,\text{m})^2}\, n^2$$

$$= (1.51 \times 10^{-18}) n^2\,\text{J} = 9.4 n^2\,\text{eV}$$

Hence, the energy difference between the states $n = 1$ and $n = 2$ is

$$\Delta E = E_2 - E_1 = 9.42(2)^2\,\text{eV} - 9.42(1)^2\,\text{eV} = \boxed{28.3\,\text{eV}}$$

(B) Atoms may be excited to higher energy states by absorbing incoming radiation. Calculate the wavelength of the photon that would cause the transition from the $n = 1$ state to the $n = 2$ state.

Solution Using the fact that $\Delta E = hf = hc/\lambda$, we obtain

$$\lambda = \frac{hc}{\Delta E} = \frac{1\,240\,\text{eV}\cdot\text{nm}}{(28.3\,\text{eV})} = \boxed{43.8\,\text{nm}}$$

This wavelength is in the far ultraviolet region, and it is interesting to note that the result is roughly correct. This oversimplified model gives a reasonable order-of-magnitude estimate for transitions between lowest-lying levels of the atom, but the estimate becomes progressively less valid for higher-energy transitions.

41.2 Reminder: Energy Belongs to a System

We often refer to the energy of a particle in commonly used language. As in Pitfall Prevention 41.1, we are actually describing the energy of the *system* of the particle and whatever environment is establishing the impenetrable walls. For the particle in a box, (Section 41.2) the only type of energy is kinetic energy belonging to the particle—this is the origin of the common description.

41.3 Quantum States Are Not Necessarily Standing Waves

While there are many similarities between the quantization of states for quantum systems and the quantization of frequencies for waves on a string (Chapter 18), quantum states are not necessarily standing waves. There may be no stationary "nodes" and no sinusoidal shape associated with a quantum wave function under boundary conditions. Systems more complicated than the particle in a box will have more complicated wave functions, and some boundary conditions will not lead to zeroes of the wave function at fixed points.

41.3 The Particle Under Boundary Conditions

The allowed wavelengths for the wave function of a particle in a box (Eq. 41.10) are identical in form to the allowed wavelengths for mechanical waves on a string fixed at both ends (Eq. 18.6, page 554). In both the string wave and the quantum particle wave, we applied **boundary conditions** to determine the allowed states of the system. For the string fixed at both ends, the boundary conditions were that the elements of the string at the boundaries must have zero displacement. For the particle in a box, the probability amplitude at the boundaries must be zero. In both cases, this results in quantized wavelengths. In the case of the vibrating string, wavelength is related to the frequency, so we have a set of harmonics, or quantized frequencies, given by Equation 18.8. In the case of the particle in a box, we also have quantized frequencies. We can go further in the quantum case, however, because the frequency of a particle is related to its energy through $E = hf$, and we generate a set of quantized *energies*.

In the model of a particle under boundary conditions, **an interaction of a particle with its environment represents one or more boundary conditions and, if the interaction restricts the particle to a finite region of space, results in quantization of the energy of the system.** Because particles have wave-like characteristics, the allowed quantum states of a system are those in which the boundary conditions on the wave function representing the system are satisfied.

In general, boundary conditions are related to the coordinates describing the problem. For the particle in a box, we require a zero value of the wave function at two values of x. In the case of a three-dimensional system such as the hydrogen atom to be discussed in Chapter 42, the problem is best presented in *spherical coordinates*. These are an extension of the plane polar coordinates introduced in Section 3.1, and consist of a radial coordinate r and two angular coordinates. Boundary conditions on the radial part of the wave function, related to r, require that the wave function must approach zero as $r \rightarrow \infty$ (so that the wave function can be normalized) and remain finite as $r \rightarrow 0$. A boundary condition on the angular part of the wave function is that adding 2π radians to the angle must return the wave function to the same value, because an addition of 2π results in the same angular position. The generation of the wave function and application of the boundary conditions for the hydrogen atom are beyond the scope of this book. However, we shall examine the behavior of some of the hydrogen-atom wave functions in Chapter 42.

Quick Quiz 41.4 Which of the following will exhibit quantized energy levels? (a) an atom in a crystal (b) an electron and a proton in a hydrogen atom (c) a proton in the nucleus of a heavy atom (d) all of the above (e) none of the above.

41.4   The Schrödinger Equation

In Section 34.2, we discussed a wave equation for electromagnetic radiation. The waves associated with particles also satisfy a wave equation. We might guess that the wave equation for material particles is different from that associated with photons due to the fact that material particles have a nonzero rest energy. The appropriate wave equation was developed by Schrödinger in 1926. In analyzing the behavior of a quantum system, the approach is to determine a solution to this equation and then apply the appropriate boundary conditions to the solution. The solution yields the allowed wave functions and energy levels of the system under consideration. Proper manipulation of the wave function then enables one to calculate all measurable features of the system.

The Schrödinger equation as it applies to a particle of mass m confined to moving along the x axis and interacting with its environment through a potential energy function $U(x)$ is

$$-\frac{\hbar^2}{2m}\frac{d^2\psi}{dx^2} + U\psi = E\psi \qquad (41.13)$$

Time-independent Schrödinger equation

where E is a constant equal to the total energy of the system (the particle and its environment). Because this equation is independent of time, it is commonly referred to as the **time-independent Schrödinger equation.** (We shall not discuss the time-dependent Schrödinger equation in this text.)

The Schrödinger equation is consistent with the concept of mechanical energy of a system. Problem 15 shows, both for a free particle and a particle in a box, that the first term in the Schrödinger equation reduces to the kinetic energy of the particle multiplied by the wave function. Therefore, Equation 41.13 tells us that the total energy is the sum of the kinetic energy and the potential energy, and that the total energy is a constant: $K + U = E =$ constant.

In principle, if the potential energy function U for the system is known, one can solve Equation 41.13 and obtain the wave functions and energies for the allowed states of the system. Because U may vary with position, it may be necessary to obtain solutions to the equation for different regions of the x axis. In the process, the wave function must be physically reasonable, as noted at the end of Section 41.1. Solutions to the Schrödinger equation in different regions must join smoothly at the boundaries—we require that $\psi(x)$ be *continuous*. In order that $\psi(x)$ obey the normalization condition, we require that $\psi(x)$ approach zero as x approaches $\pm\infty$. Finally, $d\psi/dx$ must also be continuous for finite values of the potential energy.[8]

The task of solving the Schrödinger equation may be very difficult, depending on the form of the potential energy function. As it turns out, the Schrödinger equation has been extremely successful in explaining the behavior of atomic and nuclear systems, whereas classical physics has failed to explain this behavior. Furthermore, when quantum mechanics is applied to macroscopic objects, the results agree with classical physics.

Erwin Schrödinger
Austrian Theoretical Physicist (1887–1961)

Schrödinger is best known as one of the creators of quantum mechanics. His approach to quantum mechanics was demonstrated to be mathematically equivalent to the more abstract matrix mechanics developed by Heisenberg. Schrödinger also produced important papers in the fields of statistical mechanics, color vision, and general relativity. (*AIP Emilio Segrè Visual Archives*)

The Particle in a Box Revisited

To see how the Schrödinger equation is applied to a problem, let us return to our particle in a one-dimensional box of length L (see Fig. 41.3) and analyze it with the Schrödinger equation. Figure 41.3b is the potential-energy diagram that describes this problem. A potential-energy diagram such as this is a useful representation for understanding and solving problems with the Schrödinger equation.

---

[8]   If $d\psi/dx$ were not continuous, we would not be able to evaluate $d^2\psi/dx^2$ in Equation 41.13 at the point of discontinuity.

Because of the shape of the curve in Figure 41.3b, the particle in a box is sometimes said to be in a **square well,**[9] where a **well** is an upward-facing region of the curve in a potential-energy diagram. (A downward-facing region is called a *barrier*, which we shall investigate in Section 41.6.) Figure 41.3b shows an infinite square well.

In the region $0 < x < L$, where $U = 0$, we can express the Schrödinger equation in the form

$$\frac{d^2\psi}{dx^2} = -\frac{2mE}{\hbar^2}\psi = -k^2\psi \tag{41.14}$$

where

$$k = \frac{\sqrt{2mE}}{\hbar}$$

The solution to Equation 41.14 is a function ψ whose second derivative is the negative of the same function multiplied by a constant k^2. Both the sine and cosine functions satisfy this requirement. Therefore, the most general solution to the equation is a linear combination of both solutions:

$$\psi(x) = A\sin kx + B\cos kx$$

where A and B are constants that are determined by the boundary and normalization conditions.

Our first boundary condition is that $\psi(0) = 0$:

$$\psi(0) = A\sin 0 + B\cos 0 = 0 + B = 0$$

which means that $B = 0$. Thus, our solution reduces to

$$\psi(x) = A\sin kx$$

The second boundary condition, $\psi(L) = 0$, when applied to the reduced solution, gives

$$\psi(L) = A\sin kL = 0$$

This equation could be satisfied by setting $A = 0$, but this means that $\psi = 0$ everywhere, which is not a valid wave function. We can satisfy our boundary condition only if kL is an integral multiple of π, that is, if $kL = n\pi$, where n is an integer. Because $k = \sqrt{2mE}/\hbar$, we have

$$kL = \frac{\sqrt{2mE}}{\hbar}L = n\pi$$

Each value of the integer n corresponds to a quantized energy that we call E_n. Solving for the allowed energies E_n gives

$$E_n = \left(\frac{h^2}{8mL^2}\right)n^2 \tag{41.15}$$

which are identical to the allowed energies in Equation 41.12.

Substituting the values of k in the wave function, the allowed wave functions $\psi_n(x)$ are given by

$$\psi_n(x) = A\sin\left(\frac{n\pi x}{L}\right) \tag{41.16}$$

This is the wave function (Equation 41.11) that we used in our initial discussion of the particle in a box.

---

[9]   It is called a square well even if it has a rectangular shape in a potential energy diagram.

Normalizing this wave function shows that $A = \sqrt{2/L}$. (See Problem 17.) Therefore, the normalized wave function is

$$\psi_n(x) = \sqrt{\frac{2}{L}} \sin\left(\frac{n\pi x}{L}\right) \qquad (41.17)$$

Normalized wave function for a particle in a box

Quick Quiz 41.5 Consider an electron, a proton, and an alpha particle (a helium nucleus), each trapped separately in identical infinite square wells. Which particle corresponds to the highest ground-state energy? (a) the electron (b) the proton (c) the alpha particle. (d) The ground-state energy is the same in all three cases.

Quick Quiz 41.6 Consider the three particles in Quick Quiz 41.5 again. Which particle has the longest wavelength when the system is in the ground state? (a) the electron (b) the proton (c) the alpha particle. (d) All three particles have the same wavelength.

Example 41.5 The Expectation Values for the Particle in a Box

A particle of mass m is confined to a one-dimensional box between $x = 0$ and $x = L$. Find the expectation value of the position x of the particle in the ground state.

Solution Using Equation 41.7 and the wave function in Equation 41.17, with $n = 1$, we can express the expectation value for x as follows:

$$(1) \qquad \langle x \rangle = \int_{-\infty}^{\infty} \psi_1^* x \psi_1 \, dx = \int_0^L x \left[\sqrt{\frac{2}{L}} \sin\left(\frac{\pi x}{L}\right) \right]^2 dx$$

$$= \frac{2}{L} \int_0^L x \sin^2\left(\frac{\pi x}{L}\right) dx$$

Note that we reduced the limits on the integral to 0 to L because the value of the wave function is zero elsewhere. Evaluating the integral by consulting an integral table or by mathematical integration[10] gives

$$(2) \qquad \langle x \rangle = \frac{2}{L} \left[\frac{x^2}{4} - \frac{x \sin\left(2\frac{\pi x}{L}\right)}{4\frac{\pi}{L}} - \frac{\cos\left(2\frac{\pi x}{L}\right)}{8\left(\frac{\pi}{L}\right)^2} \right]_0^L$$

$$= \frac{2}{L} \left[\frac{L^2}{4} \right] = \frac{L}{2}$$

This result shows that the expectation value of x is right at the center of the box, which we would expect from the symmetry of the square of the wave function (the probability density) about the center (Fig. 41.4b).

What If? What if we consider states of higher n? How does the expectation value of x depend on n?

Answer Because the squares of all wave functions are symmetric about the midpoint, we would not expect the expectation value to depend on n. If we write Equation (1) with a general wave function, we have

$$\langle x \rangle = \int_{-\infty}^{\infty} \psi_n^* x \psi_n \, dx = \int_0^L x \left[\sqrt{\frac{2}{L}} \sin\left(\frac{n\pi x}{L}\right) \right]^2 dx$$

$$= \frac{2}{L} \int_0^L x \sin^2\left(\frac{n\pi x}{L}\right) dx$$

which leads to a new expression for Equation (2):

$$\langle x \rangle = \frac{2}{L} \left[\frac{x^2}{4} - \frac{x \sin\left(2\frac{n\pi x}{L}\right)}{4\frac{n\pi}{L}} - \frac{\cos\left(2\frac{n\pi x}{L}\right)}{8\left(\frac{n\pi}{L}\right)^2} \right]_0^L$$

$$= \frac{2}{L} \left[\frac{L^2}{4} \right] = \frac{L}{2}$$

The presence of n in the wave function has not affected the expectation value, as we suspected.

The $n = 2$ wave function in Figure 41.4b has a value of zero at the midpoint of the box. How can the expectation value of the particle be at a position at which the particle has zero probability of existing? Remember that the expectation value is the *average* position. Therefore, the particle is as likely to be found to the right of the midpoint as to the left, so that its average position is right at the midpoint, even though its probability of being there is zero. As an analogy, consider a group of students for whom the average final examination score is 50%. There is no requirement that some student achieve a score of exactly 50% for the average of all students to be 50%.

---

[10] To integrate this function, first replace $\sin^2(\pi x/L)$ with $\frac{1}{2}(1 - \cos 2\pi x/L)$ (Table B.3 in Appendix B). This will allow $\langle x \rangle$ to be expressed as two integrals. The second integral can then be evaluated by partial integration (Section B.7 in Appendix B).

41.5 A Particle in a Well of Finite Height

Now consider a particle in a *finite* potential well, that is, a system having a potential energy that is zero when the particle is in the region $0 < x < L$ and a finite value U when the particle is outside this region, as in Figure 41.7. The system consisting of the particle and its environment determines the shape of the well. If the total energy E of the system is less than U, classically the particle is permanently bound in the potential well. If the particle were outside the well, its kinetic energy would have to be negative, an impossibility. However, according to quantum mechanics, a finite probability exists that the particle can be found outside the well even if $E < U$. That is, the wave function ψ is generally nonzero outside the well—regions I and III in Figure 41.7—so the probability density $|\psi|^2$ is also nonzero in these regions. While this may be an uncomfortable notion, the uncertainty principle tells us that the energy of the system is uncertain. This allows the particle to be outside the well as long as the apparent violation of conservation of energy does not exist in any measurable way.

In region II, where $U = 0$, the allowed wave functions are again sinusoidal because they represent solutions of Equation 41.14. However, the boundary conditions no longer require that ψ be zero at the ends of the well, as was the case with the infinite square well.

The Schrödinger equation for regions I and III may be written

$$\frac{d^2\psi}{dx^2} = \frac{2m(U - E)}{\hbar^2}\psi \tag{41.18}$$

Because $U > E$, the coefficient of ψ on the right-hand side is necessarily positive. Therefore, we can express Equation 41.18 in the form

$$\frac{d^2\psi}{dx^2} = C^2\psi \tag{41.19}$$

where $C^2 = 2m(U - E)/\hbar^2$ is a positive constant in regions I and III. As you can verify by substitution, the general solution of Equation 41.19 is

$$\psi = Ae^{Cx} + Be^{-Cx}$$

where A and B are constants.

We can use this general solution as a starting point for determining the appropriate solution for regions I and III. The function we choose for our solution must remain finite over the entire region under consideration. In region I, where $x < 0$, we must rule out the term Be^{-Cx}. In other words, we must require that $B = 0$ in region I to avoid an infinite value for ψ for large negative values of x. Likewise, in region III, where $x > L$, we must rule out the term Ae^{Cx}; this is accomplished by taking $A = 0$ in this region. This choice avoids an infinite value for ψ for large positive x values. Hence, the solutions in regions I and III are

$$\psi_I = Ae^{Cx} \qquad \text{for } x < 0$$

$$\psi_{III} = Be^{-Cx} \qquad \text{for } x > L$$

In region II the wave function is sinusoidal and has the general form

$$\psi_{II}(x) = F\sin kx + G\cos kx$$

where F and G are constants.

These results show that the wave functions outside the potential well (where classical physics forbids the presence of the particle) decay exponentially with distance. At large negative x values, ψ_I approaches zero; at large positive x values, ψ_{III} approaches zero. These functions, together with the sinusoidal solution in region II, are shown in

Figure 41.7 Potential energy diagram of a well of finite height U and length L. A particle is trapped in a well. The total energy E of the particle–well system is less than U.

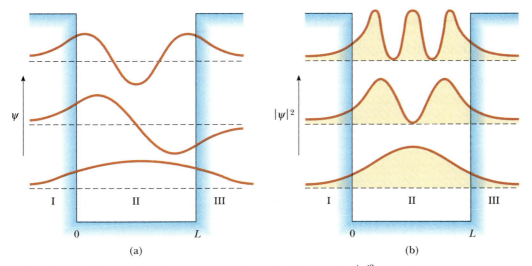

Active Figure 41.8 (a) Wave functions ψ and (b) probability densities $|\psi|^2$ for the lowest three energy states for a particle in a potential well of finite height. The states are shown superimposed on the potential energy function of Figure 41.7. The wave functions and probability densities are plotted vertically from separate axes that are offset vertically for clarity. The positions of these axes on the potential-energy function suggest the relative energies of the states, but the positions are not shown to scale.

At the Active Figures link at http://www.pse6.com, you can adjust the length of the box to see the effect on the quantized states.

Figure 41.8a for the first three energy states. In evaluating the complete wave function, we require that

$$\psi_{\text{I}} = \psi_{\text{II}} \quad \text{and} \quad \frac{d\psi_{\text{I}}}{dx} = \frac{d\psi_{\text{II}}}{dx} \quad \text{at } x = 0$$

$$\psi_{\text{II}} = \psi_{\text{III}} \quad \text{and} \quad \frac{d\psi_{\text{II}}}{dx} = \frac{d\psi_{\text{III}}}{dx} \quad \text{at } x = L$$

These four boundary conditions and the normalization condition (Eq. 41.6) are sufficient to determine the four constants A, B, F, and G, and the allowed values of the energy E. Figure 41.8b plots the probability densities for these states. Note that in each case the wave functions inside and outside the potential well join smoothly at the boundaries.

Quick Quiz 41.7 The wavelengths of the wave functions in Figure 41.8 are longer than those in Figure 41.4 because the wave function spreads out into the classically forbidden region. For an infinite and a finite square well of the same length L, the quantized energies of the particle in a finite well are (a) the same as those for a particle in an infinite well (b) higher than those for a particle in an infinite well (c) lower than those for a particle in an infinite well (d) impossible to determine.

The notion of trapping particles in potential wells is used in the burgeoning field of **nanotechnology,** which refers to the design and application of devices having dimensions ranging from 1 to 100 nm. The fabrication of these devices often involves manipulating single atoms or small groups of atoms to form structures such as the quantum corral in the opening photograph of this chapter.

One area of nanotechnology that is of interest to researchers is the **quantum dot.** The quantum dot, a small region that is grown in a silicon crystal, acts as a potential well. This region can trap electrons into states with quantized energies. The wave functions for a particle in a quantum dot look similar to those in Figure 41.8 if L is on the order of nanometers. The storage of binary information using quantum dots is an active field of research. A simple binary scheme would involve associating a one with a quantum dot containing an electron and a zero with an empty dot. Other schemes involve cells of multiple dots such that arrangements of electrons among the dots correspond to ones and zeroes. There are several research laboratories studying the properties and potential applications of quantum dots. Information should be forthcoming from these laboratories at a steady rate in the next few years.

In the next section, we shall discuss a phenomenon called *tunneling*. We will also look at how tunneling can be combined with a quantum dot to form a transistor on a nanotechnological scale.

41.6   Tunneling Through a Potential Energy Barrier

Consider the potential energy function shown in Figure 41.9. In this situation, the potential energy has a constant value of U in the region of width L and is zero in all other regions. This is called a **square barrier,** and U is called the **barrier height.** A very interesting and peculiar phenomenon occurs when a moving particle encounters such a barrier of finite height and width. Suppose a particle of energy $E < U$ is incident on the barrier from the left (Fig. 41.9). Classically, the particle is reflected by the barrier. If the particle were to exist in region II, its kinetic energy would be negative, which is not classically allowed. Consequently, region II, and therefore region III, are both classically *forbidden* to the particle incident from the left. According to quantum mechanics, however, **all regions are accessible to the particle, regardless of its energy.** (Although all regions are accessible, the probability of the particle being in a classically forbidden region is very low.) According to the uncertainty principle, the particle can be within the barrier, as long as the time interval during which it is in the barrier is short and consistent with Equation 40.24. If the barrier is relatively narrow, this short time interval can allow the particle to move through.

Let us approach this situation using a mathematical representation. The Schrödinger equation has valid solutions in all three regions I, II, and III. The solutions in regions I and III are sinusoidal like Equation 41.11. In region II, the solution is exponential. Applying the boundary conditions that the wave functions in the three regions and their derivatives must join smoothly at the boundaries, a full solution can be found, such as the one represented by the curve in Figure 41.9. Because the probability of locating the particle is proportional to $|\psi|^2$, the probability of finding the particle beyond the barrier in region III is nonzero. This result is in complete disagreement with classical physics. The movement of the particle to the far side of the barrier is called **tunneling** or **barrier penetration.**

The probability of tunneling can be described with a **transmission coefficient** T and a **reflection coefficient** R. The transmission coefficient represents the probability that the particle penetrates to the other side of the barrier, and the reflection coefficient is the probability that the particle is reflected by the barrier. Because the incident

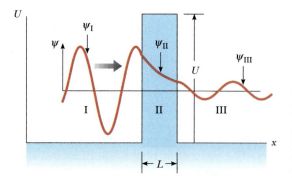

Figure 41.9 Wave function ψ for a particle incident from the left on a barrier of height U and width L. The wave function is sinusoidal in regions I and III but exponentially decaying in region II. The wave function is plotted vertically from an axis positioned at the energy of the particle.

particle is either reflected or transmitted, we require that $T + R = 1$. An approximate expression for the transmission coefficient that is obtained when $T \ll 1$ (a very wide barrier or a very high barrier, that is, $U \gg E$) is

$$T \approx e^{-2CL} \tag{41.20}$$

where

$$C = \frac{\sqrt{2m(U - E)}}{\hbar} \tag{41.21}$$

This quantum model of barrier penetration and specifically Equation 41.20 show that T can be nonzero. The fact that the phenomenon of tunneling is observed experimentally provides further confidence in the principles of quantum physics.

Quick Quiz 41.8 Which of the following changes would increase the probability of transmission of a particle through a potential barrier? (You may choose more than one answer.) (a) decreasing the width of the barrier (b) increasing the width of the barrier (c) decreasing the height of the barrier (d) increasing the height of the barrier (e) decreasing the kinetic energy of the incident particle (f) increasing the kinetic energy of the incident particle.

Example 41.6   Transmission Coefficient for an Electron

Interactive

A 30-eV electron is incident on a square barrier of height 40 eV. What is the probability that the electron will tunnel through the barrier if its width is

(A) 1.0 nm?

(B) 0.10 nm?

Solution (A) In this situation, the quantity $U - E$ has the value

$$U - E = (40 \text{ eV} - 30 \text{ eV}) = 10 \text{ eV} = 1.6 \times 10^{-18} \text{ J}$$

Using Equations 41.20 and 41.21, we find that

$$2CL = 2 \frac{\sqrt{2(9.11 \times 10^{-31} \text{ kg})(1.6 \times 10^{-18} \text{ J})}}{1.054 \times 10^{-34} \text{ J} \cdot \text{s}} (1.0 \times 10^{-9} \text{ m})$$

$$= 32.4$$

Thus, the probability of tunneling through the barrier is

$$T \approx e^{-2CL} = e^{-32.4} = \boxed{8.5 \times 10^{-15}}$$

The electron has about 1 chance in 10^{14} of tunneling through the barrier.

(B) For $L = 0.10$ nm, $2CL = 3.24$, and

$$T \approx e^{-2CL} = e^{-3.24} = \boxed{0.039}$$

Now the electron has a much higher probability (3.9%) of penetrating the barrier. Thus, reducing the width of the barrier by only one order of magnitude increases the probability of tunneling by about 12 orders of magnitude!

Investigate the tunneling of particles through barriers at the Interactive Worked Example link at **http://www.pse6.com**

Some Applications of Tunneling

As we have seen, tunneling is a quantum phenomenon, a manifestation of the wave nature of matter. Many examples exist (on the atomic and nuclear scales) for which tunneling is very important.

• **Alpha decay.** One form of radioactive decay is the emission of alpha particles (the nuclei of helium atoms) by unstable, heavy nuclei (Chapter 44). In order for an alpha particle to escape from the nucleus, it must penetrate a barrier whose height is several times larger than the energy of the nucleus–alpha particle system. The barrier is due to a combination of the attractive nuclear force (discussed in

Chapter 44) and the Coulomb repulsion (discussed in detail in Chapter 23) between the alpha particle and the rest of the nucleus. Occasionally an alpha particle tunnels through the barrier, which explains the basic mechanism for this type of decay and the large variations in the mean lifetimes of various radioactive nuclei.

*   **Nuclear fusion.** The basic reaction that powers the Sun and, indirectly, almost everything else in the solar system, is fusion, which we will study in Chapter 45. In one step of the process that occurs at the core of the Sun, protons must approach each other to within such a small distance that they fuse to form a deuterium nucleus. (See Section 45.4.) According to classical physics, these protons cannot overcome and penetrate the barrier caused by their mutual electrical repulsion. Quantum-mechanically, however, the protons are able to tunnel through the barrier and fuse together.

*   **Scanning tunneling microscopes,** discussed in Section 41.8.

As a final example of an application of tunneling, let us expand on the quantum-dot discussion in Section 41.5 by exploring the **resonant tunneling device.** Figure 41.10a shows the physical construction of such a device. The island of gallium arsenide is a quantum dot located between two barriers formed from the thin extensions of aluminum arsenide. Figure 41.10b shows the potential barriers that are encountered by electrons incident from the left as well as the quantized energy levels in the quantum dot. This situation differs from the one shown in Figure 41.9 in that there are quantized energy levels on the right of the first barrier. In Figure 41.9, an electron that tunnels through the barrier is considered a free particle and can have any energy. In contrast, in Figure 41.10b, as the electrons encounter the first barrier they have no

(a)

(b)

(c)

Active Figure 41.10 (a) The physical structure of a resonant tunneling device. Electrons travel in the gallium arsenide semiconductor and strike the barrier of the quantum dot from the left. (b) A potential energy diagram showing the double barrier representing the walls of the quantum dot. With no voltage applied to the device, none of the quantized states in the quantum dot is resonant with the incident electron energy, and the current is zero. (c) A voltage is applied across the device. The distortion of the potential energy curve causes one of the states in the quantum dot to resonate with the incident electron energy. The electrons are now able to tunnel through the barrier and produce a current in the device.

Vary the voltage across the resonant tunneling device at the Active Figures link at http://www.pse6.com.

energy levels available on the right side of the barrier, which greatly reduces the probability of tunneling.

Figure 41.10c shows the effect of applying a voltage—the potential decreases with position as we move to the right across the device. The deformation of the potential barrier results in an energy level in the quantum dot coinciding with the energy of the incident electrons. This "resonance" of energies gives the device its name. When the voltage is applied, the probability of tunneling increases tremendously and the device carries current. In this manner, the device can be used as a very fast switch on a nanotechnological scale.

Figure 41.11a shows the addition of a gate electrode at the top of the device over the quantum dot. This electrode turns the device into a **resonant tunneling transistor.** The basic function of a transistor is amplification—to convert a small varying voltage into a large varying voltage. Figure 41.11b, representing the potential-energy diagram for the tunneling transistor, has a slope at the bottom of the quantum dot due to the differing voltages at the source and drain electrodes. In this configuration, there is no resonance between the electron energies outside the quantum dot and the quantized energies within the dot. By applying a small voltage to the gate electrode as in Figure 41.11c, the quantized energies can be brought into resonance with the electron energy outside the well, and resonant tunneling occurs. The current that results causes a voltage across an external resistor that is much larger than that of the gate voltage; hence, the device amplifies the input signal to the gate electrode.

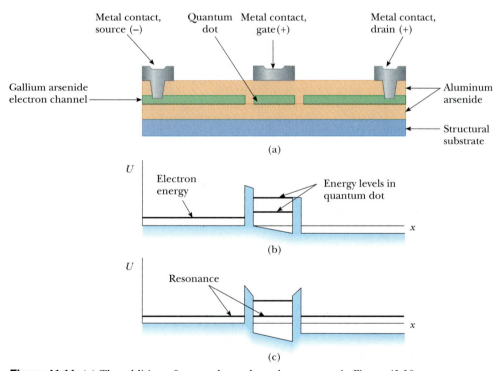

Figure 41.11 (a) The addition of a gate electrode to the structure in Figure 41.10 converts it to a resonant tunneling transistor. Electrons travel in the gallium arsenide semiconductor and strike the barrier of the quantum dot from the left. (b) A potential energy diagram showing the double barrier representing the walls of the quantum dot. With a low voltage applied across the device, none of the quantized states in the quantum dot is resonant with the incident electron energy, and the current is zero. (c) A voltage is applied to the gate electrode. The potential in the region of the quantum dot drops, along with the quantized energy levels. For very small changes in the gate voltage, the quantized level goes in and out of resonance with the electron energy, causing a large change in the current in the device and the voltage across an external resistor.

41.7 The Scanning Tunneling Microscope[11]

Figure 41.12 The surface of graphite as "viewed" with a scanning tunneling microscope. This type of microscope enables scientists to see details with a lateral resolution of about 0.2 nm and a vertical resolution of 0.001 nm. The contours seen here represent the ring-like arrangement of individual carbon atoms on the crystal surface.

One of the basic phenomena of quantum mechanics—tunneling—is at the heart of a very practical device, the scanning tunneling microscope (STM), which enables us to obtain highly detailed images of surfaces at resolutions comparable to the size of a *single atom*. Figure 41.12, showing the surface of a piece of graphite, demonstrates what the STM can do. What makes this image so remarkable is that its resolution is about 0.2 nm. For an optical microscope, the resolution is limited by the wavelength of the light used to make the image. Thus, an optical microscope has a resolution no better than 200 nm, about half the wavelength of visible light, and so could never show the detail displayed in Figure 41.12. An ideal electron microscope (Section 40.5) could have a resolution of 0.2 nm by using electron waves of this wavelength, given by the de Broglie expression $\lambda = h/p$. The linear momentum p of an electron required to give this wavelength is 10 000 eV/c, corresponding to an electron speed of 2% of the speed of light. Electrons traveling at this speed would penetrate into the interior of the graphite in Figure 41.12 and thus could not give us information about individual surface atoms.

The STM achieves its very fine resolution by using the basic idea shown in Figure 41.13. An electrically conducting probe with a very sharp tip is brought near the surface to be studied. The empty space between tip and surface represents the "barrier" we have been discussing, and the tip and surface are the two walls of the "potential well." Because electrons obey quantum rules rather than Newtonian rules, they can "tunnel" across the barrier of empty space. If a voltage is applied between surface and tip, electrons in the atoms of the surface material can be made to tunnel preferentially from surface to tip to produce a tunneling current. In this way the tip samples the distribution of electrons just above the surface.

Because of the nature of tunneling, the STM is very sensitive to the distance z from tip to surface—in other words, to the thickness of the barrier (see Example 41.6). The reason is that in the empty space between tip and surface, the electron wave function falls off exponentially (see Fig. 41.9, region II) with a decay length on the order of 0.1 nm; that is, the wave function decreases by $1/e$ over that distance. For distances $z > 1$ nm (that is, beyond a few atomic diameters), essentially no tunneling takes place. This exponential behavior causes the current of electrons tunneling from surface to tip to depend very strongly on z. This sensitivity is the basis of the operation of the STM. By monitoring the tunneling current as the tip is scanned over the surface, scientists obtain a sensitive measure of the topography of the electron distribution on the

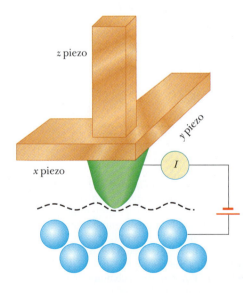

Figure 41.13 Schematic view of an STM. The tip, shown as a rounded cone, is mounted on a piezoelectric *xyz* scanner. (A piezoelectric crystal develops a voltage that is related to the amount of deformation of the crystal.) A scan of the tip over the sample can reveal contours of the surface down to the atomic level. An STM image is composed of a series of scans displaced laterally from one another. (*Based on a drawing from P. K. Hansma, V. B. Elings, O. Marti, and C. Bracker, Science 242:209, 1988.*) © *1988 by the AAAS.*

[11]   An earlier version of this section was written by Roger A. Freedman and Paul K. Hansma, University of California, Santa Barbara.

surface. The result of this scan is used to make images like that in Figure 41.12. In this way the STM can measure the height of surface features to within 0.001 nm, approximately 1/100 of an atomic diameter!

You can see just how sensitive the STM is by examining Figure 41.12. Of the six carbon atoms in each ring, three appear lower than the other three. In fact, all six atoms are at the same height, but all have slightly different electron distributions. The three atoms that appear lower are bonded to other carbon atoms directly beneath them in the underlying atomic layer; as a result, their electron distributions, which are responsible for the bonding, extend downward beneath the surface. The atoms in the surface layer that appear higher do not lie directly over subsurface atoms and hence are not bonded to any underlying atoms. For these higher-appearing atoms, the electron distribution extends upward into the space above the surface. This extra electron density is what makes these atoms appear higher in Figure 41.12, because what the STM maps is the topography of the electron distribution.

The STM has one serious limitation: Its operation depends on the electrical conductivity of the sample and the tip. Unfortunately, most materials are not electrically conductive at their surfaces. Even metals, which are usually excellent electrical conductors, are covered with nonconductive oxides. A newer microscope, the atomic force microscope (AFM), overcomes this limitation.

41.8 The Simple Harmonic Oscillator

Let us use the understanding of quantum mechanics developed in this chapter to consider again the subject of blackbody radiation, discussed in Section 40.1. In Planck's analysis, standing waves in a cavity are established by radiation from vibrating charges in the walls of the cavity. These sources of radiation have allowed energy levels equally spaced in energy (Figure 40.6). Let us see if this is consistent with a quantum treatment of the vibrating charges as simple harmonic oscillators.

We begin by considering a particle that is subject to a linear restoring force $F = -kx$, where x is the position of the particle relative to equilibrium ($x = 0$) and k is the force constant. The classical motion of a particle subject to such a force is simple harmonic motion, which was discussed in Chapter 15. The potential energy of the system is, from Equation 15.20,

$$U = \tfrac{1}{2}kx^2 = \tfrac{1}{2}m\omega^2 x^2$$

where the angular frequency of vibration is $\omega = \sqrt{k/m}$. Classically, if the particle is displaced from its equilibrium position and released, it oscillates between the points $x = -A$ and $x = A$, where A is the amplitude of the motion. Furthermore, its total energy E is, from Equation 15.21,

$$E = K + U = \tfrac{1}{2}kA^2 = \tfrac{1}{2}m\omega^2 A^2$$

In the classical model, any value of E is allowed, including $E = 0$, which is the total energy when the particle is at rest at $x = 0$.

The Schrödinger equation for this problem is obtained by substituting $U = \tfrac{1}{2}m\omega^2 x^2$ into Equation 41.13:

$$-\frac{\hbar^2}{2m}\frac{d^2\psi}{dx^2} + \tfrac{1}{2}m\omega^2 x^2\psi = E\psi \tag{41.22}$$

The mathematical technique for solving this equation is beyond the level of this text. However, it is instructive to guess at a solution. We take as our guess the following wave function:

$$\psi = Be^{-Cx^2} \tag{41.23}$$

Substituting this function into Equation 41.22, we find that it is a satisfactory solution to the Schrödinger equation, provided that

$$C = \frac{m\omega}{2\hbar} \quad \text{and} \quad E = \tfrac{1}{2}\hbar\omega$$

It turns out that the solution we have guessed corresponds to the ground state of the system, which has an energy $\tfrac{1}{2}\hbar\omega$. Because $C = m\omega/2\hbar$ it follows from Equation 41.23 that the wave function for this state is

Wave function for the ground state of a simple harmonic oscillator

$$\psi = Be^{-(m\omega/2\hbar)x^2} \tag{41.24}$$

where B is a constant to be determined from the normalization condition. This is only one solution to Equation 41.22. The remaining solutions that describe the excited states are more complicated, but all solutions include the exponential factor e^{-Cx^2}.

The energy levels of a harmonic oscillator are quantized, as we would expect because the oscillating particle is bound to stay near $x = 0$. The energy for an arbitrary quantum number n is

Quantized energies for a simple harmonic oscillator

$$E_n = (n + \tfrac{1}{2})\hbar\omega \quad n = 0, 1, 2, \ldots \tag{41.25}$$

The state $n = 0$ corresponds to the ground state, whose energy is $E_0 = \tfrac{1}{2}\hbar\omega$; the state $n = 1$ corresponds to the first excited state, whose energy is $E_1 = \tfrac{3}{2}\hbar\omega$, and so on. The energy-level diagram for this system is shown in Figure 41.14. Note that the separations

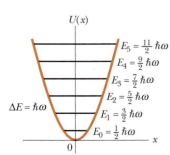

Figure 41.14 Energy-level diagram for a simple harmonic oscillator, superimposed on the potential energy function. The levels are equally spaced, with separation $\hbar\omega$. The ground-state energy is $E_0 = \tfrac{1}{2}\hbar\omega$.

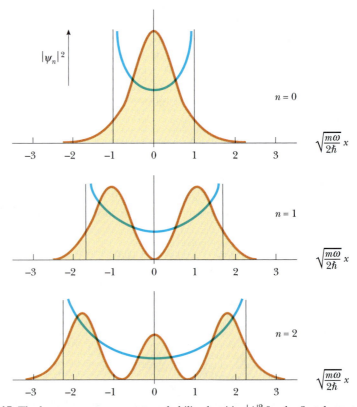

Figure 41.15 The brown curves represent probability densities $|\psi|^2$ for the first three states of a quantum simple harmonic oscillator. The blue curves represent classical probability densities corresponding to the same energies. (*From C. W. Sherwin,* Introduction to Quantum Mechanics, *New York, Holt, Rinehart and Winston, 1959. Used with permission.*)

between adjacent levels are equal and are given by

$$\Delta E = \hbar\omega \tag{41.26}$$

The brown curves in Figure 41.15 indicate the probability densities $|\psi|^2$ for the first three states of a simple harmonic oscillator. The blue curves represent the classical probability densities that correspond to the same energy and are provided for comparison. We see that as n increases, the agreement between the classical and quantum-mechanical results improves, as expected.

Notice that the energy levels for the harmonic oscillator in Figure 41.14 are equally spaced, just as Planck proposed for the oscillators in the walls of the cavity that was used in the model for blackbody radiation in Section 40.1. Planck's Equation 40.4 for the energy levels of the oscillators differs from Equation 41.25 only in the term $\frac{1}{2}$ added to n. This additional term does not affect the energy emitted in a transition, given by Equation 40.5, which is equivalent to Equation 41.26. The fact that Planck generated these concepts without the benefit of the Schrödinger equation is testimony to his genius.

Quick Quiz 41.9 For a particle undergoing simple harmonic motion in the $n = 0$ state, the most probable value of x for the particle according to quantum mechanics is (a) $x = 0$ (b) $x = \pm A$. (c) All values of x are equally likely.

Example 41.7   Molar Specific Heat of Hydrogen Gas

In Figure 21.7 (page 652), which shows the molar specific heat of hydrogen as a function of temperature, vibration does not contribute to the molar specific heat at room temperature. Explain why this is so, modeling the hydrogen molecule as a simple harmonic oscillator. The effective spring constant for the bond in the hydrogen molecule is 573 N/m.

Solution We conceptualize this problem by imagining the only mode of vibration available to a diatomic molecule. This mode (shown in Figure 21.6c) consists of the two atoms always moving in opposite directions. We categorize this as a quantum harmonic oscillator problem, with the molecule as a two-particle system. We can analyze the motion of the particles relative to the center of mass by considering a single particle with reduced mass μ given by

$$\mu = \frac{m_1 m_2}{m_1 + m_2}$$

where m_1 and m_2 are the masses of the two particles. (See Problem 37.) For the hydrogen molecule, the masses of the two particles are the same, so $\mu = \frac{1}{2}m$, where m is the mass of a hydrogen atom.

Let us assume that the molecule is in its vibrational ground state and calculate the energy necessary to excite the molecule to its first excited vibrational state. From Equation 41.26, this energy is

$$\Delta E = \hbar\omega = \hbar\sqrt{\frac{k}{\mu}}$$

where we have used Equation 15.9 for the angular frequency of the oscillator and have substituted the reduced mass μ for

the single equivalent particle. Evaluating this numerically, we find

$$\Delta E = \hbar\sqrt{\frac{k}{\frac{1}{2}m}} = \hbar\sqrt{\frac{2k}{m}}$$

$$= (1.055 \times 10^{-34}\,\text{J·s})\sqrt{\frac{2(573\,\text{N/m})}{1.67 \times 10^{-27}\,\text{kg}}}$$

$$= 8.74 \times 10^{-20}\,\text{J}$$

From Equation 21.4, we know that the average translational kinetic energy of molecules in a gas is given by $\frac{3}{2}k_B T$, where k_B is Boltzmann's constant. Let us find the temperature at which the average molecular translational kinetic energy is equal to that required to excite the first vibrational state of the molecule:

$$\frac{3}{2}k_B T = 8.74 \times 10^{-20}\,\text{J}$$

$$T = \frac{2}{3}\left(\frac{8.74 \times 10^{-20}\,\text{J}}{k_B}\right) = \frac{2}{3}\left(\frac{8.74 \times 10^{-20}\,\text{J}}{1.38 \times 10^{-23}\,\text{J/K}}\right)$$

$$= 4.22 \times 10^3\,\text{K}$$

Thus, the temperature of the gas must be more than 4 000 K in order for the translational kinetic energy to be comparable to the energy required to excite the first vibrational state. This excitation energy must come from collisions between molecules, so if the molecules do not have sufficient translational kinetic energy, they will not be excited to the first vibrational state and vibration will not contribute to the molar specific heat. This explains why the curve in Figure 21.7 does not rise to a value corresponding to the contribution of vibration until the hydrogen gas has been raised to thousands of kelvins.

To finalize this problem, we see from Figure 21.7 that rotational energy levels must be more closely spaced in energy than vibrational levels because they are excited at a lower temperature than the vibrational levels. The translational energy levels are those of a particle in a three-dimensional box, where the box is the container holding the gas. These levels are given by an expression similar to Equation 41.12. Because the box is macroscopic in size, L is very large, and the energy levels are very close together. In fact, they are so close together that translational energy levels are excited at a fraction of a kelvin.

SUMMARY

Take a practice test for this chapter by clicking on the Practice Test link at http://www.pse6.com.

In quantum mechanics, a particle in a system can be represented by a wave function $\psi(x, y, z)$. The probability per unit volume (or probability density) that a particle will be found at a point is $|\psi|^2 = \psi^*\psi$, where ψ^* is the complex conjugate of ψ. If the particle is confined to moving along the x axis, then the probability that it is located in an interval dx is $|\psi|^2 dx$. Furthermore, the sum of all these probabilities over all values of x must be 1:

$$\int_{-\infty}^{\infty} |\psi|^2 \, dx = 1 \tag{41.6}$$

This is called the **normalization condition.**

The measured position x of the particle, averaged over many trials, is called the **expectation value** of x and is defined by

$$\langle x \rangle \equiv \int_{-\infty}^{\infty} \psi^* x \psi \, dx \tag{41.7}$$

If a particle of mass m is confined to moving in a one-dimensional box of length L whose walls are impenetrable, we require that ψ be zero at the walls and outside the box. The wave functions for this system are given by

$$\psi(x) = A \sin\left(\frac{n\pi x}{L}\right) \qquad n = 1, 2, 3, \dots \tag{41.11}$$

where A is the maximum value of ψ. The particle has a well-defined wavelength λ with values such that $L = n\lambda/2$. The allowed states of a particle in a box have quantized energies given by

$$E_n = \left(\frac{h^2}{8mL^2}\right) n^2 \qquad n = 1, 2, 3, \dots \tag{41.12}$$

The wave function for a system must satisfy the **Schrödinger equation.** The time-independent Schrödinger equation for a particle confined to moving along the x axis is

$$-\frac{\hbar^2}{2m} \frac{d^2\psi}{dx^2} + U\psi = E\psi \tag{41.13}$$

where E is the total energy of the system and U is the potential energy.

The approach of quantum mechanics is to solve Equation 41.13 for ψ and E, given the potential energy $U(x)$ for the system. In doing so, we must place the following restrictions on $\psi(x)$: (1) $\psi(x)$ must be continuous, (2) $\psi(x)$ must approach zero as x approaches $\pm\infty$, (3) $\psi(x)$ must be single-valued, and (4) $d\psi/dx$ must be continuous for all finite values of $U(x)$.

QUESTIONS

1. What is the significance of the wave function ψ?

2. For a particle in a box, the probability density at certain points is zero, as seen in Figure 41.4b. Does this imply that the particle cannot move across these points? Explain.

3. Discuss the relationship between ground-state energy and the uncertainty principle.

4. As a particle of energy E is reflected from a potential barrier of height U, where $E < U$, how does the amplitude

of the reflected wave change as the barrier height is reduced?

5. A philosopher once said that "it is necessary for the very existence of science that the same conditions always produce the same results." In view of what has been discussed in this chapter, present an argument showing that this statement is false. How might the statement be reworded to make it true?

6. In quantum mechanics it is possible for the energy E of a particle to be less than the potential energy, but classically this is not possible. Explain.

7. Consider two square wells of the same length, one with finite walls and the other with infinite walls. Compare the energy and momentum of a particle trapped in the finite well with the energy and momentum of an identical particle in the infinite well.

8. Why is it impossible for the lowest-energy state of a harmonic oscillator to be zero?

9. What is the Schrödinger equation? How is it useful in describing quantum phenomena?

PROBLEMS

1, 2, 3 = straightforward, intermediate, challenging   ☐ = full solution available in the *Student Solutions Manual and Study Guide*

www = coached solution with hints available at http://www.pse6.com   🖳 = computer useful in solving problem

▨ = paired numerical and symbolic problems

Section 41.1   An Interpretation of Quantum Mechanics

1. www A free electron has a wave function

$$\psi(x) = Ae^{i(5.00 \times 10^{10}x)}$$

where x is in meters. Find (a) its de Broglie wavelength, (b) its momentum, and (c) its kinetic energy in electron volts.

2. The wave function for a particle is

$$\psi(x) = \sqrt{\frac{a}{\pi(x^2 + a^2)}}$$

for $a > 0$ and $-\infty < x < +\infty$. Determine the probability that the particle is located somewhere between $x = -a$ and $x = +a$.

Section 41.2   A Particle in a Box

3. An electron is confined to a one-dimensional region in which its ground-state $(n = 1)$ energy is 2.00 eV. (a) What is the length of the region? (b) How much energy is required to promote the electron to its first excited state?

4. An electron that has an energy of approximately 6 eV moves between rigid walls 1.00 nm apart. Find (a) the quantum number n for the energy state that the electron occupies and (b) the precise energy of the electron.

5. www An electron is contained in a one-dimensional box of length 0.100 nm. (a) Draw an energy-level diagram for the electron for levels up to $n = 4$. (b) Find the wavelengths of all photons that can be emitted by the electron in making downward transitions that could eventually carry it from the $n = 4$ state to the $n = 1$ state.

6. An alpha particle in a nucleus can be modeled as a particle moving in a box of length 1.00×10^{-14} m (the approximate diameter of a nucleus). Using this model, estimate the energy and momentum of an alpha particle in its lowest energy state. ($m_\alpha = 4 \times 1.66 \times 10^{-27}$ kg)

7. A ruby laser emits 694.3-nm light. Assume light of this wavelength is due to a transition of an electron in a box from its $n = 2$ state to its $n = 1$ state. Find the length of the box.

8. A laser emits light of wavelength λ. Assume this light is due to a transition of an electron in a box from its $n = 2$ state to its $n = 1$ state. Find the length of the box.

9. The nuclear potential energy that binds protons and neutrons in a nucleus is often approximated by a square well. Imagine a proton confined in an infinitely high square well of length 10.0 fm, a typical nuclear diameter. Calculate the wavelength and energy associated with the photon emitted when the proton moves from the $n = 2$ state to the ground state. In what region of the electromagnetic spectrum does this wavelength belong?

10. A proton is confined to move in a one-dimensional box of length 0.200 nm. (a) Find the lowest possible energy of the proton. (b) **What If?** What is the lowest possible energy of an electron confined to the same box? (c) How do you account for the great difference in your results for (a) and (b)?

11. Use the particle-in-a-box model to calculate the first three energy levels of a neutron trapped in a nucleus of diameter 20.0 fm. Do the energy-level differences have a realistic order of magnitude?

12. A photon with wavelength λ is absorbed by an electron confined to a box. As a result, the electron moves from state $n = 1$ to $n = 4$. (a) Find the length of the box. (b) What is the wavelength of the photon emitted in the transition of that electron from the state $n = 4$ to the state $n = 2$?

Section 41.3   The Particle Under Boundary Conditions
Section 41.4   The Schrödinger Equation

13. Show that the wave function $\psi = Ae^{i(kx - \omega t)}$ is a solution to the Schrödinger equation (Eq. 41.13) where $k = 2\pi/\lambda$ and $U = 0$.

14. The wave function of a particle is given by

$$\psi(x) = A\cos(kx) + B\sin(kx)$$

where A, B, and k are constants. Show that ψ is a solution of the Schrödinger equation (Eq. 41.13), assuming the particle is free $(U = 0)$, and find the corresponding energy E of the particle.

15. Prove that the first term in the Schrödinger equation, $-(\hbar^2/2m)(d^2\psi/dx^2)$, reduces to the kinetic energy of the particle multiplied by the wave function (a) for a freely moving particle, with the wave function given by Equation 41.3 and (b) for a particle in a box, with the wave function given by Equation 41.17.

16. A particle in an infinitely deep square well has a wave function given by

$$\psi_2(x) = \sqrt{\frac{2}{L}}\sin\left(\frac{2\pi x}{L}\right)$$

for $0 \le x \le L$ and zero otherwise. (a) Determine the expectation value of x. (b) Determine the probability of finding the particle near $L/2$, by calculating the probability that the particle lies in the range $0.490L \le x \le 0.510L$. (c) **What If?** Determine the probability of finding the particle near $L/4$, by calculating the probability that the particle lies in the range $0.240L \le x \le 0.260L$. (d) Argue that the result of part (a) does not contradict the results of parts (b) and (c).

17. The wave function for a particle confined to moving in a one-dimensional box is

$$\psi(x) = A\sin\left(\frac{n\pi x}{L}\right)$$

Use the normalization condition on ψ to show that

$$A = \sqrt{\frac{2}{L}}$$

Suggestion: Because the box length is L, the wave function is zero for $x < 0$ and for $x > L$, so the normalization condition (Eq. 41.6) reduces to

$$\int_0^L |\psi|^2\,dx = 1$$

18. The wave function of an electron is

$$\psi(x) = \sqrt{\frac{2}{L}}\sin\left(\frac{2\pi x}{L}\right)$$

Calculate the probability of finding the electron between $x = 0$ and $x = L/4$.

19. An electron in an infinitely deep square well has a wave function that is given by

$$\psi_2(x) = \sqrt{\frac{2}{L}}\sin\left(\frac{2\pi x}{L}\right)$$

for $0 \le x \le L$ and is zero otherwise. What are the most probable positions of the electron?

20. ⌨ A particle is in the $n = 1$ state of an infinite square well with walls at $x = 0$ and $x = L$. Let ℓ be an arbitrary value of x between $x = 0$ and $x = L$. (a) Find an expres-

sion for the probability, as a function of ℓ, that the particle will be found between $x = 0$ and $x = \ell$. (b) Sketch the probability as a function of ℓ/L. Choose values of ℓ/L ranging from 0 to 1.00 in steps of 0.100. (c) Find the value of ℓ for which the probability of finding the particle between $x = 0$ and $x = \ell$ is twice the probability of finding the particle between $x = \ell$ and $x = L$. You can solve the transcendental equation for ℓ/L numerically.

21. A particle in an infinite square well has a wave function that is given by

$$\psi_1(x) = \sqrt{\frac{2}{L}}\sin\left(\frac{\pi x}{L}\right)$$

for $0 \le x < L$ and is zero otherwise. (a) Determine the probability of finding the particle between $x = 0$ and $x = L/3$. (b) Use the result of this calculation and symmetry arguments to find the probability of finding the particle between $x = L/3$ and $x = 2L/3$. Do not re-evaluate the integral. (c) **What If?** Compare the result of part (a) with the classical probability.

22. Consider a particle moving in a one-dimensional box for which the walls are at $x = -L/2$ and $x = L/2$. (a) Write the wave functions and probability densities for $n = 1$, $n = 2$, and $n = 3$. (b) Sketch the wave functions and probability densities. (*Suggestion:* Make an analogy to the case of a particle in a box for which the walls are at $x = 0$ and $x = L$.)

23. A particle of mass m moves in a potential well of length $2L$. Its potential energy is infinite for $x < -L$ and for $x > +L$. Inside the region $-L < x < L$, its potential energy is given by

$$U(x) = \frac{-\hbar^2 x^2}{mL^2(L^2 - x^2)}$$

In addition, the particle is in a stationary state that is described by the wave function $\psi(x) = A(1 - x^2/L^2)$ for $-L < x < +L$, and by $\psi(x) = 0$ elsewhere. (a) Determine the energy of the particle in terms of $\hbar$, m, and L. (*Suggestion:* Use the Schrödinger equation, Eq. 41.13.) (b) Show that $A = (15/16L)^{1/2}$. (c) Determine the probability that the particle is located between $x = -L/3$ and $x = +L/3$.

24. In a region of space, a particle with zero total energy has a wave function

$$\psi(x) = Axe^{-x^2/L^2}$$

(a) Find the potential energy U as a function of x. (b) Make a sketch of $U(x)$ versus x.

Section 41.5 A Particle in a Well of Finite Height

25. Suppose a particle is trapped in its ground state in a box that has infinitely high walls (Fig. 41.4). Now suppose the left-hand wall is suddenly lowered to a finite height and width. (a) Qualitatively sketch the wave function for the particle a short time later. (b) If the box has a length L, what is the wavelength of the wave that penetrates the left-hand wall?

26. Sketch the wave function $\psi(x)$ and the probability density $|\psi(x)|^2$ for the $n = 4$ state of a particle in a finite potential well. (See Fig. 41.8.)

Section 41.6 Tunneling Through a Potential Energy Barrier

27. An electron with kinetic energy $E = 5.00$ eV is incident on a barrier with thickness $L = 0.200$ nm and height $U = 10.0$ eV (Fig. P41.27). What is the probability that the electron (a) will tunnel through the barrier? (b) will be reflected?

Figure P41.27 Problems 27 and 28.

28. An electron having total energy $E = 4.50$ eV approaches a rectangular energy barrier with $U = 5.00$ eV and $L = 950$ pm as shown in Figure P41.27. Classically, the electron cannot pass through the barrier because $E < U$. However, quantum-mechanically the probability of tunneling is not zero. Calculate this probability, which is the transmission coefficient.

29. **What If?** In Problem 28, by how much would the width L of the potential barrier have to be increased for the chance of an incident 4.50-eV electron tunneling through the barrier to be one in a million?

30. An electron has a kinetic energy of 12.0 eV. The electron is incident upon a rectangular barrier of height 20.0 eV and thickness 1.00 nm. By what factor would the electron's probability of tunneling through the barrier increase assuming that the electron absorbs all the energy of a photon with wavelength 546 nm (green light)?

Section 41.7 The Scanning Tunneling Microscope

31. A scanning tunneling microscope (STM) can precisely determine the depths of surface features because the current through its tip is very sensitive to differences in the width of the gap between the tip and the sample surface. Assume that in this direction the electron wave function falls off exponentially with a decay length of 0.100 nm; that is, with $C = 10.0$/nm. Determine the ratio of the current when the STM tip is 0.500 nm above a surface feature to the current when the tip is 0.515 nm above the surface.

32. The design criterion for a typical scanning tunneling microscope specifies that it must be able to detect, on the sample below its tip, surface features that differ in height by only 0.002 00 nm. What percentage change in electron transmission must the electronics of the STM be able to detect, to achieve this resolution? Assume that the electron transmission coefficient is e^{-2CL} with $C = 10.0$/nm.

Section 41.8 The Simple Harmonic Oscillator

Note: Problem 43 in Chapter 16 can be assigned with this section.

33. Show that Equation 41.24 is a solution of Equation 41.22 with energy $E = \frac{1}{2}\hbar\omega$.

34. A one-dimensional harmonic oscillator wave function is

$$\psi = Axe^{-bx^2}$$

(a) Show that ψ satisfies Equation 41.22. (b) Find b and the total energy E. (c) Is this a ground state or a first excited state?

35. A quantum simple harmonic oscillator consists of an electron bound by a restoring force proportional to its position relative to a certain equilibrium point. The proportionality constant is 8.99 N/m. What is the longest wavelength of light that can excite the oscillator?

36. (a) Normalize the wave function for the ground state of a simple harmonic oscillator. That is, apply Equation 41.6 to Equation 41.24 and find the required value for the constant B, in terms of m, ω, and fundamental constants. (b) Determine the probability of finding the oscillator in a narrow interval $-\delta/2 < x < \delta/2$ around its equilibrium position.

37. Two particles with masses m_1 and m_2 are joined by a light spring of force constant k. They vibrate along a straight line with their center of mass fixed. (a) Show that the total energy

$$\tfrac{1}{2}m_1v_1^2 + \tfrac{1}{2}m_2v_2^2 + \tfrac{1}{2}kx^2$$

can be written as $\frac{1}{2}\mu v^2 + \frac{1}{2}kx^2$ where $v = |v_1| + |v_2|$ is the *relative* speed of the particles and $\mu = m_1m_2/(m_1 + m_2)$ is the reduced mass of the system. This result demonstrates that the pair of freely vibrating particles can be precisely modeled as a single particle vibrating on one end of a spring that has its other end fixed. (b) Differentiate the equation

$$\tfrac{1}{2}\mu v^2 + \tfrac{1}{2}kx^2 = \text{constant}$$

with respect to x. Proceed to show that the system executes simple harmonic motion. Find its frequency.

38. The total energy of a particle–spring system in which the particle moves with simple harmonic motion along the x axis is

$$E = \frac{p_x^2}{2m} + \frac{kx^2}{2}$$

where p_x is the momentum of the particle and k is the spring constant. (a) Using the uncertainty principle, show that this expression can also be written

$$E \geq \frac{p_x^2}{2m} + \frac{k\hbar^2}{8p_x^2}$$

(b) Show that the minimum energy of the harmonic oscillator is

$$E_{\min} = K + U = \tfrac{1}{4}\hbar\sqrt{\frac{k}{m}} + \frac{\hbar\omega}{4} = \frac{\hbar\omega}{2}$$

Additional Problems

39. Keeping a constant speed of 0.8 m/s, a marble rolls back and forth across a shoebox. Make an order-of-magnitude

estimate of the probability of its escaping through the wall of the box by quantum tunneling. State the quantities you take as data and the values you measure or estimate for them.

40. A particle of mass 2.00×10^{-28} kg is confined to a one-dimensional box of length 1.00×10^{-10} m. For $n = 1$, what are (a) the particle's wavelength, (b) its momentum, and (c) its ground-state energy?

41. An electron is represented by the time-independent wave function

$$\psi(x) = \begin{cases} Ae^{-\alpha x} & \text{for } x > 0 \\ Ae^{+\alpha x} & \text{for } x < 0 \end{cases}$$

(a) Sketch the wave function as a function of x. (b) Sketch the probability density representing the likelihood that the electron is found between x and $x + dx$. (c) Argue that this can be a physically reasonable wave function. (d) Normalize the wave function. (e) Determine the probability of finding the electron somewhere in the range

$$x_1 = -\frac{1}{2\alpha} \quad \text{to} \quad x_2 = \frac{1}{2\alpha}$$

42. Particles incident from the left are confronted with a step in potential energy shown in Figure P41.42. Located at $x = 0$, the step has a height U. The particles have energy $E > U$. Classically, we would expect all of the particles to continue on, although with reduced speed. According to quantum mechanics, a fraction of the particles are reflected at the barrier. (a) Prove that the reflection coefficient R for this case is

$$R = \frac{(k_1 - k_2)^2}{(k_1 + k_2)^2}$$

where $k_1 = 2\pi/\lambda_1$ and $k_2 = 2\pi/\lambda_2$ are the wave numbers for the incident and transmitted particles. Proceed as follows. Show that the wave function $\psi_1 = Ae^{ik_1 x} + Be^{-ik_1 x}$ satisfies the Schrödinger equation in region 1, for $x < 0$. Here $Ae^{ik_1 x}$ represents the incident beam and $Be^{-ik_1 x}$ represents the reflected particles. Show that $\psi_2 = Ce^{ik_2 x}$ satisfies the Schrödinger equation in region 2, for $x > 0$. Impose the boundary conditions $\psi_1 = \psi_2$ and $d\psi_1/dx = d\psi_2/dx$ at $x = 0$, to find the relationship between B and A. Then evaluate $R = B^2/A^2$. (b) A particle that has kinetic energy $E = 7.00$ eV is incident from a region where the potential energy is zero onto one in which $U = 5.00$ eV. Find its probability of being reflected and its probability of being transmitted.

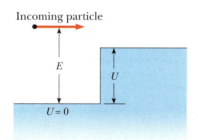

Figure P41.42 Problems 42 and 43.

43. Particles incident from the left are confronted with a step in potential energy shown in Figure P41.42. The step has a height U, and the particles have energy $E = 2U$. Classically, all the particles would pass into the region of higher potential energy at the right. However, according to quantum mechanics, a fraction of the particles are reflected at the barrier. Use the result of Problem 42 to determine the fraction of the incident particles that are reflected. (This situation is analogous to the partial reflection and transmission of light striking an interface between two different media.)

44. An electron is trapped in a quantum dot. The quantum dot may be modeled as a one-dimensional, rigid-walled box of length 1.00 nm. (a) Sketch the wave functions and probability densities for the $n = 1$ and $n = 2$ states. (b) For the $n = 1$ state, calculate the probability of finding the electron between $x_1 = 0.150$ nm and $x_2 = 0.350$ nm, where $x = 0$ is the left side of the box. (c) Repeat part (b) for the $n = 2$ state. (d) Calculate the energies in electron volts of the $n = 1$ and $n = 2$ states. *Suggestion:* For parts (b) and (c), use Equation 41.5 and note that

$$\int \sin^2 ax\, dx = \frac{1}{2}x - \frac{1}{4a}\sin 2ax$$

45. An atom in an excited state 1.80 eV above the ground state remains in that excited state 2.00 μs before moving to the ground state. Find (a) the frequency and (b) the wavelength of the emitted photon. (c) Find the approximate uncertainty in energy of the photon.

46. An electron is confined to move in the xy plane in a rectangle whose dimensions are L_x and L_y. That is, the electron is trapped in a two-dimensional potential well having lengths of L_x and L_y. In this situation, the allowed energies of the electron depend on two quantum numbers n_x and n_y. The allowed energies are given by

$$E = \frac{h^2}{8m_e}\left(\frac{n_x^2}{L_x^2} + \frac{n_y^2}{L_y^2}\right)$$

(a) Assuming $L_x = L_y = L$, find the energies of the lowest four energy levels for the electron. (b) Construct an energy-level diagram for the electron, and determine the energy difference between the second excited state and the ground state.

47. For a particle described by a wave function $\psi(x)$, the expectation value of a physical quantity $f(x)$ associated with the particle is defined by

$$\langle f(x) \rangle \equiv \int_{-\infty}^{\infty} \psi^* f(x)\, \psi\, dx$$

For a particle in a one-dimensional box extending from $x = 0$ to $x = L$, show that

$$\langle x^2 \rangle = \frac{L^2}{3} - \frac{L^2}{2n^2\pi^2}$$

48. A particle is described by the wave function

$$\psi(x) = \begin{cases} A\cos\left(\dfrac{2\pi x}{L}\right) & \text{for } -\dfrac{L}{4} \le x \le \dfrac{L}{4} \\ 0 & \text{for other values of } x \end{cases}$$

(a) Determine the normalization constant A. (b) What is the probability that the particle will be found between $x = 0$ and $x = L/8$ if its position is measured? (*Suggestion:* Use Eq. 41.5.)

49. A particle has a wave function

$$\psi(x) = \begin{cases} \sqrt{\dfrac{2}{a}}\, e^{-x/a} & \text{for } x > 0 \\ 0 & \text{for } x < 0 \end{cases}$$

(a) Find and sketch the probability density. (b) Find the probability that the particle will be at any point where $x < 0$. (c) Show that ψ is normalized, and then find the probability that the particle will be found between $x = 0$ and $x = a$.

50. A particle of mass m is placed in a one-dimensional box of length L. **What If?** Assume the box is so small that the particle's motion is *relativistic*, so that $K = p^2/2m$ is not valid. (a) Derive an expression for the kinetic energy levels of the particle. (b) Assume the particle is an electron in a box of length $L = 1.00 \times 10^{-12}$ m. Find its lowest possible kinetic energy. By what percent is the nonrelativistic equation in error? (*Suggestion:* See Eq. 39.23.)

51. Consider a "crystal" consisting of two nuclei and two electrons as shown in Figure P41.51. (a) Taking into account all the pairs of interactions, find the potential energy of the system as a function of d. (b) Assuming the electrons to be restricted to a one-dimensional box of length $3d$, find the minimum kinetic energy of the two electrons. (c) Find the value of d for which the total energy is a minimum. (d) Compare this value of d with the spacing of atoms in lithium, which has a density of 0.530 g/cm^3 and an atomic mass of 7 u. (This type of calculation can be used to estimate the density of crystals and certain stars.)

Figure P41.51

52. *The simple harmonic oscillator excited.* The wave function

$$\psi(x) = Bxe^{-(m\omega/2\hbar)x^2}$$

is a solution to the simple harmonic oscillator problem. (a) Find the energy of this state. (b) At what position are you least likely to find the particle? (c) At what positions are you most likely to find the particle? (d) Determine the value of B required to normalize the wave function. (e) **What If?** Determine the classical probability of finding the particle in an interval of small width δ centered at the position $x = 2(\hbar/m\omega)^{1/2}$. (f) What is the actual probability of finding the particle in this interval?

53. *Normalization of wave functions.* (a) Find the normalization constant A for a wave function made up of the two lowest states of a particle in a box:

$$\psi(x) = A\left[\sin\left(\frac{\pi x}{L}\right) + 4\sin\left(\frac{2\pi x}{L}\right)\right]$$

(b) A particle is described in the space $-a \le x \le a$ by the wave function

$$\psi(x) = A\cos\left(\frac{\pi x}{2a}\right) + B\sin\left(\frac{\pi x}{a}\right)$$

Determine the relationship between the values of A and B required for normalization. (*Suggestion:* Use the identity $\sin 2\theta = 2\sin\theta\cos\theta$.)

54. The normalized wave functions for the ground state, $\psi_0(x)$, and the first excited state, $\psi_1(x)$, of a quantum harmonic oscillator are

$$\psi_0(x) = \left(\frac{a}{\pi}\right)^{1/4} e^{-ax^2/2} \qquad \psi_1(x) = \left(\frac{4a^3}{\pi}\right)^{1/4} xe^{-ax^2/2}$$

where $a = m\omega/\hbar$. A mixed state, $\psi_{01}(x)$, is constructed from these states:

$$\psi_{01}(x) = \frac{1}{\sqrt{2}}\left[\psi_0(x) + \psi_1(x)\right]$$

The symbol $\langle q \rangle_s$ denotes the expectation value of the quantity q for the state $\psi_s(x)$. Calculate the following expectation values: (a) $\langle x \rangle_0$ (b) $\langle x \rangle_1$ (c) $\langle x \rangle_{01}$.

55. A two-slit electron diffraction experiment is done with slits of *unequal* widths. When only slit 1 is open, the number of electrons reaching the screen per second is 25.0 times the number of electrons reaching the screen per second when only slit 2 is open. When both slits are open, an interference pattern results in which the destructive interference is not complete. Find the ratio of the probability of an electron arriving at an interference maximum to the probability of an electron arriving at an adjacent interference minimum. (*Suggestion:* Use the superposition principle.)

Answers to Quick Quizzes

41.1 (d). The probability density for this wave function is $|\psi|^2 = \psi^*\psi = (Ae^{-ikx})(Ae^{ikx}) = A^2$, which is independent of x. Consequently, the particle is equally likely to be found at any value of x. This is consistent with the uncertainty principle—if the wavelength is known precisely (based on a specific value of k in Equation 41.3), we have no knowledge of the position of the particle.

41.2 Classically, we expect the particle to bounce back and forth between the two walls at constant speed. Thus, we are as likely to find it on the left side of the box as in the middle, on the right side, or anywhere else inside the box. Our graph of probability density versus x would therefore be a horizontal line, with a total area under the line of unity:

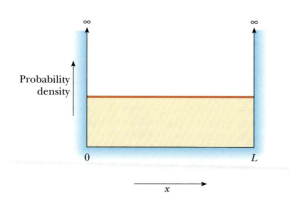

41.3 (c). According to Equation 41.12, if L is increased, all quantized energies become smaller. Thus, the energy levels move closer together. As L becomes macroscopic, the energy levels are so close together that we do not observe the quantized behavior.

41.4 (d). The particles in all three parts (a), (b), and (c) are part of a bound system.

41.5 (a). In Equation 41.15, we set $n = 1$ for the ground-state energy and see that the energy is inversely proportional to the particle mass.

41.6 (d). The wavelength is determined by the length L of the well.

41.7 (c). The longer wavelength results in a smaller value of the momentum from the de Broglie relationship. If the momentum of the particle is decreased, the energy also decreases.

41.8 (a), (c), (f). Decreasing the barrier height and increasing the particle energy both reduce the value of C in Equation 41.21, increasing the transmission coefficient in Equation 41.20. Decreasing the width L of the barrier increases the transmission coefficient in Equation 41.20.

41.9 (a). The ground state represents the largest deviation from classical behavior. Classically, the most probable values of x are $x = \pm A$ because the particle is moving the slowest near these points. As the top graph in Figure 41.15 shows, quantum mechanics predicts that the maximum probability density in the ground state is at $x = 0$.

Chapter 42

Atomic Physics

▲ *Much of the luminous matter in the Universe is hydrogen. In this photograph of the central portion of the Orion Nebula, the colors come from transitions between quantized states in hydrogen atoms. (NASA/GRIN)*

In Chapter 41 we introduced some of the basic concepts and techniques used in quantum mechanics, along with their applications to various one-dimensional systems. In this chapter, we apply quantum mechanics to atomic systems. A large portion of the chapter is focused on the application of quantum mechanics to the study of the hydrogen atom. Understanding the hydrogen atom, the simplest atomic system, is important for several reasons:

- The hydrogen atom is the only atomic system that can be solved exactly.

- Much of what was learned in the twentieth century about the hydrogen atom, with its single electron, can be extended to such single-electron ions as He^+ and Li^{2+}.

- The hydrogen atom proved to be an ideal system for performing precise tests of theory against experiment and for improving our overall understanding of atomic structure.

- The quantum numbers that are used to characterize the allowed states of hydrogen can also be used to investigate more complex atoms, and such a description enables us to understand the periodic table of the elements. This understanding is one of the greatest triumphs of quantum mechanics.

- The basic ideas about atomic structure must be well understood before we attempt to deal with the complexities of molecular structures and the electronic structure of solids.

The full mathematical solution of the Schrödinger equation applied to the hydrogen atom gives a complete and beautiful description of the atom's properties. However, because the mathematical procedures that are involved are beyond the scope of this text, many of the details are omitted. The solutions for some states of hydrogen are discussed, together with the quantum numbers used to characterize various allowed states. We also discuss the physical significance of the quantum numbers and the effect of a magnetic field on certain quantum states.

A new physical idea, the *exclusion principle,* is presented in this chapter. This principle is extremely important for understanding the properties of multielectron atoms and the arrangement of elements in the periodic table. In fact, the implications of the exclusion principle are almost as far-reaching as those of the Schrödinger equation.

Finally, we apply our knowledge of atomic structure to describe the mechanisms involved in the production of x-rays and in the operation of a laser.

42.1 Atomic Spectra of Gases

As pointed out in Section 40.1, all objects emit thermal radiation characterized by a continuous distribution of wavelengths. In sharp contrast to this continuous-distribution spectrum is the discrete **line spectrum** observed when a low-pressure gas is subject to an electric discharge. (Electric discharge occurs when the gas is subjected to a potential difference that creates an electric field greater than the dielectric strength of the gas.) Observation and analysis of these spectral lines is called **emission spectroscopy.**

When the light from a gas discharge is examined using a spectrometer (see Fig. 38.18), it is found to consist of a few bright lines of color on a generally dark background. This discrete line spectrum contrasts sharply with the continuous rainbow of colors seen when a glowing solid is viewed through the same instrument. Figure 42.1a shows that the wavelengths contained in a given line spectrum are characteristic of the element emitting the light. The simplest line spectrum is that for atomic hydrogen, and we describe this spectrum in detail. Other atoms exhibit completely different line spectra. Because no two elements have the same line spectrum, this phenomenon represents a practical and sensitive technique for identifying the elements present in unknown samples.

Another form of spectroscopy very useful in analyzing substances is **absorption spectroscopy.** An absorption spectrum is obtained by passing white light from a continuous source through a gas or a dilute solution of the element being analyzed. The absorption spectrum consists of a series of dark lines superimposed on the continuous spectrum of the light source, as shown in Figure 42.1b for atomic hydrogen.

The absorption spectrum of an element has many practical applications. For example, the continuous spectrum of radiation emitted by the Sun must pass through the cooler gases of the solar atmosphere and through the Earth's atmosphere. The various absorption lines observed in the solar spectrum have been used to identify elements in the solar atmosphere. In early studies of the solar spectrum, experimenters found some lines that did not correspond to any known element. A new element had been discovered! The new element was named helium, after the Greek

⚠ **PITFALL PREVENTION**

42.1 Why Lines?

The phrase "spectral lines" is used very often when discussing the radiation from atoms. Lines are seen because the light passes through a long and very narrow slit before being separated by wavelength. You will see many references to these "lines" in both physics and chemistry.

(a)

(b)

Figure 42.1 (a) Emission line spectra for hydrogen, mercury, and neon. (b) The absorption spectrum for hydrogen. Note that the dark absorption lines occur at the same wavelengths as the hydrogen emission lines in (a). *(K. W. Whitten, R. E. Davis, M. L. Peck, and G. G. Stanley,* General Chemistry, *7th ed.,* Belmont, CA, Brooks/Cole, *2004.)*

Figure 42.2 This street in Las Vegas is filled with neon lights of varying bright colors.

Figure 42.3 The Balmer series of spectral lines for atomic hydrogen, with several lines marked with the wavelength in nanometers. The line labeled 364.6 is the shortest-wavelength line and is in the ultraviolet region of the electromagnetic spectrum. The other labeled lines are in the visible region.

Balmer series

Lyman series

Paschen series

Brackett series

word for Sun, *helios*. Helium was subsequently isolated from subterranean gas on the Earth.

Scientists are able to examine the light from stars other than our Sun in this fashion, but elements other than those present on the Earth have never been detected. Absorption spectroscopy has also been useful in analyzing heavy-metal contamination of the food chain. For example, the first determination of high levels of mercury in tuna was made with the use of atomic absorption spectroscopy.

The discrete emissions of light from gas discharges are used in "neon" signs such as those in Figure 42.2. Neon, the first gas used in these types of signs and the gas after which they are named, emits strongly in the red region. As a result, a glass tube filled with neon gas emits bright red light when an applied voltage causes a continuous discharge. Early signs used different gases to provide different colors, although the brightness of these signs was generally very low. Many present-day "neon" signs contain mercury vapor, which emits strongly in the ultraviolet range of the electromagnetic spectrum. The inside of the glass tube is coated with a material that emits a particular color when it absorbs ultraviolet radiation from the mercury. The color of the light from the tube is due to the particular material chosen. A fluorescent light operates in the same manner, with a white-emitting material coating the inside of the glass tube.

From 1860 to 1885, scientists accumulated a great deal of data on atomic emissions using spectroscopic measurements. In 1885, a Swiss school teacher, Johann Jacob Balmer (1825–1898), found an empirical equation that correctly predicted the wavelengths of four visible emission lines of hydrogen: H_α (red), H_β (green), H_γ (blue), and H_δ (violet). Figure 42.3 shows these and other lines (in the ultraviolet) in the emission spectrum of hydrogen. The complete set of lines is called the **Balmer series.** The four visible lines occur at the wavelengths 656.3 nm, 486.1 nm, 434.1 nm, and 410.2 nm. The wavelengths of these lines can be described by the following equation, which is a modification made by Johannes Rydberg (1854–1919) of Balmer's original equation:

$$\frac{1}{\lambda} = R_H \left(\frac{1}{2^2} - \frac{1}{n^2} \right) \qquad n = 3, 4, 5, \ldots \tag{42.1}$$

where R_H is a constant now called the **Rydberg constant.** If the wavelength is in meters, R_H has the value $1.097\,373\,2 \times 10^7$ m^{-1}. In Balmer's original equation, the integer values of n varied from 3 to 6 to give the four visible lines from 656.3 nm (red) down to 410.2 nm (violet). Rydberg's modification shown in Equation 42.1 describes the ultraviolet spectral lines in the Balmer series if n is carried out beyond $n = 6$. The **series limit** is the shortest wavelength in the series and corresponds to $n \to \infty$, with a wavelength of 364.6 nm, as in Figure 42.3. The measured spectral lines agree with the empirical equation, Equation 42.1, to within 0.1%.

Other lines in the spectrum of hydrogen were found following Balmer's discovery. These spectra are called the Lyman, Paschen, and Brackett series after their discoverers. The wavelengths of the lines in these series can be calculated through the use of the following empirical equations:

$$\frac{1}{\lambda} = R_H \left(1 - \frac{1}{n^2} \right) \qquad n = 2, 3, 4, \ldots \tag{42.2}$$

$$\frac{1}{\lambda} = R_H \left(\frac{1}{3^2} - \frac{1}{n^2} \right) \qquad n = 4, 5, 6, \ldots \tag{42.3}$$

$$\frac{1}{\lambda} = R_H \left(\frac{1}{4^2} - \frac{1}{n^2} \right) \qquad n = 5, 6, 7, \ldots \tag{42.4}$$

No theoretical basis existed for these equations; they simply worked. Note that the same constant R_H appears in each equation. In Section 42.3, we discuss the remarkable achievement of a theory for the hydrogen atom that provided an explanation for these equations.

42.2   Early Models of the Atom

The model of the atom in the days of Newton was a tiny, hard, indestructible sphere. Although this model provided a good basis for the kinetic theory of gases, new models had to be devised when experiments revealed the electrical nature of atoms. In 1897, J. J. Thomson established the charge-to-mass ratio for electrons. (See Figure 29.25, page 912.) The following year, he suggested a model that describes the atom as a region in which positive charge is spread out in space with electrons embedded throughout the region, much like the seeds in a watermelon or raisins in thick pudding (Fig. 42.4). The atom as a whole would then be electrically neutral.

In 1911, Ernest Rutherford (1871–1937) and his students Hans Geiger and Ernest Marsden performed a critical experiment that showed that Thomson's model could not be correct. In this experiment, a beam of positively charged alpha particles (helium nuclei) was projected into a thin metallic foil, such as the target in Figure 42.5a. Most of the particles passed through the foil as if it were empty space. But some of the results of the experiment were astounding. Many of the particles deflected from their original direction of travel were scattered through *large* angles. Some particles were even deflected backward, completely reversing their direction of travel! When Geiger informed Rutherford that some alpha particles were scattered backward, Rutherford wrote, "It was quite the most incredible event that has ever happened to me in my life. It was almost as incredible as if you fired a 15-inch [artillery] shell at a piece of tissue paper and it came back and hit you."

Such large deflections were not expected on the basis of Thomson's model. According to that model, the positive charge of an atom in the foil is spread out over such a great volume (the entire atom) that there is no concentration of positive charge strong enough to cause any large-angle deflections of the positively charged alpha particles. Furthermore, the electrons are so much less massive than the alpha particles that they would not cause large-angle scattering either. Rutherford explained his astonishing results by developing a new atomic model, one that assumed that the positive charge in the atom was concentrated in a region that was small relative to the size of the atom. He called this concentration of positive charge the **nucleus** of the atom. Any electrons belonging to the atom were assumed to be in the relatively large volume outside the nucleus. To explain why these electrons were not pulled into the nucleus by the attractive electric force, Rutherford modeled them as moving in orbits around the nucleus in the same manner as the planets orbit the Sun (Fig. 42.5b). For this reason, this model is often referred to as the planetary model of the atom.

Two basic difficulties exist with Rutherford's planetary model. As we saw in Section 42.1, an atom emits (and absorbs) certain characteristic frequencies of electromagnetic radiation and no others; the Rutherford model cannot explain this phenomenon. A

Joseph John Thomson
English physicist (1856–1940)

The recipient of a Nobel Prize in physics in 1906, Thomson is usually considered the discoverer of the electron. He opened up the field of subatomic particle physics with his extensive work on the deflection of cathode rays (electrons) in an electric field. (*Stock Montage, Inc.*)

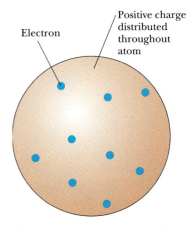

Figure 42.4 Thomson's model of the atom: negatively charged electrons in a volume of continuous positive charge.

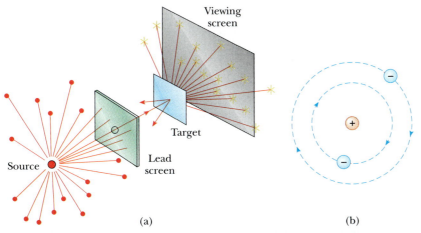

(a)   (b)

Figure 42.5 (a) Rutherford's technique for observing the scattering of alpha particles from a thin foil target. The source is a naturally occurring radioactive substance, such as radium. (b) Rutherford's planetary model of the atom.

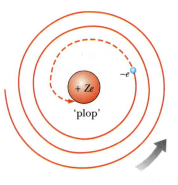

Figure 42.6 The classical model of the nuclear atom. Because the accelerating electron radiates energy, the orbit decays until the electron falls into the nucleus.

second difficulty is that Rutherford's electrons are undergoing a centripetal acceleration. According to Maxwell's theory of electromagnetism, centripetally accelerated charges revolving with frequency f should radiate electromagnetic waves of frequency f. Unfortunately, this classical model leads to a prediction of self-destruction when applied to the atom. As the electron radiates, energy is carried away from the atom, the radius of the electron's orbit steadily decreases, and its frequency of revolution increases. This would lead to an ever-increasing frequency of emitted radiation and an ultimate collapse of the atom as the electron plunges into the nucleus (Fig. 42.6).

42.3   Bohr's Model of the Hydrogen Atom

Given the situation described at the end of Section 42.2, the stage was set for Niels Bohr in 1913 when he presented a new model of the hydrogen atom that circumvented the erroneous deductions of Rutherford's planetary model. Bohr applied Planck's ideas of quantized energy levels (Section 40.1) to orbiting atomic electrons. Bohr's theory was historically important to the development of quantum physics, and it appears to explain the four spectral line series described by Equations 42.1 through 42.4. Though his model is now considered obsolete and has been completely replaced by a probabilistic quantum-mechanical theory, we can use his model to develop the notions of energy quantization and angular momentum quantization as applied to atomic-sized systems. We can also use the model to calculate certain properties of the hydrogen atom and then compare these predictions to experimental observations.

Bohr combined ideas from Planck's original quantum theory, Einstein's concept of the photon, Rutherford's planetary model of the atom, and Newtonian mechanics to arrive at a semiclassical model based on some revolutionary postulates. The basic ideas of the Bohr theory as it applies to the hydrogen atom are as follows:

1. The electron moves in circular orbits around the proton under the influence of the electric force of attraction, as shown in Figure 42.7.

2. Only certain electron orbits are stable. When in one of these **stationary states,** as Bohr called them, the electron does not emit energy in the form of radiation. Hence, the total energy of the atom remains constant, and classical mechanics can be used to describe the electron's motion. Note that this representation claims that the centripetally accelerated electron does not continuously emit radiation, losing energy and eventually spiraling into the nucleus, as predicted by classical physics in the form of Rutherford's planetary model.

3. Radiation is emitted by the atom when the electron makes a transition from a more energetic initial orbit to a lower-energy orbit. This transition cannot be visualized or treated classically. In particular, the frequency f of the photon emitted in the transition is related to the change in the atom's energy and *is independent of the frequency of the electron's orbital motion.* The frequency of the emitted radiation is found from the energy-conservation expression

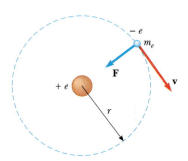

Figure 42.7 Diagram representing Bohr's model of the hydrogen atom. The orbiting electron is allowed to be only in specific orbits of discrete radii.

$$E_i - E_f = hf \qquad (42.5)$$

where E_i is the energy of the initial state, E_f is the energy of the final state, and $E_i > E_f$. In addition, energy of an incident photon can be absorbed by the atom, but only if the photon has an energy that exactly matches the difference in energy between an allowed state of the atom and its existing state upon incidence of the photon. Upon absorption, the photon disappears and the atom makes a transition to the higher-energy state.

4. The size of an allowed electron orbit is determined by a condition imposed on the electron's orbital angular momentum: the allowed orbits are those for which the electron's orbital angular momentum about the nucleus is quantized and equal to an integral multiple of $\hbar = h/2\pi$,

$$m_e v r = n\hbar \qquad n = 1, 2, 3, \ldots \qquad (42.6)$$

Niels Bohr
Danish Physicist (1885–1962)

Bohr was an active participant in the early development of quantum mechanics and provided much of its philosophical framework. During the 1920s and 1930s, he headed the Institute for Advanced Studies in Copenhagen. The institute was a magnet for many of the world's best physicists and provided a forum for the exchange of ideas. When Bohr visited the United States in 1939 to attend a scientific conference, he brought news that the fission of uranium had been observed by Hahn and Strassman in Berlin. The results were the foundations of the nuclear weapon developed in the United States during World War II. Bohr was awarded the 1922 Nobel Prize in physics for his investigation of the structure of atoms and the radiation emanating from them. (*Photo courtesy of AIP Niels Bohr Library, Margarethe Bohr Collection*)

Note that assumption 3 implies the existence of a characteristic discrete emission line spectrum *and also* a corresponding absorption line spectrum of the kind shown in Figure 42.1a and b for hydrogen. Using these four assumptions, we can calculate the allowed energy levels and emission wavelengths of the hydrogen atom. We can find the electric potential energy of the system shown in Figure 42.7 from Equation 25.13, $U = k_e q_1 q_2 / r = -k_e e^2 / r$, where k_e is the Coulomb constant and the negative sign arises from the charge $-e$ on the electron. Thus, the total energy of the atom, which consists of the kinetic energy of the electron and the potential energy of the system, is

$$E = K + U = \tfrac{1}{2} m_e v^2 - k_e \frac{e^2}{r} \qquad (42.7)$$

Applying Newton's second law to the electron, we see that the electric force $k_e e^2 / r^2$ exerted on the electron must equal the product of its mass and its centripetal acceleration ($a_c = v^2/r$):

$$\frac{k_e e^2}{r^2} = \frac{m_e v^2}{r}$$

From this expression, we see that the kinetic energy of the electron is

$$K = \tfrac{1}{2} m_e v^2 = \frac{k_e e^2}{2r} \qquad (42.8)$$

Substituting this value of K into Equation 42.7, we find that the total energy of the atom is[1]

$$E = -\frac{k_e e^2}{2r} \qquad (42.9)$$

Note that the total energy is *negative*, indicating a bound electron–proton system. This means that energy in the amount of $k_e e^2 / 2r$ must be added to the atom to remove the electron and make the total energy of the system zero.

We can obtain an expression for r, the radius of the allowed orbits, by solving Equations 42.6 and 42.8 for v^2 and equating the results:

$$v^2 = \frac{n^2 \hbar^2}{m_e^2 r^2} = \frac{k_e e^2}{m_e r}$$

$$r_n = \frac{n^2 \hbar^2}{m_e k_e e^2} \qquad n = 1, 2, 3, \ldots \qquad (42.10)$$

This equation shows that the radii of the allowed orbits have discrete values—they are quantized. The result is based on the *assumption* that the electron can exist only in certain allowed orbits determined by the integer n.

[1] Compare Equation 42.9 to its gravitational counterpart, Equation 13.18.

The orbit with the smallest radius, called the **Bohr radius** a_0, corresponds to $n = 1$ and has the value

Bohr radius

$$a_0 = \frac{\hbar^2}{m_e k_e e^2} = 0.052\,9 \text{ nm} \tag{42.11}$$

We obtain a general expression for the radius of any orbit in the hydrogen atom by substituting Equation 42.11 into Equation 42.10:

Radii of Bohr orbits in hydrogen

$$r_n = n^2 a_0 = n^2 (0.052\,9 \text{ nm}) \tag{42.12}$$

Bohr's theory predicts a value for the radius of a hydrogen atom on the right order of magnitude, based on experimental measurements. This result was a striking triumph for Bohr's theory. The first three Bohr orbits are shown to scale in Figure 42.8.

The quantization of orbit radii immediately leads to energy quantization. We can see this by substituting $r_n = n^2 a_0$ into Equation 42.9, obtaining for the allowed energy levels

Allowed energies of the Bohr hydrogen atom

$$E_n = -\frac{k_e e^2}{2a_0}\left(\frac{1}{n^2}\right) \qquad n = 1, 2, 3, \ldots \tag{42.13}$$

Inserting numerical values into this expression, we find

$$E_n = -\frac{13.606}{n^2} \text{ eV} \qquad n = 1, 2, 3, \ldots \tag{42.14}$$

Only energies satisfying this equation are permitted. The lowest allowed energy level, the ground state, has $n = 1$ and energy $E_1 = -13.606$ eV. The next energy level, the first excited state, has $n = 2$ and energy $E_2 = E_1/2^2 = -3.401$ eV. Figure 42.9 is an energy-level diagram showing the energies of these discrete energy states and the corresponding quantum numbers n. The uppermost level, corresponding to $n = \infty$ (or $r = \infty$) and $E = 0$, represents the state for which the electron is removed from the atom.

The minimum energy required to ionize the atom in its ground state (that is, to completely remove an electron from the proton's influence) is called the **ionization energy.** As can be seen from Figure 42.9, the ionization energy for hydrogen in the ground state, based on Bohr's calculation, is 13.6 eV. This constituted another major

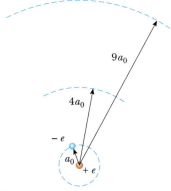

Active Figure 42.8 The first three circular orbits predicted by the Bohr model of the hydrogen atom.

At the Active Figures link at http://www.pse6.com, choose the initial and final states of the hydrogen atom and observe the transitions in this figure and in Figure 42.9.

At the Active Figures link at http://www.pse6.com, choose the initial and final states of the hydrogen atom and observe the transitions in this figure and in Figure 42.8.

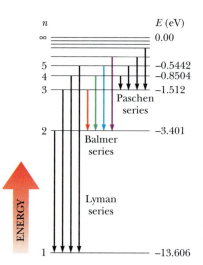

Active Figure 42.9 An energy-level diagram for the hydrogen atom. Quantum numbers are given on the left and energies (in electron volts) on the right. Vertical arrows represent the four lowest-energy transitions for each of the spectral series shown. The colored arrows for the Balmer series indicate that this series results in visible light.

achievement for the Bohr theory because the ionization energy for hydrogen had already been measured to be 13.6 eV.

Equations 42.5 and 42.13 can be used to calculate the frequency of the photon emitted when the electron makes a transition from an outer orbit to an inner orbit:

$$f = \frac{E_i - E_f}{h} = \frac{k_e e^2}{2a_0 h}\left(\frac{1}{n_f^2} - \frac{1}{n_i^2}\right) \qquad (42.15)$$

Frequency of a photon emitted from hydrogen

Because the quantity measured experimentally is wavelength, it is convenient to use $c = f\lambda$ to express Equation 42.15 in terms of wavelength:

$$\frac{1}{\lambda} = \frac{f}{c} = \frac{k_e e^2}{2a_0 hc}\left(\frac{1}{n_f^2} - \frac{1}{n_i^2}\right) \qquad (42.16)$$

The remarkable fact is that this expression, which is purely theoretical, is *identical* to the general form of the empirical relationships discovered by Balmer and Rydberg and given by Equations 42.1 to 42.4,

$$\frac{1}{\lambda} = R_H\left(\frac{1}{n_f^2} - \frac{1}{n_i^2}\right) \qquad (42.17)$$

provided the constant $k_e e^2/2a_0 hc$ is equal to the experimentally determined Rydberg constant $R_H = 1.097\,373\,2 \times 10^7\ \text{m}^{-1}$. Soon after Bohr demonstrated that these two quantities agree to within approximately 1%, this work was recognized as the crowning achievement of his new quantum theory of the hydrogen atom. Furthermore, Bohr showed that all of the spectral series for hydrogen have a natural interpretation in his theory. Figure 42.9 shows the origin of these spectral series as transitions between energy levels.

Bohr immediately extended his model for hydrogen to other elements in which all but one electron had been removed. These systems have the same structure as the hydrogen atom except that the nuclear charge is larger. Ionized elements such as He^+, Li^{2+}, and Be^{3+} were suspected to exist in hot stellar atmospheres, where atomic collisions frequently have enough energy to completely remove one or more atomic electrons. Bohr showed that many mysterious lines observed in the spectra of the Sun and several other stars could not be due to hydrogen but were correctly predicted by his theory if attributed to singly ionized helium. In general, to describe a single electron orbiting a fixed nucleus of charge $+Ze$, where Z is the atomic number of the element (the number of protons in the nucleus), Bohr's theory gives

$$r_n = (n^2)\frac{a_0}{Z} \qquad (42.18)$$

$$E_n = -\frac{k_e e^2}{2a_0}\left(\frac{Z^2}{n^2}\right) \qquad n = 1, 2, 3, \ldots \qquad (42.19)$$

⚠ **PITFALL PREVENTION**

42.2 The Bohr Model Is Great, But . . .

The Bohr model correctly predicts the ionization energy for hydrogen, but it cannot account for the spectra of more complex atoms and is unable to predict many subtle spectral details of hydrogen and other simple atoms. The notion of electrons in well-defined orbits around the nucleus is *not* consistent with current models of the atom.

While the Bohr theory was triumphant in its agreement with some experimental results on the hydrogen atom, it suffered from some difficulties. One of the first indications that the Bohr theory needed to be modified arose when improved spectroscopic techniques were used to examine the spectral lines of hydrogen. It was found that many of the lines in the Balmer and other series were not single lines at all. Instead, each was a group of lines spaced very close together. An additional difficulty arose when it was observed that, in some situations, certain single spectral lines were split into three closely spaced lines when the atoms were placed in a strong magnetic field. Efforts to explain these and other deviations from the Bohr model led to modifications in the theory and ultimately to a replacement theory that we will discuss in Section 42.4.

Bohr's Correspondence Principle

In our study of relativity, we found that Newtonian mechanics is a special case of relativistic mechanics and is usable only when v is much less than c. Similarly, **quantum physics agrees with classical physics when the difference between quantized levels becomes vanishingly small.** This principle, first set forth by Bohr, is called the **correspondence principle.**[2]

For example, consider an electron orbiting the hydrogen atom with $n > 10\ 000$. For such large values of n, the energy differences between adjacent levels approach zero, and therefore the levels are nearly continuous. Consequently, the classical model is reasonably accurate in describing the system for large values of n. According to the classical picture, the frequency of the light emitted by the atom is equal to the frequency of revolution of the electron in its orbit about the nucleus. Calculations show that for $n > 10\ 000$, this frequency is different from that predicted by quantum mechanics by less than 0.015%.

Quick Quiz 42.1 A hydrogen atom is in its ground state. Incident on the atom are many photons each having an energy of 10.5 eV. The result is that (a) the atom is excited to a higher allowed state (b) the atom is ionized (c) the photons pass by the atom without interaction.

Quick Quiz 42.2 When electrons collide with an atom, they can transfer some or all of their energy to the atom. Suppose a hydrogen atom in its ground state is struck by many electrons each having a kinetic energy of 10.5 eV. The result is that (a) the atom is excited to a higher allowed state (b) the atom is ionized (c) the electrons pass by the atom without interaction.

Quick Quiz 42.3 A hydrogen atom makes a transition from the $n = 3$ level to the $n = 2$ level. It then makes a transition from the $n = 2$ level to the $n = 1$ level. Which transition results in emission of the longest-wavelength photon? (a) the first transition (b) the second transition (c) neither, because the wavelengths are the same for both transitions.

Example 42.1 Spectral Lines from the Star ξ-Puppis

Some mysterious lines observed in 1896 in the emission spectrum of the star ξ-Puppis (ξ is the Greek letter xi) fit the empirical equation

$$\frac{1}{\lambda} = R_{H} \left(\frac{1}{(n_f/2)^2} - \frac{1}{(n_i/2)^2} \right)$$

Show that these lines can be explained by the Bohr theory as originating from He$^+$.

Solution The ion He$^+$ has $Z = 2$. Thus, the allowed energy levels are given by Equation 42.19 as

$$E_n = -\frac{k_e e^2}{2a_0} \left(\frac{4}{n^2} \right)$$

Similar to Equation 42.15, we find

$$f = \frac{E_i - E_f}{h} = \frac{k_e e^2}{2a_0 h} \left(\frac{4}{n_f^2} - \frac{4}{n_i^2} \right)$$

$$f = \frac{k_e e^2}{2a_0 h} \left(\frac{1}{(n_f/2)^2} - \frac{1}{(n_i/2)^2} \right)$$

$$\frac{1}{\lambda} = \frac{f}{c} = \frac{k_e e^2}{2a_0 hc} \left(\frac{1}{(n_f/2)^2} - \frac{1}{(n_i/2)^2} \right)$$

This is the desired solution when we recognize that $R_{H} \equiv k_e e^2/2a_0 hc$. (See text discussion immediately following Eq. 42.17.)

---

[2] The correspondence principle is in reality the starting point for Bohr's postulate 4 on angular momentum quantization. To see how postulate 4 arises from the correspondence principle, see pages 353–356 of J. W. Jewett, *Physics Begins With Another M . . . Mysteries, Magic, Myth, and Modern Physics,* Allyn & Bacon, Boston, 1996.

Example 42.2   Electronic Transitions in Hydrogen

(A) The electron in a hydrogen atom makes a transition from the $n = 2$ energy level to the ground level ($n = 1$). Find the wavelength and frequency of the emitted photon.

Solution We can use Equation 42.17 directly to obtain λ, with $n_i = 2$ and $n_f = 1$:

$$\frac{1}{\lambda} = R_H \left(\frac{1}{n_f^2} - \frac{1}{n_i^2} \right)$$

$$\frac{1}{\lambda} = R_H \left(\frac{1}{1^2} - \frac{1}{2^2} \right) = \frac{3R_H}{4}$$

$$\lambda = \frac{4}{3R_H} = \frac{4}{3(1.097 \times 10^7 \text{ m}^{-1})}$$

$$= 1.215 \times 10^{-7} \text{ m} = \boxed{121.5 \text{ nm}} \qquad \text{(ultraviolet)}$$

Because $c = f\lambda$, the frequency of the photon is

$$f = \frac{c}{\lambda} = \frac{3.00 \times 10^8 \text{ m/s}}{1.215 \times 10^{-7} \text{ m}} = \boxed{2.47 \times 10^{15} \text{ Hz}}$$

(B) In interstellar space, highly excited hydrogen atoms called Rydberg atoms have been observed. Find the wavelength to which radio astronomers must tune to detect signals from electrons dropping from the $n = 273$ level to $n = 272$.

Solution We can again use Equation 42.17, this time with $n_i = 273$ and $n_f = 272$:

$$\frac{1}{\lambda} = R_H \left(\frac{1}{n_f^2} - \frac{1}{n_i^2} \right)$$

$$\lambda = \boxed{0.922 \text{ m}}$$

(C) What is the radius of the electron orbit for a Rydberg atom for which $n = 273$?

Solution Using Equation 42.12, we find

$$r_{273} = (273)^2 (0.052\ 9 \text{ nm}) = \boxed{3.94\ \mu\text{m}}$$

This is large enough that the atom is on the verge of becoming macroscopic!

(D) How fast is the electron moving in a Rydberg atom for which $n = 273$?

Solution Using Equation 42.8, we find

$$v = \sqrt{\frac{k_e e^2}{m_e r}} = \sqrt{\frac{(8.99 \times 10^9 \text{ N·m}^2/\text{C}^2)(1.60 \times 10^{-19} \text{ C})^2}{(9.11 \times 10^{-31} \text{ kg})(3.94 \times 10^{-6} \text{ m})}}$$

$$= \boxed{8.01 \times 10^3 \text{ m/s}}$$

What If? What if radiation from the Rydberg atom in part (B) is treated classically? What is the wavelength of radiation emitted by the atom in the $n = 273$ level?

Answer Classically, the frequency of the emitted radiation is that of the rotation of the electron around the nucleus. We can calculate this using the period defined in Equation 4.16:

$$f = \frac{1}{T} = \frac{v}{2\pi r}$$

We have the radius and speed from parts (C) and (D), so

$$f = \frac{v}{2\pi r} = \frac{8.01 \times 10^3 \text{ m/s}}{2\pi (3.94 \times 10^{-6} \text{ m})} = 3.24 \times 10^8 \text{ Hz}$$

The wavelength of the radiation is

$$\lambda = \frac{c}{f} = \frac{3.00 \times 10^8 \text{ m/s}}{3.24 \times 10^8 \text{ Hz}} = 0.926 \text{ m}$$

This is less than half of one percent different from the wavelength calculated in part (B). As indicated in the discussion of Bohr's correspondence principle, this difference becomes even smaller for higher values of n.

Investigate transitions of the atom between states at the Interactive Worked Example link at **http://www.pse6.com.**

42.4   The Quantum Model of the Hydrogen Atom

In the preceding section we described how the Bohr model views the electron as a particle orbiting the nucleus in nonradiating, quantized energy levels. This approach leads to an analysis that combines both classical and quantum concepts. While the model demonstrates excellent agreement with some experimental results, it cannot explain others such as the "splitting" of spectral lines, as noted in Section 42.3. These difficulties are removed when a full quantum model involving the Schrödinger equation is used to describe the hydrogen atom.

The potential energy function for the hydrogen atom is

$$U(r) = -k_e \frac{e^2}{r} \qquad (42.20)$$

where $k_e = 8.99 \times 10^9$ N·m^2/C^2 is the Coulomb constant and r is the radial distance from the proton (situated at $r = 0$) to the electron.

The formal procedure for solving the problem of the hydrogen atom is to substitute $U(r)$ into the Schrödinger equation and find appropriate solutions to the equation, as we did for the particle in a box in Chapter 41. The present problem is more complicated, however, because it is three-dimensional and because U depends on the radial coordinate r. If the time-independent Schrödinger equation (Equation 41.13) is extended to three-dimensional rectangular coordinates, the result is:

$$-\frac{\hbar^2}{2m}\left(\frac{\partial^2 \psi}{\partial x^2} + \frac{\partial^2 \psi}{\partial y^2} + \frac{\partial^2 \psi}{\partial z^2}\right) + U\psi = E\psi$$

It is easier to solve this equation for the hydrogen atom if rectangular coordinates are converted to spherical polar coordinates, an extension of the plane polar coordinates we introduced in Section 3.1. In spherical polar coordinates, a point in space is represented by the three variables r, θ, and ϕ where r is the radial distance from the origin, $r = \sqrt{x^2 + y^2 + z^2}$. With the point represented at the end of a position vector **r** as shown in Figure 42.10, the angular coordinate θ specifies its angular position relative to the z axis. Once that position vector is projected onto the xy plane, the angular coordinate ϕ specifies the projection's (and therefore the point's) angular position relative to the x axis.

The conversion of the three-dimensional time-independent Schrödinger equation for $\psi(x, y, z)$ to the equivalent form for $\psi(r, \theta, \phi)$ is straightforward but very tedious, and we will omit the details.[3] As we did in Chapter 41 when we separated the time dependence from the space dependence in solutions to the one-dimensional Schrödinger equation, in this case we can separate the three space variables by writing $\psi(r, \theta, \phi)$ as a product of functions of each single variable:

$$\psi(r, \theta, \phi) = R(r)\,f(\theta)\,g(\phi)$$

In this way, the three-dimensional partial differential equation can be transformed into three separate ordinary differential equations: one for $R(r)$, one for $f(\theta)$, and one for $g(\phi)$. Each of these functions is subject to boundary conditions. For example, $R(r)$ must remain finite as $r \rightarrow 0$ and $r \rightarrow \infty$, and $g(\phi)$ must have the same value as $g(\phi + 2\pi)$.

When the full set of boundary conditions is applied to all three functions, we are led to three different quantum numbers for each allowed state of the hydrogen atom. These are restricted to integer values and correspond to the three independent degrees of freedom (three space dimensions).

The first quantum number, associated with the radial function $R(r)$ of the full wave function, is called the **principal quantum number** and is assigned the symbol n. The radial wave equation will lead to functions giving the probability of finding the electron at a certain radial distance from the nucleus. In Section 42.5, we will display two of these radial wave functions. Note that the potential-energy function given in Equation 42.20 depends *only* on the radial coordinate r and not on either of the angular coordinates. The energies of the allowed states for the hydrogen atom are found to be

$$E_n = -\left(\frac{k_e e^2}{2a_0}\right)\frac{1}{n^2} = -\frac{13.606 \text{ eV}}{n^2} \qquad n = 1, 2, 3, \ldots \qquad (42.21)$$

This result is in exact agreement with that obtained in the Bohr theory (Eqs. 42.13 and 42.14).

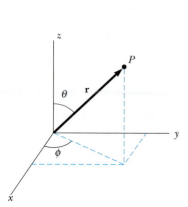

Figure 42.10 A point P in space is located by means of a position vector **r**. In Cartesian coordinates, the components of this vector are x, y, and z. In spherical polar coordinates, the point is described by r, the distance from the origin, θ, the angle between **r** and the z axis, and ϕ, the angle between the x axis and a projection of **r** onto the xy plane.

⚠ **PITFALL PREVENTION**

42.3 Energy Depends on n Only for Hydrogen

This statement that the energy depends only on the quantum number n is only true for the hydrogen atom. For more complicated atoms, we will use the same quantum numbers that we develop here for hydrogen. The energy levels for these atoms will depend primarily on n, but will also depend to a lesser degree on other quantum numbers.

Allowed energies of the quantum hydrogen atom

---

[3]   Descriptions of the solutions to the Schrödinger equation for the hydrogen atom are available in modern physics textbooks, such as R. A. Serway, C. Moses, and C. A. Moyer, *Modern Physics*, 2nd ed., Saunders College Publishing, 1998.

The **orbital quantum number**, symbolized ℓ, is associated with the orbital angular momentum of the electron, as is the **orbital magnetic quantum number** m_ℓ. Both ℓ and m_ℓ are integers. We will expand our discussion of these two quantum numbers in Section 42.6, and also introduce a fourth (non-integral) quantum number, resulting from a relativistic treatment of the hydrogen atom. Each set of appropriate values for these four quantum numbers corresponds to an allowed state of the atom.

The application of boundary conditions on the three parts of the full wave function leads to important relationships among the three quantum numbers, as well as certain restrictions on their values:

> The values of n can range from 1 to ∞.
>
> The values of ℓ can range from 0 to $n - 1$.
>
> The values of m_ℓ can range from $-\ell$ to ℓ.

Restrictions on the values of hydrogen-atom quantum numbers

For example, if $n = 1$, only $\ell = 0$ and $m_\ell = 0$ are permitted. If $n = 2$, ℓ may be 0 or 1; if $\ell = 0$, then $m_\ell = 0$; but if $\ell = 1$, then m_ℓ may be 1, 0, or -1. Table 42.1 summarizes the rules for determining the allowed values of ℓ and m_ℓ for a given n.

For historical reasons, **all states having the same principal quantum number are said to form a shell.** Shells are identified by the letters K, L, M, . . . , which designate the states for which $n = 1, 2, 3, ...$. Likewise, **all states having the same values of n and ℓ are said to form a subshell.** The letters[4] $s, p, d, f, g, h, . . .$ are used to designate the subshells for which $\ell = 0, 1, 2, 3, ...$. For example, the state designated by $3p$ has the quantum numbers $n = 3$ and $\ell = 1$; the $2s$ state has the quantum numbers $n = 2$ and $\ell = 0$. These notations are summarized in Table 42.2.

⚠ **PITFALL PREVENTION**

42.4   Quantum Numbers Describe a System

It is common to assign the quantum numbers to an electron. Remember, however, that these quantum numbers arise from the Schrödinger equation, which involves a potential energy function for the *system* of the electron and the nucleus. Thus, it is more *proper* to assign the quantum numbers to the atom, but it is more *popular* to assign them to an electron. We will follow this latter usage because it is so common.

Table 42.1

Three Quantum Numbers for the Hydrogen Atom			
Quantum Number	Name	Allowed Values	Number of Allowed States
n	Principal quantum number	1, 2, 3, . . .	Any number
ℓ	Orbital quantum number	0, 1, 2, . . . , $n - 1$	n
m_ℓ	Orbital magnetic quantum number	$-\ell, -\ell + 1, . . . , 0, . . . , \ell - 1, \ell$	$2\ell + 1$

Table 42.2

Atomic Shell and Subshell Notations			
n	Shell Symbol	ℓ	Subshell Symbol
1	K	0	s
2	L	1	p
3	M	2	d
4	N	3	f
5	O	4	g
6	P	5	h
.		.	
.		.	
.		.	

[4] The first four of these letters come from early classifications of spectral lines: sharp, principal, diffuse, and fundamental. The remaining letters are in alphabetical order.

States that violate the rules given in Table 42.1 do not exist. (They do not satisfy the boundary conditions on the wave function.) For instance, the 2*d* state, which would have $n = 2$ and $\ell = 2$, cannot exist because the highest allowed value of ℓ is $n - 1$, which in this case is 1. Thus, for $n = 2$, the 2*s* and 2*p* states are allowed but 2*d*, 2*f*, . . . are not. For $n = 3$, the allowed subshells are 3*s*, 3*p*, and 3*d*.

Quick Quiz 42.4 How many possible subshells are there for the $n = 4$ level of hydrogen? (a) 5   (b) 4   (c) 3   (d) 2   (e) 1.

Quick Quiz 42.5 When the principal quantum number is $n = 5$, how many different values of (a) ℓ and (b) m_ℓ are possible?

Quick Quiz 42.6 In the hydrogen atom, the quantum number n can increase without limit. True or False: because of this, the frequency of possible spectral lines from hydrogen also increases without limit.

Example 42.3   The $n = 2$ Level of Hydrogen

For a hydrogen atom, determine the number of allowed states corresponding to the principal quantum number $n = 2$, and calculate the energies of these states.

Solution When $n = 2$, ℓ can be 0 or 1. If $\ell = 0$, the only value that m_ℓ can have is 0; for $\ell = 1$, m_ℓ can be -1, 0, or 1. Hence, we have one state, designated as the 2*s* state, that is associated with the quantum numbers $n = 2$, $\ell = 0$, and $m_\ell = 0$, and three states, designated as 2*p* states, for which

the quantum numbers are $n = 2$, $\ell = 1$, $m_\ell = -1$; $n = 2$, $\ell = 1$, $m_\ell = 0$; and $n = 2$, $\ell = 1$, $m_\ell = 1$.

Because all four of these states have the same principal quantum number $n = 2$, they all have the same energy, according to Equation 42.21:

$$E_2 = -\frac{13.606 \text{ eV}}{2^2} = \boxed{-3.401 \text{ eV}}$$

42.5   The Wave Functions for Hydrogen

Because the potential energy of the hydrogen atom depends only on the radial distance r between nucleus and electron, some of the allowed states for this atom can be represented by wave functions that depend only on r. (For these states, $f(\theta)$ and $g(\phi)$ are constants.) The simplest wave function for hydrogen is the one that describes the 1*s* state and is designated $\psi_{1s}(r)$:

Wave function for hydrogen in its ground state

$$\psi_{1s}(r) = \frac{1}{\sqrt{\pi a_0^3}}\, e^{-r/a_0} \tag{42.22}$$

where a_0 is the Bohr radius. (In Problem 21, you can verify that this function satisfies the Schrödinger equation.) Note that ψ_{1s} approaches zero as r approaches ∞ and is normalized as presented (see Eq. 41.6). Furthermore, because ψ_{1s} depends only on r, it is *spherically symmetric*. This symmetry exists for all *s* states.

Recall that the probability of finding a particle in any region is equal to an integral of the probability density $|\psi|^2$ for the particle over the region. The probability density for the 1*s* state is

$$|\psi_{1s}|^2 = \left(\frac{1}{\pi a_0^3}\right) e^{-2r/a_0} \tag{42.23}$$

Because we imagine the nucleus to be fixed in space at $r = 0$, we can assign this probability density to the question of locating the electron. The probability of finding the electron in a volume element dV is $|\psi|^2 \, dV$. It is convenient to define the *radial probability density function* $P(r)$ as the probability per unit radial length of finding the electron in a spherical shell of radius r and thickness dr. Thus, $P(r) \, dr$ is the probability of finding the electron in this shell. The volume dV of such an infinitesimally thin shell equals its surface area $4\pi r^2$ multiplied by the shell thickness dr (Fig. 42.11), so we can write this probability as

$$P(r) \, dr = |\psi|^2 \, dV = |\psi|^2 \, 4\pi r^2 dr$$

Thus, the radial probability density function is

$$P(r) = 4\pi r^2 |\psi|^2 \qquad (42.24)$$

Substituting Equation 42.23 into Equation 42.24 gives the radial probability density function for the hydrogen atom in its ground state:

$$P_{1s}(r) = \left(\frac{4r^2}{a_0^3}\right) e^{-2r/a_0} \qquad (42.25)$$

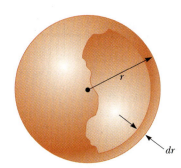

Figure 42.11 A spherical shell of radius r and thickness dr has a volume equal to $4\pi r^2 dr$.

Radial probability density for the 1s state of hydrogen

A plot of the function $P_{1s}(r)$ versus r is presented in Figure 42.12a. The peak of the curve corresponds to the most probable value of r for this particular state. We show in Example 42.4 that this peak occurs at the Bohr radius, the radial position of the electron when the hydrogen atom is in its ground state (in the Bohr theory)! This agreement, along with that of the energies given by Equation 42.21, is remarkable because the Bohr theory and the full quantum theory arrive at these results from completely different starting points.

We also find that the average value of r for the ground state of hydrogen is $\frac{3}{2}a_0$, which is 50% greater than the most probable value. The reason the average value is so much greater is the asymmetry in the radial probability density function (Fig. 42.12a), which has more area to the right of the peak. According to quantum mechanics, the atom has no sharply defined boundary as suggested by the Bohr theory. The probability distribution in Figure 42.12a suggests that the charge of the electron is extended throughout a diffuse region of space, commonly referred to as an *electron cloud*. Figure 42.12b shows the probability density of the electron in a hydrogen atom in the 1s state as a function of position in the xy plane. The darkest portion of the distribution appears at $r = a_0$, corresponding to the most probable value of r for the electron.

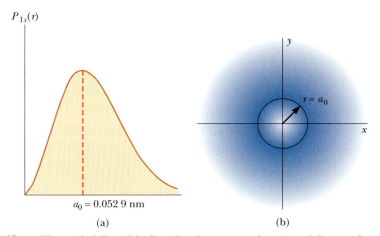

(a) (b)

Figure 42.12 (a) The probability of finding the electron as a function of distance from the nucleus for the hydrogen atom in the 1s (ground) state. Note that the probability has its maximum value when r equals the Bohr radius a_0. (b) The cross section in the xy plane of the spherical electronic charge distribution for the hydrogen atom in its 1s state.

Example 42.4   The Ground State of Hydrogen

Calculate the most probable value of r for an electron in the ground state of the hydrogen atom.

Solution We conceptualize a hydrogen atom as having a single electron and proton. Because the statement of the problem asks for the "most probable value of r," we categorize this as a problem in which we use the quantum approach. (In the Bohr atom, the electron moves in an orbit with an *exact* value of r.) Thus, our conceptualization should include the electron cloud image of the electron rather than the well-defined orbits of the Bohr model. To analyze the problem, we note that the most probable value of r corresponds to the peak of the plot of $P_{1s}(r)$ versus r. Because the slope of the curve at this point is zero, we can evaluate the most probable value of r by setting $dP_{1s}/dr = 0$ and solving for r. Using Equation 42.25, we obtain

$$\frac{dP_{1s}}{dr} = \frac{d}{dr}\left[\left(\frac{4r^2}{a_0^3}\right)e^{-2r/a_0}\right] = 0$$

Carrying out the derivative operation and simplifying the expression, we obtain

$$e^{-2r/a_0}\frac{d}{dr}(r^2) + r^2\frac{d}{dr}(e^{-2r/a_0}) = 0$$

$$2re^{-2r/a_0} + r^2(-2/a_0)e^{-2r/a_0} = 0$$

$$(1) \qquad 2r[1 - (r/a_0)]e^{-2r/a_0} = 0$$

This expression is satisfied if

$$1 - \frac{r}{a_0} = 0$$

$$r = \boxed{a_0}$$

The most probable value of r is the Bohr radius!

To finalize the problem, note that Equation (1) is also satisfied at $r = 0$ and as $r \to \infty$. These are points of *minimum* probability, which is equal to zero, as seen in Figure 42.12a.

What If? What if you were asked for the *average* value of r for the electron in the ground state rather than the most probable value?

Answer From the discussion before Example 42.4, we expect the value to be $\frac{3}{2}a_0$. The average value of r is the same as the expectation value for r. From Equation 41.7,

$$r_{av} = \langle r \rangle = \int_0^\infty rP(r)\,dr = \int_0^\infty r\left[\left(\frac{4r^2}{a_0^3}\right)e^{-2r/a_0}\right]dr$$

$$= \left(\frac{4}{a_0^3}\right)\int_0^\infty r^3 e^{-2r/a_0}\,dr$$

We can evaluate the integral with the help of the first integral listed in Table B.6 in Appendix B:

$$r_{av} = \left(\frac{4}{a_0^3}\right)\int_0^\infty r^3 e^{-2r/a_0}\,dr = \left(\frac{4}{a_0^3}\right)\left(\frac{3!}{(2/a_0)^4}\right) = \frac{3}{2}a_0$$

Example 42.5   Probabilities for the Electron in Hydrogen

Calculate the probability that the electron in the ground state of hydrogen will be found outside the first Bohr radius.

Solution The probability is found by integrating the radial probability density function $P_{1s}(r)$ for this state from the Bohr radius a_0 to ∞. Using Equation 42.25, we obtain

$$P = \int_{a_0}^\infty P_{1s}(r)\,dr = \frac{4}{a_0^3}\int_{a_0}^\infty r^2 e^{-2r/a_0}\,dr$$

We can put the integral in dimensionless form by changing variables from r to $z = 2r/a_0$. Noting that $z = 2$ when $r = a_0$

and that $dr = (a_0/2)\,dz$, we obtain

$$P = \frac{1}{2}\int_2^\infty z^2 e^{-z}\,dz = -\frac{1}{2}(z^2 + 2z + 2)e^{-z}\Big|_2^\infty$$

where we have used partial integration (see Appendix B.7). Evaluating between the limits, we have

$$P = 5e^{-2} = \boxed{0.677 \text{ or } 67.7\%}$$

This probability is larger than 50%, which is consistent with the fact that, as shown in Example 42.4, the average value of r is *greater* than the most probable value of r, the Bohr radius.

The next-simplest wave function for the hydrogen atom is the one corresponding to the 2s state ($n = 2$, $\ell = 0$). The normalized wave function for this state is

Wave function for hydrogen in the 2s state

$$\psi_{2s}(r) = \frac{1}{4\sqrt{2\pi}}\left(\frac{1}{a_0}\right)^{3/2}\left(2 - \frac{r}{a_0}\right)e^{-r/2a_0} \qquad (42.26)$$

Again we see that ψ_{2s} depends only on r and is spherically symmetric. The energy corresponding to this state is $E_2 = -(13.606/4)$ eV $= -3.401$ eV. This energy level represents the first excited state of hydrogen. A plot of the radial probability density function for this state in comparison to the 1s state is shown in Figure 42.13. The plot for the 2s state has two peaks. In this case, the most probable value corresponds

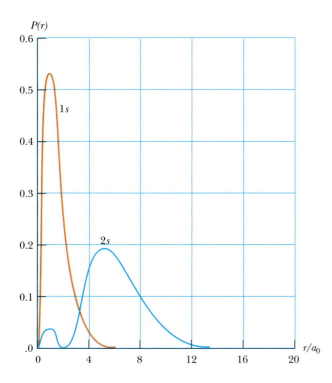

Active Figure 42.13 The radial probability density function versus r/a_0 for the $1s$ and $2s$ states of the hydrogen atom. *(From E. U. Condon and G. H. Shortley,* The Theory of Atomic Spectra, *Cambridge, England, Cambridge University Press, 1953. Used with permission.)*

🌐 *At the Active Figures link at* **http://www.pse6.com,** *choose values of r/a_0 and find the probability that the electron is located between two values.*

to that value of r that has the highest value of P ($\approx 5a_0$). An electron in the $2s$ state would be much farther from the nucleus (on the average) than an electron in the $1s$ state.

42.6 Physical Interpretation of the Quantum Numbers

The energy of a particular state in the hydrogen atom depends on the principal quantum number n (Eq. 42.21). Now let us see what the other quantum numbers contribute to our atomic model.

The Orbital Quantum Number ℓ

We begin this discussion by returning briefly to the Bohr model of the atom. If the electron moves in a circle of radius r, the magnitude of its angular momentum relative to the center of the circle is $L = m_e v r$. The direction of **L** is perpendicular to the plane of the circle and is given by a right-hand rule. According to classical physics, the magnitude L of the orbital angular momentum can have any value. However, the Bohr model of hydrogen postulates that the magnitude of the angular momentum of the electron is restricted to multiples of $\hbar$; that is, $m_e v r = n\hbar$. This model must be modified because it predicts (incorrectly) that the ground state of hydrogen has one unit of angular momentum. Furthermore, if L is taken to be zero in the Bohr model, we are forced to accept a picture of the electron as a particle oscillating along a straight line through the nucleus, a physically unacceptable situation.

These difficulties are resolved with the quantum-mechanical model of the atom, although we must give up the mental representation of an electron orbiting in a well-defined circular path. Despite the absence of this representation, the atom does indeed possess an angular momentum and we still call it orbital angular momentum. According to quantum mechanics, an atom in a state whose principal quantum

Allowed values of L

number is n can take on the following *discrete* values of the magnitude of the orbital angular momentum[5]:

$$L = \sqrt{\ell(\ell + 1)}\,\hbar \qquad \ell = 0, 1, 2, \ldots, n - 1 \qquad (42.27)$$

Given these allowed values of ℓ, we see that $L = 0$ (corresponding to $\ell = 0$) is an acceptable value of the magnitude of the angular momentum. The fact that L can be zero in this model serves to point out the inherent difficulties in any attempt to describe results based on quantum mechanics in terms of a purely particle-like (classical) model. In the quantum-mechanical interpretation, the electron cloud for the $L = 0$ state is spherically symmetric and has no fundamental rotation axis.

Example 42.6   Calculating L for a p State

Calculate the magnitude of the orbital angular momentum of an electron in a p state of hydrogen.

Solution Because we know that $\hbar = 1.054 \times 10^{-34}$ J·s, we can use Equation 42.27 to calculate L. With $\ell = 1$ for a p state (Table 42.2), we have

$$L = \sqrt{1(1 + 1)}\,\hbar = \sqrt{2}\,\hbar = \boxed{1.49 \times 10^{-34}\,\text{J·s}}$$

This number is extremely small relative to, say, the orbital angular momentum of the Earth orbiting the Sun, which is approximately 2.7×10^{40} J·s. The quantum number that describes L for macroscopic objects such as the Earth is so large that the separation between adjacent states cannot be measured. Once again, the correspondence principle is upheld.

The Orbital Magnetic Quantum Number m_ℓ

Because angular momentum is a vector, its direction must be specified. Recall from Chapter 29 that a current loop has a corresponding magnetic moment $\boldsymbol{\mu} = I\mathbf{A}$ (Eq. 29.10), where I is the current in the loop and $\mathbf{A}$ is a vector perpendicular to the loop whose magnitude is the area of the loop. Such a moment placed in a magnetic field $\mathbf{B}$ interacts with the field. Suppose a weak magnetic field applied along the z axis defines a direction in space. According to classical physics, the energy of the loop–field system depends on the direction of the magnetic moment of the loop with respect to the magnetic field, as described by Equation 29.12, $U = -\boldsymbol{\mu} \cdot \mathbf{B}$. Any energy between $-\mu B$ and $+\mu B$ is allowed by classical physics.

In the Bohr theory, the circulating electron represents a current loop. In the quantum-mechanical approach to the hydrogen atom, we abandon the circular orbit viewpoint of the Bohr theory, but the atom still possesses an orbital angular momentum. Thus, there is some sense of rotation of the electron around the nucleus, so that a magnetic moment is present due to this angular momentum.

As mentioned in Section 42.3, spectral lines from some atoms are observed to split into groups of three closely spaced lines when the atoms are placed in a magnetic field. Suppose we place the hydrogen atom in a magnetic field. According to quantum mechanics, **there are discrete directions allowed for the magnetic moment vector $\boldsymbol{\mu}$ with respect to the magnetic field vector B.** This is very different from the situation in classical physics, in which all directions are allowed.

Because the magnetic moment $\boldsymbol{\mu}$ of the atom can be related[6] to the angular momentum vector $\mathbf{L}$, the discrete directions of $\boldsymbol{\mu}$ translate to the fact that the direction of $\mathbf{L}$ is quantized. This quantization means that L_z (the projection of $\mathbf{L}$ along the z axis) can have

---

[5]   Equation 42.27 is a direct result of the mathematical solution of the Schrödinger equation and the application of angular boundary conditions. This development, however, is beyond the scope of this text and will not be presented.

[6]   See Equation 30.25 for this relationship, as derived from a classical viewpoint. Quantum mechanics arrives at the same result.

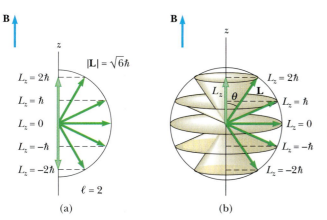

Figure 42.14 A vector model for $\ell = 2$. (a) The allowed projections of the orbital angular momentum **L**. (b) The orbital angular momentum vector **L** lies on the surface of a cone and precesses about the z axis when a magnetic field **B** is applied in the z direction.

only discrete values. The orbital magnetic quantum number m_ℓ specifies the allowed values of the z component of the orbital angular momentum according to the expression

$$L_z = m_\ell \hbar \tag{42.28}$$

Allowed values of L_z

The quantization of the possible orientations of **L** with respect to an external magnetic field is often referred to as **space quantization.**

Let us look at the possible orientations of **L** for a given value of ℓ. Recall that m_ℓ can have values ranging from $-\ell$ to ℓ. If $\ell = 0$, then $m_\ell = 0$ and $L_z = 0$. If $\ell = 1$, then the possible values of m_ℓ are -1, 0, and 1, so that L_z may be $-\hbar$, 0, or $\hbar$. If $\ell = 2$, then m_ℓ can be $-2, -1, 0, 1,$ or 2, corresponding to L_z values of $-2\hbar, -\hbar, 0, \hbar,$ or $2\hbar$, and so on.

Figure 42.14a shows what is commonly called a **vector model** that describes space quantization for the case $\ell = 2$. Note that **L** can never be aligned parallel or antiparallel to **B** because L_z must be less than the total angular momentum L. For L_z to be zero, **L** must be perpendicular to **B**.

It can be shown that **L** does not point in one specific direction, even though its z component is fixed. If **L** were known exactly, then all three components L_x, L_y, and L_z would be specified, which is inconsistent with the uncertainty principle. How can the magnitude and z component of a vector be specified, but the vector not be completely specified? To answer this, we can imagine that **L** must lie anywhere on the surface of a cone that makes an angle θ with the z axis, as shown in Figure 42.14b. From the figure, we see that θ is also quantized and that its values are specified through the relationship

$$\cos \theta = \frac{L_z}{|\mathbf{L}|} = \frac{m_\ell}{\sqrt{\ell(\ell + 1)}} \tag{42.29}$$

Note that m_ℓ is never greater than ℓ, and therefore θ can never be zero.

If the atom is placed in a magnetic field, the energy $U = -\boldsymbol{\mu} \cdot \mathbf{B}$ is additional energy for the atom beyond that described in Equation 42.21. Because the directions of $\boldsymbol{\mu}$ are quantized, there are discrete total energies for the atom corresponding to different values of m_ℓ. Figure 42.15a shows a transition between two atomic levels in the absence of a magnetic field. In Figure 42.15b, a magnetic field is applied, and the upper level, with $\ell = 1$, splits into three levels corresponding to the different directions of $\boldsymbol{\mu}$. There are now three possible transitions from the $\ell = 1$ subshell to the $\ell = 0$ subshell. Thus, in a collection of atoms, the single spectral line in Figure 42.15a will split into three spectral lines when the atoms are placed in a magnetic field. This phenomenon is called the *Zeeman effect.*

The Zeeman effect can be used to measure extraterrestrial magnetic fields. For example, the splitting of spectral lines in light from hydrogen atoms in the surface of the Sun can be used to calculate the magnitude of the magnetic field at that location. The Zeeman effect is one of many phenomena that cannot be explained with the Bohr model but are successfully explained by the quantum model of the atom.

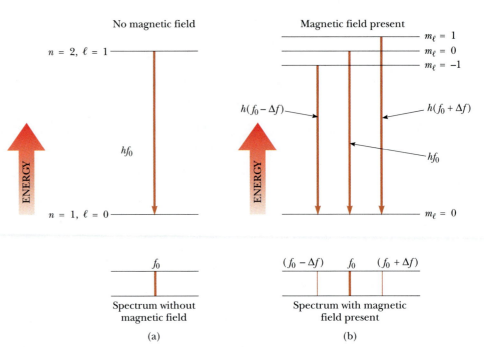

Figure 42.15 The Zeeman effect. (a) Energy levels for the ground and first excited states of a hydrogen atom. When **B** = 0, only a single spectral line at f_0 is observed. (b) When the atom is immersed in a magnetic field **B**, the state with $\ell = 1$ splits into three states. Atoms in the excited states decay to the ground state with the emission of photons with three different energies, giving rise to emission lines at f_0, $f_0 + \Delta f$, and $f_0 - \Delta f$, where Δf is the frequency shift of the emission caused by the magnetic field.

> **Quick Quiz 42.7** Sketch a vector model (shown in Figure 42.14 for $\ell = 2$) for $\ell = 1$.

Example 42.7   Space Quantization for Hydrogen   Interactive

Consider the hydrogen atom in the $\ell = 3$ state. Calculate the magnitude of **L**, the allowed values of L_z, and the corresponding angles θ that $|\mathbf{L}|$ makes with the z axis.

Solution We can calculate the magnitude of the orbital angular momentum using Equation 42.27:

$$L = \sqrt{\ell(\ell + 1)}\,\hbar = \sqrt{3(3 + 1)}\,\hbar = \boxed{2\sqrt{3}\,\hbar}$$

The allowed values of L_z can be calculated using $L_z = m_\ell \hbar$ with $m_\ell = -3, -2, -1, 0, 1, 2,$ and 3:

$$L_z = \boxed{-3\hbar, -2\hbar, -\hbar, 0, \hbar, 2\hbar, 3\hbar}$$

Finally, we calculate the allowed values of θ using Equation 42.29:

$$\cos \theta = \frac{m_\ell}{2\sqrt{3}}$$

Substituting the allowed values of m_ℓ gives

$$\cos \theta = \pm 0.866, \pm 0.577, \pm 0.289, \text{ and } 0$$

$$\theta = \boxed{30.0°, 54.8°, 73.2°, 90.0°, 107°, 125°, \text{ and } 150°}$$

What If? What if the value of ℓ is an arbitrary integer? For an arbitrary value of ℓ, how many values of m_ℓ are allowed?

Answer For a given value of ℓ, the values of m_ℓ range from $-\ell$ to $+\ell$ in steps of 1. Thus, there are 2ℓ nonzero values of m_ℓ (specifically, $\pm 1, \pm 2, \ldots, \pm \ell$). In addition, one more value of $m_\ell = 0$ is possible. This makes a total of $2\ell + 1$ values of m_ℓ. This result is critical in the understanding of the results of the Stern–Gerlach experiment described below with regard to spin.

🌀 *At the Interactive Worked Example link at* **http://www.pse6.com,** *practice evaluating the angular momentum for various quantum states of the hydrogen atom.*

The Spin Magnetic Quantum Number m_s

The three quantum numbers n, ℓ, and m_ℓ that we have discussed so far are generated by applying boundary conditions to solutions of the Schrödinger equation, and we can assign a physical interpretation to each of the quantum numbers. Let us now consider **electron spin,** which does *not* come from the Schrödinger equation.

In Example 42.3, we found four quantum states corresponding to $n = 2$. In reality, however, eight such states occur. The additional four states can be explained by requiring a fourth quantum number for each state—the **spin magnetic quantum number m_s.**

The need for this new quantum number arises because of an unusual feature that is observed in the spectra of certain gases, such as sodium vapor. Close examination of one prominent line in the emission spectrum of sodium reveals that the line is, in fact, two closely spaced lines called a *doublet*.[7] The wavelengths of these lines occur in the yellow region of the electromagnetic spectrum at 589.0 nm and 589.6 nm. In 1925, when this doublet was first observed, it could not be explained with the existing atomic theory. To resolve this dilemma, Samuel Goudsmit (1902–1978) and George Uhlenbeck (1900–1988), following a suggestion made by the Austrian physicist Wolfgang Pauli (1900–1958), proposed the spin quantum number.

To describe this new quantum number, it is convenient (but technically incorrect) to think of the electron as spinning about its axis as it orbits the nucleus, as described in Section 30.8. Only two directions exist for the electron spin, as illustrated in Figure 42.16. If the direction of spin is as shown in Figure 42.16a, the electron is said to have *spin up*. If the direction of spin is as shown in Figure 42.16b, the electron is said to have *spin down*. In the presence of a magnetic field, the energy of the electron is slightly different for the two spin directions, and this energy difference accounts for the sodium doublet.

The classical description of electron spin—as resulting from a spinning electron— is incorrect. More recent theory indicates that the electron is a point particle, without spatial extent. Thus, the electron cannot be considered to be spinning. Despite this conceptual difficulty, all experimental evidence supports the idea that an electron does have some intrinsic angular momentum that can be described by the spin magnetic quantum number. Paul Dirac (1902–1984) showed that this fourth quantum number originates in the relativistic properties of the electron.

In 1921, Otto Stern (1888–1969) and Walter Gerlach (1889–1979) performed an experiment that demonstrated space quantization. However, their results were not in quantitative agreement with the atomic theory that existed at that time. In their experiment, a beam of silver atoms sent through a nonuniform magnetic field was split into two discrete components (Fig. 42.17). They repeated the experiment using other atoms, and in each case the beam split into two or more components. The classical argument is as follows: If the z direction is chosen to be the direction of the maximum nonuniformity of **B**, the net magnetic force on the atoms is along the z axis and is proportional to the component of the magnetic moment $\boldsymbol{\mu}$ of the atom in the z direction. Classically, $\boldsymbol{\mu}$ can have any orientation, so the deflected beam should be spread out continuously. According to quantum mechanics, however, the deflected beam has an integral number of discrete components, and the number of components determines the number of possible values of μ_z. Therefore, because the Stern–Gerlach experiment showed split beams, space quantization was at least qualitatively verified.

For the moment, let us assume that the magnetic moment $\boldsymbol{\mu}$ of the atom is due to the orbital angular momentum. Because μ_z is proportional to m_ℓ, the number of possible values of μ_z is $2\ell + 1$, as found in the **What If?** section of Example 42.7. Furthermore, because ℓ is an integer, the number of values of μ_z is always odd. This prediction is not consistent with Stern and Gerlach's observation of two components (an *even*

Wolfgang Pauli and Niels Bohr watch a spinning top. The spin of the electron is analogous to the spin of the top but different in many ways.

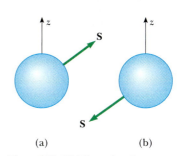

Figure 42.16 The spin of an electron can be either (a) up or (b) down relative to a specified z axis. The spin can never be aligned with the axis.

⚠️ **PITFALL PREVENTION**

42.5 The Electron Is Not Spinning

While the concept of a spinning electron is conceptually useful, it should not be taken literally. The spin of the Earth is a physical rotation. Electron spin is a purely quantum effect that gives the electron an angular momentum as if it were physically spinning.

---

[7] This is a Zeeman effect for spin and is identical in nature to the Zeeman effect for orbital angular momentum discussed before Example 42.7. The magnetic field for this Zeeman effect is internal to the atom and arises due to the relative motion of the electron and the nucleus.

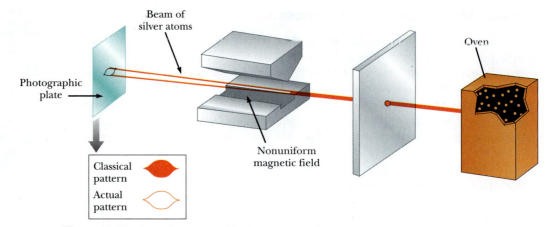

Figure 42.17 The technique used by Stern and Gerlach to verify space quantization. A beam of silver atoms is split in two by a nonuniform magnetic field.

number) in the deflected beam of silver atoms. Hence, we are forced to conclude that either quantum mechanics is incorrect or the model is in need of refinement.

In 1927, Phipps and Taylor repeated the Stern–Gerlach experiment using a beam of hydrogen atoms. This experiment was important because it involved an atom containing a single electron in its ground state, for which the quantum theory makes reliable predictions. Recall that $\ell = 0$ for hydrogen in its ground state, so $m_\ell = 0$. Therefore, we would not expect the beam to be deflected by the field because the magnetic moment $\boldsymbol{\mu}$ of the atom is zero. However, the beam in the Phipps–Taylor experiment was again split into two components. On the basis of that result, we can come to only one conclusion: something other than the electron's orbital motion is contributing to the atomic magnetic moment.

As we learned earlier, Goudsmit and Uhlenbeck had proposed that the electron has an intrinsic angular momentum, spin, apart from its orbital angular momentum. In other words, the total angular momentum of the electron in a particular electronic state contains both an orbital contribution **L** and a spin contribution **S**. The Phipps–Taylor result confirmed the hypothesis of Goudsmit and Uhlenbeck.

In 1929, Dirac used the relativistic form of the total energy of a system to solve the relativistic wave equation for the electron in a potential well. His analysis confirmed the fundamental nature of electron spin. (Spin, like mass and charge, is an *intrinsic* property of a particle, independent of its surroundings.) Furthermore, the analysis showed that electron spin[8] can be described by a single quantum number s, whose value can be only $s = \frac{1}{2}$. The spin angular momentum of the electron *never changes*. This notion contradicts classical laws, which dictate that a rotating charge slows down in the presence of an applied magnetic field because of the Faraday emf that accompanies the changing field. Furthermore, if the electron is viewed as a spinning ball of charge subject to classical laws, parts of it near its surface would be rotating with speeds exceeding the speed of light. Thus, the classical picture must not be pressed too far; ultimately, the spinning electron is a quantum entity defying any simple classical description.

Because spin is a form of angular momentum, it must follow the same quantum rules as orbital angular momentum. In accordance with Equation 42.27, the magnitude of the **spin angular momentum S** for the electron is

Magnitude of the spin angular momentum of an electron

$$S = \sqrt{s(s+1)}\,\hbar = \frac{\sqrt{3}}{2}\,\hbar \qquad (42.30)$$

[8] Physicists often use the word *spin* when referring to the spin angular momentum quantum number. For example, it is common to make the statement "The electron has a spin of one half."

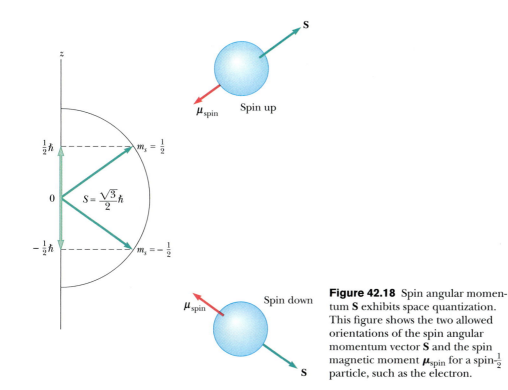

Figure 42.18 Spin angular momentum **S** exhibits space quantization. This figure shows the two allowed orientations of the spin angular momentum vector **S** and the spin magnetic moment $\boldsymbol{\mu}_{\text{spin}}$ for a spin-$\frac{1}{2}$ particle, such as the electron.

Like orbital angular momentum **L**, spin angular momentum **S** is quantized in space, as described in Figure 42.18. It can have two orientations relative to a z axis, specified by the spin magnetic quantum number $m_s = \pm\frac{1}{2}$. Similar to Equation 42.28 for orbital angular momentum, the z component of spin angular momentum is

$$S_z = m_s \hbar = \pm\tfrac{1}{2}\hbar \qquad (42.31)$$

Allowed values of S_z

The two values $\pm\hbar/2$ for S_z correspond to the two possible orientations for **S** shown in Figure 42.18. The value $m_s = +\frac{1}{2}$ refers to the spin-up case, and $m_s = -\frac{1}{2}$ refers to the spin-down case. Notice that Equations 42.30 and 42.31 do not allow the spin vector to lie along the z axis. The actual direction of **S** is at a relatively large angle with respect to the z axis as shown in Figures 42.16 and 42.18.

The spin magnetic moment $\boldsymbol{\mu}_{\text{spin}}$ of the electron is related to its spin angular momentum **S** by the expression

$$\boldsymbol{\mu}_{\text{spin}} = -\frac{e}{m_e}\,\mathbf{S} \qquad (42.32)$$

where e is the electronic charge and m_e is the mass of the electron. Because $S_z = \pm\frac{1}{2}\hbar$, the z component of the spin magnetic moment can have the values

$$\mu_{\text{spin},z} = \pm\frac{e\hbar}{2m_e} \qquad (42.33)$$

As we learned in Section 30.8, the quantity $e\hbar/2m_e$ is the Bohr magneton $\mu_{\text{B}} = 9.27 \times 10^{-24}$ J/T. Note that the ratio of magnetic moment to angular momentum is twice as great for spin angular momentum (Eq. 42.32) as it is for orbital angular momentum (Eq. 30.25). The factor of 2 is explained in a relativistic treatment first carried out by Dirac.

Today physicists explain the Stern–Gerlach experiment as follows. The observed magnetic moments for both silver and hydrogen are due to spin angular momentum only, with no contribution from orbital angular momentum. A single-electron atom such as hydrogen has its electron spin quantized in the magnetic field in such a way that the z component of spin angular momentum is either $\frac{1}{2}\hbar$ or $-\frac{1}{2}\hbar$, corresponding to $m_s = \pm\frac{1}{2}$. Electrons with spin $+\frac{1}{2}$ are deflected downward, and those with spin $-\frac{1}{2}$ are deflected upward.

The Stern–Gerlach experiment provided two important results. First, it verified the concept of space quantization. Second, it showed that spin angular momentum exists—even though this property was not recognized until four years after the experiments were performed.

Example 42.8 Putting Some Spin on Hydrogen

For a hydrogen atom, determine the quantum numbers associated with the possible states that correspond to the principal quantum number $n = 2$.

Solution With the addition of the spin quantum number, we have the possibilities given in the accompanying table.

n	ℓ	m_ℓ	m_s	Subshell	Shell	Number of States in Subshell
2	0	0	$\frac{1}{2}$			
2	0	0	$-\frac{1}{2}$	$2s$	L	2
2	1	1	$\frac{1}{2}$			
2	1	1	$-\frac{1}{2}$			
2	1	0	$\frac{1}{2}$			
2	1	0	$-\frac{1}{2}$	$2p$	L	6
2	1	-1	$\frac{1}{2}$			
2	1	-1	$-\frac{1}{2}$			

42.7 The Exclusion Principle and the Periodic Table

We have found that the state of a hydrogen atom is specified by four quantum numbers: n, ℓ, m_ℓ, and m_s. As it turns out, the number of states available to other atoms may also be predicted by this same set of quantum numbers. In fact, these four quantum numbers can be used to describe all the electronic states of an atom regardless of the number of electrons in its structure.

For the present discussion of atoms with many electrons, it is often easiest to assign the quantum numbers to the electrons in the atom as opposed to the entire atom. An obvious question that arises here is: "How many electrons can be in a particular quantum state?" Pauli answered this important question in 1925, in a statement known as the **exclusion principle:**

> No two electrons can ever be in the same quantum state; therefore, no two electrons in the same atom can have the same set of quantum numbers.

If this principle were not valid, an atom could radiate energy until every electron in the atom is in the lowest possible energy state, and the chemical behavior of the elements would be grossly modified. Nature as we know it would not exist!

In reality, we can view the electronic structure of complex atoms as a succession of filled levels increasing in energy. As a general rule, the order of filling of an atom's subshells is as follows. Once a subshell is filled, the next electron goes into the lowest-energy vacant subshell. We can understand this behavior by recognizing that if the atom were not in the lowest energy state available to it, it would radiate energy until it reached this state.

▲ **PITFALL PREVENTION**

42.6 The Exclusion Principle Is More General

A more general form of the exclusion principle will be discussed in Chapter 46. It states that no two *fermions* can be in the same quantum state. Fermions are particles with half-integral spin ($\frac{1}{2}$, $\frac{3}{2}$, $\frac{5}{2}$, etc.).

Table 42.3

Allowed Quantum States for an Atom up to $n = 3$												
n	1	2			3								
ℓ	0	0	1			0	1			2				
m_ℓ	0	0	1	0	−1	0	1	0	−1	2	1	0	−1	−2
m_s	⇅	⇅	⇅	⇅	⇅	⇅	⇅	⇅	⇅	⇅	⇅	⇅	⇅	⇅

Before we discuss the electronic configuration of various elements, it is convenient to define an *orbital* as the atomic state characterized by the quantum numbers n, ℓ, and m_ℓ. From the exclusion principle, we see that **only two electrons can be present in any orbital.** One of these electrons has a spin magnetic quantum number $m_s = +\frac{1}{2}$, and the other has $m_s = -\frac{1}{2}$. Because each orbital is limited to two electrons, the number of electrons that can occupy the various shells is also limited.

Table 42.3 shows the allowed quantum states for an atom up to $n = 3$. The arrows pointing upward indicate an electron described by $m_s = +\frac{1}{2}$, and those pointing downward indicate that $m_s = -\frac{1}{2}$. The $n = 1$ shell can accommodate only two electrons because $m_\ell = 0$ means that only one orbital is allowed. (The three quantum numbers describing this orbital are $n = 1$, $\ell = 0$, and $m_\ell = 0$.) The $n = 2$ shell has two subshells, one for $\ell = 0$ and one for $\ell = 1$. The $\ell = 0$ subshell is limited to two electrons because $m_\ell = 0$. The $\ell = 1$ subshell has three allowed orbitals, corresponding to $m_\ell = 1$, 0, and −1. Because each orbital can accommodate two electrons, the $\ell = 1$ subshell can hold six electrons. Thus, the $n = 2$ shell can contain eight electrons, as shown in Example 42.8. The $n = 3$ shell has three subshells ($\ell = 0, 1, 2$) and nine orbitals, accommodating up to 18 electrons. In general, each shell can accommodate up to $2n^2$ electrons.

The exclusion principle can be illustrated by an examination of the electronic arrangement in a few of the lighter atoms. The **atomic number** Z of any element is the number of protons in the nucleus of an atom of that element. A neutral atom of that element has Z electrons. Hydrogen ($Z = 1$) has only one electron—which, in the ground state of the atom, can be described by either of two sets of quantum numbers n, ℓ, m_ℓ, m_s: $1, 0, 0, \frac{1}{2}$ or $1, 0, 0, -\frac{1}{2}$. This electronic configuration is often written $1s^1$. The notation $1s$ refers to a state for which $n = 1$ and $\ell = 0$, and the superscript indicates that one electron is present in the s subshell.

Neutral helium ($Z = 2$) has two electrons. In the ground state, their quantum numbers are $1, 0, 0, \frac{1}{2}$ and $1, 0, 0, -\frac{1}{2}$. No other possible combinations of quantum numbers exist for this level, and we say that the K shell is filled. This electronic configuration is written $1s^2$.

Neutral lithium ($Z = 3$) has three electrons. In the ground state, two of these are in the $1s$ subshell. The third is in the $2s$ subshell because this subshell is slightly lower in energy than the $2p$ subshell.[9] Hence, the electronic configuration for lithium is $1s^2 2s^1$.

The electronic configurations of lithium and the next several elements are provided in Figure 42.19. The electronic configuration of beryllium ($Z = 4$), with its four electrons, is $1s^2 2s^2$, and boron ($Z = 5$) has a configuration of $1s^2 2s^2 2p^1$. The $2p$ electron in boron may be described by any of the six equally probable sets of quantum numbers listed in

Wolfgang Pauli
Austrian Theoretical Physicist (1900–1958)

An extremely talented theoretician who made important contributions in many areas of modern physics, Pauli gained public recognition at the age of 21 with a masterful review article on relativity that is still considered one of the finest and most comprehensive introductions to the subject. His other major contributions were the discovery of the exclusion principle, the explanation of the connection between particle spin and statistics, theories of relativistic quantum electrodynamics, the neutrino hypothesis, and the hypothesis of nuclear spin. (*CERN, courtesy of AIP Emilio Segré Visual Archive*)

---

[9] To a first approximation, energy depends only on the quantum number n, as we have discussed. However, because of the effect of the electronic charge shielding the nuclear charge, energy depends on ℓ also in multielectron atoms. We shall discuss these shielding effects in Section 42.8.

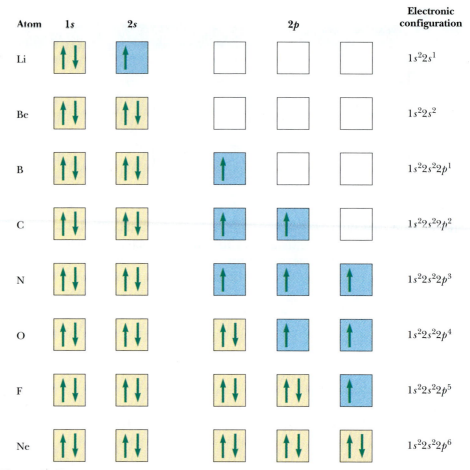

Figure 42.19 The filling of electronic states must obey both the exclusion principle and Hund's rule.

Example 42.8. In Figure 42.19, we show this electron in the leftmost $2p$ box with spin up, but it is equally likely to be in any $2p$ box with spin either up or down.

Carbon ($Z = 6$) has six electrons, giving rise to a question concerning how to assign the two $2p$ electrons. Do they go into the same orbital with paired spins ($\uparrow\downarrow$), or do they occupy different orbitals with unpaired spins ($\uparrow\uparrow$)? Experimental data show that the most stable configuration (that is, the one that is energetically preferred) is the latter, in which the spins are unpaired. Hence, the two $2p$ electrons in carbon and the three $2p$ electrons in nitrogen ($Z = 7$) have unpaired spins, as Figure 42.19 shows. The general rule that governs such situations, called **Hund's rule,** states that

> when an atom has orbitals of equal energy, the order in which they are filled by electrons is such that a **maximum number of electrons have unpaired spins.**

Some exceptions to this rule occur in elements having subshells that are close to being filled or half-filled.

A complete list of electronic configurations is provided in Table 42.4. In 1871, long before quantum mechanics was developed, the Russian chemist Dmitri Mendeleev (1834–1907) made an early attempt at finding some order among the chemical elements. He was trying to organize the elements for the table of contents of a book he was writing. He arranged the atoms in a table similar to that shown in Appendix C, according to their atomic masses and chemical similarities. The first table Mendeleev proposed contained many blank spaces, and he boldly stated that the gaps were there only because the elements had not yet been discovered. By noting the columns in

Table 42.4

Electronic Configuration of the Elements			
Atomic Number Z	Symbol	Ground-State Configuration	Ionization Energy (eV)
1	H	$1s^1$	13.595
2	He	$1s^2$	24.581
3	Li	[He] $2s^1$	5.39
4	Be	$2s^2$	9.320
5	B	$2s^2 2p^1$	8.296
6	C	$2s^2 2p^2$	11.256
7	N	$2s^2 2p^3$	14.545
8	O	$2s^2 2p^4$	13.614
9	F	$2s^2 2p^5$	17.418
10	Ne	$2s^2 2p^6$	21.559
11	Na	[Ne] $3s^1$	5.138
12	Mg	$3s^2$	7.644
13	Al	$3s^2 3p^1$	5.984
14	Si	$3s^2 3p^2$	8.149
15	P	$3s^2 3p^3$	10.484
16	S	$3s^2 3p^4$	10.357
17	Cl	$3s^2 3p^5$	13.01
18	Ar	$3s^2 3p^6$	15.755
19	K	[Ar] $4s^1$	4.339
20	Ca	$4s^2$	6.111
21	Sc	$3d^1 4s^2$	6.54
22	Ti	$3d^2 4s^2$	6.83
23	V	$3d^3 4s^2$	6.74
24	Cr	$3d^5 4s^1$	6.76
25	Mn	$3d^5 4s^2$	7.432
26	Fe	$3d^6 4s^2$	7.87
27	Co	$3d^7 4s^2$	7.86
28	Ni	$3d^8 4s^2$	7.633
29	Cu	$3d^{10} 4s^1$	7.724
30	Zn	$3d^{10} 4s^2$	9.391
31	Ga	$3d^{10} 4s^2 4p^1$	6.00
32	Ge	$3d^{10} 4s^2 4p^2$	7.88
33	As	$3d^{10} 4s^2 4p^3$	9.81
34	Se	$3d^{10} 4s^2 4p^4$	9.75
35	Br	$3d^{10} 4s^2 4p^5$	11.84
36	Kr	$3d^{10} 4s^2 4p^6$	13.996
37	Rb	[Kr] $5s^1$	4.176
38	Sr	$5s^2$	5.692
39	Y	$4d^1 5s^2$	6.377
40	Zr	$4d^2 5s^2$	
41	Nb	$4d^4 5s^1$	6.881
42	Mo	$4d^5 5s^1$	7.10
43	Tc	$4d^6 5s^1$	7.228
44	Ru	$4d^7 5s^1$	7.365
45	Rh	$4d^8 5s^1$	7.461
46	Pd	$4d^{10}$	8.33
47	Ag	$4d^{10} 5s^1$	7.574
48	Cd	$4d^{10} 5s^2$	8.991

continued

Table 42.4

	Electronic Configuration of the Elements *continued*		
Atomic Number Z	**Symbol**	**Ground-State Configuration**	**Ionization Energy (eV)**
49	In	$4d^{10}5s^25p^1$	
50	Sn	$4d^{10}5s^25p^2$	7.342
51	Sb	$4d^{10}5s^25p^3$	8.639
52	Te	$4d^{10}5s^25p^4$	9.01
53	I	$4d^{10}5s^25p^5$	10.454
54	Xe	$4d^{10}5s^25p^6$	12.127
55	Cs	[Xe] $6s^1$	3.893
56	Ba	$6s^2$	5.210
57	La	$5d^16s^2$	5.61
58	Ce	$4f^15d^16s^2$	6.54
59	Pr	$4f^36s^2$	5.48
60	Nd	$4f^46s^2$	5.51
61	Pm	$4f^56s^2$	
62	Fm	$4f^66s^2$	5.6
63	Eu	$4f^76s^2$	5.67
64	Gd	$4f^75d^16s^2$	6.16
65	Tb	$4f^96s^2$	6.74
66	Dy	$4f^{10}6s^2$	
67	Ho	$4f^{11}6s^2$	
68	Er	$4f^{12}6s^2$	
69	Tm	$4f^{13}6s^2$	
70	Yb	$4f^{14}6s^2$	6.22
71	Lu	$4f^{14}5d^16s^2$	6.15
72	Hf	$4f^{14}5d^26s^2$	7.0
73	Ta	$4f^{14}5d^36s^2$	7.88
74	W	$4f^{14}5d^46s^2$	7.98
75	Re	$4f^{14}5d^56s^2$	7.87
76	Os	$4f^{14}5d^66s^2$	8.7
77	Ir	$4f^{14}5d^76s^2$	9.2
78	Pt	$4f^{14}5d^96s^1$	8.88
79	Au	[Xe, $4f^{14}5d^{10}$] $6s^1$	9.22
80	Hg	$6s^2$	10.434

which some missing elements should be located, he was able to make rough predictions about their chemical properties. Within 20 years of this announcement, most of these elements were indeed discovered.

The elements in the **periodic table** (Appendix C) are arranged so that all those in a column have similar chemical properties. For example, consider the elements in the last column, which are all gases at room temperature: He (helium), Ne (neon), Ar (argon), Kr (krypton), Xe (xenon), and Rn (radon). The outstanding characteristic of all these elements is that they do not normally take part in chemical reactions—that is, they do not join with other atoms to form molecules. They are therefore called *inert gases*.

We can partially understand this behavior by looking at the electronic configurations in Table 42.4. The chemical behavior of an element depends on the outermost shell that contains electrons. The electronic configuration for helium is $1s^2$—the $n = 1$ shell (which is the outermost shell because it is the only shell) is filled. Additionally, the energy of the atom in this configuration is considerably lower than the energy for the configuration in which an electron is in the next available level, the $2s$ subshell. Next, look at the electronic configuration for neon, $1s^22s^22p^6$. Again, the outermost

Table 42.4

Atomic Number Z	Symbol	Ground-State Configuration	Ionization Energy (eV)
		Electronic Configuration of the Elements *continued*	
81	Tl	$6s^2 6p^1$	6.106
82	Pb	$6s^2 6p^2$	7.415
83	Bi	$6s^2 6p^3$	7.287
84	Po	$6s^2 6p^4$	8.43
85	At	$6s^2 6p^5$	
86	Rn	$6s^2 6p^6$	10.745
87	Fr	[Rn] $7s^1$	
88	Ra	$7s^2$	5.277
89	Ac	$6d^1 7s^2$	6.9
90	Th	$6d^2 7s^2$	
91	Pa	$5f^2 6d^1 7s^2$	
92	U	$5f^3 6d^1 7s^2$	4.0
93	Np	$5f^4 6d^1 7s^2$	
94	Pu	$5f^6 7s^2$	
95	Am	$5f^7 7s^2$	
96	Cm	$5f^7 6d^1 7s^2$	
97	Bk	$5f^9 7s^2$	
98	Cf	$5f^{10} 7s^2$	
99	Es	$5f^{11} 7s^2$	
100	Fm	$5f^{12} 7s^2$	
101	Md	$5f^{13} 7s^2$	
102	No	$5f^{14} 7s^2$	
103	Lr	$5f^{14} 7s^2 7p^1$	
104	Rf	$5f^{14} 6d^2 7s^2$	
105	Db	$5f^{14} 6d^3 7s^2$	
106	Sg	$5f^{14} 6d^4 7s^2$	
107	Bh	$5f^{14} 6d^5 7s^2$	
108	Hs	$5f^{14} 6d^6 7s^2$	
109	Mt	$5f^{14} 6d^7 7s^2$	
110	Ds	$5f^{14} 6d^9 7s^1$	

Note: The bracket notation is used as a shorthand method to avoid repetition in indicating inner-shell electrons. Thus, [He] represents $1s^2$, [Ne] represents $1s^2 2s^2 2p^6$, [Ar] represents $1s^2 2s^2 2p^6 3s^2 3p^6$, and so on. Configurations for elements above $Z = 102$ are tentative.

shell ($n = 2$ in this case) is filled, and a wide gap in energy occurs between the filled $2p$ subshell and the next available one, the $3s$ subshell. Argon has the configuration $1s^2 2s^2 2p^6 3s^2 3p^6$. Here, it is only the $3p$ subshell that is filled, but again a wide gap in energy occurs between the filled $3p$ subshell and the next available one, the $3d$ subshell. This pattern continues through all the inert gases. Krypton has a filled $4p$ subshell, xenon a filled $5p$ subshell, and radon a filled $6p$ subshell.

If we consider the column to the left of the inert gases in the periodic table, we find a group of elements called the *halogens*: fluorine, chlorine, bromine, iodine, and astatine. At room temperature, fluorine and chlorine are gases, bromine is a liquid, and iodine and astatine are solids. In each of these atoms, the outer subshell is one electron short of being filled. As a result, the halogens are chemically very active, readily accepting an electron from another atom to form a closed shell. The halogens tend to form strong ionic bonds with atoms at the other side of the periodic table. (We will discuss ionic bonds in Chapter 43.) In a halogen lightbulb, bromine or iodine atoms combine with tungsten atoms evaporated from the filament and return them to the filament, resulting in a longer-lasting bulb. In addition, the filament can

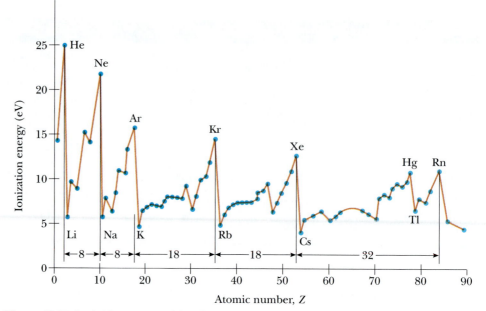

Figure 42.20 Ionization energy of the elements versus atomic number.

be operated at a higher temperature than in ordinary bulbs, giving a brighter and whiter light.

At the left side of the periodic table, the Group I elements consist of hydrogen and the *alkali metals*, lithium, sodium, potassium, rubidium, cesium, and francium. Each of these atoms contains one electron in a subshell outside of a closed subshell. Thus, these elements easily form positive ions because the lone electron is bound with a relatively low energy and is easily removed. Because of this, the alkali metal atoms are chemically active and form very strong bonds with halogen atoms. For example, table salt, NaCl, is a combination of an alkali metal and a halogen. Because the outer electron is weakly bound, pure alkali metals tend to be good electrical conductors, although, because of their high chemical activity, pure alkali metals are not generally found in nature.

It is interesting to plot ionization energy (see page 1358) versus atomic number Z, as in Figure 42.20. Note the pattern of $\Delta Z = 2, 8, 8, 18, 18, 32$ for the various peaks. This pattern follows from the exclusion principle and helps explain why the elements repeat their chemical properties in groups. For example, the peaks at $Z = 2, 10, 18$, and 36 correspond to the inert gases helium, neon, argon, and krypton, which, as we have mentioned, all have filled outermost shells. These elements have relatively high ionization energies and similar chemical behavior.

Quick Quiz 42.8 Rank the energy necessary to remove the outermost electron from the following three elements, smallest to largest: helium, neon, argon.

42.8 More on Atomic Spectra: Visible and X-Ray

In Section 42.1 we discussed the observation and early interpretation of visible spectral lines from gases. These spectral lines have their origin in transitions between quantized atomic states. We shall investigate these transitions more deeply in these final three sections of this chapter.

A modified energy-level diagram for hydrogen is shown in Figure 42.21. In this diagram, the allowed values of ℓ for each shell are separated horizontally. Figure 42.21 shows only those states up to $\ell = 2$; the shells from $n = 4$ upward would have more sets

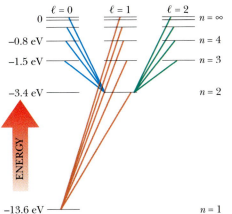

Figure 42.21 Some allowed electronic transitions for hydrogen, represented by the colored lines. These transitions must obey the selection rule $\Delta\ell = \pm 1$.

of states to the right which are not shown. We find that transitions for which ℓ does not change are very unlikely to occur and are called *forbidden transitions*. (Such transitions actually can occur, but their probability is very low relative to the probability of "allowed" transitions.) The various diagonal lines represent allowed transitions between stationary states. Whenever an atom makes a transition from a higher energy state to a lower one, a photon of light is emitted. The frequency of this photon is $f = \Delta E/h$, where ΔE is the energy difference between the two states and h is Planck's constant. The **selection rules** for the *allowed transitions* are

$$\Delta\ell = \pm 1 \qquad \text{and} \qquad \Delta m_\ell = 0, \pm 1 \tag{42.34}$$

Selection rules for allowed atomic transitions

Because the orbital angular momentum of an atom changes when a photon is emitted or absorbed (that is, as a result of a transition between states) and because angular momentum of the atom–photon system must be conserved, we conclude that **the photon involved in the process must carry angular momentum.** In fact, the photon has an angular momentum equivalent to that of a particle having a spin of 1. Therefore, a photon has energy, linear momentum, and angular momentum.

Recall from Equation 42.19 that the allowed energies for one-electron atoms and ions, such as hydrogen and He$^+$, are

$$E_n = -\frac{k_e e^2}{2a_0}\left(\frac{Z^2}{n^2}\right) = -\frac{13.6Z^2}{n^2}\text{ eV} \tag{42.35}$$

This equation was developed from the Bohr theory, but it serves as a good first approximation in quantum theory as well. For multielectron atoms, the positive nuclear charge Ze is largely shielded by the negative charge of the inner-shell electrons. Therefore, the outer electrons interact with a net charge that is smaller than the nuclear charge. The expression for the allowed energies for multielectron atoms has the same form as Equation 42.35 with Z replaced by an effective atomic number Z_{eff}:

$$E_n = -\frac{13.6Z_{\text{eff}}^2}{n^2}\text{ eV} \tag{42.36}$$

where Z_{eff} depends on n and ℓ.

X-Ray Spectra

X-rays are emitted when high-energy electrons or any other charged particles bombard a metal target. The x-ray spectrum typically consists of a broad continuous band containing a series of sharp lines, as shown in Figure 42.22. In Section 34.5, we mentioned the fact that an accelerated electric charge emits electromagnetic radiation. The x-rays we see in Figure 42.22 are the result of the slowing down of high-energy electrons as they strike the target. It may take several interactions with the atoms of the target before the

Figure 42.22 The x-ray spectrum of a metal target consists of a broad continuous spectrum (*bremsstrahlung*) plus a number of sharp lines, which are due to *characteristic x-rays*. The data shown were obtained when 37-keV electrons bombarded a molybdenum target.

electron loses all of its kinetic energy. (We explored this idea in Quick Quiz 42.2.) The amount of kinetic energy lost in any given interaction can vary from zero up to the entire kinetic energy of the electron. Thus, the wavelength of radiation from these interactions lies in a continuous range from some minimum value up to infinity. It is this general slowing down of the electrons that provides the continuous curve in Figure 42.22, which shows the cutoff of x-rays below a minimum wavelength value that depends on the kinetic energy of the incoming electrons. X-ray radiation with its origin in the slowing down of electrons is called **bremsstrahlung,** the German word for "braking radiation."

The discrete lines in Figure 42.22, called **characteristic x-rays** and discovered in 1908, have a different origin. Their origin remained unexplained until the details of atomic structure were understood. The first step in the production of characteristic x-rays occurs when a bombarding electron collides with a target atom. The electron must have sufficient energy to remove an inner-shell electron from the atom. The vacancy created in the shell is filled when an electron in a higher level drops down into the level containing the vacancy. The time interval for this to happen is very short, less than 10^{-9} s. As usual, this transition is accompanied by the emission of a photon whose energy equals the difference in energy between the two levels. Typically, the energy of such transitions is greater than 1 000 eV, and the emitted x-ray photons have wavelengths in the range of 0.01 nm to 1 nm.

Let us assume that the incoming electron has dislodged an atomic electron from the innermost shell—the K shell. If the vacancy is filled by an electron dropping from the next higher shell—the L shell—the photon emitted has an energy corresponding to the K_α characteristic x-ray line on the curve of Figure 42.22. In this notation, K refers to the final level of the electron and the subscript α, as the *first* letter of the Greek alphabet, refers to the initial level as the *first* one above the final level. If the vacancy in the K shell is filled by an electron dropping from the M shell, the K_β line in Figure 42.22 is produced.

Other characteristic x-ray lines are formed when electrons drop from upper levels to vacancies other than those in the K shell. For example, L lines are produced when vacancies in the L shell are filled by electrons dropping from higher shells. An L_α line is produced as an electron drops from the M shell to the L shell, and an L_β line is produced by a transition from the N shell to the L shell.

Although multielectron atoms cannot be analyzed exactly with either the Bohr model or the Schrödinger equation, we can apply our knowledge of Gauss's law from Chapter 24 to make some surprisingly accurate estimates of expected x-ray energies and wavelengths. Consider an atom of atomic number Z in which one of the two electrons in the K shell has been ejected. Imagine that a gaussian sphere is drawn just inside the most probable radius of the L electrons. The electric field at the position of the L electrons is a combination of the fields created by the nucleus, the single K electron, the other L electrons, and the outer electrons. The wave functions of the outer electrons are such that the electrons have a very high probability of being farther from the nucleus than the L electrons are. Therefore, they are much more likely to be outside the gaussian surface than inside and, on the average, do not contribute significantly to the electric field at the position of the L electrons. The effective charge inside the gaussian surface is the positive nuclear charge and one negative charge due to the single K electron. If we ignore the interactions between L electrons, a single L electron behaves as if it experiences an electric field due to a charge $(Z - 1)e$ enclosed by the gaussian surface. The nuclear charge is shielded by the electron in the K shell such that Z_{eff} in Equation 42.36 is $Z - 1$. For higher-level shells, the nuclear charge is shielded by electrons in all of the inner shells.

We can now use Equation 42.36 to estimate the energy associated with an electron in the L shell:

$$E_{\text{L}} = -(Z - 1)^2 \frac{13.6 \text{ eV}}{2^2}$$

After the atom makes the transition, there are two electrons in the K shell. Using a similar argument for a gaussian surface drawn just inside the most probable radius for a K electron, we find that the energy associated with one of these electrons is approximately that of a one-electron atom with the nuclear charge reduced by the negative charge of the other electron. That is,

$$E_K \approx -(Z-1)^2 (13.6 \text{ eV}) \tag{42.37}$$

As Example 42.9 shows, the energy of the atom with an electron in an M shell can be estimated in a similar fashion. Taking the energy difference between the initial and final levels, we can then calculate the energy and wavelength of the emitted photon.

In 1914, Henry G. J. Moseley (1887–1915) plotted the Z values for a number of elements versus $\sqrt{1/\lambda}$, where λ is the wavelength of the K_α line of each element. He found that the plot is a straight line, as in Figure 42.23. This is consistent with rough calculations of the energy levels given by Equation 42.37. From this plot, Moseley determined the Z values of elements that had not yet been discovered and produced a periodic table in excellent agreement with the known chemical properties of the elements. Until that experiment, atomic numbers had been merely placeholders for the elements that appeared in the periodic table, the elements being ordered according to mass.

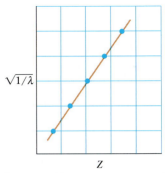

Figure 42.23 A Moseley plot of $\sqrt{1/\lambda}$ versus Z, where λ is the wavelength of the K_α x-ray line of the element of atomic number Z.

Quick Quiz 42.9 What are the initial and final shells for an M_β line in an x-ray spectrum?

Quick Quiz 42.10 In an x-ray tube, as you increase the energy of the electrons striking the metal target, the wavelengths of the characteristic x-rays (a) increase (b) decrease (c) do not change.

Quick Quiz 42.11 True or false: it is possible for an x-ray spectrum to show the continuous spectrum of x-rays without the presence of the characteristic x-rays.

Example 42.9 **Estimating the Energy of an X-Ray**

Estimate the energy of the characteristic x-ray emitted from a tungsten target when an electron drops from an M shell ($n = 3$ state) to a vacancy in the K shell ($n = 1$ state).

Solution The atomic number for tungsten is $Z = 74$. Using Equation 42.37, we see that the energy associated with the electron in the K shell is approximately

$$E_K \approx -(74-1)^2 (13.6 \text{ eV}) = -7.2 \times 10^4 \text{ eV}$$

An electron in the M shell is subject to an effective nuclear charge that depends on the number of electrons in the $n = 1$ and $n = 2$ states, which shield the nucleus. Because there are eight electrons in the $n = 2$ state and one electron in the $n = 1$ state, nine electrons shield the nucleus, and so

$Z_{\text{eff}} = Z - 9$. Hence, the energy of the M shell, following Equation 42.36, is approximately

$$E_M \approx -\frac{(13.6 \text{ eV})(74-9)^2}{(3)^2} \approx -6.4 \times 10^3 \text{ eV}$$

Therefore, the emitted x-ray has an energy equal to $hf = E_M - E_K \approx -6.4 \times 10^3 \text{ eV} - (-7.2 \times 10^4 \text{ eV}) \approx 6.6 \times 10^4 \text{ eV} = $ 66 keV.

Consultation of x-ray tables shows that the M–K transition energies in tungsten vary from 66.9 keV to 67.7 keV, where the range of energies is due to slightly different energy values for states of different ℓ. Thus, our estimate differs from the midpoint of this experimentally measured range by about 2%.

42.9 Spontaneous and Stimulated Transitions

We have seen that an atom absorbs and emits electromagnetic radiation only at frequencies that correspond to the energy differences between allowed states. Let us now examine more details of these processes. Consider an atom having the allowed energy levels labeled $E_1, E_2, E_3, \ldots$. When radiation is incident on the atom, only those photons whose energy hf matches the energy separation ΔE between two

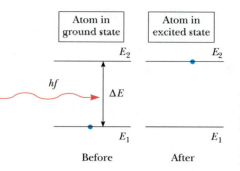

Active Figure 42.24 Stimulated absorption of a photon. The blue dot represents an electron. The electron is transferred from the ground state to the excited state when the atom absorbs a photon of energy $hf = E_2 - E_1$.

energy levels can be absorbed by the atom, as represented in Figure 42.24. This process is called **stimulated absorption** because the photon stimulates the atom to make the upward transition. At ordinary temperatures, most of the atoms in a sample are in the ground state. If a vessel containing many atoms of a gaseous element is illuminated with radiation of all possible photon frequencies (that is, a continuous spectrum), only those photons having energy $E_2 - E_1$, $E_3 - E_1$, $E_4 - E_1$, and so on are absorbed by the atoms. As a result of this absorption, some of the atoms are raised to excited states.

Once an atom is in an excited state, the excited atom can make a transition back to a lower energy level, emitting a photon in the process, as in Figure 42.25. This process is known as **spontaneous emission** because it happens naturally, without requiring an event to trigger the transition. Typically, an atom remains in an excited state for only about 10^{-8} s.

In addition to spontaneous emission, **stimulated emission** occurs. Suppose an atom is in an excited state E_2, as in Figure 42.26. If the excited state is a *metastable state*—that is, if its lifetime is much longer than the typical 10^{-8} s lifetime of excited states—then the time interval until spontaneous emission occurs will be relatively long. Let us imagine that during that interval a photon of energy $hf = E_2 - E_1$ is incident on the atom. One possibility is that the photon energy will be sufficient for the photon to ionize the atom. Another possibility is that the interaction between the incoming photon and the atom will cause the atom to return to the ground state[10] and thereby emit a second photon with energy $hf = E_2 - E_1$. In this process the incident photon is not absorbed; thus, after the stimulated emission, two photons with identical energy exist—the incident photon and the emitted photon. The two are in phase and travel in the same direction—an important consideration in lasers, which we shall discuss in the next section.

Active Figure 42.25 Spontaneous emission of a photon by an atom that is initially in the excited state E_2. When the atom falls to the ground state, it emits a photon of energy $hf = E_2 - E_1$.

[10] This is fundamentally a *resonance* phenomenon. The incoming photon is considered to be driving the system of the atom. Because the driving frequency matches that associated with a transition between states—one of the natural frequencies of the atom—there is a large response: the atom makes the transition.

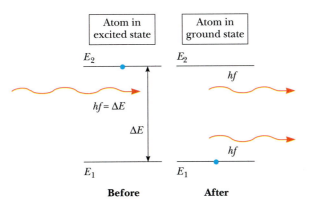

Active Figure 42.26 Stimulated emission of a photon by an incoming photon of energy $hf = E_2 - E_1$. Initially, the atom is in the excited state. The incoming photon stimulates the atom to emit a second photon of energy $hf = E_2 - E_1$.

Adjust the energy difference between states at the Active Figures link at http://www.pse6.com and observe stimulated emission.

42.10 Lasers

In this section, we shall explore the nature of laser light and a variety of applications of lasers in our technological society. The primary properties of laser light that make it useful in these technological applications are these:

- Laser light is coherent. The individual rays of light in a laser beam maintain a fixed phase relationship with each other, resulting in no destructive interference.

- Laser light is monochromatic. Light in a laser beam has a very narrow range of wavelengths.

- Laser light has a small angle of divergence. The beam spreads out very little, even over large distances.

In order to understand the origin of these properties, let us combine our knowledge of atomic energy levels from this chapter with some special requirements for the atoms that emit laser light.

We have described how an incident photon can cause atomic energy transitions either upward (stimulated absorption) or downward (stimulated emission). The two processes are equally probable. When light is incident on a collection of atoms, a net absorption of energy usually occurs because, when the system is in thermal equilibrium, many more atoms are in the ground state than in excited states. However, if the situation can be inverted so that more atoms are in an excited state than in the ground state, a net emission of photons can result. Such a condition is called **population inversion.**

This, in fact, is the fundamental principle involved in the operation of a **laser**—an acronym for *l*ight *a*mplification by *s*timulated *e*mission of *r*adiation. The full name indicates one of the requirements for laser light—the process of stimulated emission must occur in order to achieve laser action.

Suppose an atom is in the excited state E_2, as in Figure 42.26, and a photon with energy $hf = E_2 - E_1$ is incident on it. As described in Section 42.9, the incoming photon can stimulate the excited atom to return to the ground state and thereby emit a second photon having the same energy hf and traveling in the same direction. The incident photon is not absorbed, so after the stimulated emission, there are two identical photons—the incident photon and the emitted photon. The emitted photon is in phase with the incident photon. These photons can stimulate other atoms to emit photons in a chain of similar processes. The many photons produced in this fashion are the source of the intense, coherent light in a laser.

Figure 42.27 Schematic diagram of a laser design. The tube contains the atoms that are the active medium. An external source of energy (for example, an optical or electrical device) "pumps" the atoms to excited states. The parallel end mirrors confine the photons to the tube, but mirror 2 is only partially reflective.

In order that the stimulated emission result in laser light, we must have a buildup of photons in the system. The following three conditions must be satisfied in order to achieve this buildup:

- The system must be in a state of population inversion—there must be more atoms in an excited state than in the ground state. This must be true because the number of photons emitted must be greater than the number absorbed.

- The excited state of the system must be a *metastable state*, which means its lifetime must be long compared with the usually short lifetimes of excited states, which are typically 10^{-8} s. In this case, the population inversion can be established and stimulated emission is likely to occur before spontaneous emission.

- The emitted photons must be confined in the system long enough to enable them to stimulate further emission from other excited atoms. This is achieved by using reflecting mirrors at the ends of the system. One end is made totally reflecting, and the other is partially reflecting. A fraction of the light intensity passes through the partially reflecting end, forming the beam of laser light (Fig. 42.27).

One device that exhibits stimulated emission of radiation is the helium–neon gas laser. Figure 42.28 is an energy-level diagram for the neon atom in this system. The mixture of helium and neon is confined to a glass tube that is sealed at the ends by mirrors. A voltage applied across the tube causes electrons to sweep through the tube, colliding with the atoms of the gases and raising them into excited states. Neon atoms are excited to state E_3^* through this process (the asterisk * indicates a metastable state) and also as a result of collisions with excited helium atoms. Stimulated emission occurs, causing neon atoms to make transitions to state E_2. Neighboring excited atoms are also stimulated. This results in the production of coherent light at a wavelength of 632.8 nm.

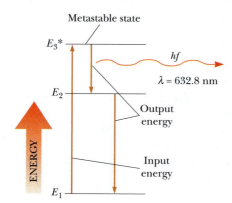

Figure 42.28 Energy-level diagram for a neon atom in a helium–neon laser. The atom emits 632.8-nm photons through stimulated emission in the transition $E_3^* - E_2$. This is the source of coherent light in the laser.

Applications

Since the development of the first laser in 1960, tremendous growth has occurred in laser technology. Lasers that cover wavelengths in the infrared, visible, and ultraviolet regions are now available. Applications include surgical "welding" of detached retinas, precision surveying and length measurement, precision cutting of metals and other materials (such as the fabric in Figure 42.29), and telephone communication along optical fibers. These and other applications are possible because of the unique characteristics of laser light. In addition to being highly monochromatic, laser light is also highly directional and can be sharply focused to produce regions of extremely intense light energy (with energy densities 10^{12} times that in the flame of a typical cutting torch).

Lasers are used in precision long-range distance measurement (range finding). In recent years it has become important, for astronomical and geophysical purposes, to measure as precisely as possible the distances from various points on the surface of the Earth to a point on the Moon's surface. To facilitate this, the Apollo astronauts set up a 0.5-m square of reflector prisms on the Moon, which enables laser pulses directed from an Earth station to be retroreflected to the same station (see Fig. 35.8a). Using the known speed of light and the measured round-trip travel time of a 1-ns pulse, the Earth–Moon distance can be determined to a precision of better than 10 cm.

Medical applications use the fact that various laser wavelengths can be absorbed in specific biological tissues. For example, certain laser procedures have greatly reduced blindness in glaucoma and diabetes patients. Glaucoma is a widespread eye condition characterized by a high fluid pressure in the eye, a condition that can lead to destruction of the optic nerve. A simple laser operation (iridectomy) can "burn" open a tiny hole in a clogged membrane, relieving the destructive pressure. A serious side effect of diabetes is neovascularization, the proliferation of weak blood vessels, which often leak blood. When this occurs in the retina, vision deteriorates (diabetic retinopathy) and finally is destroyed. It is now possible to direct the green light from an argon ion laser through the clear eye lens and eye fluid, focus on the retina edges, and photocoagulate the leaky vessels. Even people who have only minor vision defects such as nearsightedness are benefiting from the use of lasers to reshape the cornea, changing its focal length and reducing the need for eyeglasses.

Laser surgery is now an everyday occurrence at hospitals around the world. Infrared light at 10 μm from a carbon dioxide laser can cut through muscle tissue, primarily by vaporizing the water contained in cellular material. Laser power of about 100 W is required in this technique. The advantage of the "laser knife" over conventional

Philippe Plailly/SPL/Photo Researchers, Inc.

Figure 42.29 This robot carrying laser scissors, which can cut up to 50 layers of fabric at a time, is one of the many applications of laser technology.

methods is that laser radiation cuts tissue and coagulates blood at the same time, leading to a substantial reduction in blood loss. In addition, the technique virtually eliminates cell migration, an important consideration when tumors are being removed.

A laser beam can be trapped in fine optical-fiber light guides (endoscopes) by means of total internal reflection. An endoscope can be introduced through natural orifices, conducted around internal organs, and directed to specific interior body locations, eliminating the need for invasive surgery. For example, bleeding in the gastrointestinal tract can be optically cauterized by endoscopes inserted through the mouth.

In biological and medical research, it is often important to isolate and collect unusual cells for study and growth. A laser cell separator exploits the fact that specific cells can be tagged with fluorescent dyes. All cells are then dropped from a tiny charged nozzle and laser-scanned for the dye tag. If triggered by the correct light-emitting tag, a small voltage applied to parallel plates deflects the falling electrically charged cell into a collection beaker.

An exciting area of research and technological applications arose in the 1990s with the development of *laser trapping* of atoms. One scheme, called *optical molasses* and developed by Steven Chu, of Stanford University, and his colleagues, involves focusing six laser beams onto a small region in which atoms are to be trapped. Each pair of lasers is along one of the *x*, *y*, and *z* axes and emits light in opposite directions (Fig. 42.30). The frequency of the laser light is tuned to be just below the absorption frequency of the subject atom. Imagine that an atom has been placed into the trap region and moves along the positive *x* axis toward the laser which is emitting light toward it (the rightmost laser in Figure 42.30). Because the atom is moving, the light from the laser appears Doppler-shifted upward in frequency in the reference frame of the atom. This creates a match between the Doppler-shifted laser frequency and the absorption frequency of the atom, and the atom absorbs photons.[11] The momentum carried by these photons results in the atom being pushed back to the center of the trap. By incorporating six lasers, the atoms are pushed back into the trap regardless of which way they move along any axis.

In 1986, Chu developed *optical tweezers*, a device that uses a single tightly focused laser beam to trap and manipulate small particles. In combination with microscopes, optical tweezers have opened up many new possibilities for biologists. Optical tweezers

Figure 42.30 An optical trap for atoms is formed at the intersection point of six counterpropagating laser beams along mutually perpendicular axes. The frequency of the laser light is tuned to be just below that for absorption by the trapped atoms. If an atom moves away from the trap, it absorbs the Doppler-shifted laser light, and the momentum of the light pushes the atom back into the trap.

---

[11] The laser light traveling in the same direction as the atom is Doppler-shifted further downward in frequency, so that there is no absorption. Thus, the atom is not pushed out of the trap by the diametrically opposed laser.

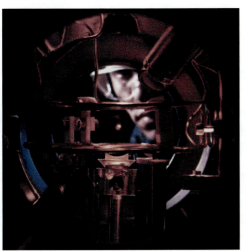

Figure 42.31 A staff member of the National Institute of Standards and Technology views a sample of trapped sodium atoms (the small yellow dot in the center of the vacuum chamber) cooled to a temperature of less than 1 mK.

Courtesy of Mark Helfer/National Institute of Standards and Technology

have been used to manipulate live bacteria without damage, move chromosomes within a cell nucleus, and measure the elastic properties of a single DNA molecule. Steven Chu shared the 1997 Nobel Prize in physics with two of his colleagues for the development of the techniques of optical trapping.

An extension of laser trapping, *laser cooling,* is possible because the normal high speeds of the atoms are reduced when they are restricted to the region of the trap. As a result, the temperature of the collection of atoms can be reduced to a few microkelvins. The technique of laser cooling allows scientists to study the behavior of atoms at extremely low temperatures (Figure 42.31).

SUMMARY

The wavelengths of spectral lines from hydrogen, called the **Balmer series,** can be described by the following equation:

$$\frac{1}{\lambda} = R_H \left(\frac{1}{2^2} - \frac{1}{n^2} \right) \qquad n = 3, 4, 5, \ldots \tag{42.1}$$

where R_H is the **Rydberg constant.** The spectral lines corresponding to values of n from 3 to 6 are in the visible range of the electromagnetic spectrum. Values of n higher than 6 correspond to spectral lines in the ultraviolet region of the spectrum.

The Bohr model of the atom is successful in describing the spectra of atomic hydrogen and hydrogen-like ions. One of the basic assumptions of the model is that the electron can exist only in discrete orbits such that the angular momentum of the electron $m_e v r$ is an integral multiple of $h/2\pi = \hbar$. When we assume circular orbits and a simple Coulomb attraction between electron and proton, the energies of the quantum states for hydrogen are calculated to be

$$E_n = -\frac{k_e e^2}{2a_0} \left(\frac{1}{n^2} \right) \qquad n = 1, 2, 3, \ldots \tag{42.13}$$

where k_e is the Coulomb constant, e is the electronic charge, n is an integer called the **quantum number,** and $a_0 = 0.052\,9$ nm is the **Bohr radius.**

If the electron in a hydrogen atom makes a transition from an orbit whose quantum number is n_i to one whose quantum number is n_f, where $n_f < n_i$, a photon is emitted by the atom, and the frequency of this photon is

$$f = \frac{k_e e^2}{2a_0 h} \left(\frac{1}{n_f^2} - \frac{1}{n_i^2} \right) \tag{42.15}$$

Take a practice test for this chapter by clicking on the Practice Test link at http://www.pse6.com.

Quantum mechanics can be applied to the hydrogen atom by the use of the potential energy function $U(r) = -k_e e^2/r$ in the Schrödinger equation. The solution to this equation yields wave functions for allowed states and allowed energies:

$$E_n = -\left(\frac{k_e e^2}{2a_0}\right)\frac{1}{n^2} = -\frac{13.606 \text{ eV}}{n^2} \qquad n = 1, 2, 3, \ldots \qquad (42.21)$$

where n is the **principal quantum number.** The allowed wave functions depend on three quantum numbers: n, ℓ, and m_ℓ, where ℓ is the **orbital quantum number** and m_ℓ is the **orbital magnetic quantum number.** The restrictions on the quantum numbers are

$$n = 1, 2, 3, \ldots$$
$$\ell = 0, 1, 2, \ldots, n-1$$
$$m_\ell = -\ell, -\ell+1, \ldots, \ell-1, \ell$$

All states having the same principal quantum number n form a **shell,** identified by the letters K, L, M, . . . (corresponding to $n = 1, 2, 3, \ldots$). All states having the same values of n and ℓ form a **subshell,** designated by the letters s, p, d, f, . . . (corresponding to $\ell = 0, 1, 2, 3, \ldots$).

An atom in a state characterized by a specific value of n can have the following values of L, the magnitude of the atom's orbital angular momentum $\mathbf{L}$:

$$L = \sqrt{\ell(\ell+1)}\,\hbar \qquad (42.27)$$

The allowed values of the projection of $\mathbf{L}$ along the z axis are

$$L_z = m_\ell \hbar \qquad (42.28)$$

Only discrete values of L_z are allowed, as determined by the restrictions on m_ℓ. This quantization of L_z is referred to as **space quantization.**

The electron has an intrinsic angular momentum called the **spin angular momentum.** That is, the total angular momentum of an electron in an atom has two contributions, one arising from the spin of the electron ($\mathbf{S}$) and one arising from the orbital motion of the electron ($\mathbf{L}$). Electron spin can be described by a single quantum number $s = \frac{1}{2}$. To completely describe a quantum state, it is necessary to include a fourth quantum number m_s, called the **spin magnetic quantum number.** This quantum number can have only two values, $\pm\frac{1}{2}$. The magnitude of the spin angular momentum is

$$S = \frac{\sqrt{3}}{2}\,\hbar \qquad (42.30)$$

and the z component of $\mathbf{S}$ is

$$S_z = m_s \hbar = \pm\frac{1}{2}\hbar \qquad (42.31)$$

That is, the spin angular momentum is also quantized in space, as specified by the spin magnetic quantum number $m_s = \pm\frac{1}{2}$.

The magnetic moment $\boldsymbol{\mu}_{\text{spin}}$ associated with the spin angular momentum of an electron is

$$\boldsymbol{\mu}_{\text{spin}} = -\frac{e}{m_e}\mathbf{S} \qquad (42.32)$$

The z component of $\boldsymbol{\mu}_{\text{spin}}$ can have the values

$$\mu_{\text{spin},z} = \pm\frac{e\hbar}{2m_e} \qquad (42.33)$$

The **exclusion principle** states that **no two electrons in an atom can be in the same quantum state.** In other words, no two electrons can have the same set of quantum numbers n, ℓ, m_ℓ, and m_s. Using this principle, the electronic configurations of the elements can be determined. This serves as a basis for understanding atomic structure and the chemical properties of the elements. The allowed electronic transitions between any two levels in an atom are governed by the **selection rules**

$$\Delta \ell = \pm 1 \qquad \text{and} \qquad \Delta m_\ell = 0, \pm 1 \qquad (42.34)$$

The x-ray spectrum of a metal target consists of a set of sharp characteristic lines superimposed on a broad continuous spectrum. **Bremsstrahlung** is x-radiation with its origin in the slowing down of high-energy electrons as they encounter the target. **Characteristic x-rays** are emitted by atoms when an electron undergoes a transition from an outer shell to a vacancy in an inner shell.

Atomic transitions can be described with three processes: **stimulated absorption,** in which an incoming photon raises the atom to a higher energy state; **spontaneous emission**, in which the atom makes a transition to a lower energy state, emitting a photon; and **stimulated emission,** in which an incident photon causes an excited atom to make a downward transition, emitting a photon identical to the incident one.

QUESTIONS

1. Does the light emitted by a neon sign constitute a continuous spectrum or only a few colors? Defend your answer.

2. The Bohr theory of the hydrogen atom is based upon several assumptions. Discuss these assumptions and their significance. Do any of them contradict classical physics?

3. Suppose that the electron in the hydrogen atom obeyed classical mechanics rather than quantum mechanics. Why should such a hypothetical atom emit a continuous spectrum rather than the observed line spectrum?

4. Can a hydrogen atom in the ground state absorb a photon of energy (a) less than 13.6 eV and (b) greater than 13.6 eV?

5. Explain why, in the Bohr model, the total energy of the atom is negative.

6. Let $-E$ represent the energy of a hydrogen atom. What is the kinetic energy of the electron? What is the potential energy of the atom?

7. According to Bohr's model of the hydrogen atom, what is the uncertainty in the radial coordinate of the electron? What is the uncertainty in the radial component of the velocity of the electron? In what way does the model violate the uncertainty principle?

8. Why are three quantum numbers needed to describe the state of a one-electron atom (ignoring spin)?

9. Compare the Bohr theory and the Schrödinger treatment of the hydrogen atom. Comment on the total energy and orbital angular momentum.

10. Discuss why the term *electron cloud* is used to describe the electronic arrangement in the quantum-mechanical model of an atom.

11. Why is the direction of the orbital angular momentum of an electron opposite to that of its magnetic moment?

12. Could the Stern–Gerlach experiment be performed with ions rather than neutral atoms? Explain.

13. Why is a nonuniform magnetic field used in the Stern–Gerlach experiment?

14. Discuss some of the consequences of the exclusion principle.

15. Describe some experiments that support the conclusion that the spin magnetic quantum number for electrons can only have the values $\pm\frac{1}{2}$.

16. Why do lithium, potassium, and sodium exhibit similar chemical properties?

17. Explain why a photon must have a spin of 1.

18. An energy of about 21 eV is required to excite an electron in a helium atom from the $1s$ state to the $2s$ state. The same transition for the He$^+$ ion requires approximately twice as much energy. Explain.

19. The absorption or emission spectrum of a gas consists of lines that broaden as the density of gas molecules increases. Why do you suppose this occurs?

20. How is it possible that electrons, whose positions are described by a probability distribution around a nucleus, can exist in states of *definite* energy (e.g., $1s$, $2p$, $3d$, . . .)?

21. It is easy to understand how two electrons (one spin up, one spin down) can fill the $1s$ shell for a helium atom. How is it possible that eight more electrons can fit into the $2s$, $2p$ level to complete the $1s^2 2s^2 2p^6$ shell for a neon atom?

22. In 1914, Henry Moseley was able to define the atomic number of an element from its characteristic x-ray spectrum. How was this possible? (*Suggestion:* See Figure 42.23.)

23. Does the intensity of light from a laser fall off as $1/r^2$?

24. Why is stimulated emission so important in the operation of a laser?

25. (a) "As soon as I define a particular direction as the z axis, precisely one half of the electrons in this part of the Universe have their magnetic moment vectors oriented at 54.735 61° to that axis, and all the rest have their magnetic moments at 125.264 39°." Argue for or against this statement. (b) "The Universe is not simply stranger than we suppose; it is stranger than we *can* suppose." Argue for or against this statement.

PROBLEMS

1, **2**, **3** = straightforward, intermediate, challenging   ☐ = full solution available in the *Student Solutions Manual and Study Guide*

🌐 = coached solution with hints available at http://www.pse6.com   💻 = computer useful in solving problem

▭ = paired numerical and symbolic problems

Section 42.1   Atomic Spectra of Gases

1. (a) What value of n_i is associated with the 94.96-nm spectral line in the Lyman series of hydrogen? (b) **What If?** Could this wavelength be associated with the Paschen or Balmer series?

2. (a) Compute the shortest wavelength in each of these hydrogen spectral series: Lyman, Balmer, Paschen, and Brackett. (b) Compute the energy (in electron volts) of the highest-energy photon produced in each series.

Section 42.2   Early Models of the Atom

3. 🌐 According to classical physics, a charge e moving with an acceleration a radiates at a rate

$$\frac{dE}{dt} = -\frac{1}{6\pi\epsilon_0}\frac{e^2 a^2}{c^3}$$

(a) Show that an electron in a classical hydrogen atom (see Fig. 42.6) spirals into the nucleus at a rate

$$\frac{dr}{dt} = -\frac{e^4}{12\pi^2\epsilon_0^2 r^2 m_e^2 c^3}$$

(b) Find the time interval over which the electron will reach $r = 0$, starting from $r_0 = 2.00 \times 10^{-10}$ m.

4. In the Rutherford scattering experiment, 4.00-MeV alpha particles (^{4}He nuclei containing 2 protons and 2 neutrons) scatter off gold nuclei (containing 79 protons and 118 neutrons). Assume that a particular alpha particle makes a direct head-on collision with the gold nucleus and scatters backward at 180°. Determine (a) the distance of closest approach of the alpha particle to the gold nucleus, and (b) the maximum force exerted on the alpha particle. Assume that the gold nucleus remains fixed throughout the entire process.

Section 42.3   Bohr's Model of the Hydrogen Atom

5. For a hydrogen atom in its ground state, use the Bohr model to compute (a) the orbital speed of the electron, (b) the kinetic energy of the electron, and (c) the electric potential energy of the atom.

6. Four possible transitions for a hydrogen atom are as follows:

(i) $n_i = 2$; $n_f = 5$ (ii) $n_i = 5$; $n_f = 3$

(iii) $n_i = 7$; $n_f = 4$ (iv) $n_i = 4$; $n_f = 7$

(a) In which transition is light of the shortest wavelength emitted? (b) In which transition does the atom gain the most energy? (c) In which transition(s) does the atom lose energy?

7. 🌐 A hydrogen atom is in its first excited state ($n = 2$). Using the Bohr theory of the atom, calculate (a) the radius of the orbit, (b) the linear momentum of the electron, (c) the angular momentum of the electron, (d) the kinetic energy of the electron, (e) the potential energy of the system, and (f) the total energy of the system.

8. How much energy is required to ionize hydrogen (a) when it is in the ground state? (b) when it is in the state for which $n = 3$?

9. A photon is emitted as a hydrogen atom undergoes a transition from the $n = 6$ state to the $n = 2$ state. Calculate (a) the energy, (b) the wavelength, and (c) the frequency of the emitted photon.

10. Show that the speed of the electron in the nth Bohr orbit in hydrogen is given by

$$v_n = \frac{k_e e^2}{n\hbar}$$

11. Two hydrogen atoms collide head-on and end up with zero kinetic energy. Each atom then emits light with a wavelength of 121.6 nm ($n = 2$ to $n = 1$ transition). At what speed were the atoms moving before the collision?

12. A monochromatic beam of light is absorbed by a collection of ground-state hydrogen atoms in such a way that six different wavelengths are observed when the hydrogen relaxes back to the ground state. What is the wavelength of the incident beam?

13. (a) Construct an energy-level diagram for the He$^+$ ion, for which $Z = 2$. (b) What is the ionization energy for He$^+$?

14. In a hot star, because of the high temperature, an atom can absorb sufficient energy to remove several electrons from the atom. Consider such a multiply ionized atom with a single remaining electron. The ion produces a series of spectral lines as described by the Bohr model. The series corresponds to electronic transitions that terminate in the same final state. The longest and shortest wavelengths of the series are 63.3 nm and 22.8 nm, respectively. (a) What is the ion? (b) Find the wavelengths of the next three spectral lines nearest to the line of longest wavelength.

15. (a) Calculate the angular momentum of the Moon due to its orbital motion about the Earth. In your calculation, use 3.84×10^8 m as the average Earth–Moon distance and 2.36×10^6 s as the period of the Moon in its orbit. (b) Assume the Moon's angular momentum is described by Bohr's assumption $mvr = n\hbar$. Determine the corresponding quantum number. (c) By what fraction would the Earth–Moon distance have to be increased to raise the quantum number by 1?

Section 42.4   The Quantum Model of the Hydrogen Atom

16. A general expression for the energy levels of one-electron atoms and ions is

$$E_n = -\frac{\mu k_e^2 q_1^2 q_2^2}{2\hbar^2 n^2}$$

where k_e is the Coulomb constant, q_1 and q_2 are the charges of the electron and the nucleus, and μ is the reduced mass, given by $\mu = m_1 m_2 / (m_1 + m_2)$. The wavelength for the $n = 3$ to $n = 2$ transition of the hydrogen atom is 656.3 nm (visible red light). **What If?** What are the wavelengths for this same transition in (a) positronium, which consists of an electron and a positron, and (b) singly ionized helium? (*Note:* A positron is a positively charged electron.)

17. An electron of momentum p is at a distance r from a stationary proton. The electron has kinetic energy $K = p^2/2m_e$. The atom has potential energy $U = -k_e e^2/r$, and total energy $E = K + U$. If the electron is bound to the proton to form a hydrogen atom, its average position is at the proton, but the uncertainty in its position is approximately equal to the radius r of its orbit. The electron's average vector momentum is zero, but its average squared momentum is approximately equal to the squared uncertainty in its momentum, as given by the uncertainty principle. Treating the atom as a one-dimensional system, (a) estimate the uncertainty in the electron's momentum in terms of r. (b) Estimate the electron's kinetic, potential, and total energies in terms of r. (c) The actual value of r is the one that *minimizes the total energy*, resulting in a stable atom. Find that value of r and the resulting total energy. Compare your answer with the predictions of the Bohr theory.

Section 42.5 The Wave Functions for Hydrogen

18. Plot the wave function $\psi_{1s}(r)$ (see Eq. 42.22) and the radial probability density function $P_{1s}(r)$ (see Eq. 42.25) for hydrogen. Let r range from 0 to $1.5a_0$, where a_0 is the Bohr radius.

19. The ground-state wave function for the electron in a hydrogen atom is

$$\psi(r) = \frac{1}{\sqrt{\pi a_0^3}} e^{-r/a_0}$$

where r is the radial coordinate of the electron and a_0 is the Bohr radius. (a) Show that the wave function as given is normalized. (b) Find the probability of locating the electron between $r_1 = a_0/2$ and $r_2 = 3a_0/2$.

20. The wave function for an electron in the $2p$ state of hydrogen is

$$\psi_{2p} = \frac{1}{\sqrt{3}(2a_0)^{3/2}} \frac{r}{a_0} e^{-r/2a_0}$$

What is the most likely distance from the nucleus to find an electron in the $2p$ state?

21. For a spherically symmetric state of a hydrogen atom, the Schrödinger equation in spherical coordinates is

$$-\frac{\hbar^2}{2m}\left(\frac{d^2\psi}{dr^2} + \frac{2}{r}\frac{d\psi}{dr}\right) - \frac{k_e e^2}{r}\psi = E\psi$$

Show that the $1s$ wave function for an electron in hydrogen,

$$\psi(r) = \frac{1}{\sqrt{\pi a_0^3}} e^{-r/a_0}$$

satisfies the Schrödinger equation.

22. In an experiment, electrons are fired at a sample of neutral hydrogen atoms and observations are made of how the incident particles scatter. A large set of trials can be thought of as containing 1 000 observations of the electron in the ground state of a hydrogen atom being momentarily at a distance $a_0/2$ from the nucleus. How many times is the atomic electron observed at a distance $2a_0$ from the nucleus in this set of trials?

Section 42.6 Physical Interpretation of the Quantum Numbers

23. List the possible sets of quantum numbers for electrons in (a) the $3d$ subshell and (b) the $3p$ subshell.

24. Calculate the angular momentum for an electron in (a) the $4d$ state and (b) the $6f$ state.

25. If an electron has orbital angular momentum equal to 4.714×10^{-34} J·s, what is the orbital quantum number for the state of the electron?

26. A hydrogen atom is in its fifth excited state, with principal quantum number 6. The atom emits a photon with a wavelength of 1 090 nm. Determine the maximum possible orbital angular momentum of the electron after emission.

27. How many sets of quantum numbers are possible for an electron for which (a) $n = 1$, (b) $n = 2$, (c) $n = 3$, (d) $n = 4$, and (e) $n = 5$? Check your results to show that they agree with the general rule that the number of sets of quantum numbers for a shell is equal to $2n^2$.

28. Find all possible values of L, L_z, and θ for an electron in a $3d$ state of hydrogen.

29. (a) Find the mass density of a proton, modeling it as a solid sphere of radius 1.00×10^{-15} m. (b) **What If?** Consider a classical model of an electron as a solid sphere with the same density as the proton. Find its radius. (c) Imagine that this electron possesses spin angular momentum $I\omega = \hbar/2$ because of classical rotation about the z axis. Determine the speed of a point on the equator of the electron and (d) compare this speed to the speed of light.

30. An electron is in the N shell. Determine the maximum value the z component of its angular momentum could have.

31. The ρ-meson has a charge of $-e$, a spin quantum number of 1, and a mass 1 507 times that of the electron. **What If?** Imagine that the electrons in atoms were replaced by ρ-mesons. List the possible sets of quantum numbers for ρ-mesons in the $3d$ subshell.

Section 42.7 The Exclusion Principle and the Periodic Table

32. (a) Write out the electronic configuration for the ground state of oxygen ($Z = 8$). (b) Write out a set of possible values for the quantum numbers n, ℓ, m_ℓ, and m_s for each electron in oxygen.

33. As we go down the periodic table, which subshell is filled first, the $3d$ or the $4s$ subshell? Which electronic configuration has a lower energy: $[Ar]3d^4 4s^2$ or $[Ar]3d^5 4s^1$? Which has the greater number of unpaired spins? Identify this element and discuss Hund's rule in this case. (*Note:* The notation [Ar] represents the filled configuration for argon.)

34. Devise a table similar to that shown in Figure 42.19 for atoms containing 11 through 19 electrons. Use Hund's rule and educated guesswork.

35. A certain element has its outermost electron in a $3p$ state. It has valence $+3$, since it has 3 more electrons than a certain noble gas. What element is it?

36. Two electrons in the same atom both have $n = 3$ and $\ell = 1$. (a) List the quantum numbers for the possible states of the atom. (b) **What If?** How many states would be possible if the exclusion principle were inoperative?

37. (a) Scanning through Table 42.4 in order of increasing atomic number, note that the electrons fill the subshells in such a way that those subshells with the lowest values of $n + \ell$ are filled first. If two subshells have the same value of $n + \ell$, the one with the lower value of n is filled first. Using these two rules, write the order in which the subshells are filled through $n + \ell = 7$. (b) Predict the chemical valence for the elements that have atomic numbers 15, 47, and 86, and compare your predictions with the actual valences (which may be found in a chemistry text).

38. For a neutral atom of element 110, what would be the probable ground-state electronic configuration?

39. Review problem. For an electron with magnetic moment $\boldsymbol{\mu}_S$ in a magnetic field **B**, Section 29.3 showed the following. The electron can be in a higher energy state with the z component of its magnetic moment opposite to the field, or in a lower energy state with the z component of its magnetic moment in the direction of the field. The difference in energy between the two states is $2\mu_B B$.

Under high resolution, many spectral lines are observed to be doublets. The most famous of these are the two yellow lines in the spectrum of sodium (the D lines), with wavelengths of 588.995 nm and 589.592 nm. Their existence was explained in 1925 by Goudsmit and Uhlenbeck, who postulated that an electron has intrinsic spin angular momentum. When the sodium atom is excited with its outermost electron in a $3p$ state, the orbital motion of the outermost electron creates a magnetic field. The atom's energy is somewhat different depending on whether the electron is itself spin-up or spin-down in this field. Then the photon energy the atom radiates as it falls back into its ground state depends on the energy of the excited state. Calculate the magnitude of the internal magnetic field mediating this so-called *spin-orbit coupling.*

Section 42.8   More on Atomic Spectra: Visible and X-Ray

40. (a) Determine the possible values of the quantum numbers ℓ and m_ℓ for the He^+ ion in the state corresponding to $n = 3$. (b) What is the energy of this state?

41. If you wish to produce 10.0-nm x-rays in the laboratory, what is the minimum voltage you must use in accelerating the electrons?

42 In x-ray production, electrons are accelerated through a high voltage ΔV and then decelerated by striking a target. Show that the shortest wavelength of an x-ray that can be produced is

$$\lambda_{min} = \frac{1\,240\ nm \cdot V}{\Delta V}$$

43. Use the method illustrated in Example 42.9 to calculate the wavelength of the x-ray emitted from a molybdenum target $(Z = 42)$ when an electron moves from the L shell $(n = 2)$ to the K shell $(n = 1)$.

44. The K series of the discrete x-ray spectrum of tungsten contains wavelengths of 0.018 5 nm, 0.020 9 nm, and 0.021 5 nm. The K-shell ionization energy is 69.5 keV. Determine the ionization energies of the L, M, and N shells. Draw a diagram of the transitions.

45. The wavelength of characteristic x-rays in the K_β line is 0.152 nm. Determine the material in the target.

Section 42.9   Spontaneous and Stimulated Transitions

Section 42.10   Lasers

46. Figure P42.46 shows portions of the energy-level diagrams of the helium and neon atoms. An electrical discharge excites the He atom from its ground state to its excited state of 20.61 eV. The excited He atom collides with a Ne atom in its ground state and excites this atom to the state at 20.66 eV. Lasing action takes place for electron transitions from E_3^* to E_2 in the Ne atoms. From the data in the figure, show that the wavelength of the red He–Ne laser light is approximately 633 nm.

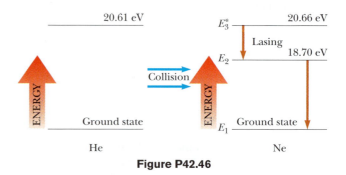

Figure P42.46

47. The carbon dioxide laser is one of the most powerful developed. The energy difference between the two laser levels is 0.117 eV. Determine the frequency and wavelength of the radiation emitted by this laser. In what portion of the electromagnetic spectrum is this radiation?

48. A Nd:YAG laser used in eye surgery emits a 3.00-mJ pulse in 1.00 ns, focused to a spot 30.0 μm in diameter on the retina. (a) Find (in SI units) the power per unit area at the retina. (This quantity is called the *irradiance* in the optics industry.) (b) What energy is delivered to an area of molecular size, taken as a circular area 0.600 nm in diameter?

49. A ruby laser delivers a 10.0-ns pulse of 1.00 MW average power. If the photons have a wavelength of 694.3 nm, how many are contained in the pulse?

50. The number N of atoms in a particular state is called the population of that state. This number depends on the energy of that state and the temperature. In thermal equilibrium the population of atoms in a state of energy E_n is given by a Boltzmann distribution expression

$$N = N_g e^{-(E_n - E_g)/k_B T}$$

where T is the absolute temperature and N_g is the population of the ground state, of energy E_g. For simplicity, we assume that each energy level has only one quantum state associated with it. (a) Before the power is switched on, the neon atoms in a laser are in thermal equilibrium at 27.0°C. Find the equilibrium ratio of the populations of the states $E_3{}^*$ and E_2 shown in Figure 42.28. Lasers operate by a clever artificial production of a "population inversion" between the upper and lower atomic energy states involved in the lasing transition. This means that more atoms are in the upper excited state then in the lower one. Consider the helium–neon laser transition at 632.8 nm. Assume that 2% more atoms occur in the upper state than in the lower. (b) To demonstrate how unnatural such a situation is, find the temperature for which the Boltzmann distribution describes a 2.00% population inversion. (c) Why does such a situation not occur naturally?

51. **Review problem.** A helium–neon laser can produce a green laser beam instead of red. Refer to Figure 42.28, which omits some energy levels between E_2 and E_1. After a population inversion is established, neon atoms will make a variety of downward transitions in falling from the state labeled $E_3{}^*$ down eventually to level E_1. The atoms will emit both red light with a wavelength of 632.8 nm and also green light with a wavelength of 543 nm in a competing transition. Assume the atoms are in a cavity between mirrors designed to reflect the green light with high efficiency but to allow the red light to leave the cavity immediately. Then stimulated emission can lead to the buildup of a collimated beam of green light between the mirrors having a greater intensity than does the red light. A small fraction of the green light is permitted to escape by transmission through one mirror, to constitute the radiated laser beam. The mirrors forming the resonant cavity can be made of layers of silicon dioxide and titanium dioxide. (a) How thick a layer of silicon dioxide, between layers of titanium dioxide, would minimize reflection of the red light? (b) What should be the thickness of a similar but separate layer of silicon dioxide to maximize reflection of the green light?

Additional Problems

52. As the Earth moves around the Sun, its orbits are quantized. (a) Follow the steps of Bohr's analysis of the hydrogen atom to show that the allowed radii of the Earth's orbit are given by

$$r = \frac{n^2 \hbar^2}{G M_S M_E{}^2}$$

where M_S is the mass of the Sun, M_E is the mass of the Earth, and n is an integer quantum number. (b) Calculate the numerical value of n. (c) Find the distance between the orbit for quantum number n and the next orbit out from the Sun corresponding to the quantum number $n + 1$. Discuss the significance of your results.

53. **LENINGRAD, 1930**—Four years after publication of the Schrödinger equation, Lev Davidovich Landau, age 23,

solved the equation for a charged particle moving in a uniform magnetic field. A single electron moving perpendicular to a field **B** can be considered as a model atom without a nucleus, or as the irreducible quantum limit of the cyclotron. Landau proved that its energy is quantized in uniform steps of $e\hbar B/m_e$. **HARVARD, 1999**—Gerald Gabrielse traps a single electron in an evacuated centimeter-size metal can cooled to a temperature of 80 mK. In a magnetic field of magnitude 5.26 T, the electron circulates for hours in its lowest energy level, generating a measurable signal as it moves. (a) Evaluate the size of a quantum jump in the electron's energy. (b) For comparison, evaluate $k_B T$ as a measure of the energy available to the electron in blackbody radiation from the walls of its container. (c) Microwave radiation can be introduced to excite the electron. Calculate the frequency and wavelength of the photon that the electron absorbs as it jumps to its second energy level. Measurement of the resonant absorption frequency verifies the theory and permits precise determination of properties of the electron.

54. Example 42.4 calculates the most probable value and the average value for the radial coordinate r of the electron in the ground state of a hydrogen atom. **What If?** For comparison with these modal and mean values, find the median value of r. Proceed as follows. (a) Derive an expression for the probability, as a function of r, that the electron in the ground state of hydrogen will be found outside a sphere of radius r centered on the nucleus. (b) Make a graph of the probability as a function of r/a_0. Choose values of r/a_0 ranging from 0 to 4.00 in steps of 0.250. (c) Find the value of r for which the probability of finding the electron outside a sphere of radius r is equal to the probability of finding the electron inside this sphere. You must solve a transcendental equation numerically, and your graph is a good starting point.

55. The positron is the antiparticle to the electron. It has the same mass and a positive electric charge of the same magnitude as that of the electron. Positronium is a hydrogen-like atom consisting of a positron and an electron revolving around each other. Using the Bohr model, find the allowed distances between the two particles and the allowed energies of the system.

56. **Review problem.** (a) How much energy is required to cause an electron in hydrogen to move from the $n = 1$ state to the $n = 2$ state? (b) Suppose the electron gains this energy through collisions among hydrogen atoms at a high temperature. At what temperature would the average atomic kinetic energy $3k_B T/2$, where k_B is the Boltzmann constant, be great enough to excite the electron?

57. *An example of the correspondence principle.* Use Bohr's model of the hydrogen atom to show that when the electron moves from the state n to the state $n - 1$, the frequency of the emitted light is

$$f = \left(\frac{2\pi^2 m_e k_e{}^2 e^4}{h^3 n^2} \right) \frac{2n - 1}{(n - 1)^2}$$

Show that as $n \rightarrow \infty$, this expression varies as $1/n^3$ and reduces to the classical frequency one expects the atom to emit. (*Suggestion:* To calculate the classical frequency, note that the frequency of revolution is $v/2\pi r$, where r is given by Eq. 42.10.)

58. Astronomers observe a series of spectral lines in the light from a distant galaxy. On the hypothesis that the lines form the Lyman series for a (new?) one-electron atom, they start to construct the energy-level diagram shown in Figure P42.58, which gives the wavelengths of the first four lines and the short-wavelength limit of this series. Based on this information, calculate (a) the energies of the ground state and first four excited states for this one-electron atom, and (b) the wavelengths of the first three lines and the short-wavelength limit in the Balmer series for this atom. (c) Show that the wavelengths of the first four lines and the short wavelength limit of the Lyman series for the hydrogen atom are all 60.0% of the wavelengths for the Lyman series in the one-electron atom described in part (b). (d) Based on this observation, explain why this atom could be hydrogen.

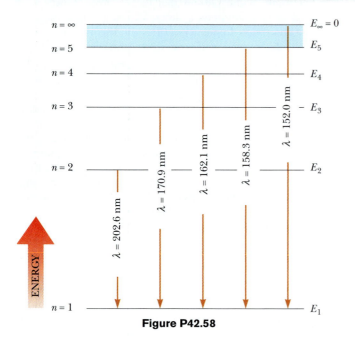

Figure P42.58

59. Suppose a hydrogen atom is in the 2s state, with its wave function given by Equation 42.26. Taking $r = a_0$, calculate values for (a) $\psi_{2s}(a_0)$, (b) $|\psi_{2s}(a_0)|^2$, and (c) $P_{2s}(a_0)$.

60. The states of matter are solid, liquid, gas, and plasma. Plasma can be described as a gas of charged particles, or a gas of ionized atoms. Most of the matter in the Solar System is plasma (throughout the interior of the Sun). In fact, most of the matter in the Universe is plasma; so is a candle flame. Use the information in Figure 42.20 to make an order-of-magnitude estimate for the temperature to which a typical chemical element must be raised to turn into plasma by ionizing most of the atoms in a sample. Explain your reasoning.

61. A pulsed ruby laser emits light at 694.3 nm. For a 14.0-ps pulse containing 3.00 J of energy, find (a) the physical length of the pulse as it travels through space and (b) the number of photons in it. (c) The beam has a circular cross section of diameter 0.600 cm. Find the number of photons per cubic millimeter.

62. A pulsed laser emits light of wavelength λ. For a pulse of duration Δt having energy E, find (a) the physical length of the pulse as it travels through space and (b) the number of photons in it. (c) The beam has a circular cross section having diameter d. Find the number of photons per unit volume.

63. Assume that three identical uncharged particles of mass m and spin 1/2 are contained in a one-dimensional box of length L. What is the ground-state energy of this system?

64. The force on a magnetic moment μ_z in a nonuniform magnetic field B_z is given by $F_z = \mu_z (dB_z / dz)$. If a beam of silver atoms travels a horizontal distance of 1.00 m through such a field and each atom has a speed of 100 m/s, how strong must be the field gradient dB_z / dz in order to deflect the beam 1.00 mm?

65. (a) Show that the most probable radial position for an electron in the 2s state of hydrogen is $r = 5.236a_0$. (b) Show that the wave function given by Equation 42.26 is normalized.

66. Suppose the ionization energy of an atom is 4.10 eV. In the spectrum of this same atom, we observe emission lines with wavelengths 310 nm, 400 nm, and 1 377.8 nm. Use this information to construct the energy-level diagram with the fewest levels. Assume that the higher levels are closer together.

67. An electron in chromium moves from the $n = 2$ state to the $n = 1$ state without emitting a photon. Instead, the excess energy is transferred to an outer electron (one in the $n = 4$ state), which is then ejected by the atom. (This is called an Auger [pronounced 'ohjay'] process, and the ejected electron is referred to as an Auger electron.) Use the Bohr theory to find the kinetic energy of the Auger electron.

68. In interstellar space, atomic hydrogen produces the sharp spectral line called the 21-cm radiation, which astronomers find most helpful in detecting clouds of hydrogen between stars. This radiation is useful because it is the only signal cold hydrogen emits and because interstellar dust that obscures visible light is transparent to these radio waves. The radiation is not generated by an electron transition between energy states characterized by different values of n. Instead, in the ground state ($n = 1$), the electron and proton spins may be parallel or antiparallel, with a resultant slight difference in these energy states. (a) Which condition has the higher energy? (b) More precisely, the line has wavelength 21.11 cm. What is the energy difference between the states? (c) The average lifetime in the excited state is about 10^7 yr. Calculate the associated uncertainty in energy of the excited energy level.

69. For hydrogen in the 1s state, what is the probability of finding the electron farther than $2.50a_0$ from the nucleus?

70. All atoms have the same size, to an order of magnitude. (a) To show this, estimate the diameters for aluminum (with molar mass 27.0 g/mol and density 2.70 g/cm^3) and uranium (molar mass 238 g/mol and density 18.9 g/cm^3). (b) What do the results imply about the wave functions for inner-shell electrons as we progress to higher and higher atomic mass atoms? (*Suggestion:* The molar volume is approximately $D^3 N_A$, where D is the atomic diameter and N_A is Avogadro's number.)

71. In the technique known as electron spin resonance (ESR), a sample containing unpaired electrons is placed in a magnetic field. Consider the simplest situation, in which only one electron is present and therefore only two energy

states are possible, corresponding to $m_s = \pm\frac{1}{2}$. In ESR, the absorption of a photon causes the electron's spin magnetic moment to flip from the lower energy state to the higher energy state. According to Section 29.3, the change in energy is $2\mu_B B$. (The lower energy state corresponds to the case where the z component of the magnetic moment $\boldsymbol{\mu}_{spin}$ is aligned with the magnetic field, and the higher energy state is the case where the z component of $\boldsymbol{\mu}_{spin}$ is aligned opposite to the field.) What is the photon frequency required to excite an ESR transition in a 0.350-T magnetic field?

72. Show that the wave function for an electron in the $2s$ state in hydrogen

$$\psi_{2s}(r) = \frac{1}{4\sqrt{2\pi}}\left(\frac{1}{a_0}\right)^{3/2}\left(2 - \frac{r}{a_0}\right)e^{-r/2a_0}$$

satisfies the spherically symmetric Schrödinger equation given in Problem 21.

73. Review problem. Steven Chu, Claude Cohen-Tannoudji, and William Phillips received the 1997 Nobel Prize in physics for "the development of methods to cool and trap atoms with laser light." One part of their work was with a beam of atoms (mass $\sim 10^{-25}$ kg) that move at a speed on the order of 1 km/s, similar to the speed of molecules in air at room temperature. An intense laser light beam tuned to a visible atomic transition (assume 500 nm) is directed straight into the atomic beam. That is, the atomic beam and light beam are traveling in opposite directions. An atom in the ground state immediately absorbs a photon. Total system momentum is conserved in the absorption process. After a lifetime on the order of 10^{-8} s, the excited atom radiates by spontaneous emission. It has an equal probability of emitting a photon in any direction. Thus, the average "recoil" of the atom is zero over many absorption and emission cycles. (a) Estimate the average deceleration of the atomic beam. (b) What is the order of magnitude of the distance over which the atoms in the beam will be brought to a halt?

74. Find the average (expectation) value of $1/r$ in the $1s$ state of hydrogen. It is given by

$$\langle 1/r \rangle = \int_{\text{all space}} |\psi|^2\,(1/r)\,dV = \int_0^\infty P(r)\,(1/r)\,4\pi r^2\,dr$$

Is the result equal to the inverse of the average value of r?

Answers to Quick Quizzes

42.1 (c). Because the energy of 10.5 eV does not correspond to raising the atom from the ground state to an allowed excited state, there is no interaction between the photon and the atom.

42.2 (a), (c). As the electrons strike the atom, they can give up any amount of energy between 0 and 10.5 eV, unlike the photons in Quick Quiz 42.1, which must give up all of their energy in one interaction. Thus, those electrons that undergo the appropriate collision with the atom can transfer 13.606 eV − 3.401 eV = 10.205 eV to the atom and excite it to the $n = 2$ state. Those electrons that do not make the appropriate collision will transfer only enough kinetic energy to the atom as a whole to satisfy conservation of momentum in the collision, without raising the atom to an excited state.

42.3 (a). The longest-wavelength photon is associated with the lowest energy transition, which is $n = 3$ to $n = 2$.

42.4 (b). The number of subshells is the same as the number of allowed values of ℓ. The allowed values of ℓ for $n = 4$ are $\ell = 0, 1, 2,$ and 3, so there are four subshells.

42.5 (a). Five values (0, 1, 2, 3, 4) of ℓ and (b) nine different values $(-4, -3, -2, -1, 0, 1, 2, 3, 4)$ of m_ℓ as follows:

ℓ	m_ℓ
0	0
1	−1, 0, 1
2	−2, −1, 0, 1, 2
3	−3, −2, −1, 0, 1, 2, 3
4	−4, −3, −2, −1, 0, 1, 2, 3, 4

42.6 False. If the energy of the hydrogen atom were proportional to n (or any power of n), the energy would become infinite as n grows to infinity. But the energy of the atom is *inversely* proportional to n^2. Thus, as n increases to very large values, the energy of the atom approaches zero from the negative side. As a result, the maximum frequency of emitted radiation approaches a value determined by the difference in energy between zero and the (negative) energy of the ground state.

42.7 The vector model for $\ell = 1$ is shown below:

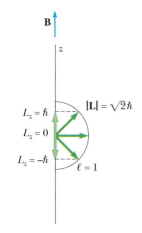

42.8 Argon, neon, helium. The higher the value of Z, the closer to zero is the energy associated with the outermost electron and the smaller is the ionization energy.

42.9 Final: M. Initial: O (because the subscript β indicates that the initial shell is the second shell higher than M).

42.10 (c). The wavelengths of the characteristic x-rays are determined by the separation between energy levels in the atoms of the target, which is unrelated to the energy with which electrons are fired at the target. The only dependence is that the incoming electrons must have enough energy to eject an atomic electron from an inner shell.

42.11 True. If the electrons arrive at the target with very low energy, atomic electrons cannot be ejected and characteristic x-rays do not appear. Because the incoming electrons experience accelerations, the continuous spectrum appears.

Chapter 43

Molecules and Solids

▲ *An understanding of the physics of solids has led to the technology of integrated circuits, found in countless electronic devices used by consumers in today's society. In this photograph, the microchip sitting on a fingertip contains millions of electrical components. (Bruce Dale/Getty Images)*

The most random atomic arrangement, that of a gas, was well understood in the 1800s, as we discussed in Chapter 21. In a crystalline solid, the atoms are not randomly arranged, but form a regular array. The symmetry of the arrangement of atoms both stimulated and allowed rapid progress in the field of solid-state physics in the twentieth century. Recently, our understanding of liquids and amorphous solids has advanced. (In an amorphous solid, the atoms do not form a regular array.) The recent interest in the physics of low-cost amorphous materials has been driven by their use in such devices as solar cells, memory elements, and fiber-optic waveguides.

In this chapter, we begin by studying the aggregates of atoms known as molecules. We describe the bonding mechanisms in molecules, the various modes of molecular excitation, and the radiation emitted or absorbed by molecules. We then take the next logical step and show how molecules combine to form solids. Then, by examining their energy-level structure, we explain the differences between insulating, conducting, semiconducting, and superconducting materials. The chapter also includes discussions of semiconducting junctions and several semiconductor devices.

43.1   Molecular Bonds

The bonding mechanisms in a molecule are fundamentally due to electric forces between atoms (or ions), but these forces are more complex than the simple Coulomb attraction or repulsion between single charges that we studied in Chapter 23. If the atoms in a molecule are considered as a system, the forces between components of the system are related to a potential energy function. A stable molecule would be expected at a configuration for which the potential energy function has its minimum value. (See Section 8.6.)

A potential energy function that can be used to model a molecule should account for two known features of molecular bonding:

- The force between atoms is repulsive at very small separation distance. When two atoms are brought close to each other, some of their electron shells overlap, resulting in repulsion between the shells. This repulsion is partly electrostatic in origin and partly the result of the exclusion principle. Because all electrons must obey the exclusion principle, some electrons in the overlapping shells are forced into higher energy states, and the system energy increases, as if a repulsive force existed between the atoms.

- At somewhat larger separations, the force between atoms is attractive. If this were not true, the atoms in a molecule would not be bound together. For many molecules, the attractive force is due to dipole–dipole interaction between charge distributions within the atoms of the molecule. We found in Example 23.6 that an electric dipole possesses an electric field, despite the fact that it carries no net charge. The electric fields of two dipoles will interact, resulting in a force between the dipoles.

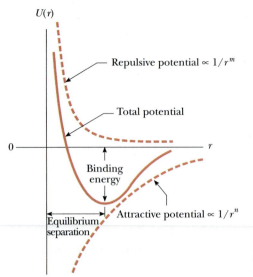

Figure 43.1 Total potential energy as a function of internuclear separation distance for a system of two atoms.

Taking into account these two features, the potential energy for a system of two atoms can be represented by an expression of the form

$$U(r) = -\frac{A}{r^n} + \frac{B}{r^m} \tag{43.1}$$

where r is the internuclear separation distance between the two atoms, and n and m are small integers. The parameter A is associated with the attractive force and B with the repulsive force. Example 8.11 on page 237 gives one common model for such a potential energy function, the Lennard–Jones potential.

Potential energy versus internuclear separation distance for a two-atom system is graphed in Figure 43.1. At large separation distances between the two atoms, the slope of the curve is positive, corresponding to a net attractive force. At the equilibrium separation distance, the attractive and repulsive forces just balance. At this point the potential energy has its minimum value, and the slope of the curve is zero.

A complete description of the bonding mechanisms in molecules is highly complex because bonding involves the mutual interactions of many particles. In this section, therefore, we discuss only some simplified models: ionic bonding, covalent bonding, van der Waals bonding, and hydrogen bonding.

Ionic Bonding

When two atoms combine in such a way that one or more outer electrons are transferred from one atom to the other, the bond formed is called an **ionic bond.** Ionic bonds are fundamentally caused by the Coulomb attraction between oppositely charged ions.

A familiar example of an ionically bonded solid is sodium chloride, NaCl, which is common table salt. Sodium, which has the electronic configuration $1s^2 2s^2 2p^6 3s^1$, is ionized relatively easily, giving up its $3s$ electron to form a Na^+ ion. The energy required to ionize the atom to form Na^+ is 5.1 eV. Chlorine, which has the electronic configuration $1s^2 2s^2 2p^5$, is one electron short of the filled-shell structure of argon. If we compare the energy of a system of a free electron and a Cl atom to one in which the electron joins the atom to make the Cl^- ion, we find that the energy of the ion is lower. When the electron makes a transition from the $E = 0$ state to the negative energy state associated with the available shell in the atom, energy is released. This amount of energy is called the **electron affinity** of the atom. For chlorine, the electron affinity is 3.7 eV. Therefore, the energy required to form Na^+ and Cl^- from isolated atoms is $5.1 - 3.7 = 1.4$ eV. It costs 5.1 eV to remove the electron from the Na atom but you gain 3.7 eV of it back when that electron is allowed to join with the Cl atom.

Figure 43.2 Total energy versus internuclear separation distance for Na^+ and Cl^- ions. The horizontal axis is labeled Na + Cl because we define zero energy as that for the system of neutral sodium and chlorine *atoms.* The asymptote of the curve for large values of r is marked $Na^+ + Cl^-$ because that is the energy of the system of sodium and chloride *ions.*

Now imagine that these two charged ions interact with one another to form a NaCl "molecule."[1] The total energy of the NaCl molecule versus internuclear separation distance is graphed in Figure 43.2. At very large separation distances, the energy of the system of ions is 1.4 eV, as calculated above. The total energy has a minimum value of -4.2 eV at the equilibrium separation distance, which is about 0.24 nm. This means that the energy required to break the Na^+–Cl^- bond and form neutral sodium and chlorine atoms, called the **dissociation energy,** is 4.2 eV. Notice that the energy of the molecule is lower than that of the system of two neutral atoms. Consequently, we say that it is **energetically favorable** for the molecule to form—if a lower energy state of a system exists, the system will tend to emit energy in order to achieve this lower energy state. The system of neutral sodium and chlorine atoms can reduce its total energy by transferring energy out of the system (by electromagnetic radiation, for example) and forming the NaCl molecule.

Covalent Bonding

A **covalent bond** between two atoms is one in which electrons supplied by either one or both atoms are shared by the two atoms. Many diatomic molecules, such as H_2, F_2, and CO, owe their stability to covalent bonds. We will describe covalent bonds by means of atomic wave functions, focusing on the bonds between hydrogen atoms, for which we saw the ground-state wave function in Chapter 42:

$$\psi_{1s}(r) = \frac{1}{\sqrt{\pi a_0^3}}\, e^{-r/a_0}$$

This wave function is graphed in Figure 43.3a for two hydrogen atoms that are far apart. Notice that there is very little overlap of the wave functions $\psi_1(r)$ for atom 1, located at $r = 0$, and $\psi_2(r)$ for atom 2, located some distance away. Suppose now we bring the two atoms close together. As this happens, their wave functions overlap and form the compound wave function $\psi_1(r) + \psi_2(r)$ shown in Figure 43.3b. Notice that the probability amplitude is larger between the atoms than it is on either side of the combination of atoms. As a result, the probability is higher that the electrons associated with the atoms will be located between the atoms than on the outer regions of the system. Consequently, the average position of negative charge in the system is halfway

▲ **PITFALL PREVENTION**

43.1   Ionic and Covalent Bonds

In practice, these descriptions of ionic and covalent bonds represent extreme ends of a spectrum of bonds involving electron transfer. In a real bond, the electron may not be *completely* transferred as in an ionic bond or *equally* shared as in a covalent bond. Thus, real bonds lie somewhere between these extremes.

---

[1]   NaCl does not exist as an isolated molecule. In the solid state, NaCl forms a crystalline array of ions as described in Section 43.3. In the liquid state, or in solution with water, the Na^+ and Cl^- ions dissociate and are free to move relative to each other.

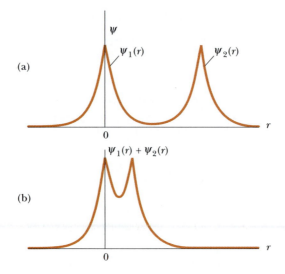

(a)

(b)

Active Figure 43.3 Ground-state wave functions $\psi_1(r)$ and $\psi_2(r)$ for two atoms making a covalent bond. (a) The atoms are far apart and their wave functions overlap minimally. (b) The atoms are close together, forming a composite wave function $\psi_1(r) + \psi_2(r)$ for the system. The probability amplitude for an electron to be between the atoms is high.

between the atoms. This can be modeled as if there were a fixed negative charge between the atoms, exerting attractive Coulomb forces on both nuclei. The result is an overall attractive force between the atoms, resulting in a covalent bond.

Because of the exclusion principle, the two electrons in the ground state of H_2 must have antiparallel spins. Also because of the exclusion principle, if a third H atom is brought near the H_2 molecule, the third electron would have to occupy a higher energy level, which is an energetically unfavorable situation. For this reason, the H_3 molecule is not stable and does not form.

Van der Waals Bonding

Ionic and covalent bonds occur between atoms to form molecules or ionic solids, so they can be described as bonds *within* molecules. Two additional types of bonds, van der Waals bonds and hydrogen bonds, can occur *between* molecules.

We might expect that two neutral molecules would not interact by means of the electric force because they each have zero net charge. We find, however, that they are attracted to each other by weak electrostatic forces called **van der Waals forces.** Likewise, atoms that do not form ionic or covalent bonds are attracted to each other by van der Waals forces. Inert gases, for example, because of their filled shell structure, do not generally form molecules. Because of van der Waals forces, however, at sufficiently low temperatures at which thermal excitations are negligible, inert gases first condense to liquids and then solidify (with the exception of helium, which does not solidify at atmospheric pressure).

The van der Waals force is due to the fact that, while being electrically neutral, a molecule has a charge distribution with positive and negative centers at different positions in the molecule. As a result, the molecule may act as an electric dipole. Because of the dipole electric fields, two molecules can interact such that there is an attractive force between them.

There are three types of van der Waals forces. The first type, called the *dipole–dipole force,* is an interaction between two molecules each having a permanent electric dipole moment—for example, polar molecules such as HCl have permanent electric dipole moments and attract other polar molecules.

The second type, the *dipole–induced dipole force,* results when a polar molecule having a permanent electric dipole moment induces a dipole moment in a nonpolar molecule. In this case, the electric field of the polar molecule creates the dipole moment in the nonpolar molecule, which then results in an attractive force between the molecules.

The third type is called the *dispersion force,* an attractive force that occurs between two nonpolar molecules. In this case, the interaction results from the fact that, although the average dipole moment of a nonpolar molecule is zero, the average of the square of the dipole moment is nonzero because of charge fluctuations. Two non-polar molecules near each other tend to have dipole moments that are correlated in time so as to produce an attractive van der Waals force.

Hydrogen Bonding

Because hydrogen has only one electron, it is expected to form a covalent bond with only one other atom within a molecule. A hydrogen atom in a given molecule can also form a second type of bond between molecules called a **hydrogen bond.** Let us use the water molecule H_2O as an example. In the two covalent bonds in this molecule, the electrons from the hydrogen atoms are more likely to be found near the oxygen atom than near the hydrogen atoms. This leaves essentially bare protons at the positions of the hydrogen atoms. This unshielded positive charge can be attracted to the negative end of another polar molecule. Because the proton is unshielded by electrons, the negative end of the other molecule can come very close to the proton to form a bond that is strong enough to form a solid crystalline structure, such as that of ice. The bonds within a water molecule are covalent, but the bonds between water molecules in ice are hydrogen bonds.

The hydrogen bond is relatively weak compared with other chemical bonds—it can be broken with an input energy of about 0.1 eV. Because of this, ice melts at the low temperature of 0°C. Despite the fact that this bond is very weak, however, hydrogen bonding is a critical mechanism responsible for the linking of biological molecules and polymers. For example, in the case of the DNA (deoxyribonucleic acid) molecule, which has a double-helix structure (Fig. 43.4), hydrogen bonds formed by the sharing of a proton between two atoms create linkages between the turns of the helix.

Figure 43.4 DNA molecules are held together by hydrogen bonds.

> **Quick Quiz 43.1**   For each of the following, identify the most likely type of bonding that will occur between the atoms or between the molecules. Choose from the following list: ionic, covalent, van der Waals, hydrogen. (a) atoms of krypton (b) potassium and chlorine atoms (c) hydrogen fluoride (HF) molecules (d) chlorine and oxygen atoms in a hypochlorite ion (ClO^-).

43.2 Energy States and Spectra of Molecules

As in the case of atoms, we can study the structure and properties of molecules by examining the radiation they emit or absorb. Before we describe these processes, it is important to understand the various ways of exciting a molecule.

Consider an individual molecule in the gaseous phase of a substance. The energy of the molecule can be divided into four categories: (1) electronic energy, due to the interactions between the molecule's electrons and nuclei; (2) translational energy, due to the motion of the molecule's center of mass through space; (3) rotational energy, due to the rotation of the molecule about its center of mass; and (4) vibrational energy, due to the vibration of the molecule's constituent atoms:

$$E = E_{el} + E_{trans} + E_{rot} + E_{vib}$$

Total energy of a molecule

We explored the role of translational, rotational, and vibrational energy of molecules in determining the molar specific heats of gases in Sections 21.2 and 21.4. Because the translational energy is unrelated to internal structure, this molecular energy is unimportant in interpreting molecular spectra. The electronic energy of a molecule is very complex because it involves the interaction of many charged particles, but various techniques have been developed to approximate its values. While the electronic energies can be studied, significant information about a molecule can be determined by analyzing its rotational and vibrational energy states, which give spectral lines in the microwave and infrared regions of the electromagnetic spectrum, respectively.

Rotational Motion of Molecules

Let us consider the rotation of a molecule around its center of mass, confining our discussion to the diatomic case (Fig. 43.5a) but noting that the same ideas can be extended to polyatomic molecules. A diatomic molecule aligned along an x axis has only two rotational degrees of freedom, corresponding to rotations about the y and z axes. If ω is the angular frequency of rotation about one of these axes, the rotational kinetic energy of the molecule about that axis can be expressed in the form

$$E_{rot} = \tfrac{1}{2} I \omega^2 \tag{43.2}$$

In this equation, I is the moment of inertia of the molecule, given by

Moment of inertia for a diatomic molecule

$$I = \left(\frac{m_1 m_2}{m_1 + m_2} \right) r^2 = \mu r^2 \tag{43.3}$$

where m_1 and m_2 are the masses of the atoms that form the molecule, r is the atomic separation, and μ is the **reduced mass** of the molecule (see Example 41.7):

Reduced mass of a diatomic molecule

$$\mu = \frac{m_1 m_2}{m_1 + m_2} \tag{43.4}$$

The magnitude of the molecule's angular momentum is $L = I\omega$, which classically can have any value. Quantum mechanics, however, restricts the molecule to certain quantized rotational frequencies such that the angular momentum of the molecule has the values[2]

Allowed values of rotational angular momentum

$$L = \sqrt{J(J+1)}\, \hbar \qquad J = 0, 1, 2, \ldots \tag{43.5}$$

where J is an integer called the **rotational quantum number.** Combining Equations 43.5 and 43.2, we obtain an expression for the allowed values of the rotational

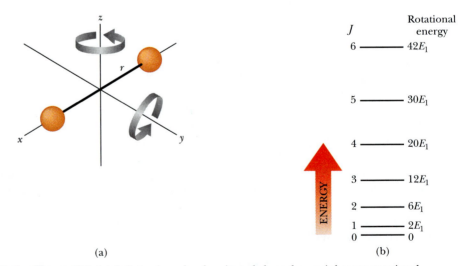

(a)

(b)

Active Figure 43.5 (a) A diatomic molecule oriented along the x axis has two rotational degrees of freedom, corresponding to rotation about the y and z axes. (b) Allowed rotational energies of a diatomic molecule calculated with Equation 43.6, where $E_1 = \hbar^2/2I$.

At the Active Figures link at http://www.pse6.com, you can adjust the distance between the atoms and choose the initial rotational energy state of the molecule. Then observe transitions of the molecule to lower energy states.

---

[2] Equation 43.5 is similar to Equation 42.27 for orbital angular momentum in an atom. In fact, the relationship between the magnitude of the angular momentum of a system and the associated quantum number is the same as it is in these equations for *any* system that exhibits rotation, as long as the potential energy function for the system is spherically symmetric.

kinetic energy of the molecule:

$$E_{rot} = \tfrac{1}{2} I \omega^2 = \frac{1}{2I}(I\omega)^2 = \frac{L^2}{2I} = \frac{\left(\sqrt{J(J+1)}\,\hbar\right)^2}{2I}$$

$$E_{rot} = \frac{\hbar^2}{2I} J(J+1) \qquad J = 0, 1, 2, \ldots \qquad (43.6)$$

Allowed values of rotational energy

This result shows that **the rotational energy of the molecule is quantized and depends on its moment of inertia.** The allowed rotational energies of a diatomic molecule are plotted in Figure 43.5b. As the quantum number J goes up, the states become farther apart, as displayed earlier for rotational energy levels in Figure 21.8.

For most molecules, transitions between adjacent rotational energy levels result in radiation that lies in the microwave range of frequencies ($f \sim 10^{11}$ Hz). When a molecule absorbs a microwave photon, the molecule jumps from a lower rotational energy level to a higher one. The allowed rotational transitions of linear molecules are regulated by the selection rule $\Delta J = \pm 1$. Given this selection rule, all absorption lines in the spectrum of a linear molecule correspond to energy separations equal to $E_J - E_{J-1}$, where $J = 1, 2, 3, \ldots$. From Equation 43.6, we see that the energies of the allowed transitions are given by the condition

$$\Delta E = E_J - E_{J-1} = \frac{\hbar^2}{2I}[J(J+1) - (J-1)J]$$

$$\Delta E = \frac{\hbar^2}{I}J = \frac{h^2}{4\pi^2 I}J \qquad J = 1, 2, 3, \ldots \qquad (43.7)$$

Energy separation between adjacent rotational levels

where J is the rotational quantum number of the higher energy state. Because $\Delta E = hf$, where f is the frequency of the absorbed photon, we see that the allowed frequency for the transition $J = 0$ to $J = 1$ is $f_1 = h/4\pi^2 I$. The frequency corresponding to the $J = 1$ to $J = 2$ transition is $2f_1$, and so on. These predictions are in excellent agreement with the observed frequencies.

The wavelengths and frequencies for the microwave absorption spectrum of the carbon monoxide molecule are given in Table 43.1. From these data, we can evaluate the moment of inertia and bond length of the molecule (Example 43.1).

Quick Quiz 43.2 A gas of diatomic molecules is absorbing electromagnetic radiation over a wide range of frequencies. Molecule 1 is in the $J = 0$ rotation state and makes a transition to the $J = 1$ state. Molecule 2 is in the $J = 2$ state and makes a transition to the $J = 3$ state. The ratio of the frequency of the photon that excited molecule 2 to that of the photon that excited molecule 1 is (a) 1 (b) 2 (c) 3 (d) 4 (e) impossible to determine.

Table 43.1

Several Rotational Transitions of the CO Molecule		
Rotational Transition	Wavelength of Absorbed Photon (m)	Frequency of Absorbed Photon (Hz)
$J = 0 \rightarrow J = 1$	2.60×10^{-3}	1.15×10^{11}
$J = 1 \rightarrow J = 2$	1.30×10^{-3}	2.30×10^{11}
$J = 2 \rightarrow J = 3$	8.77×10^{-4}	3.46×10^{11}
$J = 3 \rightarrow J = 4$	6.50×10^{-4}	4.61×10^{11}

From G. M. Barrows, *The Structure of Molecules,* New York, W. A. Benjamin, 1963.

Example 43.1   Rotation of the CO Molecule

The $J = 0$ to $J = 1$ rotational transition of the CO molecule occurs at a frequency of 1.15×10^{11} Hz.

(A) Use this information to calculate the moment of inertia of the molecule.

Solution From Equation 43.7, we see that the energy difference between the $J = 0$ and $J = 1$ rotational levels is $h^2/4\pi^2 I$. Equating this ΔE value to the energy of the absorbed photon, we have

$$\Delta E = \frac{h^2}{4\pi^2 I} = hf$$

Solving for I gives

$$I = \frac{h}{4\pi^2 f} = \frac{6.626 \times 10^{-34}\,\text{J} \cdot \text{s}}{4\pi^2 (1.15 \times 10^{11}\,\text{s}^{-1})}$$

$$= \boxed{1.46 \times 10^{-46}\,\text{kg} \cdot \text{m}^2}$$

(B) Calculate the bond length of the molecule.

Solution We can use Equation 43.3 to calculate the bond length, but we first need to know the value for the reduced mass μ of the CO molecule:

$$\mu = \frac{m_1 m_2}{m_1 + m_2} = \frac{(12\,\text{u})(16\,\text{u})}{12\,\text{u} + 16\,\text{u}} = 6.86\,\text{u}$$

$$\mu = (6.86\,\text{u})\left(\frac{1.66 \times 10^{-27}\,\text{kg}}{1\,\text{u}}\right) = 1.14 \times 10^{-26}\,\text{kg}$$

where we have used the fact that $1\,\text{u} = 1.66 \times 10^{-27}$ kg.

Substituting this value and the result of part (A) into Equation 43.3 and solving for r, we obtain

$$r = \sqrt{\frac{I}{\mu}} = \sqrt{\frac{1.46 \times 10^{-46}\,\text{kg} \cdot \text{m}^2}{1.14 \times 10^{-26}\,\text{kg}}}$$

$$= 1.13 \times 10^{-10}\,\text{m} = \boxed{0.113\,\text{nm}}$$

What If? What if another photon of frequency 1.15×10^{11} Hz is incident on the CO molecule while it is in the $J = 1$ state? What happens?

Answer Because the rotational quantum states are not equally spaced in energy, the $J = 1$ to $J = 2$ transition does not have the same energy as the $J = 0$ to $J = 1$ transition. Thus, the molecule will *not* be excited to the $J = 2$ state. Two possibilities are available. First, the photon could pass by the molecule with no interaction. Second, the photon could induce a stimulated emission, similar to that for atoms and discussed in Section 42.9. In this case, the molecule makes a transition back to the $J = 0$ state, and the original photon and a second identical photon leave the scene of the interaction.

🌀 *At the Interactive Worked Example link at* **http://www.pse6.com,** *practice evaluating the moment of inertia and the bond length from spectroscopic data of diatomic molecules.*

Vibrational Motion of Molecules

If we consider a molecule to be a flexible structure in which the atoms are bonded together by "effective springs" as shown in Figure 43.6a, we can model the molecule as a simple harmonic oscillator, as long as the atoms in the molecule are not too far from their equilibrium positions. Figure 43.6b shows a plot of potential energy versus atomic separation for the molecule in Figure 43.6a, where r_0 is the equilibrium atomic separation. For separations close to r_0, the shape of the potential energy curve closely

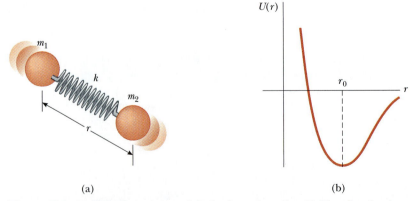

🌀 *At the Active Figures link at* **http://www.pse6.com,** *adjust the spring constant and choose the initial vibrational energy state of the molecule. Then observe transitions of the molecule to lower energy states in Figure 43.7.*

(a) (b)

Active Figure 43.6 (a) Effective-spring model of a diatomic molecule. The vibration is along the molecular axis. (b) Plot of the potential energy of a diatomic molecule versus atomic separation distance, where r_0 is the equilibrium separation distance of the atoms. Compare with Figure 15.12, page 464.

resembles a parabola. Recall from Section 15.3 that the potential energy function for a simple harmonic oscillator is indeed parabolic, varying as the square of the displacement from equilibrium. (See Equation 15.20.)

According to classical mechanics, the frequency of vibration for the system shown in Figure 43.6a is given by Equation 15.14:

$$f = \frac{1}{2\pi} \sqrt{\frac{k}{\mu}} \qquad (43.8)$$

where k is the effective spring constant and μ is the reduced mass given by Equation 43.4.

Quantum mechanics predicts that a molecule will vibrate in quantized states as described in Section 41.8. The vibrational motion and quantized vibrational energy can be altered if the molecule acquires energy of the proper value to cause a transition between quantized vibrational states. As discussed in Section 41.8, the allowed vibrational energies are

$$E_{\text{vib}} = (v + \tfrac{1}{2})hf \qquad v = 0, 1, 2, \ldots \qquad (43.9)$$

where v is an integer called the **vibrational quantum number.** (We used n in Section 41.8 for a general harmonic oscillator, but v is often used for the quantum number when discussing molecular vibrations.) If the system is in the lowest vibrational state, for which $v = 0$, its ground-state energy is $\tfrac{1}{2}hf$. The accompanying vibration is always present, even if the molecule is not excited. In the first excited state, $v = 1$ and the vibrational energy is $\tfrac{3}{2}hf$, and so on.

Substituting Equation 43.8 into Equation 43.9, we obtain the following expression for the allowed vibrational energies:

$$E_{\text{vib}} = (v + \tfrac{1}{2})\frac{h}{2\pi}\sqrt{\frac{k}{\mu}} \qquad v = 0, 1, 2, \ldots \qquad (43.10)$$

The selection rule for the allowed vibrational transitions is $\Delta v = \pm 1$. From Equation 43.10, we see that the energy difference between any two successive vibrational levels is

$$\Delta E_{\text{vib}} = \frac{h}{2\pi}\sqrt{\frac{k}{\mu}} = hf \qquad (43.11)$$

where f is the frequency of the oscillator.

The vibrational energies of a diatomic molecule are plotted in Figure 43.7. At ordinary temperatures, most molecules have vibrational energies corresponding to the $v = 0$ state because the spacing between vibrational states is much greater than $k_{\text{B}}T$ where k_{B} is Boltzmann's constant and T is the temperature. The molecules are not thermally excited into higher states, as discussed in Example 41.7.

Transitions between vibrational levels are caused by absorption of photons in the infrared region of the spectrum. The photon frequencies corresponding to the $v = 0$ to $v = 1$ transition are listed in Table 43.2 for several diatomic molecules, together with the force constants of the effective springs holding the molecules together. The latter values were calculated using Equation 43.11. The "stiffness" of a bond can be measured by the size of the effective force constant k.

Allowed values of vibrational energy

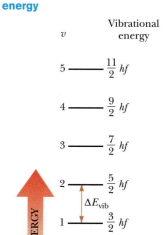

Active Figure 43.7 Allowed vibrational energies of a diatomic molecule, where f is the frequency of vibration of the molecule, given by Equation 43.8. The spacings between adjacent vibrational levels are equal if the molecule behaves as a harmonic oscillator.

At the Active Figures link at http://www.pse6.com, *adjust the spring constant in Figure 43.6 and choose the initial vibrational energy state of the molecule. Then observe transitions of the molecule to lower energy states.*

Quick Quiz 43.3 A gas of diatomic molecules is absorbing electromagnetic radiation over a wide range of frequencies. Molecule 1, initially in the $v = 0$ vibrational state, makes a transition to the $v = 1$ state. Molecule 2, initially in the $v = 2$ state, makes a transition to the $v = 3$ state. The ratio of the frequency of the photon that excited molecule 2 to that of the photon that excited molecule 1 is (a) 1 (b) 2 (c) 3 (d) 4 (e) impossible to determine.

Table 43.2

Photon Frequency and Effective Spring Force Constant for $v = 0$ to $v = 1$ Transition in Some Diatomic Molecules		
Molecule	**Photon Frequency (Hz)**	**Force Constant k (N/m)**
HF	8.72×10^{13}	970
HCl	8.66×10^{13}	480
HBr	7.68×10^{13}	410
HI	6.69×10^{13}	320
CO	6.42×10^{13}	1 850
NO	5.63×10^{13}	1 530

From G. M. Barrows, *The Structure of Molecules,* New York, W. A. Benjamin, 1963. The k values were calculated from Equation 43.11.

Example 43.2 Vibration of the CO Molecule Interactive

The frequency of the photon that causes the $v = 0$ to $v = 1$ transition in the CO molecule is 6.42×10^{13} Hz. We will ignore any changes in the rotational energy for this example.

(A) Calculate the force constant k for this molecule.

Solution We can use Equation 43.11 and the value $\mu = 1.14 \times 10^{-26}$ kg that we calculated in Example 43.1b:

$$\frac{h}{2\pi}\sqrt{\frac{k}{\mu}} = hf$$

$$k = 4\pi^2 \mu f^2$$

$$= 4\pi^2 (1.14 \times 10^{-26} \text{ kg})(6.42 \times 10^{13} \text{ s}^{-1})^2$$

$$= \boxed{1.85 \times 10^3 \text{ N/m}}$$

(B) What is the maximum classical amplitude of vibration for this molecule in the $v = 0$ vibrational state?

Solution The maximum elastic potential energy in the molecule is $\frac{1}{2}kA^2$, where A is the classical amplitude of

vibration. Equating this maximum energy to the vibrational energy given by Equation 43.10 with $v = 0$, we have

$$\frac{1}{2}kA^2 = \frac{h}{4\pi}\sqrt{\frac{k}{\mu}}$$

$$A = \sqrt{\frac{h}{2\pi}}\left(\frac{1}{\mu k}\right)^{1/4}$$

Substituting the value for k from part (A) and the value for μ gives

$$A = \sqrt{\frac{6.626 \times 10^{-34}\,\text{J}\cdot\text{s}}{2\pi}}$$

$$\times \left(\frac{1}{(1.14 \times 10^{-26}\text{ kg})(1.85 \times 10^3 \text{ N/m})}\right)^{1/4}$$

$$= 4.79 \times 10^{-12} \text{ m} = \boxed{0.004\ 79 \text{ nm}}$$

Comparing this result with the bond length of 0.113 nm we calculated in Example 43.1b, we see that the classical amplitude of vibration is about 4% of the bond length.

At the Interactive Worked Example link at **http://www.pse6.com,** *practice evaluating the effective force constant and the classical vibration amplitude from spectroscopic data of diatomic molecules.*

Molecular Spectra

In general, a molecule vibrates and rotates simultaneously. To a first approximation, these motions are independent of each other, and so the total energy of the molecule for these motions is the sum of Equations 43.6 and 43.9:

$$E = (v + \tfrac{1}{2})hf + \frac{\hbar^2}{2I}J(J + 1) \tag{43.12}$$

The energy levels of any molecule can be calculated from this expression, and each level is indexed by the two quantum numbers v and J. From these calculations, an energy-level diagram like the one shown in Figure 43.8a can be constructed. For each allowed value of the vibrational quantum number v, there is a complete set of rotational

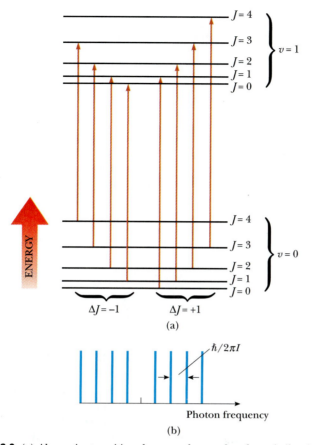

Active Figure 43.8 (a) Absorptive transitions between the $v = 0$ and $v = 1$ vibrational states of a diatomic molecule. The transitions obey the selection rule $\Delta J = \pm 1$ and fall into two sequences, those for $\Delta J = +1$ and those for $\Delta J = -1$. The transition energies are given by Equations 43.13 and 43.14. Compare the energy levels in this figure with those in Figure 21.8 on page 653. (b) Expected lines in the absorption spectrum of a molecule. The lines to the right of the center mark correspond to transitions in which J changes by $+1$; the lines to the left of the center mark correspond to transitions for which J changes by -1. These same lines appear in the emission spectrum.

At the Active Figures link at http://www.pse6.com, adjust the spring constant and the moment of inertia of the molecule, and observe the effect on the energy levels and the spectral lines.

levels corresponding to $J = 0, 1, 2, \ldots$. The energy separation between successive rotational levels is much smaller than the separation between successive vibrational levels. As noted earlier, most molecules at ordinary temperatures are in the $v = 0$ vibrational state; these molecules can be in various rotational states, as Figure 43.8a shows.

When a molecule absorbs a photon with the appropriate energy, the vibrational quantum number v increases by one unit while the rotational quantum number J either increases or decreases by one unit, as can be seen in Figure 43.8. Therefore, the molecular absorption spectrum consists of two groups of lines: one group to the right of center and satisfying the selection rules $\Delta J = +1$ and $\Delta v = +1$, and the other group to the left of center and satisfying the selection rules $\Delta J = -1$ and $\Delta v = +1$.

The energies of the absorbed photons can be calculated from Equation 43.12:

$$\Delta E = hf + \frac{\hbar^2}{I}(J + 1) \qquad J = 0, 1, 2, \ldots \qquad (\Delta J = +1) \qquad \textbf{(43.13)}$$

$$\Delta E = hf - \frac{\hbar^2}{I}J \qquad J = 1, 2, 3, \ldots \qquad (\Delta J = -1) \qquad \textbf{(43.14)}$$

where J is the rotational quantum number of the *initial* state. Equation 43.13 generates the series of equally spaced lines *higher* than the frequency f, whereas Equation 43.14 generates the series *lower* than this frequency. Adjacent lines are separated in frequency by the fundamental unit $\hbar/2\pi I$. Figure 43.8b shows the expected frequencies in the absorption spectrum of the molecule; these same frequencies appear in the emission spectrum.

Figure 43.9 Absorption spectrum of the HCl molecule. Each line is split into a doublet because the sample contains two chlorine isotopes that have different masses and therefore different moments of inertia.

The experimental absorption spectrum of the HCl molecule shown in Figure 43.9 follows this pattern very well and reinforces our model. However, one peculiarity is apparent: each line is split into a doublet. This doubling occurs because two chlorine isotopes (see Section 44.1) were present in the sample used to obtain this spectrum. Because the isotopes have different masses, the two HCl molecules have different values of I.

The intensity of the spectral lines in Figure 43.9 follows an interesting pattern, rising first as one moves away from the central gap (at about 8.65×10^{13} Hz, corresponding to the forbidden $J = 0$ to $J = 0$ transition) and then falling. This intensity is determined by a product of two functions of J. The first function corresponds to the number of available states for a given value of J. This function is $2J + 1$, corresponding to the number of values of m_J, the molecular rotation analog to m_ℓ for atomic states. For example, the $J = 2$ state has five substates with five values of m_J ($m_J = -2, -1, 0, 1, 2$), while the $J = 1$ state has only three substates ($m_J = -1, 0, 1$). Thus, on the average and without regard for the second function described below, $\frac{5}{3}$ as many molecules will make the transition from the $J = 2$ state as from the $J = 1$ state.

The second function determining the envelope of the intensity of the spectral lines is the Boltzmann factor, introduced in Section 21.5. The number of molecules in an excited rotational state is given by

$$n = n_0 e^{-\hbar^2 J(J+1)/(2Ik_BT)}$$

where n_0 is the number of molecules in the $J = 0$ state.

Multiplying these factors together indicates that the intensity of spectral lines should be described by a function of J as follows:

Intensity variation in the vibration-rotation spectrum of a molecule

$$I \propto (2J + 1)e^{-\hbar^2 J(J+1)/(2Ik_BT)} \tag{43.15}$$

The factor $(2J + 1)$ increases with J while the exponential second factor decreases. The product of the two factors gives a behavior that describes very well the envelope of the spectral lines in Figure 43.9.

The excitation of rotational and vibrational energy levels is an important consideration in current models for the phenomenon of global warming. For CO_2 molecules, most of the absorption lines are in the infrared portion of the spectrum. Thus, visible light from the Sun is not absorbed by atmospheric CO_2 but instead strikes the Earth's surface, warming it. In turn, the surface of the Earth, being at a much lower temperature than the Sun, emits thermal radiation that peaks in the infrared portion of the electromagnetic spectrum (Section 40.1). This infrared radiation is absorbed by the CO_2 molecules in the air instead of radiating out into space. Thus, atmospheric CO_2 acts like a one-way valve for energy from the Sun and is responsible, along with some other atmospheric molecules, for raising the temperature of the Earth's surface above its value in the absence of an atmosphere. This phenomenon is commonly called the "greenhouse effect." The burning of fossil fuels in today's industrialized society adds more CO_2 to the atmosphere. Many scientists fear that this will increase the absorption of infrared radiation, raising the Earth's temperature further, and may cause substantial climatic changes.

Quick Quiz 43.4 In Figure 43.9, there is a gap between the two sets of peaks. This is because (a) there is no pair of energy levels that are separated by an energy corresponding to this frequency (b) the transition between energy levels that are separated by an energy corresponding to this frequency is not allowed (c) molecules cannot rotate and vibrate simultaneously.

Quick Quiz 43.5 A spectrum like that in Figure 43.9 is taken when the HCl gas is at a much higher temperature. Compared to Figure 43.9, in this new spectrum, the absorption peak with the highest intensity (a) is the same as the one in Figure 43.9 (b) is farther from the gap than in Figure 43.9 (c) is closer to the gap than in Figure 43.9.

Conceptual Example 43.3   Comparing Figures 43.8 and 43.9

In Figure 43.8a, the transitions indicated correspond to spectral lines that are equally spaced, as shown in Figure 43.8b. The actual spectrum in Figure 43.9, however, shows lines that move closer together as the frequency increases. Why does the spacing of the actual spectral lines differ from the diagram in Figure 43.8?

Solution In Figure 43.8, we modeled the rotating diatomic molecule as a rigid object (Chapter 10). In reality, however, as the molecule rotates faster and faster, the effective spring in Figure 43.6a stretches, in order to provide the increased force associated with the larger centripetal acceleration of each atom. As the molecule stretches along its length, its moment of inertia I increases. Thus, in Equation 43.12, the rotational part of the energy expression has an extra dependence on J in the moment of inertia I. Because the increasing moment of inertia is in the denominator, as J increases, the energies do not increase as rapidly with J as indicated in Equation 43.12. With each higher energy level being lower than indicated by Equation 43.12, the energy associated with a transition to that level is smaller, as is the frequency of the absorbed photon. This destroys the even spacing of the spectral lines and gives the uneven spacing seen in Figure 43.9.

43.3   Bonding in Solids

A crystalline solid consists of a large number of atoms arranged in a regular array, forming a periodic structure. Two of the bonding mechanisms described in Section 43.1–ionic and covalent–are appropriate for describing bonds in solids. For example, the ions in the NaCl crystal are ionically bonded, as already noted, and the carbon atoms in the crystal that we call diamond form covalent bonds with one another. The metallic bond described at the end of this section is responsible for the cohesion of copper, silver, sodium, and other solid metals.

Ionic Solids

Many crystals are formed by ionic bonding, in which the dominant interaction between ions is through the Coulomb force. Consider the NaCl crystal in Figure 43.10. Each Na^+ ion has six nearest-neighbor Cl^- ions, and each Cl^- ion has six nearest-neighbor Na^+ ions. Each Na^+ ion is attracted to its six Cl^- neighbors. The corresponding potential

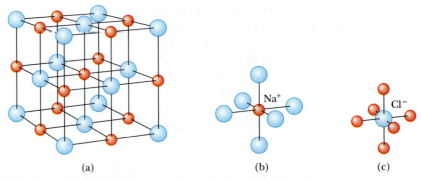

Figure 43.10 (a) Crystalline structure of NaCl. (b) Each positive sodium ion (orange spheres) is surrounded by six negative chloride ions (blue spheres). (c) Each chloride ion is surrounded by six sodium ions.

energy is $-6k_e e^2/r$, where k_e is the Coulomb constant and r is the separation distance between each Na$^+$ and Cl$^-$. In addition, there are 12 next nearest-neighbor Na$^+$ ions at a distance of $\sqrt{2}\,r$ from the Na$^+$ ion, and these 12 positive ions exert weaker repulsive forces on the central Na$^+$. Furthermore, beyond these 12 Na$^+$ ions are more Cl$^-$ ions that exert an attractive force, and so on. The net effect of all these interactions is a resultant negative electric potential energy

$$U_{\text{attractive}} = -\alpha k_e \frac{e^2}{r} \tag{43.16}$$

where α is a dimensionless number known as the **Madelung constant.** The value of α depends only on the particular crystalline structure of the solid. For example, $\alpha = 1.747\,6$ for the NaCl structure. When the constituent ions of a crystal are brought close together, a repulsive force exists because of electrostatic forces and the exclusion principle, as discussed in Section 43.1. The potential energy term B/r^m in Equation 43.1 accounts for this repulsive force. We do not include neighbors other than nearest neighbors here because the repulsive forces occur only for ions that are very close together. (Electron shells must overlap for exclusion-principle effects to become important.) Therefore, we can express the total potential energy of the crystal as

$$U_{\text{total}} = -\alpha k_e \frac{e^2}{r} + \frac{B}{r^m} \tag{43.17}$$

where m in this expression is some small integer.

A plot of total potential energy versus ion separation distance is shown in Figure 43.11. The potential energy has its minimum value U_0 at the equilibrium separation, when $r = r_0$. It is left as a problem for you (Problem 53) to show that

$$U_0 = -\alpha k_e \frac{e^2}{r_0} \left(1 - \frac{1}{m} \right) \tag{43.18}$$

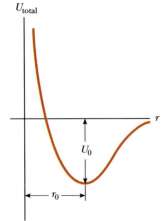

Figure 43.11 Total potential energy versus ion separation distance for an ionic solid, where U_0 is the ionic cohesive energy and r_0 is the equilibrium separation distance between ions.

This minimum energy U_0 is called the **ionic cohesive energy** of the solid, and its absolute value represents the energy required to separate the solid into a collection of isolated positive and negative ions. Its value for NaCl is -7.84 eV per ion pair.

To calculate the **atomic cohesive energy,** which is the binding energy relative to the energy of the neutral atoms, we must add 5.14 eV to the ionic cohesive energy value to account for the transition from Na$^+$ to Na, and we must subtract 3.61 eV to account for the conversion of Cl$^-$ to Cl. Therefore, the atomic cohesive energy of NaCl is

$$-7.84 \text{ eV} + 5.14 \text{ eV} - 3.61 \text{ eV} = -6.31 \text{ eV}$$

In other words, it requires 6.31 eV of energy per ion pair to separate the solid into isolated neutral atoms of Na and Cl.

Ionic crystals have the following general properties:

- They form relatively stable, hard crystals.

- They are poor electrical conductors because they contain no free electrons. This means that each electron in the solid is bound tightly to one of the ions, so it is not sufficiently mobile to carry current.

- They have high melting points. (The melting point of NaCl is 801°C.)

- They are transparent to visible radiation but absorb strongly in the infrared region. No visible light is absorbed because the shells formed by the electrons in ionic solids are so tightly bound that visible radiation does not possess sufficient energy to promote electrons to the next allowed shell. Infrared radiation is absorbed strongly because the vibrations of the ions have natural resonant frequencies in the low-energy infrared region.

- Many are quite soluble in polar liquids, such as water. The polar solvent molecules exert an attractive electric force on the charged ions, which breaks the ionic bonds and dissolves the solid.

Covalent Solids

Solid carbon, in the form of diamond, is a crystal whose atoms are covalently bonded. Because atomic carbon has the electronic configuration $1s^2 2s^2 2p^2$, it is four electrons short of filling its $n = 2$ shell, which can accommodate eight electrons. Hence, two carbon atoms have a strong attraction for each other, with a cohesive energy of 7.37 eV.

In the diamond structure, each carbon atom is covalently bonded to four other carbon atoms located at four corners of a cube, as shown in Figure 43.12a. To form such a configuration of bonds, one $2s$ electron of each atom must be promoted to the $2p$ subshell so that the electronic configuration becomes $1s^2 2s^1 2p^3$, which corresponds to a half-filled p subshell. The promotion of this electron requires an energy input of about 4 eV.

The crystalline structure of diamond is shown in Figure 43.12b. Note that each carbon atom forms covalent bonds with four nearest-neighbor atoms. The basic structure of diamond is called tetrahedral (each carbon atom is at the center of a regular tetrahedron), and the angle between the bonds is 109.5°. Other crystals such as silicon and germanium have the same structure.

A cylinder of nearly pure crystalline silicon (Si), approximately 25 cm long. Such crystals are cut into wafers and processed to make various semiconductor devices.

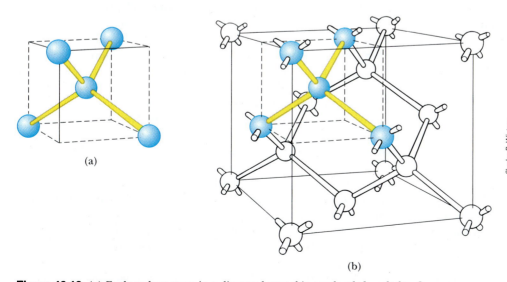

(a)

(b)

Figure 43.12 (a) Each carbon atom in a diamond crystal is covalently bonded to four other carbon atoms and forms a tetrahedral structure. (b) The crystal structure of diamond, showing the tetrahedral bond arrangement.

Charles D. Winters

Figure 43.13 Computer rendering of a "buckyball," short for the molecule buckminsterfullerene. These nearly spherical molecular structures that look like soccer balls were named for the inventor of the geodesic dome. This form of carbon, C_{60}, was discovered by astrophysicists while investigating the carbon gas that exists between stars. Scientists are actively studying the properties and potential uses of buckminsterfullerene and related molecules.

Table 43.3

Atomic Cohesive Energies of Some Covalent Solids	
Solid	Cohesive Energy (eV per ion pair)
C (diamond)	7.37
Si	4.63
Ge	3.85
InAs	5.70
SiC	6.15
ZnS	6.32
CuCl	9.24

Carbon is interesting in that it can form several different types of structures. In addition to the diamond structure, it forms graphite, with completely different properties. In this form, the carbon atoms form flat layers with hexagonal arrays of atoms. There is a very weak interaction between the layers. This allows the layers to be removed easily under friction, as occurs in the graphite used in pencil lead.

Carbon atoms can also form a large hollow structure; in this case, the compound is called **buckminsterfullerene** after the famous architect R. Buckminster Fuller, who invented the geodesic dome. The unique shape of this molecule (Fig. 43.13) provides a "cage" to hold other atoms or molecules. Related structures, called "buckytubes" because of their long, narrow cylindrical arrangements of carbon atoms, may provide the basis for extremely strong, yet lightweight materials.

The atomic cohesive energies of some covalent solids are given in Table 43.3. The large energies account for the hardness of covalent solids. Diamond is particularly hard and has an extremely high melting point (about 4 000 K). Covalently bonded solids are usually very hard, have high bond energies and high melting points, and are good electrical insulators.

Metallic Solids

Metallic bonds are generally weaker than ionic or covalent bonds. The outer electrons in the atoms of a metal are relatively free to move throughout the material, and the number of such mobile electrons in a metal is large. The metallic structure can be viewed as a "sea" or a "gas" of nearly free electrons surrounding a lattice of positive ions (Fig. 43.14). The bonding mechanism in a metal is the attractive force between the entire collection of positive ions and the electron gas. Metals have a cohesive energy in the range of 1 to 3 eV per atom, which is less than the cohesive energies of ionic or covalent solids.

Light interacts strongly with the free electrons in metals. Hence, visible light is absorbed and re-emitted quite close to the surface of a metal, which accounts for the shiny nature of metal surfaces. In addition to the high electrical conductivity of metals produced by the free electrons, the nondirectional nature of the metallic bond allows many different types of metal atoms to be dissolved in a host metal in varying amounts. The resulting *solid solutions,* or *alloys,* may be designed to have particular properties, such as tensile strength, ductility, electrical and thermal conductivity, and resistance to corrosion. Such properties are usually controllable and in many cases predictable.

Because the bonding in metals is between all of the electrons and all of the positive ions, metals tend to bend when stressed. This is in contrast to nonmetallic solids, which tend to fracture when stressed. This latter property is due to the fact that bonding in nonmetallic solids is primarily with nearest-neighbor ions or atoms. When the distortion causes sufficient stress between some set of nearest neighbors, fracture occurs.

Metal ion

Electron gas

Figure 43.14 Highly schematic diagram of a metal. The blue area represents the electron gas, and the orange circles represent the positive metal ions.

43.4 Free-Electron Theory of Metals

In the remaining sections of this chapter, we discuss various aspects of electrical conduction in solids. This important area of study has led to many technological developments, a few of which we will discuss.

In Section 27.3 we described a classical free-electron theory of electrical conduction in metals that led to Ohm's law. According to this theory, a metal is modeled as a classical gas of conduction electrons moving through a fixed lattice of ion cores. Although this theory predicts the correct functional form of Ohm's law, it does not predict the correct values of electrical and thermal conductivities.

This section introduces a quantum-based free-electron theory of metals, which remedies the shortcomings of the classical model by taking into account the wave nature of the electrons. In this model, one imagines that the outer-shell electrons are free to move through the metal but are trapped within a three-dimensional box formed by the metal surfaces. Therefore, each electron is represented as a particle in a box. As we discussed in Section 41.2, particles in a box are restricted to quantized energy levels.

Statistical physics can be applied to a collection of particles in an effort to relate microscopic properties to macroscopic properties, as we saw with kinetic theory of gases in Chapter 21. In the case of electrons, it is necessary to use *quantum statistics,* with the requirement that each state of the system can be occupied by only two electrons (one with spin up and the other with spin down) as a consequence of the exclusion principle. The probability that a particular state having energy E is occupied by one of the electrons in a solid is given by

$$f(E) = \frac{1}{e^{(E-E_F)/k_B T} + 1} \tag{43.19}$$

Fermi–Dirac distribution function

where $f(E)$ is called the **Fermi–Dirac distribution function** and E_F is called the **Fermi energy.** A plot of $f(E)$ versus E at $T = 0$ K is shown in Figure 43.15a. Note that $f(E) = 1$ for $E < E_F$ and $f(E) = 0$ for $E > E_F$. That is, at 0 K all states having energies less than the Fermi energy are occupied, and all states having energies greater than the Fermi energy are vacant. A plot of $f(E)$ versus E at some temperature $T > 0$ K is shown in Figure 43.15b. This curve shows that as T increases, the distribution rounds off slightly, with states near and below E_F losing population and states near and above E_F gaining population, due to thermal excitation. The Fermi energy E_F also depends on temperature, but the dependence is weak in metals.

Let us now follow up on our discussion of the particle in a box in Chapter 41 in order to generalize the results to a three-dimensional box. We found that if a particle of mass m is confined to move in a one-dimensional box of length L, the allowed states

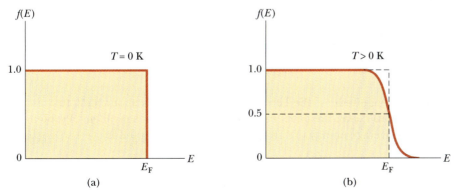

Active Figure 43.15 Plot of the Fermi–Dirac distribution function $f(E)$ versus energy at (a) $T = 0$ K and (b) $T > 0$ K. The energy E_F is the Fermi energy.

At the Active Figures link at http://www.pse6.com, *adjust the temperature and observe the effect on the Fermi–Dirac distribution function.*

have quantized energy levels given by Equation 41.12:

$$E_n = \frac{h^2}{8mL^2}n^2 = \frac{\hbar^2\pi^2}{2mL^2}n^2 \qquad n = 1, 2, 3, \ldots$$

Now imagine a piece of metal in the shape of a solid cube of sides L and volume L^3, and let us focus on one electron that is free to move anywhere in this volume. Thus, the electron is modeled as a particle in a three-dimensional box. In this model, we require that $\psi(x, y, z) = 0$ at the boundaries of the metal. It can be shown (see Problem 34) that the energy for such an electron is

$$E = \frac{\hbar^2\pi^2}{2m_eL^2}(n_x^2 + n_y^2 + n_z^2) \tag{43.20}$$

where m_e is the mass of the electron and n_x, n_y, and n_z are quantum numbers. As we expect, the energies are quantized, and each allowed value of the energy is characterized by this set of three quantum numbers (one for each degree of freedom) and the spin quantum number m_s. For example, the ground state, corresponding to $n_x = n_y = n_z = 1$, has an energy equal to $3\hbar^2\pi^2/2m_eL^2$, and can be occupied by two electrons, corresponding to spin-up and spin-down.

Because of the macroscopic size L of the box, the energy levels for the electrons are very close together. As a result, we can treat the quantum numbers as continuous variables. Under this assumption, the number of allowed states per unit volume that have energies between E and $E + dE$ is

$$g(E)\,dE = \frac{8\sqrt{2}\pi m_e^{3/2}}{h^3}E^{1/2}\,dE \tag{43.21}$$

(See Example 43.5 on page 1417.) The function $g(E)$ is called the **density-of-states function.**

For a metal in thermal equilibrium, the number of electrons per unit volume $N(E)\,dE$ that have energy between E and $E + dE$ is equal to the product of the number of allowed states and the probability that a state is occupied, that is, $N(E)\,dE = f(E)g(E)\,dE$:

$$N(E)\,dE = \frac{8\sqrt{2}\pi m_e^{3/2}}{h^3}\frac{E^{1/2}\,dE}{e^{(E-E_F)/k_BT} + 1} \tag{43.22}$$

Plots of $N(E)$ versus E for two temperatures are given in Figure 43.16.

If n_e is the total number of electrons per unit volume, we require that

$$n_e = \int_0^\infty N(E)\,dE = \frac{8\sqrt{2}\pi m_e^{3/2}}{h^3}\int_0^\infty \frac{E^{1/2}\,dE}{e^{(E-E_F)/k_BT} + 1} \tag{43.23}$$

We can use this condition to calculate the Fermi energy. At $T = 0$ K, the Fermi-Dirac distribution function $f(E) = 1$ for $E < E_F$ and $f(E) = 0$ for $E > E_F$. Therefore, at $T = 0$ K, Equation 43.23 becomes

$$n_e = \frac{8\sqrt{2}\pi m_e^{3/2}}{h^3}\int_0^{E_F} E^{1/2}\,dE = \frac{2}{3}\frac{8\sqrt{2}\pi m_e^{3/2}}{h^3}E_F^{3/2} \tag{43.24}$$

Solving for the Fermi energy at 0 K, we obtain

$$E_F(0) = \frac{h^2}{2m_e}\left(\frac{3n_e}{8\pi}\right)^{2/3} \tag{43.25}$$

The order of magnitude of the Fermi energy for metals is about 5 eV. Representative values for various metals are given in Table 43.4. It is left as a problem (Problem 33) to show that the average energy of a free electron in a metal at 0 K is

$$E_{av} = \tfrac{3}{5}E_F \tag{43.26}$$

In summary, we can consider a metal to be a system comprising a very large number of energy levels available to the free electrons. These electrons fill these levels in accordance with the Pauli exclusion principle, beginning with $E = 0$ and ending with E_F. At $T = 0$ K, all levels below the Fermi energy are filled and all levels above the

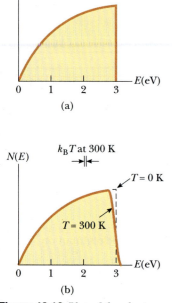

Figure 43.16 Plot of the electron distribution function versus energy in a metal at (a) $T = 0$ K and (b) $T = 300$ K. The Fermi energy E_F is 3 eV.

Fermi energy at $T = 0$ K

Table 43.4

Calculated Values of the Fermi Energy for Metals at 300 K Based on the Free-Electron Theory		
Metal	**Electron Concentration (m^{-3})**	**Fermi Energy (eV)**
Li	4.70×10^{28}	4.72
Na	2.65×10^{28}	3.23
K	1.40×10^{28}	2.12
Cu	8.49×10^{28}	7.05
Ag	5.85×10^{28}	5.48
Au	5.90×10^{28}	5.53

Fermi energy are empty. At 300 K, a very small fraction of the free electrons are excited above the Fermi energy. In Section 43.6, we will combine our understanding of the Fermi energy with the information that we study in Section 43.5 on band theory of solids to explain the high electrical conductivity of metals relative to other materials in the solid state.

Quick Quiz 43.6 In a piece of silver, free-electron energy levels are measured near 2 eV and near 6 eV. (The Fermi energy for silver is 5.48 eV.) Near which of these energies are the energy levels closer together? (a) 2 eV (b) 6 eV (c) The spacing of energy levels is the same near both energies.

Quick Quiz 43.7 Consider the situation in Quick Quiz 43.6 again. Near which of these energies are there more electrons occupying quantum states? (a) 2 eV (b) 6 eV (c) The number of electrons is the same near both energies.

Example 43.4 The Fermi Energy of Gold

Each atom of gold (Au) contributes one free electron to the metal. Compute the Fermi energy for gold.

Solution The concentration of free electrons in gold is 5.90×10^{28} m^{-3} (see Table 43.4). Substitution of this value into Equation 43.25 gives

$$E_F(0) = \frac{h^2}{2m_e}\left(\frac{3n_e}{8\pi}\right)^{2/3}$$

$$= \frac{(6.626 \times 10^{-34}\,\text{J·s})^2}{2(9.11 \times 10^{-31}\,\text{kg})}\left(\frac{3(5.90 \times 10^{28}\,\text{m}^{-3})}{8\pi}\right)^{2/3}$$

$$= 8.85 \times 10^{-19}\,\text{J} = \boxed{5.53\ \text{eV}}$$

Example 43.5 Deriving Equation 43.21

Based on the allowed states of a particle in a three-dimensional box, derive Equation 43.21.

Solution We begin by conceptualizing a particle confined to a three-dimensional box, subject to boundary conditions in three dimensions. We categorize this problem as that of a quantum system in which the energies of the particle are quantized. As noted previously, the allowed states of the particle in a three-dimensional box are described by three quantum numbers n_x, n_y, and n_z. To analyze the problem, imagine a three-dimensional quantum number space whose axes represent n_x, n_y, and n_z. The allowed states in this space can be represented as dots located at integral values of the three quantum numbers, as in Figure 43.17.

Equation 43.20 can be written as

$$(1) \qquad n_x^2 + n_y^2 + n_z^2 = \frac{E}{E_0} = n^2$$

where

$$(2) \qquad E_0 = \frac{\hbar^2 \pi^2}{2m_e L^2}$$

and $n = (E/E_0)^{1/2}$. In this space, Equation (1) is the equation of a sphere of radius n. Thus, the number of allowed states having energies between E and $E + dE$ is equal to the number of points in a spherical shell of radius n and thickness dn. The "volume" of this shell is

$$(3) \qquad G(E)\,dE = \tfrac{1}{8}(4\pi n^2\,dn) = \tfrac{1}{2}\pi n^2\,dn$$

where we have taken one eighth of the total volume because we are restricted to the octant of a three-dimensional space in which all three quantum numbers are positive. We use $G(E)$ here to represent the total number of states, while

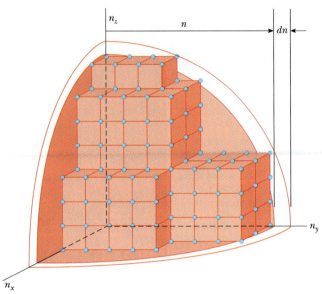

Figure 43.17 (Example 43.5) The allowed states of particles in a three-dimensional box can be represented by dots (blue circles) in a quantum number space. This is not traditional space in which a location is specified by coordinates x, y, and z. It is a space in which allowed states can be specified by coordinates representing the quantum numbers. The dots representing the allowed states are located at integer values of n_x, n_y, and n_z and are therefore at the corners of cubes with sides of "length" 1. The number of allowed states having energies between E and $E + dE$ corresponds to the number of dots in the spherical shell of radius n and thickness dn.

$g(E)$ represents the number of states per unit volume V (in normal space). Thus, $g(E) = G(E)/V$.

Now, we replace n in Equation (3) with its equivalent in terms of E using the relation $n^2 = E/E_0$ from Equation (1):

$$G(E)\, dE = \tfrac{1}{2}\, \pi \left(\frac{E}{E_0} \right) d\left[\left(\frac{E}{E_0} \right)^{1/2} \right]$$

$$= \tfrac{1}{2}\, \pi \left(\frac{E}{E_0} \right) E_0^{-1/2} \tfrac{1}{2} E^{-1/2}\, dE = \tfrac{1}{4}\, \pi E_0^{-3/2} E^{1/2}\, dE$$

Substituting for E_0 from Equation (2) gives

$$G(E)\, dE = \tfrac{1}{4}\, \pi \left(\frac{\hbar^2 \pi^2}{2m_e L^2} \right)^{-3/2} E^{1/2}\, dE$$

$$= \frac{\sqrt{2}}{2}\, \frac{m_e^{3/2} L^3}{\pi^2 \hbar^3} E^{1/2}\, dE$$

Recognizing L^3 as the volume V of the cubical box in normal space, and recalling that $g(E) = G(E)/V$, we find

$$g(E)\, dE = \frac{G(E)}{V}\, dE = \frac{\sqrt{2}}{2}\, \frac{m_e^{3/2}}{\pi^2 \hbar^3} E^{1/2}\, dE$$

Substituting $\hbar = h/2\pi$ gives

$$g(E)\, dE = \frac{4\sqrt{2}\, \pi m_e^{3/2}}{h^3} E^{1/2}\, dE$$

Finally, we multiply by 2 for the two possible spin states in each particle-in-a-box state:

$$g(E)\, dE = \frac{8\sqrt{2}\, \pi m_e^{3/2}}{h^3} E^{1/2}\, dE$$

To finalize this problem, note that this is Equation 43.21, which is what we set out to derive.

43.5 Band Theory of Solids

In the preceding section, we modeled the electrons in a metal as particles free to move around inside a three-dimensional box and we paid no attention to the influence of the parent atoms. In this section, we make the model more sophisticated by incorporating the contribution of the parent atoms that form the crystal.

Recall from Section 41.1 that the probability density $|\psi|^2$ for a system is physically significant while the probability amplitude ψ is not. Let us consider as an example an atom that has a single s electron outside of a closed shell. Both of the following wave functions are valid for such an atom with atomic number Z:

$$\psi_s^+(r) = +Af(r)\, e^{-Zr/na_0} \qquad \psi_s^-(r) = -Af(r)\, e^{-Zr/na_0}$$

where A is the normalization constant and $f(r)$ is a function[3] of r that varies with the value of n. Choosing either of these wave functions leads to the same value of $|\psi|^2$, so both choices are equivalent. A difference arises, however, when we combine two atoms.

If two identical atoms are very far apart, they do not interact and their electronic energy levels can be considered to be those of isolated atoms. Suppose that the two atoms

---

[3]   The functions $f(r)$ are called *Laguerre polynomials*. They can be found in the quantum treatment of the hydrogen atom in modern physics textbooks.

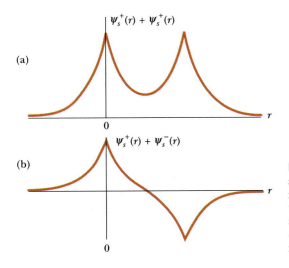

(a)

$|\psi_s^+(r) + \psi_s^+(r)$

0 r

(b)

$\psi_s^+(r) + \psi_s^-(r)$

0 r

Figure 43.18 The wave functions of two atoms combine to form a composite wave function for the two-atom system when the atoms are close together. (a) Two atoms with wave functions $\psi_s^+(r)$ combine. (b) Two atoms with wave functions $\psi_s^+(r)$ and $\psi_s^-(r)$ combine.

are sodium, each having a lone $3s$ electron that is in a well-defined quantum state. As the two sodium atoms are brought closer together, their wave functions begin to overlap, as we discussed for covalent bonding in Section 43.1. The properties of the combined system differ depending on whether we combine two atoms with wave functions $\psi_s^+(r)$, as in Figure 43.18a, or one with wave function $\psi_s^+(r)$ and the other with $\psi_s^-(r)$, as in Figure 43.18b. (The choice of two atoms with wave function $\psi_s^-(r)$ is physically equivalent to that with two positive wave functions, so we do not consider it separately.) When two wave functions $\psi_s^+(r)$ are combined, the result is a composite wave function in which the probability amplitudes add between the atoms. If $\psi_s^+(r)$ combines with $\psi_s^-(r)$, however, the wave functions between the nuclei subtract. Thus, the composite probability amplitudes for the two possibilities are different. **These two possible combinations of wave functions represent two possible states of the two-atom system.** We interpret these curves as representing the probability amplitude of finding an electron. The positive–positive curve shows some probability of finding the electron at the midpoint between the atoms. The positive–negative function shows no such probability. A state with a high probability of an electron *between* two positive nuclei (resulting in a covalent binding force from the Coulomb attraction of the electron on both nuclei) must have a different energy than a state with a high probability of the electron being elsewhere! Thus, the states are *split* into two energy levels due to the two ways of combining the wave functions. The energy difference is relatively small, so the two states are close together on an energy scale.

Figure 43.19a shows this splitting effect as a function of separation distance. For large separations r, the electron clouds do not overlap and there is no splitting. As the

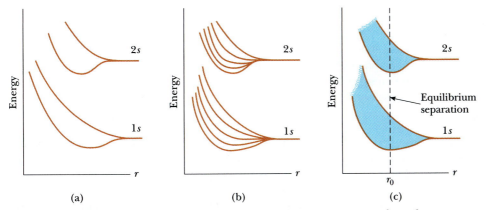

(a)   (b)   (c)

Figure 43.19 (a) Splitting of the $1s$ and $2s$ levels when two sodium atoms are brought together. (b) Splitting of the $1s$ and $2s$ levels when five sodium atoms are brought together. (c) Formation of energy bands when a large number of sodium atoms are assembled to form a solid.

atoms are brought closer so that r decreases, the electron clouds overlap so that we need to consider the system of two atoms.

When a large number of atoms are brought together to form a solid, a similar phenomenon occurs. The individual wave functions can be brought together in various combinations of $\psi_s^+(r)$ and $\psi_s^-(r)$, each possible combination corresponding to a different energy. As the atoms are brought close together, the various isolated-atom energy levels split into multiple energy levels for the composite system. This splitting in levels for five atoms in close proximity is shown in Figure 43.19b. In this case, there are five energy levels corresponding to five different combinations of isolated-atom wave functions.

If we extend this argument to the large number of atoms found in solids (on the order of 10^{23} atoms per cm^3), we obtain a large number of levels of varying energy so closely spaced that they may be regarded as a continuous **band** of energy levels, as shown in Figure 43.19c. In the case of sodium, it is customary to refer to the continuous distributions of allowed energy levels as s bands because the bands originate from the s levels of the individual sodium atoms.

In general, a crystalline solid has a large number of allowed energy bands that arise from the various atomic energy levels. Figure 43.20 shows the allowed energy bands of sodium. Note that energy gaps, corresponding to *forbidden energies,* occur between the allowed bands. In addition, some bands exhibit sufficient spreading in energy that there is an overlap between bands arising from different quantum states ($3s$ and $3p$).

The $1s$, $2s$, and $2p$ bands of sodium are each full of electrons, as indicated by the blue-shaded areas in Figure 43.20. This is because the $1s$, $2s$, and $2p$ states of each atom are full. An energy level in which the orbital angular momentum is ℓ can hold $2(2\ell + 1)$ electrons. The factor 2 arises from the two possible electron spin orientations, and the factor $2\ell + 1$ corresponds to the number of possible orientations of the orbital angular momentum. The capacity of each band for a system of N atoms is $2(2\ell + 1)N$ electrons. Therefore, the $1s$ and $2s$ bands each contain $2N$ electrons ($\ell = 0$), and the $2p$ band contains $6N$ electrons ($\ell = 1$). Because sodium has only one $3s$ electron and there are a total of N atoms in the solid, the $3s$ band contains only N electrons and is partially full, as indicated by the blue coloring in Figure 43.20. The $3p$ band, which is the higher region of the overlapping bands, is completely empty (all gold in the figure).

Band theory allows us to build simple models in order to understand the behavior of conductors, insulators, and semiconductors, as well as that of semiconductor devices, as we see in the following sections.

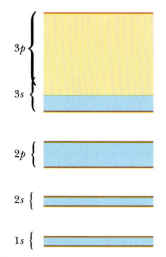

Figure 43.20 Energy bands of a sodium crystal. Note the energy gaps (white regions) between the allowed bands; electrons cannot occupy states that lie in these gaps. Blue represents energy bands occupied by the sodium electrons when the atom is in its ground state. Gold represents energy bands that are empty.

43.6 Electrical Conduction in Metals, Insulators, and Semiconductors

Good electrical conductors contain a high density of free charge carriers, and the density of free charge carriers in insulators is nearly zero. **Semiconductors** are a class of technologically important materials in which charge-carrier densities are intermediate between those of insulators and those of conductors. In this section, we discuss the mechanisms of conduction in these three classes of materials in terms of a model based on energy bands.

Metals

If a material is to be a good electrical conductor, the charge carriers in the material must be free to move in response to an applied electric field. Let us consider the electrons in a metal as the charge carriers we shall investigate. The motion of the electrons in response to an electric field represents an increase in energy of the system of the metal lattice and the free electrons corresponding to the additional kinetic energy of the moving electrons. Therefore, when an electric field is applied to a conductor, electrons must move upward to an available higher energy state on an energy-level diagram.

Figure 43.21 shows a half-filled band in a metal at $T = 0$ K, where the blue region represents levels filled with electrons. Because electrons obey Fermi–Dirac

Metal

Figure 43.21 Half-filled band of a metal, an electrical conductor. At $T = 0$ K, the Fermi energy lies in the middle of the band.

statistics, all levels below the Fermi energy are filled with electrons and all levels above the Fermi energy are empty. The Fermi energy lies in the band at the highest filled state. At temperatures slightly greater than 0 K, some electrons are thermally excited to levels above E_F, but overall there is little change from the 0 K case. However, **if a potential difference is applied to the metal, electrons having energies near the Fermi energy require only a small amount of additional energy from the applied field to reach nearby empty energy states above the Fermi energy.** Therefore, electrons in a metal experiencing only a small applied electric field are free to move because there are many empty levels available close to the occupied energy levels. The model of metals based on band theory shows that metals are excellent electrical conductors.

Insulators

Now consider the two outermost energy bands of a material in which the lower band is filled with electrons and the higher band is empty at 0 K (Fig. 43.22). The lower, filled band is called the **valence band,** and the upper, empty band is the **conduction band.** (The conduction band is the one that is partially filled in a metal.) It is common to refer to the energy separation between the valence and conduction bands as the **energy gap** E_g of the material. The Fermi energy lies somewhere in the energy gap,[4] as shown in Figure 43.22.

Suppose a material has a relatively large energy gap, for example, about 5 eV. At 300 K (room temperature), $k_B T = 0.025$ eV, which is much smaller than the energy gap. At such temperatures, the Fermi–Dirac distribution predicts that very few electrons are thermally excited into the conduction band. There are no available states that lie close in energy and into which electrons can move upward in order to account for the extra kinetic energy associated with motion through the material in response to an electric field. Consequently, the electrons do not move—the material is an insulator. Although an insulator has many vacant states in its conduction band that can accept electrons, these states are separated from the filled states by a large energy gap. Only a few electrons occupy these states, so the overall electrical conductivity of insulators is very small.

Semiconductors

Semiconductors have the same type of band structure as an insulator but the energy gap is much smaller—on the order of 1 eV. Table 43.5 shows the energy gaps for some representative materials. The band structure of a semiconductor is shown in Figure 43.23. Because the Fermi level is located near the middle of the gap for a semiconductor and because E_g is small, appreciable numbers of electrons are thermally excited from the valence band to the conduction band. There are many empty levels above the thermally filled levels in the conduction band; therefore, a small applied potential difference can easily raise the energy of the electrons in the conduction band, resulting in a moderate current.

At $T = 0$ K, all electrons in these materials are in the valence band, and no energy is available to excite them across the energy gap. Thus, semiconductors are poor conductors at very low temperatures. Because the thermal excitation of electrons across the narrow gap is more probable at higher temperatures, the conductivity of semiconductors increases rapidly with temperature. This contrasts sharply with the conductivity of metals, which decreases slowly with increasing temperature.

Charge carriers in a semiconductor can be negative or positive, or both. When an electron moves from the valence band into the conduction band, it leaves behind a vacant site, called a **hole,** in the otherwise filled valence band. This hole

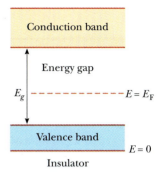

Figure 43.22 An electrical insulator at $T = 0$ K has a filled valence band and an empty conduction band. The Fermi level lies somewhere between these bands in the region known as the energy gap.

Table 43.5

Energy-Gap Values for Some Semiconductors[a]		
	E_g(eV)	
Crystal	0 K	300 K
Si	1.17	1.14
Ge	0.744	0.67
InP	1.42	1.35
GaP	2.32	2.26
GaAs	1.52	1.43
CdS	2.582	2.42
CdTe	1.607	1.45
ZnO	3.436	3.2
ZnS	3.91	3.6

[a] From C. Kittel, *Introduction to Solid State Physics*, 5th ed., New York, John Wiley & Sons, 1976.

Figure 43.23 Band structure of a semiconductor at ordinary temperatures ($T \approx 300$ K). The energy gap is much smaller than in an insulator, and some electrons from the valence band occupy states in the conduction band.

---

[4] We defined the Fermi energy as the energy of the highest filled state at $T = 0$. This might suggest that the Fermi energy should be at the top of the valence band in Figure 43.22. A more sophisticated general treatment of the Fermi energy, however, shows that it is located at that energy at which the probability of occupation is $\frac{1}{2}$ (see Fig. 43.15b). According to this definition, the Fermi energy lies in the energy gap between the bands.

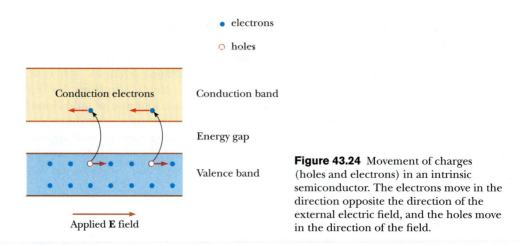

Figure 43.24 Movement of charges (holes and electrons) in an intrinsic semiconductor. The electrons move in the direction opposite the direction of the external electric field, and the holes move in the direction of the field.

(electron-deficient site) acts as a charge carrier in the sense that a free electron from a nearby site can transfer into the hole. Whenever an electron does so, it creates a new hole at the site it abandoned. Thus, the net effect can be viewed as the hole migrating through the material in the direction opposite the direction of electron movement. The hole behaves as if it were a particle with a positive charge $+ e$.

A pure semiconductor crystal containing only one element or one compound is called an **intrinsic semiconductor.** In such a semiconductor, there are equal numbers of conduction electrons and holes. Such combinations of charges are called **electron–hole pairs.** In the presence of an external electric field, the holes move in the direction of the field and the conduction electrons move in the direction opposite the field (Fig. 43.24). Because the electrons and holes have opposite signs, both of these motions correspond to a current in the same direction.

Quick Quiz 43.8   Consider the data on three materials in the table below:

Material	Conduction Band	E_g
A	Empty	1.2 eV
B	Half full	1.2 eV
C	Empty	8.0 eV

Identify these materials as a conductor, an insulator, or a semiconductor.

Doped Semiconductors

When impurities are added to a semiconductor, both the band structure of the semiconductor and its resistivity are modified. The process of adding impurities, called **doping,** is important in controlling the conductivity of semiconductors. For example, when an atom containing five outer-shell electrons, such as arsenic, is added to a Group IV semiconductor, four of the electrons form covalent bonds with atoms of the semiconductor and one is left over (Fig. 43.25a). This extra electron is nearly free of its parent atom and can be modeled as having an energy level that lies in the energy gap, just below the conduction band (Fig. 43.25b). Such a pentavalent atom in effect donates an electron to the structure and hence is referred to as a **donor atom.** Because the spacing between the energy level of the electron of the donor atom and the bottom of the conduction band is very small (typically, about 0.05 eV), only a small amount of thermal excitation is needed to cause this electron to move into the conduction band. (Recall that the average energy of an electron at room temperature is about $k_B T \approx 0.025$ eV.) Semiconductors doped with

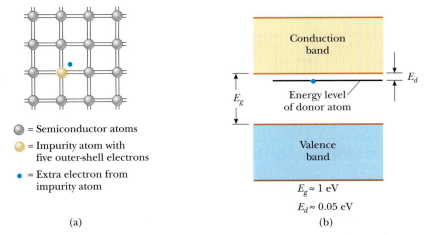

= Semiconductor atoms

= Impurity atom with
   five outer-shell electrons

= Extra electron from
   impurity atom

(a)

$E_g \approx 1 \text{ eV}$

$E_d \approx 0.05 \text{ eV}$

(b)

Figure 43.25 (a) Two-dimensional representation of a semiconductor consisting of Group IV atoms (gray) and an impurity atom (yellow) that has five outer-shell electrons. Each double line between atoms represents a covalent bond in which two electrons are shared. (b) Energy-band diagram for a semiconductor in which the nearly free electron of the impurity atom lies in the energy gap, just below the bottom of the conduction band. A small amount of energy can excite the electron into the conduction band.

donor atoms are called **n-type semiconductors** because the majority of charge carriers are electrons, which are **n**egatively charged.

If a Group IV semiconductor is doped with atoms containing three outer-shell electrons, such as indium and aluminum, the three electrons form covalent bonds with neighboring semiconductor atoms, leaving an electron deficiency—a hole—where the fourth bond would be if an impurity-atom electron were available to form it (Fig. 43.26a). This can be modeled by placing an energy level in the energy gap, just above the valence band, as in Figure 43.26b. An electron from the valence band has enough energy at room temperature to fill this impurity level, leaving behind a hole in the valence band. This hole can carry current in the presence of an electric field. Because a trivalent atom in effect accepts an electron from the valence band, such impurities are referred to as **acceptor atoms.** A semiconductor doped with trivalent (acceptor)

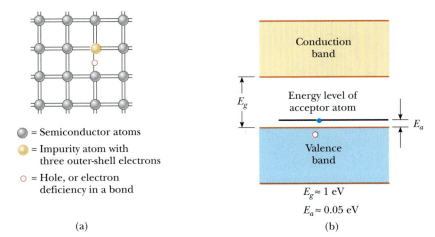

= Semiconductor atoms

= Impurity atom with
   three outer-shell electrons

= Hole, or electron
   deficiency in a bond

(a)

$E_g \approx 1 \text{ eV}$

$E_a \approx 0.05 \text{ eV}$

(b)

Figure 43.26 (a) Two-dimensional representation of a semiconductor consisting of Group IV atoms (gray) and an impurity atom (yellow) having three outer-shell electrons. The single line between the impurity atom and the semiconductor atom below it represents the fact that there is only one electron shared in this bond. (b) Energy-band diagram for a semiconductor in which the hole resulting from the trivalent impurity atom lies in the energy gap, just above the top of the valence band. This diagram shows an electron excited into the energy level of the acceptor atom, leaving a hole in the valence band.

impurities is known as a **p-type semiconductor** because the majority of charge carriers are **p**ositively charged holes.

When conduction in a semiconductor is the result of acceptor or donor impurities, the material is called an **extrinsic semiconductor.** The typical range of doping densities for extrinsic semiconductors is 10^{13} to 10^{19} cm^{-3}, whereas the electron density in a typical semiconductor is roughly 10^{21} cm^{-3}.

43.7 Semiconductor Devices

The electronics of the first half of the twentieth century was based on vacuum tubes, in which electrons pass through empty space between a cathode and an anode. The number of electrons in a vacuum tube can be controlled by a small voltage applied to a grid electrode, which repels electrons in their path to the anode. We have seen vacuum tubes in Figure 23.27 (the cathode-ray tube), Figure 29.5 (television picture tube), Figure 29.20 (circular electron beam), Figure 29.25 (Thomson's apparatus for measuring e/m_e for the electron), and Figure 40.9 (photoelectric effect apparatus).

The transistor was invented in 1948, leading to a shift away from vacuum tubes and toward semiconductors as the basis of electronic devices. This phase of electronics has been underway for several decades. As discussed in Chapter 41, there may be a new phase of electronics in the near future using nanotechnological devices employing quantum dots and other nanoscale structures.

In this section, we discuss electronic devices based on semiconductors, which are still in wide use and will be for many years to come.

The Junction Diode

A fundamental unit of a semiconductor device is formed when a p-type semiconductor is joined to an n-type semiconductor to form a **p–n junction.** A **junction diode** is a device that is based on a single p–n junction. The role of a diode of any type is to pass current in one direction but not the other. Thus, it acts as a one-way valve for current.

Figure 43.27 (a) Physical arrangement of a p–n junction. (b) Internal electric field versus x for the p–n junction. (c) Internal electric potential difference ΔV versus x for the p–n junction. The potential difference ΔV_0 represents the potential difference across the junction in the absence of an applied electric field.

The p–n junction is shown in Figure 43.27a. We can identify three distinct regions: a p region, an n region, and a small area that extends several micrometers to either side of the interface, called a *depletion region*.

The depletion region may be visualized as arising when the two halves of the junction are brought together. The mobile n-side donor electrons nearest the junction (deep-blue area in Fig. 43.27a) diffuse to the p side and fill holes located there, leaving behind immobile positive ions. While this happens, we can model the holes that are being filled as diffusing to the n side, leaving behind a region (brown area in Fig. 43.27a) of fixed negative ions.

Because the two sides of the depletion region each carry a net charge, an internal electric field on the order of 10^4 to 10^6 V/cm exists in the depletion region (see Fig. 43.27b). This field produces an electric force on any remaining mobile charge carriers that sweeps them out of the depletion region. The depletion region is so named because it is depleted of mobile charge carriers. This internal electric field creates an internal potential difference ΔV_0 that prevents further diffusion of holes and electrons across the junction and thereby ensures zero current in the junction when no potential difference is applied.

The operation of the junction as a diode is easiest to understand in terms of the potential-difference graph shown in Figure 43.27c. If a voltage ΔV is applied to the junction such that the p side is connected to the positive terminal of a voltage source, as shown in Figure 43.28a, the internal potential difference ΔV_0 across the junction is decreased; the decrease results in a current that increases exponentially with increasing forward voltage, or *forward bias*. For *reverse bias* (where the n side of the junction is connected to the positive terminal of a voltage source), the internal potential difference ΔV_0 increases with increasing reverse bias; the increase results in a very small reverse current that quickly reaches a saturation value I_0. The current–voltage relationship for an ideal diode is

$$I = I_0 \left(e^{e \Delta V / k_B T} - 1 \right) \qquad (43.27)$$

where the first e is the base of the natural logarithm, the second e represents the magnitude of the electron charge, k_B is Boltzmann's constant, and T is the absolute temperature. Figure 43.28b shows a circuit diagram for a diode under forward bias and Figure 43.28c shows an I–ΔV plot characteristic of a real p–n junction, demonstrating the diode behavior.

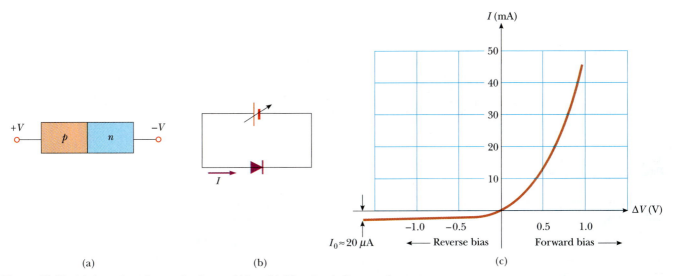

(a)   (b)   (c)

Figure 43.28 (a) A p–n junction under forward bias. (b) The circuit diagram for a diode under forward bias, showing a battery with an adjustable voltage. Both positive and negative voltages can be applied to the diode to study its nonlinear behavior. (c) The characteristic curve for a real p–n junction.

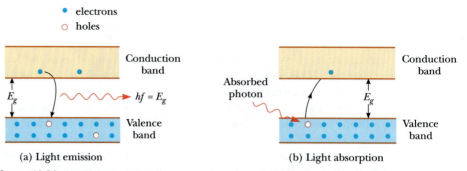

Figure 43.29 (a) Light emission from a semiconductor. (b) Light absorption by a semiconductor.

Light-Emitting and Light-Absorbing Diodes

Light-emitting diodes (LEDs) and semiconductor lasers are common examples of devices that depend on the behavior of semiconductors. LEDs are commonly used in traffic signals, in electronic displays, and as indicator lights for electronic equipment. Semiconductor lasers are often used for pointers in presentations and in compact disc and DVD playback equipment.

Light emission and absorption in semiconductors is similar to light emission and absorption by gaseous atoms, except that in the discussion of semiconductors, we must incorporate the concept of energy bands rather than the discrete energy levels in single atoms. As shown in Figure 43.29a, an electron excited electrically into the conduction band can easily recombine with a hole (especially if the electron is injected into a *p* region). As this recombination takes place, a photon of energy E_g is emitted. With proper design of the semiconductor and the associated plastic envelope or mirrors, the light from a large number of these transitions will serve as the source of an LED or a semiconductor laser.

Conversely, an electron in the valence band may absorb an incoming photon of light and be promoted to the conduction band, leaving a hole behind (Fig. 43.29b). This absorbed energy can be used to operate an electrical circuit. One device that operates on this principle is the photovoltaic solar cell, which appears in many handheld calculators. Arrays of solar cells are used in generating electric power in space vehicles and in remote areas on the Earth.

Quick Quiz 43.9 Figure 43.29a shows a photon of frequency $f = E_g/h$ being emitted from a semiconductor. In addition to photons of this frequency, (a) photons with no other frequencies are emitted (b) photons with lower frequencies are emitted (c) photons with higher frequencies are emitted.

Example 43.6   Where's the Remote?

Estimate the band gap of the semiconductor in the infrared LED of a typical television remote control.

Solution In Chapter 34 we learned that the wavelength of infrared light ranges from 700 nm to 1 mm. Let us pick a number that is easy to work with, such as 1 000 nm. (This is not a bad estimate. Remote controls typically operate in the range of 880 to 950 nm.)

The energy of a photon is given by $E = hc/\lambda$, and so the energy of the photons from the remote control is

$$E = \frac{hc}{\lambda} = \frac{1\ 240\ \text{eV} \cdot \text{nm}}{1\ 000\ \text{nm}} = \boxed{1.2\ \text{eV}}$$

This corresponds to an energy gap E_g of approximately 1.2 eV in the LED's semiconductor.

The Transistor

The invention of the transistor by John Bardeen (1908–1991), Walter Brattain (1902–1987), and William Shockley (1910–1989) in 1948 totally revolutionized the world of electronics. For this work, these three men shared a Nobel Prize in physics in 1956. By 1960, the transistor had replaced the vacuum tube in many electronic applications. The advent of the transistor created a multitrillion-dollar industry that produces such popular devices as portable radios, CD players, handheld calculators, computers, television receivers, wireless telephones, and electronic games.

A **junction transistor** consists of a semiconducting material in which a very narrow *n* region is sandwiched between two *p* regions or a *p* region is sandwiched between two *n* regions. In either case, the transistor is formed from two *p–n* junctions. These types of transistors were used widely in the early days of semiconductor electronics.

During the 1960s, the electronics industry converted many electronic applications from the junction transistor to the **field-effect transistor,** which is much easier to manufacture and just as effective. Figure 43.30a shows the structure of a very common device, the **MOSFET,** or **metal-oxide-semiconductor field-effect transistor.** You are likely using millions of MOSFET devices when you are working on your computer.

There are three metal connections (the M in MOSFET) to the transistor—the *source, drain,* and *gate.* The source and drain are connected to *n*-type semiconductor regions (the S in MOSFET) at either end of the structure. These regions are connected by a narrow channel of additional *n*-type material, the *n* channel. The source and drain regions and the *n* channel are embedded in a *p*-type substrate material. This forms a depletion region, as in the junction diode, along the bottom of the *n* channel. (Depletion regions also exist at the junctions underneath the source and drain regions, but we will ignore these because the operation of the device depends primarily on the behavior in the channel.)

The gate is separated from the *n* channel by a layer of insulating silicon dioxide (the O in MOSFET, for oxide). Thus, it does not make electrical contact with the rest of the semiconducting material.

Figure 43.30 (a) The structure of a metal-oxide-semiconductor field-effect transistor (MOSFET). (b) A source–drain voltage is applied, with the result that current exists in the circuit. (c) A gate voltage is applied. The gate voltage can be used to control the source–drain current, so that the MOSFET acts as an amplifier.

Imagine that a voltage source ΔV_{SD} is applied across the source and drain as shown in Figure 43.30b. With this situation, electrons will flow through the upper region of the n channel. Electrons cannot flow through the depletion region in the lower part of the n channel because this region is depleted of charge carriers. Now a second voltage ΔV_{SG} is applied across the source and gate as in Figure 43.30c. The positive potential on the gate electrode results in an electric field below the gate that is directed downward in the n channel. (This is the field in "field-effect.") This electric field exerts upward forces on electrons in the region below the gate, causing them to move into the n channel. This causes the depletion region to become smaller, widening the area through which there is current between the top of the n channel and the depletion region. As the area becomes wider, the value of the current increases.

If a varying voltage, such as that generated from music stored on a compact disc, is applied to the gate, the area through which the source–drain current exists will vary in size according to the varying gate voltage. A small variation in gate voltage results in a large variation in current and a correspondingly large voltage across the resistor in Figure 43.30c. Therefore, the MOSFET acts as a voltage amplifier. A circuit consisting of a chain of such transistors can result in a very small initial signal from a microphone being amplified enough to drive the powerful speakers at an outdoor concert.

The MOSFET can be used as a switch by reversing the potential difference ΔV_{SG} in Figure 43.30c. In this case, increasing the voltage causes the n channel to decrease in size. If this voltage is large enough to completely block the area through which electrons pass, the current falls to zero. This on–off behavior is used in many applications.

The MOSFET is used widely in current electronic equipment. As mentioned in Chapter 41, a new era of transistors based on quantum dots and other nanotechnologies may await us in the future.

The Integrated Circuit

Invented independently by Jack Kilby (b. 1923, Nobel Prize, 2000) at Texas Instruments in late 1958 and by Robert Noyce (b. 1927) at Fairchild Camera and Instrument in early 1959, the integrated circuit has been justly called "the most remarkable technology ever to hit mankind." Kilby's first device is shown in Figure 43.31. Integrated circuits have indeed started a "second industrial revolution" and are found at the heart of computers, watches, cameras, automobiles, aircraft, robots, space vehicles, and all sorts of communication and switching networks.

In simplest terms, an **integrated circuit** is a collection of interconnected transistors, diodes, resistors, and capacitors fabricated on a single piece of silicon known as a *chip*. Contemporary electronic devices often contain many integrated circuits (Fig. 43.32). State-of-the-art chips easily contain several million components within a 1-cm^2 area with the number of components per square inch having increased steadily since the integrated circuit was invented. Figure 43.33a illustrates the dramatic advances made in chip technology since Intel introduced the first microprocessor in 1971. Figure 43.33b is a

Courtesy of Texas Instruments, Inc.

Figure 43.31 Jack Kilby's first integrated circuit, tested on September 12, 1958.

Figure 43.32 Integrated circuits are prevalent in many electronic devices. All of the flat circuit elements with black-topped surfaces in this photograph are integrated circuits.

graph of the logarithm of the number of transistors in a chip as a function of the year in which the chip was introduced. Because this growth follows an approximately straight line, we conclude that that the growth is exponential. "Moore's law," proposed in 1965 by Gordon Moore, a co-founder of Intel, claims that the number of transistors per square inch on integrated circuits should double every 18 months. The doubling time in Figure 43.33b is longer than 18 months because this graph shows only the total number of transistors and does not take into account the shrinking size of the integrated circuit over the years. When the density of transistors per square centimeter is graphed, the results are similar to those predicted by Moore's law.

Integrated circuits were invented partly to solve the interconnection problem spawned by the transistor. In the era of vacuum tubes, power and size considerations of individual components set modest limits on the number of components that could be interconnected in a given circuit. With the advent of the tiny, low-power, highly reliable transistor, design limits on the number of components disappeared and were replaced by the problem of wiring together hundreds of thousands of components. The magnitude of this problem can be appreciated when we consider that second-generation computers (consisting of discrete transistors rather than integrated circuits) contained several hundred thousand components requiring more than a million joints that had to be hand-soldered and tested.

In addition to solving the interconnection problem, integrated circuits possess the advantages of miniaturization and fast response, two attributes critical for high-speed computers. The fast response results from the miniaturization and close packing of components, because the response time of a circuit depends on the time interval required for electrical signals traveling at the speed of light to pass from one component to another.

Figure 43.33 Dramatic advances in chip technology related to computer microchips manufactured by Intel. (a) A plot of the number of transistors on a single computer chip versus year of manufacture. (b) A plot of the logarithm of the number of transistors in part (a). The fact that this plot is approximately a straight line shows that the growth is exponential.

Table 43.6

Critical Temperatures for Various Superconductors	
Material	**T_c (K)**
Zn	0.88
Al	1.19
Sn	3.72
Hg	4.15
Pb	7.18
Nb	9.46
Nb_3Sn	18.05
Nb_3Ge	23.2
$YBa_2Cu_3O_7$	92
Bi–Sr–Ca–Cu–O	105
Tl–Ba–Ca–Cu–O	125
$HgBa_2Ca_2Cu_3O_8$	134

43.8 Superconductivity

We learned in Section 27.5 that there is a class of metals and compounds known as **superconductors** whose electrical resistance decreases to virtually zero below a certain temperature T_c called the *critical temperature* (Table 43.6). Let us now look at these amazing materials in greater detail, using what we have just learned about the properties of solids to help us understand the behavior of superconductors.

Let us start by examining the Meissner effect, described in Section 30.8 as the exclusion of magnetic flux from the interior of superconductors. Simple arguments based on the laws of electricity and magnetism can be used to show that the magnetic field inside a superconductor cannot change with time. According to Equation 27.8, $R = \Delta V/I$, and because the potential difference ΔV across a conductor is proportional to the electric field inside the conductor, we see that the electric field is proportional to the resistance of the conductor. Thus, because $R = 0$ for a superconductor at or below its critical temperature, *the electric field in its interior must be zero*. Now recall that Faraday's law of induction can be expressed in the form shown in Equation 31.9:

$$\oint \mathbf{E} \cdot d\mathbf{s} = -\frac{d\Phi_B}{dt} \qquad (43.28)$$

That is, the line integral of the electric field around any closed loop is equal to the negative rate of change in the magnetic flux Φ_B through the loop. Because **E** is zero everywhere inside the superconductor, the integral over any closed path inside the superconductor is zero. Hence, $d\Phi_B/dt = 0$; this tells us that **the magnetic flux in the superconductor cannot change.** From this information, we conclude that $B (= \Phi_B/A)$ must remain constant inside the superconductor.

Before 1933, it was assumed that superconductivity was a manifestation of perfect conductivity. If a perfect conductor is cooled below its critical temperature in the presence of an applied magnetic field, the field should be trapped in the interior of the conductor even after the external field is removed. In addition, the final state of the perfect conductor should depend on which occurs first, the application of the field or cooling the material below T_c. If the field is applied after the material has been cooled, the field should be expelled from the superconductor. If the field is applied before the material is cooled, the field should not be expelled once the material has been cooled. In 1933, however, W. Hans Meissner and Robert Ochsenfeld discovered that, when a metal becomes superconducting in the presence of a weak magnetic field, the field is expelled. Thus, the same final state **B** = 0 is achieved whether the field is applied before or after the material is cooled below its critical temperature.

The Meissner effect is illustrated in Figure 43.34 for a superconducting material in the shape of a long cylinder. Note that the field penetrates the cylinder when its temperature is greater than T_c (Fig. 43.34a). As the temperature is lowered to below T_c, however, the field lines are spontaneously expelled from the interior of the superconductor (Fig. 43.34b). Thus, a superconductor is more than a perfect conductor (resistivity $\rho = 0$); it is also a perfect diamagnet (**B** = 0). The property that **B** = 0 in the interior of a superconductor is as fundamental as the property of zero resistance. If the magnitude of the applied magnetic field exceeds a critical value B_c, defined as the value of B that destroys a material's superconducting properties, the field again penetrates the sample.

Because a superconductor is a perfect diamagnet having a *negative* magnetic susceptibility, it repels a permanent magnet. In fact, one can perform a demonstration of the Meissner effect by floating a small permanent magnet above a superconductor and achieving magnetic levitation, as seen in Figure 30.35 on page 952.

Recall from our study of electricity that a good conductor expels static electric fields by moving charges to its surface. In effect, the surface charges produce an electric field that exactly cancels the externally applied field inside the conductor. In a similar manner, a superconductor expels magnetic fields by forming surface currents. To see why this happens, consider again the superconductor shown in Figure 43.34. Let us assume that the sample is initially at a temperature $T > T_c$ as

$T > T_c$

$T < T_c$

i

(a)

(b)

Figure 43.34 A superconductor in the form of a long cylinder in the presence of an external magnetic field. (a) At temperatures above T_c, the field lines penetrate the cylinder because it is in its normal state. (b) When the cylinder is cooled to $T < T_c$ and becomes superconducting, magnetic flux is excluded from its interior by the induction of surface currents.

illustrated in Figure 43.34a, so that the magnetic field penetrates the cylinder. As the cylinder is cooled to a temperature $T < T_c$, the field is expelled, as shown in Figure 43.34b. Surface currents induced on the superconductor's surface produce a magnetic field that exactly cancels the externally applied field inside the superconductor. As you would expect, the surface currents disappear when the external magnetic field is removed.

A successful theory for superconductivity in metals was published in 1957 by J. Bardeen, L. N. Cooper, and J. R. Schrieffer, and is generally called BCS theory, based on the first letters of their last names. This theory led to a Nobel Prize in physics for the three scientists in 1972. In this theory, two electrons can interact via distortions in the array of lattice ions so that there is a net attractive force between the electrons.[5] As a result, the two electrons are bound into an entity called a *Cooper pair*. The Cooper pair behaves like a particle with integral spin. Particles with integral spin are called *bosons*. (As noted in Pitfall Prevention 42.6, *fermions* make up another class of particles, those with half-integral spin.) An important feature of bosons is that they do not obey the Pauli exclusion principle. Consequently, at very low temperatures, it is possible for all bosons in a collection of such particles to be in the lowest quantum state. As a result, the entire collection of Cooper pairs in the metal is described by a single wave function. Above the energy level associated with this wave function is an energy gap equal to the binding energy of a Cooper pair. Under the action of an applied electric field, the Cooper pairs experience an electric force and move through the metal. A random scattering event of a Cooper pair from a lattice ion would represent resistance to the electric current. Such a collision would change the energy of the Cooper pair because some energy would be transferred to the lattice ion. But there are no available energy levels below that of the Cooper pair (it is already in the lowest state) and none available above, because of the energy gap. As a result, collisions do not occur and there is no resistance to the movement of Cooper pairs.

An important development in physics that elicited much excitement in the scientific community was the discovery of high-temperature copper oxide–based superconductors. The excitement began with a 1986 publication by J. Georg Bednorz (b. 1950) and K. Alex Müller (b. 1927), scientists at the IBM Zurich Research Laboratory in Switzerland. In their seminal paper,[6] Bednorz and Müller reported strong evidence for superconductivity at 30 K in an oxide of barium, lanthanum, and copper. They were awarded

---

[5] A highly simplified explanation of this attraction between electrons is as follows. Around one electron, the attractive Coulomb force causes surrounding positively charged lattice ions to move inward slightly toward the electron. As a result, there is a higher concentration of positive charge in this region than elsewhere in the lattice. A second electron is attracted to the higher concentration of positive charge.

[6] J. G. Bednorz and K. A. Müller, *Z. Phys. B* 64:189, 1986.

the Nobel Prize for physics in 1987 for their remarkable discovery. Shortly thereafter, a new family of compounds was open for investigation, and research activity in the field of superconductivity proceeded vigorously. In early 1987, groups at the University of Alabama at Huntsville and the University of Houston announced superconductivity at about 92 K in an oxide of yttrium, barium, and copper ($YBa_2Cu_3O_7$). Later that year, teams of scientists from Japan and the United States reported superconductivity at 105 K in an oxide of bismuth, strontium, calcium, and copper. More recently, scientists have reported superconductivity at temperatures as high as 150 K in an oxide containing mercury. Today, one cannot rule out the possibility of room-temperature superconductivity, and the mechanisms responsible for the behavior of high-temperature superconductors are still under investigation. The search for novel superconducting materials continues both for scientific reasons and because practical applications become more probable and widespread as the critical temperature is raised.

While BCS theory was very successful in explaining superconductivity in metals, there is currently no widely accepted theory for high-temperature superconductivity. This remains an area of active research.

SUMMARY

Take a practice test for this chapter by clicking on the Practice Test link at http://www.pse6.com.

Two or more atoms combine to form molecules because of a net attractive force between the atoms. The mechanisms responsible for molecular bonding can be classified as follows:

- **Ionic bonds** form primarily because of the Coulomb attraction between oppositely charged ions. Sodium chloride (NaCl) is one example.

- **Covalent bonds** form when the constituent atoms of a molecule share electrons. For example, the two electrons of the H_2 molecule are equally shared between the two nuclei.

- **Van der Waals bonds** are weak electrostatic bonds between molecules or between atoms that do not form ionic or covalent bonds. These bonds are responsible for the condensation of inert gas atoms and nonpolar molecules into the liquid phase.

- **Hydrogen bonds** form between the center of positive charge in a polar molecule that includes one or more hydrogen atoms and the center of negative charge in another polar molecule.

The energy of a gas molecule consists of contributions from the electronic energy in the bonds and from the translational, rotational, and vibrational motions of the molecule.

The allowed values of the rotational energy of a diatomic molecule are

$$E_{\text{rot}} = \frac{\hbar^2}{2I}J(J+1) \qquad J = 0, 1, 2, \ldots \qquad (43.6)$$

where I is the moment of inertia of the molecule and J is an integer called the **rotational quantum number.** The selection rule for transitions between rotational states is given by $\Delta J = \pm 1$.

The allowed values of the vibrational energy of a diatomic molecule are

$$E_{\text{vib}} = (v + \tfrac{1}{2}) \frac{h}{2\pi} \sqrt{\frac{k}{\mu}} \qquad v = 0, 1, 2, \ldots \qquad (43.10)$$

where v is the **vibrational quantum number,** k is the force constant of the "effective spring" bonding the molecule, and μ is the **reduced mass** of the molecule. The selection rule for allowed vibrational transitions is $\Delta v = \pm 1$, and the energy difference between any two adjacent levels is the same regardless of which two levels are involved.

Bonding mechanisms in solids can be classified in a manner similar to the schemes for molecules. For example, the Na^+ and Cl^- ions in NaCl form **ionic bonds,** while the carbon atoms in diamond form **covalent bonds.** The **metallic bond** is characterized by a net attractive force between positive ion cores and the mobile free electrons of a metal.

In the **free-electron theory of metals,** the free electrons fill the quantized levels in accordance with the Pauli exclusion principle. The number of states per unit volume available to the conduction electrons having energies between E and $E + dE$ is

$$N(E)\, dE = \frac{8\sqrt{2}\,\pi m_e^{3/2}}{h^3} \frac{E^{1/2}\, dE}{e^{(E-E_F)/k_B T} + 1} \qquad (43.22)$$

where E_F is the **Fermi energy.** At $T = 0$ K, all levels below E_F are filled, all levels above E_F are empty, and

$$E_F(0) = \frac{h^2}{2m_e} \left(\frac{3n_e}{8\pi} \right)^{2/3} \qquad (43.25)$$

where n_e is the total number of conduction electrons per unit volume. Only those electrons having energies near E_F can contribute to the electrical conductivity of the metal.

In a crystalline solid, the energy levels of the system form a set of **bands.** Electrons occupy the lowest energy states, with no more than one electron per state. Energy gaps are present between the bands of allowed states.

A **semiconductor** is a material having an energy gap of approximately 1 eV and a valence band that is filled at $T = 0$ K. Because of the small energy gap, a significant number of electrons can be thermally excited from the valence band into the conduction band. The band structures and electrical properties of a Group IV semiconductor can be modified by the addition of either donor atoms containing five outer-shell electrons (such as arsenic) or acceptor atoms containing three outer-shell electrons (such as indium). A semiconductor **doped** with donor impurity atoms is called an ***n*-type semiconductor,** and one doped with acceptor impurity atoms is called a ***p*-type semiconductor.** The energy levels of these impurity atoms fall within the energy gap of the material.

QUESTIONS

Note: Questions 14 and 15 in Chapter 27 can be assigned wih this chapter.

1. Discuss the three major forms of excitation of a molecule (other than translational motion) and the relative energies associated with these three forms.

2. Explain the role of the Pauli exclusion principle in describing the electrical properties of metals.

3. Discuss the properties of a material that determine whether it is a good electrical insulator or a good conductor.

4. Table 43.5 shows that the energy gaps for semiconductors decrease with increasing temperature. What do you suppose accounts for this behavior?

5. The resistivity of metals increases with increasing temperature, whereas the resistivity of an intrinsic semiconductor decreases with increasing temperature. Explain.

6. Discuss the differences in the band structures of metals, insulators, and semiconductors. How does the band-structure model enable you to better understand the electrical properties of these materials?

7. Discuss models for the different types of bonds that form stable molecules.

8. Discuss the electrical, physical, and optical properties of ionically bonded solids. Compare your expectations with tabulated properties for such solids.

9. Discuss the electrical and physical properties of covalently bonded solids. Compare your expectations with tabulated properties for such solids.

10. Discuss the electrical and physical properties of metals.

11. When a photon is absorbed by a semiconductor, an electron–hole pair is created. Give a physical explanation of this statement using the energy-band model as the basis for your description.

12. Pentavalent atoms such as arsenic are donor atoms in a semiconductor such as silicon, while trivalent atoms such as indium are acceptors. Inspect the periodic table in Appendix C, and determine what other elements might make good donors or acceptors.

13. What are the essential assumptions made in the free-electron theory of metals? How does the energy-band model differ from the free-electron theory in describing the properties of metals?

14. How do the vibrational and rotational levels of heavy hydrogen (D_2) molecules compare with those of H_2 molecules?

15. Which is easier to excite in a diatomic molecule, rotational or vibrational motion?

16. The energies of photons of visible light range between 1.8 and 3.1 eV. Does this explain why silicon, with an energy gap of 1.1 eV (see Table 43.5), appears opaque, whereas diamond, with an energy gap of 5.5 eV, appears transparent?

17. How can the analysis of the rotational spectrum of a molecule lead to an estimate of the size of that molecule?

PROBLEMS

1, 2, 3 = straightforward, intermediate, challenging ☐ = full solution available in the *Student Solutions Manual and Study Guide*

= coached solution with hints available at http://www.pse6.com 💻 = computer useful in solving problem

= paired numerical and symbolic problems

Section 43.1 Molecular Bonds

1. **Review problem.** A K^+ ion and a Cl^- ion are separated by a distance of 5.00×10^{-10} m. Assuming the two ions act like point charges, determine (a) the force each ion exerts on the other and (b) the potential energy of the two-ion system in electron volts.

2. Potassium chloride is an ionically bonded molecule, sold as a salt substitute for use in a low-sodium diet. The electron affinity of chlorine is 3.6 eV. An energy input of 0.7 eV is required to form separate K^+ and Cl^- ions from separate K and Cl atoms. What is the ionization energy of K?

3. One description of the potential energy of a diatomic molecule is given by the Lennard–Jones potential,

$$U = \frac{A}{r^{12}} - \frac{B}{r^6}$$

where A and B are constants. Find, in terms of A and B, (a) the value r_0 at which the energy is a minimum and (b) the energy E required to break up a diatomic molecule. (c) Evaluate r_0 in meters and E in electron volts for the H_2 molecule. In your calculations, take $A = 0.124 \times 10^{-120}$ eV·m^{12} and $B = 1.488 \times 10^{-60}$ eV·m^6. (*Note:* Although this potential is widely used for modeling, it is known to have serious defects. For example, its behavior at both small and large values of r is significantly in error.)

4. Potassium iodide can be taken as a medicine to reduce radiation dosage to the thyroid gland before or after exposure to radioactive iodine. In the potassium iodide molecule, assume that the K and I atoms bond ionically by the transfer of one electron from K to I. (a) The ionization energy of K is 4.34 eV, and the electron affinity of I is 3.06 eV. What energy is needed to transfer an electron from K to I, to form K^+ and I^- ions from neutral atoms? This is sometimes called the activation energy E_a. (b) A model potential energy function for the KI molecule is the Lennard–Jones potential:

$$U(r) = 4\epsilon\left[\left(\frac{\sigma}{r}\right)^{12} - \left(\frac{\sigma}{r}\right)^6\right] + E_a$$

where r is the internuclear separation distance, and σ and ϵ are adjustable parameters. The E_a term is added to ensure the correct asymptotic behavior at large r. At the equilibrium separation distance, $r = r_0 = 0.305$ nm, $U(r)$ is a minimum, and $dU/dr = 0$. Now $U(r_0)$ is the negative of the dissociation energy: $U(r_0) = -3.37$ eV. Evaluate σ and ϵ. (c) Calculate the force needed to break up a KI molecule. (d) Calculate the force constant for small oscillations about $r = r_0$. (*Suggestion:* Set $r = r_0 + s$ where $s/r_0 \ll 1$, and expand $U(r)$ in powers of s/r_0 up to second-order terms.)

5. A van der Waals dispersion force between helium atoms produces a very shallow potential well, with a depth on the order of 1 meV. At about what temperature would you expect helium to condense?

Section 43.2 Energy States and Spectra of Molecules

6. The cesium iodide (CsI) molecule has an atomic separation of 0.127 nm. (a) Determine the energy of the lowest excited rotational state and the frequency of the photon absorbed in the $J = 0$ to $J = 1$ transition. (b) **What If?** What would be the fractional change in this frequency if the estimate of the atomic separation is off by 10%?

7. An HCl molecule is excited to its first rotational energy level, corresponding to $J = 1$. If the distance between its nuclei is 0.127 5 nm, what is the angular speed of the molecule about its center of mass?

8. The CO molecule makes a transition from the $J = 1$ to the $J = 2$ rotational state when it absorbs a photon of frequency 2.30×10^{11} Hz. Find the moment of inertia of this molecule from these data.

9. A diatomic molecule consists of two atoms having masses m_1 and m_2 and separated by a distance r. Show that the moment of inertia about an axis through the center of mass of the molecule is given by Equation 43.3, $I = \mu r^2$.

10. (a) Calculate the moment of inertia of an NaCl molecule about its center of mass. The atoms are separated by a

distance $r = 0.28$ nm. (b) Calculate the wavelength of radiation emitted when an NaCl molecule undergoes a transition from the $J = 2$ state to the $J = 1$ state.

11. The rotational spectrum of the HCl molecule contains lines with wavelengths of 0.060 4, 0.069 0, 0.080 4, 0.096 4, and 0.120 4 mm. What is the moment of inertia of the molecule?

12. Use the data in Table 43.2 to calculate the minimum amplitude of vibration for (a) the HI molecule and (b) the HF molecule. Which has the weaker bond?

13. 🌿 Taking the effective force constant of a vibrating HCl molecule as $k = 480$ N/m, find the energy difference between the ground state and the first excited vibrational level.

14. The nuclei of the O_2 molecule are separated by 1.20×10^{-10} m. The mass of each oxygen atom in the molecule is 2.66×10^{-26} kg. (a) Determine the rotational energies of an oxygen molecule in electron volts for the levels corresponding to $J = 0$, 1, and 2. (b) The effective force constant k between the atoms in the oxygen molecule is 1 177 N/m. Determine the vibrational energies (in electron volts) corresponding to $v = 0$, 1, and 2.

15. Figure P43.15 is a model of a benzene molecule. All atoms lie in a plane, and the carbon atoms form a regular hexagon, as do the hydrogen atoms. The carbon atoms are 0.110 nm apart center-to-center. Determine the allowed energies of rotation about an axis perpendicular to the plane of the paper through the center point O. Hydrogen and carbon atoms have masses of 1.67×10^{-27} kg and 1.99×10^{-26} kg, respectively.

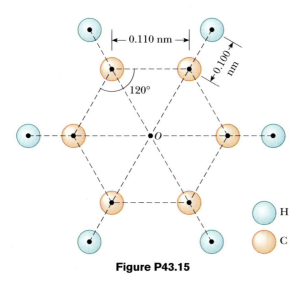

Figure P43.15

16. Calculate the longest wavelength in the rotational spectrum of HCl. Take the Cl atom to be the isotope ^{35}Cl. The equilibrium separation of the H and Cl atoms is 0.127 46 nm. The atomic mass of the H atom is 1.007 825 u, and that of the ^{35}Cl atom is 34.968 853 u. (b) **What If?** Repeat the calculation in (a), but take the Cl atom to be the isotope ^{37}Cl, which has atomic mass 36.965 903 u. The equilibrium separation distance is the same as in (a). (c) Naturally occurring chlorine contains approximately three parts of ^{35}Cl to one part of ^{37}Cl. Because of the two different Cl

masses, each line in the microwave rotational spectrum of HCl is split into a doublet. Calculate the doublet separation for the longest wavelength. (Figure 43.9 shows the doublets in the infrared vibrational spectrum).

17. If the CO molecule were rigid, the rotational transition into what J state would absorb the same wavelength photon as the 0 to 1 vibrational transition? (Use information given in Table 43.2.)

18. Calculate the moment of inertia of an HCl molecule from its infrared absorption spectrum shown in Figure 43.9.

19. An H_2 molecule is in its vibrational and rotational ground states. It absorbs a photon of wavelength 2.211 2 μm and jumps to the $v = 1, J = 1$ energy level. It then drops to the $v = 0, J = 2$ energy level, while emitting a photon of wavelength 2.405 4 μm. Calculate (a) the moment of inertia of the H_2 molecule about an axis through its center of mass and perpendicular to the H–H bond (b) the vibrational frequency of the H_2 molecule (c) the equilibrium separation distance for this molecule.

20. Photons of what frequencies can be spontaneously emitted by CO molecules in the state with $v = 1$ and $J = 0$?

21. Most of the mass of an atom is in its nucleus. Model the mass distribution in a diatomic molecule as two spheres, each of radius 2.00×10^{-15} m and mass 1.00×10^{-26} kg, located at points along the x axis in Figure 43.5a, and separated by 2.00×10^{-10} m. Rotation about the axis joining the nuclei in the diatomic molecule is ordinarily ignored because the first excited state would have an energy that is too high to access. To see why, calculate the ratio of the energy of the first excited state for rotation about the x axis to the energy of the first excited state for rotation about the y axis.

Section 43.3 Bonding in Solids

22. Use a magnifying glass to look at the table salt that comes out of a salt shaker. Compare what you see to Figure 43.10(a). The distance between a sodium ion and a nearest-neighbor chlorine ion is 0.261 nm. (a) Make an order-of-magnitude estimate of the number N of atoms in a typical grain of salt. (b) **What If?** Suppose that you had a number of grains of salt equal to this number N. What would be the volume of this quantity of salt?

23. Use Equation 43.18 to calculate the ionic cohesive energy for NaCl. Take $\alpha = 1.747\ 6$, $r_0 = 0.281$ nm, and $m = 8$.

24. The distance between the K^+ and Cl^- ions in a KCl crystal is 0.314 nm. Calculate the distances from one K^+ ion to its nearest-neighbor K^+ ions, to its second-nearest-neighbor K^+ ions, and to its third-nearest-neighbor K^+ ions.

25. Consider a one-dimensional chain of alternating positive and negative ions. Show that the potential energy associated with one of the ions and its interactions with the rest of this hypothetical crystal is

$$U(r) = -k_e \alpha \frac{e^2}{r}$$

where the Madelung constant is $\alpha = 2 \ln 2$ and r is the interionic spacing. [*Suggestion:* Use the series expansion for $\ln(1 + x)$.]

Section 43.4   Free-Electron Theory of Metals

Section 43.5   Band Theory of Solids

26. Show that Equation 43.25 can be expressed as $E_F = (3.65 \times 10^{-19}) n_e^{2/3}$ eV where E_F is in electron volts when n_e is in electrons per cubic meter.

27. The Fermi energy for silver is 5.48 eV. Silver has a density of 10.6×10^3 kg/m^3 and an atomic mass of 108. Use this information to show that silver has one free electron per atom.

28. (a) Find the typical speed of a conduction electron in copper, taking its kinetic energy as equal to the Fermi energy, 7.05 eV. (b) How does this compare with a drift speed of 0.1 mm/s?

29. Sodium is a monovalent metal having a density of 0.971 g/cm^3 and a molar mass of 23.0 g/mol. Use this information to calculate (a) the density of charge carriers and (b) the Fermi energy.

30. When solid silver starts to melt, what is the approximate fraction of the conduction electrons that are thermally excited above the Fermi level?

31. Calculate the energy of a conduction electron in silver at 800 K, assuming the probability of finding an electron in that state is 0.950. The Fermi energy is 5.48 eV at this temperature.

32. Consider a cube of gold 1.00 mm on an edge. Calculate the approximate number of conduction electrons in this cube whose energies lie in the range 4.000 to 4.025 eV.

33. Show that the average kinetic energy of a conduction electron in a metal at 0 K is $E_{av} = \frac{3}{5}E_F$. (*Suggestion:* In general, the average kinetic energy is

$$E_{av} = \frac{1}{n_e} \int E\, N(E)\, dE$$

where n_e is the density of particles, $N(E)\, dE$ is given by Equation 43.22, and the integral is over all possible values of the energy.)

34. **Review problem.** An electron moves in a three-dimensional box of edge length L and volume L^3. The wave function of the particle is $\psi = A \sin(k_x x) \sin(k_y y) \sin(k_z z)$. Show that its energy is given by Equation 43.20,

$$E = \frac{\hbar^2 \pi^2}{2 m_e L^2} (n_x^2 + n_y^2 + n_z^2)$$

where the quantum numbers (n_x, n_y, n_z) are integers ≥ 1. (*Suggestions:* The Schrödinger equation in three dimensions may be written

$$\frac{\hbar^2}{2m}\left(\frac{\partial^2 \psi}{\partial x^2} + \frac{\partial^2 \psi}{\partial y^2} + \frac{\partial^2 \psi}{\partial z^2}\right) = (U - E)\psi$$

To confine the electron inside the box, take $U = 0$ inside and $U = \infty$ outside.)

35. (a) Consider a system of electrons confined to a three-dimensional box. Calculate the ratio of the number of allowed energy levels at 8.50 eV to the number at 7.00 eV. (b) **What If?** Copper has a Fermi energy of 7.0 eV at 300 K. Calculate the ratio of the number of occupied levels at an energy of 8.50 eV to the number at the Fermi energy. Compare your answer with that obtained in part (a).

Section 43.6   Electrical Conduction in Metals, Insulators, and Semiconductors

36. The energy gap for silicon at 300 K is 1.14 eV. (a) Find the lowest-frequency photon that will promote an electron from the valence band to the conduction band. (b) What is the wavelength of this photon?

37. Light from a hydrogen discharge tube is incident on a CdS crystal. Which spectral lines from the Balmer series are absorbed and which are transmitted?

38. A light-emitting diode (LED) made of the semiconductor GaAsP emits red light ($\lambda = 650$ nm). Determine the energy-band gap E_g in the semiconductor.

39. Most solar radiation has a wavelength of 1 μm or less. What energy gap should the material in a solar cell have in order to absorb this radiation? Is silicon appropriate (see Table 43.5)?

40. Assume you are to build a scientific instrument that is thermally isolated from its surroundings, but such that you can use an external laser to raise the temperature of a target inside it. (It might be a calorimeter, but these design criteria could apply to other devices as well.) Since you know that diamond is transparent and a good thermal insulator, you decide to use a diamond window in the apparatus. Diamond has an energy gap of 5.5 eV between its valence and conduction bands. What is the shortest laser wavelength you can use to warm the sample inside?

41. **Review problem.** Silicon is a semiconductor widely used in computer chips and other electronic devices. Its most important properties result from doping it with impurities in order to control its electrical conductivity. Phosphorus, which is adjacent to silicon in the periodic table, has five outer valence electrons as compared to four for silicon. When a phosphorus atom is substituted for a silicon atom in a crystal, four of the phosphorus valence electrons form bonds with neighboring atoms and the remaining electron is much more loosely bound. You can model the electron as free to move through the crystal lattice. The phosphorus nucleus has one more positive charge than does the silicon nucleus, however, so the extra electron provided by the phosphorus atom is attracted to this single nuclear charge $+ e$. The energy levels of the extra electron are similar to those of the electron in the Bohr hydrogen atom with two important exceptions. First, the Coulomb attraction between the electron and the positive charge on the phosphorus nucleus is reduced by a factor of $1/\kappa$ from what it would be in free space (see Eq. 26.21), so the orbit radii are greatly increased. Here κ is the dielectric constant of the crystal, with a value of 11.7 in silicon. Second, the influence of the periodic electric potential of the lattice causes the electron to move as if it had an effective mass m^*, quite different from the mass m_e of a free electron. One can use the Bohr model of hydrogen to obtain fairly accurate values for the allowed energy levels of the extra electron. These energy levels, called donor states, play an important role in semiconductor devices. Assume that $m^* = 0.220\, m_e$. Calculate the energy and the radius for an extra electron in the first Bohr orbit around a donor atom in silicon.

Section 43.7   Semiconductor Devices

> *Note:* Problem 74 in Chapter 27 can be assigned with this section.

42. For what value of the bias voltage ΔV in Equation 43.27 does (a) $I = 9.00I_0$? (b) **(What If?)** $I = -0.900I_0$? Assume $T = 300$ K.

43. The diode shown in Figure 43.28 is connected in series with a battery and a 150-Ω resistor. What battery emf is required for a current of 25.0 mA?

44. You put a diode in a microelectronic circuit to protect the system in case an untrained person installs the battery backward. In the correct forward-bias situation, the current is 200 mA with a potential difference of 100 mV across the diode at room temperature (300 K). If the battery were reversed, what would be the magnitude of the current through the diode?

45. A diode, a resistor, and a battery are connected in a series circuit. The diode is at a temperature for which $k_B T = 25.0$ meV, and the saturation value I_0 of the current is 1.00 μA. The resistance of the resistor is 745 Ω, and the battery maintains a constant potential difference between its terminals of 2.42 V. (a) Find graphically the current in the loop. Proceed as follows. On the same axes, draw graphs of the diode current I_D and the current in the wire I_W versus the voltage across the diode ΔV. Choose values of ΔV ranging from 0 to 0.250 V in steps of 0.025 V. Determine the value of ΔV at the intersection of the two graph lines, and calculate the corresponding currents I_D and I_W. Do they agree? (b) Find the ohmic resistance of the diode, which is defined as the ratio $\Delta V/I_D$. (c) Find the dynamic resistance of the diode, which is defined as the derivative $d(\Delta V)/dI_D$.

Section 43.8   Superconductivity

> *Note:* Problem 28 in Chapter 30 and Problems 76 through 79 in Chapter 32 can be assigned wih this section.

46. A thin rod of superconducting material 2.50 cm long is placed into a 0.540-T magnetic field with its cylindrical axis along the magnetic field lines. (a) Sketch the directions of the applied field and the induced surface current. (b) Find the magnitude of the surface current on the curved surface of the rod.

47. Determine the current generated in a superconducting ring of niobium metal 2.00 cm in diameter when a 0.020 0-T magnetic field directed perpendicular to the ring is suddenly decreased to zero. The inductance of the ring is 3.10×10^{-8} H.

48. *A convincing demonstration of zero resistance.* A direct and relatively simple demonstration of zero DC resistance can be carried out using the four-point probe method. The probe shown in Figure P43.48 consists of a disk of $YBa_2Cu_3O_7$ (a high-T_c superconductor) to which four wires are attached by indium solder or some other suitable contact material. Current is maintained through

the sample by applying a DC voltage between points a and b, and it is measured with a DC ammeter. The current can be varied with the variable resistance R. The potential difference ΔV_{cd} between c and d is measured with a digital voltmeter. When the probe is immersed in liquid nitrogen, the sample quickly cools to 77 K, below the critical temperature of the material, 92 K. The current remains approximately constant, but ΔV_{cd} *drops abruptly to zero.* (a) Explain this observation on the basis of what you know about superconductors. (b) The data in Table P43.48 represent actual values of ΔV_{cd} for different values of I taken on the sample at room temperature. A 6-V battery in series with a variable resistor R supplied the current. The values of R ranged from 10 Ω to 100 Ω. The data are from one author's laboratory. Make an I-ΔV plot of the data, and determine whether the sample behaves in a linear manner. From the data obtain a value for the DC resistance of the sample at room temperature. (c) At room temperature it is found that $\Delta V_{cd} = 2.234$ mV for $I = 100.3$ mA, but after the sample is cooled to 77 K, $\Delta V_{cd} = 0$ and $I = 98.1$ mA. What do you think might cause the slight decrease in current?

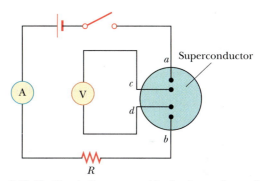

Figure P43.48 Circuit diagram used in the four-point probe measurement of the DC resistance of a sample. A DC digital ammeter is used to measure the current, and the potential difference between c and d is measured with a DC digital voltmeter. Note that there is no voltage source in the inner loop circuit where ΔV_{cd} is measured.

Table P43.48

Current Versus Potential Difference ΔV_{cd} Measured in a Bulk Ceramic Sample of $YBa_2Cu_3O_{7-\delta}$ at Room Temerature	
I (mA)	ΔV_{cd} (mV)
57.8	1.356
61.5	1.441
68.3	1.602
76.8	1.802
87.5	2.053
102.2	2.398
123.7	2.904
155	3.61

Additional Problems

49. As you will learn in Chapter 44, carbon-14 (^{14}C) is an isotope of carbon. It has the same chemical properties and electronic structure as the much more abundant isotope carbon-12 (^{12}C) but has different nuclear properties. Its mass is 14 u, greater because it has two extra neutrons in its nucleus. Assume that the CO molecular potential is the same for both isotopes of carbon, and that the tables and examples in Section 43.2 refer to carbon monoxide with carbon-12 atoms. (a) What is the vibrational frequency of ^{14}CO? (b) What is the moment of inertia of ^{14}CO? (c) What wavelengths of light can be absorbed by ^{14}CO in the ($v = 0$, $J = 10$) state that will cause it to end up in the $v = 1$ level?

50. The effective spring constant associated with bonding in the N_2 molecule is $2\,297\,N/m$. The nitrogen atoms each have a mass of 2.32×10^{-26} kg, and their nuclei are 0.120 nm apart. Assume that the molecule is rigid and in the ground vibrational state. Calculate the J value of the rotational state that has the same energy as the first excited vibrational state.

51. The hydrogen molecule comes apart (dissociates) when it is excited internally by 4.5 eV. Assuming that this molecule behaves like a harmonic oscillator having classical angular frequency $\omega = 8.28 \times 10^{14}$ rad/s, find the highest vibrational quantum number for a state below the 4.5-eV dissociation energy.

52. Under pressure, liquid helium can solidify as each atom bonds with four others, and each bond has an average energy of 1.74×10^{-23} J. Find the latent heat of fusion for helium in joules per gram. (The molar mass of He is 4.00 g/mol.)

53. Show that the ionic cohesive energy of an ionically bonded solid is given by Equation 43.18. (*Suggestion:* Start with Equation 43.17, and note that $dU/dr = 0$ at $r = r_0$.)

54. The dissociation energy of ground-state molecular hydrogen is 4.48 eV, while it only takes 3.96 eV to dissociate it when it starts in the first excited vibrational state with $J = 0$. Using this information, determine the depth of the H_2 molecular potential-energy function.

55. A particle moves in one-dimensional motion through a field for which the potential energy of the particle–field system is

$$U(x) = \frac{A}{x^3} - \frac{B}{x}$$

where $A = 0.150$ eV·nm^3 and $B = 3.68$ eV·nm. The general shape of this function is shown in Figure 43.11, where x replaces r. (a) Find the static equilibrium position x_0 of the particle. (b) Determine the depth U_0 of this potential well. (c) In moving along the x axis, what maximum force toward the negative x direction does the particle experience?

56. A particle of mass m moves in one-dimensional motion through a field for which the potential energy of the particle–field system is

$$U(x) = \frac{A}{x^3} - \frac{B}{x}$$

where A and B are constants with appropriate units. The general shape of this function is shown in Figure 43.11, where x replaces r. (a) Find the static equilibrium position x_0 of the particle in terms of m, A, and B. (b) Determine the depth U_0 of this potential well. (c) In moving along the x axis, what maximum force toward the negative x direction does the particle experience?

57. As an alternative to Equation 43.1, another useful model for the potential energy of a diatomic molecule is the Morse potential

$$U(r) = B[e^{-a(r-r_0)} - 1]^2$$

where B, a, and r_0 are parameters used to adjust the shape of the potential and its depth. (a) What is the equilibrium separation of the nuclei? (b) What is the depth of the potential well, i.e., the difference in energy between the potential's minimum value and its asymptote as r approaches infinity? (c) If μ is the reduced mass of the system of two nuclei, what is the vibrational frequency of the diatomic molecule in its ground state? (Assume that the potential is nearly parabolic about the well minimum). (d) What amount of energy needs to be supplied to the ground-state molecule to separate the two nuclei to infinity?

58. The Fermi-Dirac distribution function can be written as

$$f(E) = \frac{1}{e^{(E-E_F)/k_B T} + 1} = \frac{1}{e^{(E/E_F - 1)T_F/T} + 1}$$

where T_F is the *Fermi temperature*, defined according to

$$k_B T_F \equiv E_F$$

Write a spreadsheet to calculate and plot $f(E)$ versus E/E_F at a fixed temperature T. Examine the curves obtained for $T = 0.1T_F$, $0.2T_F$, and $0.5T_F$.

59. The Madelung constant for sodium chloride may be found by summing an infinite alternating series of terms giving the electric potential energy between a Na^+ ion and its six nearest Cl^- neighbors, its twelve next-nearest Na^+ neighbors, and so on (Fig. 43.10a). (a) From this expression, show that the first three terms of the series yield $\alpha = 2.13$ for the NaCl structure. (b) **What If?** Does this series converge rapidly? Calculate the fourth term as a check.

Answers to Quick Quizzes

43.1 (a) van der Waals (b) ionic (c) hydrogen (d) covalent

43.2 (c). Equation 43.7 shows that the energy spacing is proportional to J, the quantum number of the higher state in the transition. Because the frequency of the absorbed photon is proportional to the energy separation of the states, the frequencies are in the same ratio as the energy separations.

43.3 (a). This is similar to Quick Quiz 43.2, except that the vibrational states are all separated by the same energy difference.

43.4 (b). The $J = 0$ to $J = 0$ transition is forbidden according to the selection rules.

43.5 (b). At a higher temperature, transitions from higher excited states will be more likely, so that peaks farther from the gap will be more intense. In Equation 43.15, at a higher temperature, the exponential factor does not decrease as rapidly with J if the temperature is higher.

43.6 (b). In Equation 43.21, the density of states increases with energy. Thus, at the higher energy of 6 eV, the states are closer together.

43.7 (a). Because the energy of 6 eV is higher than the Fermi energy, there are essentially no electrons in states near 6 eV.

43.8 A: semiconductor; B: conductor; C: insulator

43.9 (c). Photons can be emitted when electrons make a transition from anywhere in the conduction band above its bottom to anywhere in the valence band below its top. The lowest-energy transition, resulting in the lowest-frequency photon, is that shown in Figure 43.29a.

Chapter 44

Nuclear Structure

▲ *The Ice Man, discovered in 1991 when an Italian glacier melted enough to expose his remains. His possessions, particularly his tools, have shed light on the way people lived in the Bronze Age. Radioactivity was used to determine how long ago this person lived. (Paul Hanny/Gamma Liaison)*

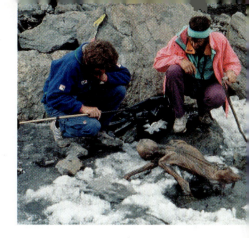

The year 1896 marks the birth of nuclear physics when the French physicist Henri Becquerel (1852–1908) discovered radioactivity in uranium compounds. This discovery prompted scientists to investigate the details of radioactivity and, ultimately, the structure of the nucleus. Pioneering work by Rutherford showed that the radiation emitted from radioactive substances is of three types: alpha, beta, and gamma rays, classified according to the nature of their electric charge and their ability to penetrate matter and ionize air. Later experiments showed that alpha rays are helium nuclei, beta rays are electrons, and gamma rays are high-energy photons.

In 1911, Rutherford, Geiger, and Marsden performed the alpha-particle scattering experiments described in Section 42.2. These experiments established that (a) the nucleus of an atom can be regarded as essentially a point mass and point charge and (b) most of the atomic mass is contained in the nucleus. Subsequent studies revealed the presence of a new type of force, the short-range nuclear force, which is predominant at particle separation distances less than approximately 10^{-14} m and is zero for large distances.

Other milestones in the development of nuclear physics include

- The observation of nuclear reactions in 1930 by Cockroft and Walton using artificially accelerated nuclei;
- The discovery of the neutron in 1932 by Chadwick and the conclusion that neutrons make up about half of the nucleus;
- The discovery of artificial radioactivity in 1933 by Irène and Frédéric Joliot-Curie;
- The discovery of nuclear fission in 1938 by Hahn and Strassmann;
- The development of the first controlled fission reactor in 1942 by Fermi and his collaborators.

In this chapter we discuss the properties and structure of the atomic nucleus. We start by describing the basic properties of nuclei, followed by a discussion of nuclear forces and binding energy, nuclear models, and the phenomenon of radioactivity. We then discuss various processes by which nuclei decay, and describe nuclear reactions.

44.1 Some Properties of Nuclei

All nuclei are composed of two types of particles: protons and neutrons. The only exception is the ordinary hydrogen nucleus, which is a single proton. We describe the atomic nucleus by the number of protons and neutrons it contains, using the following quantities:

- The **atomic number** Z, which equals the number of protons in the nucleus (sometimes called the *charge number*);

44.1 Mass Number Is Not Atomic Mass

The mass number A should not be confused with the atomic mass, defined in Section 1.3. Mass number is an integer specific to an isotope and has no units—it is simply a count of the number of nucleons. Atomic mass is generally not an integer, because it is an average of the masses of the naturally occurring isotopes of a given element, and has units of u.

- The **neutron number** N, which equals the number of neutrons in the nuclcus;
- The **mass number** $A = Z + N$, which equals the number of **nucleons** (neutrons plus protons) in the nucleus.

In representing nuclei, it is convenient to use the symbol $^{A}_{Z}X$ to show how many protons and neutrons are present, where X represents the chemical symbol of the element. For example, $^{56}_{26}Fe$ (iron) has mass number 56 and atomic number 26; therefore, it contains 26 protons and 30 neutrons. When no confusion is likely to arise, we omit the subscript Z because the chemical symbol can always be used to determine Z. Thus, $^{56}_{26}Fe$ is the same as ^{56}Fe and can also be expressed as "iron-56."

The nuclei of all atoms of a particular element contain the same number of protons but often contain different numbers of neutrons. Nuclei that are related in this way are called **isotopes.** The isotopes of an element have the same Z value but different N and A values.

The natural abundance of isotopes can differ substantially. For example $^{11}_{6}C$, $^{12}_{6}C$, $^{13}_{6}C$, and $^{14}_{6}C$ are four isotopes of carbon. The natural abundance of the $^{12}_{6}C$ isotope is approximately 98.9%, whereas that of the $^{13}_{6}C$ isotope is only about 1.1%. Some isotopes, such as $^{11}_{6}C$ and $^{14}_{6}C$, do not occur naturally but can be produced by nuclear reactions in the laboratory or by cosmic rays.

Even the simplest element, hydrogen, has isotopes: $^{1}_{1}H$, the ordinary hydrogen nucleus; $^{2}_{1}H$, deuterium; and $^{3}_{1}H$, tritium.

Quick Quiz 44.1 Consider the following three nuclei: ^{12}C, ^{13}N, ^{14}O. These nuclei have the same (a) number of protons (b) number of neutrons (c) number of nucleons.

Quick Quiz 44.2 Consider the following three nuclei: ^{12}N, ^{13}N, ^{14}N. These nuclei have the same (a) number of protons (b) number of neutrons (c) number of nucleons.

Quick Quiz 44.3 Consider the following three nuclei: ^{14}C, ^{14}N, ^{14}O. These nuclei have the same (a) number of protons (b) number of neutrons (c) number of nucleons.

Charge and Mass

The proton carries a single positive charge e, equal in magnitude to the charge $-e$ on the electron ($e = 1.6 \times 10^{-19}$ C). The neutron is electrically neutral, as its name implies. Because the neutron has no charge, it was difficult to detect with early experimental apparatus and techniques. Today we can detect neutrons relatively easily with devices such as plastic scintillators.

Nuclear masses can be measured with great precision with the use of a mass spectrometer (see Section 29.5) and by the analysis of nuclear reactions. The proton is approximately 1 836 times as massive as the electron, and the masses of the proton and the neutron are almost equal. The atomic mass unit u, first introduced in Section 1.3, is defined in such a way that the mass of one atom of the isotope ^{12}C is exactly 12 u, where 1 u $= 1.660\,539 \times 10^{-27}$ kg. According to this definition, the proton and

Table 44.1

Masses of Selected Particles in Various Units			
	Mass		
Particle	**kg**	**u**	**MeV/c^2**
Proton	$1.672\,62 \times 10^{-27}$	$1.007\,276$	938.28
Neutron	$1.674\,93 \times 10^{-27}$	$1.008\,665$	939.57
Electron	$9.109\,39 \times 10^{-31}$	$5.485\,79 \times 10^{-4}$	$0.510\,999$
$_1^1$H atom	$1.673\,53 \times 10^{-27}$	$1.007\,825$	938.783
$_2^4$He nucleus	$6.644\,66 \times 10^{-27}$	$4.001\,506$	$3\,727.38$
$_6^{12}$C atom	$1.992\,65 \times 10^{-27}$	$12.000\,000$	$11\,177.9$

neutron each have a mass of approximately 1 u, and the electron has a mass that is only a small fraction of this value:

$$\text{Mass of proton} = 1.007\,276 \text{ u}$$

$$\text{Mass of neutron} = 1.008\,665 \text{ u}$$

$$\text{Mass of electron} = 0.000\,548\,6 \text{ u}$$

One might wonder how six protons and six neutrons, each having a mass larger than 1 u, can be combined with six electrons to form a carbon-12 atom having a mass of exactly 12 u. The bound system of ^{12}C has a lower rest energy (Section 39.8) than that of six separate protons and six separate neutrons. According to Equation 39.24, $E_R = mc^2$, this lower rest energy corresponds to a smaller mass for the bound system. The extra mass of the separated particles is equal to the binding energy when the particles are combined to form the nucleus. We shall discuss this point in more detail in Section 44.3.

Because the rest energy of a particle is given by $E_R = mc^2$, it is often convenient to express the atomic mass unit in terms of its *rest energy equivalent*. For one atomic mass unit, we have

$$E_R = mc^2 = (1.660\,539 \times 10^{-27} \text{ kg})(2.997\,92 \times 10^8 \text{ m/s})^2 = 931.494 \text{ MeV}$$

where we have used the conversion 1 eV $= 1.602\,176 \times 10^{-19}$ J.

Based on the rest-energy expression in Equation 39.24, nuclear physicists often express mass in terms of the unit MeV/c^2. The masses of several nuclei and atoms are given in Table 44.1. The masses and some other properties of selected isotopes are provided in Appendix A.3.

Example 44.1   The Atomic Mass Unit

Use Avogadro's number to show that 1 u $= 1.66 \times 10^{-27}$ kg.

Solution From the definition of the mole given in Section 19.5, we know that exactly 12 g ($=1$ mol) of ^{12}C contains Avogadro's number of atoms, where $N_A = 6.02 \times 10^{23}$ atoms/mol. Thus, the mass of one carbon atom is

$$\text{Mass of one } ^{12}\text{C atom} = \frac{0.012 \text{ kg}}{6.02 \times 10^{23} \text{ atoms}}$$

$$= 1.99 \times 10^{-26} \text{ kg}$$

Because one atom of ^{12}C is defined to have a mass of 12.0 u, we find that

$$1 \text{ u} = \frac{1.99 \times 10^{-26} \text{ kg}}{12.0} = 1.66 \times 10^{-27} \text{ kg}$$

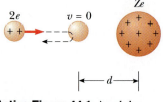

Active Figure 44.1 An alpha particle on a head-on collision course with a nucleus of charge Ze. Because of the Coulomb repulsion between the charges of the same sign, the alpha particle approaches to a distance d from the nucleus, called the distance of closest approach.

At the Active Figures link at http://www.pse6.com, adjust the atomic number of the target nucleus and the kinetic energy of the alpha particle. Then observe the approach of the alpha particle toward the nucleus.

Figure 44.2 A nucleus can be modeled as a cluster of tightly packed spheres, where each sphere is a nucleon.

Nuclear radius

The Size and Structure of Nuclei

In Rutherford's scattering experiments, positively charged nuclei of helium atoms (alpha particles) were directed at a thin piece of metallic foil. As the alpha particles moved through the foil, they often passed near a metal nucleus. Because of the positive charge on both the incident particles and the nuclei, the particles were deflected from their straight-line paths by the Coulomb repulsive force. Some particles were even deflected straight backward! These particles apparently were moving directly toward a nucleus, on a head-on collision course.

Rutherford used conservation of energy for an isolated system to find an expression for the separation distance d at which an alpha particle approaching a nucleus head-on is turned around by Coulomb repulsion. In such a head-on collision, he reasoned, the kinetic energy of the incoming particle must be converted completely to electric potential energy of the alpha-particle–nucleus system when the particle stops momentarily at the point of closest approach before moving back along the same path (Fig. 44.1). If we equate the initial and final energies of the system, we have

$$\frac{1}{2} mv^2 = k_e \frac{q_1 q_2}{r} = k_e \frac{(2e)(Ze)}{d}$$

where m is the mass of the alpha particle and Z is the atomic number of the target nucleus. Solving for d, the distance of closest approach, we obtain

$$d = \frac{4 k_e Z e^2}{m v^2}$$

From this expression, Rutherford found that the alpha particles approached nuclei to within 3.2×10^{-14} m when the foil was made of gold. Thus, the radius of the gold nucleus must be less than this value. For silver atoms, the distance of closest approach was found to be 2×10^{-14} m. From the results of his scattering experiments, Rutherford concluded that the positive charge in an atom is concentrated in a small sphere, which he called the nucleus, whose radius is no greater than about 10^{-14} m.

Because such small lengths are common in nuclear physics, an often-used convenient length unit is the femtometer (fm), which is sometimes called the **fermi** and is defined as

$$1 \text{ fm} \equiv 10^{-15} \text{ m}$$

In the early 1920s it was known that the nucleus of an atom contains Z protons and has a mass nearly equivalent to that of A protons, where on the average $A \approx 2Z$ for lighter nuclei ($Z \leq 20$) and $A > 2Z$ for heavier nuclei. To account for the nuclear mass, Rutherford proposed that each nucleus must also contain $A - Z$ neutral particles that he called neutrons. In 1932, the British physicist James Chadwick (1891–1974) discovered the neutron and was awarded the Nobel Prize for this important work.

Since the time of Rutherford's scattering experiments, a multitude of other experiments have shown that most nuclei are approximately spherical and have an average radius given by

$$r = r_0 A^{1/3} \tag{44.1}$$

where r_0 is a constant equal to 1.2×10^{-15} m and A is the mass number. Because the volume of a sphere is proportional to the cube of its radius, it follows from Equation 44.1 that the volume of a nucleus (assumed to be spherical) is directly proportional to A, the total number of nucleons. This proportionality suggests that *all nuclei have nearly the same density.* When nucleons combine to form a nucleus, they combine as though they were tightly packed spheres (Fig. 44.2). This fact has led to an analogy between the nucleus and a drop of liquid, in which the density of the drop is independent of its size. We shall discuss the liquid-drop model of the nucleus in Section 44.3.

Example 44.2   The Volume and Density of a Nucleus

For a nucleus of mass number A, find

(A) an approximate expression for the mass of the nucleus,

(B) an expression for the volume of this nucleus in terms of A, and

(C) a numerical value for its density.

Solution (A) The mass of the proton is approximately equal to that of the neutron. Thus, if the mass of one of these particles is m, the mass of the nucleus is approximately Am.

(B) Assuming the nucleus is spherical and using Equation 44.1, we find that the volume is

$$V_{\text{nucleus}} = \tfrac{4}{3}\pi r^3 = \boxed{\tfrac{4}{3}\pi r_0^3 A}$$

(C) The nuclear density is

$$\rho = \frac{m_{\text{nucleus}}}{V_{\text{nucleus}}} = \frac{Am}{\tfrac{4}{3}\pi r_0^3 A} = \frac{3m}{4\pi r_0^3}$$

Taking $r_0 = 1.2 \times 10^{-15}$ m and $m = 1.67 \times 10^{-27}$ kg, we find that

$$\rho = \frac{3(1.67 \times 10^{-27}\ \text{kg})}{4\pi(1.2 \times 10^{-15}\ \text{m})^3} = \boxed{2.3 \times 10^{17}\ \text{kg/m}^3}$$

The nuclear density is approximately 2.3×10^{14} times as great as the density of water ($\rho_{\text{water}} = 1.0 \times 10^3$ kg/m^3).

What If? What if the Earth were compressed until it had this density? How large would it be?

Answer Because this density is so large, we predict that an Earth of this density would be very small. Using Equation 1.1 and the mass of the Earth, we can find the volume of the compressed Earth:

$$V = \frac{M_E}{\rho} = \frac{5.98 \times 10^{24}\ \text{kg}}{2.3 \times 10^{17}\ \text{kg/m}^3} = 2.6 \times 10^7\ \text{m}^3$$

From this volume we find the radius:

$$V = \tfrac{4}{3}\pi r^3$$

$$r = \left(\frac{3}{4}\frac{V}{\pi}\right)^{1/3} = \left[\frac{3}{4}\left(\frac{2.6 \times 10^7\ \text{m}^3}{\pi}\right)\right]^{1/3} = 1.8 \times 10^2\ \text{m}$$

This is indeed a small Earth!

Nuclear Stability

Because the nucleus is viewed as a closely packed collection of protons and neutrons, you might be surprised that it can exist. Because charges of the same sign (the protons) in proximity exert very large repulsive Coulomb forces on each other, these forces should cause the nucleus to fly apart. Because this doesn't happen, there must be a counteracting attractive force. The **nuclear force** is a very-short-range (about 2 fm) attractive force that acts between all nuclear particles. The protons attract each other by means of the nuclear force, and, at the same time, they repel each other through the Coulomb force. The nuclear force also acts between pairs of neutrons and between neutrons and protons. The nuclear force dominates the Coulomb repulsive force within the nucleus (at short ranges), so that stable nuclei can exist.

The nuclear force is independent of charge. In other words, the forces associated with the proton–proton, proton–neutron, and neutron–neutron interactions are the same, apart from the additional repulsive Coulomb force for the proton–proton interaction.

Evidence for the limited range of nuclear forces comes from scattering experiments and from studies of nuclear binding energies. The short range of the nuclear force is shown in the neutron–proton (n–p) potential energy plot of Figure 44.3a obtained by scattering neutrons from a target containing hydrogen. The depth of the n–p potential energy well is 40 to 50 MeV, and there is a strong repulsive component that prevents the nucleons from approaching much closer than 0.4 fm.

The nuclear force does not affect electrons, enabling energetic electrons to serve as point-like probes of nuclei. The charge independence of the nuclear force also means that the main difference between the n–p and p–p interactions is that the p–p potential energy consists of a *superposition* of nuclear and Coulomb interactions as shown in Figure 44.3b. At distances less than 2 fm, both p–p and n–p potential energies are nearly identical, but for distances greater than this, the p–p potential has a positive energy barrier with a maximum at 4 fm.

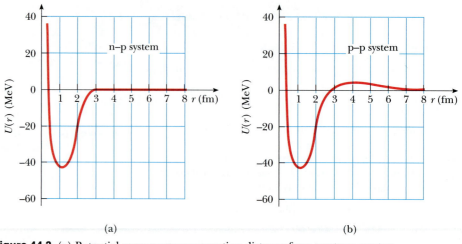

(a)   (b)

Figure 44.3 (a) Potential energy versus separation distance for a neutron–proton system. (b) Potential energy versus separation distance for a proton–proton system. The difference in the two curves is due to the large Coulomb repulsion in the case of the proton–proton interaction. The height of the peak for the proton–proton curve has been exaggerated by a factor of 10 in order to display the difference in the curves on this scale.

The general features of the nuclear force responsible for stable nuclei have been revealed in a wide variety of experiments and are summarized as follows:

- The nuclear force is attractive and is the strongest force in nature.

- It is a short-range force that falls to zero value when the separation between nucleons exceeds several fermis.

Figure 44.4 Neutron number N versus atomic number Z for stable nuclei (blue dots). These nuclei lie in a narrow band called the line of stability. The dashed line corresponds to the condition $N = Z$.

- The magnitude of the nuclear force depends on the relative spin orientations of the nucleons.

- Scattering experiments and other indirect evidence show that the nuclear force is independent of the charge of the interacting nucleons.

The existence of the nuclear force results in approximately 270 stable nuclei; hundreds of other nuclei have been observed, but they are unstable. A plot of neutron number N versus atomic number Z for a number of stable nuclei is given in Figure 44.4. The stable nuclei are represented by the blue dots, which lie in a narrow range called the *line of stability*. Note that the light stable nuclei contain an equal number of protons and neutrons; that is, $N = Z$. Also, note that in heavy stable nuclei the number of neutrons exceeds the number of protons—above $Z = 20$, the line of stability deviates upward from the line representing $N = Z$. This can be understood by recognizing that, as the number of protons increases, the strength of the Coulomb force increases, which tends to break the nucleus apart. As a result, more neutrons are needed to keep the nucleus stable because neutrons experience only the attractive nuclear force. Eventually, the repulsive Coulomb forces between protons cannot be compensated by the addition of more neutrons. This occurs when $Z = 83$, meaning that elements that contain more than 83 protons do not have stable nuclei.

44.2 Nuclear Binding Energy

As mentioned in the discussion of ^{12}C in the preceding section, the total mass of a nucleus is less than the sum of the masses of its individual nucleons. Because mass is a measure of energy, **the total energy of the bound system (the nucleus) is less than the combined energy of the separated nucleons.** This difference in energy is called the **binding energy** of the nucleus and can be interpreted as the energy that must be added to a nucleus to break it apart into its components. Therefore, in order to separate a nucleus into protons and neutrons, energy must be delivered to the system.

Conservation of energy and the Einstein mass–energy equivalence relationship show that the binding energy E_b of any nucleus of mass M_A is

$$E_b = (Zm_p + Nm_n - M_A) \times 931.494 \text{ MeV/u} \qquad (44.2)$$

Binding energy of a nucleus

where m_p is the mass of the proton, m_n is the mass of the neutron, and the masses are all expressed in atomic mass units. In practice, it is often more convenient to use the mass of neutral atoms (nuclear mass plus mass of electrons) in computing binding energy because mass spectrometers generally measure atomic masses.[1]

A plot of binding energy per nucleon E_b/A as a function of mass number A for various stable nuclei is shown in Figure 44.5. Note that the curve in Figure 44.5 peaks in the vicinity of $A = 60$. That is, nuclei having mass numbers either greater or less than 60 are not as strongly bound as those near the middle of the periodic table. The higher values of binding energy per nucleon near $A = 60$ imply that energy is released when a heavy nucleus splits, or *fissions*, into two lighter nuclei. Energy is released in fission because the nucleons in each product nucleus are more tightly bound to one another than are the nucleons in the original nucleus. The important process of fission and a second important process of *fusion*, in which energy is released as light nuclei combine, are considered in detail in Chapter 45.

Another important feature of Figure 44.5 is that the binding energy per nucleon is approximately constant at around 8 MeV per nucleon for all nuclei with $A > 50$. For these nuclei, the nuclear forces are said to be *saturated*, meaning that, in the closely packed structure shown in Figure 44.2, a particular nucleon can form attractive bonds with only a limited number of other nucleons.

⚠ PITFALL PREVENTION

44.2 Binding Energy

When separate nucleons are combined to form a nucleus, the energy of the system is reduced. Thus, the change in energy is negative. The absolute value of this change is called the binding energy. This difference in sign may be a source of confusion. For example, an *increase* in binding energy corresponds to a *decrease* in the energy of the system.

---

[1] It is possible to use atomic masses rather than nuclear masses because electron masses cancel in the calculations.

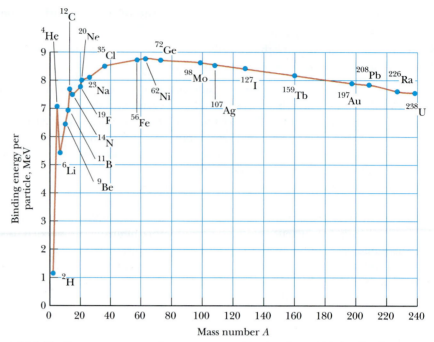

Figure 44.5 Binding energy per nucleon versus mass number for nuclei that lie along the line of stability in Figure 44.4. Some representative nuclei appear as blue dots with labels. (Nuclei to the right of ^{208}Pb are unstable. The curve represents the binding energy for the most stable isotopes.)

Figure 44.5 provides insight into fundamental questions about the origin of the chemical elements. In the early life of the Universe, the only elements that existed were hydrogen and helium. Clouds of cosmic gas coalesced under gravitational forces to form stars. As a star ages, it produces heavier elements from the lighter elements contained within it, beginning by fusing hydrogen atoms to form helium. This process continues as the star becomes older, generating atoms having larger and larger atomic numbers, up through the isotope of iron having $A = 56$, which is near the peak of the curve shown in Figure 44.5.

The nuclide $^{62}_{28}$Ni has the largest binding energy per nucleon of 8.794 5 MeV per nucleon. It takes additional energy to create elements with mass numbers larger than 62 because of their lower binding energies per nucleon. This energy comes from the supernova explosion that occurs at the end of some large stars' lives. Thus, all the heavy atoms in your body were produced from the explosions of ancient stars. You are literally made of stardust!

44.3 Nuclear Models

The details of the nuclear force are still an area of active research. Several nuclear models have been proposed, and these are useful in understanding general features of nuclear experimental data and the mechanisms responsible for binding energy. Two such models are discussed below. The liquid-drop model provides good agreement with the observed behavior of nuclear binding energy, and the shell model predicts the existence of stable isotopes.

Liquid-Drop Model

In 1936, Bohr proposed treating nucleons like molecules in a drop of liquid. In this **liquid-drop model,** the nucleons interact strongly with one another and undergo frequent collisions as they jiggle around within the nucleus. This jiggling motion is analogous to the thermally agitated motion of molecules in a drop of liquid.

Four major effects influence the binding energy of the nucleus in the liquid-drop model:

- **The volume effect.** Figure 44.5 shows that, for $A > 50$, the binding energy per nucleon is approximately constant. This indicates that the nuclear force on a given nucleon is due only to a few nearest neighbors and not to all the other nucleons in the nucleus. On the average, then, the binding energy associated with the nuclear force for each nucleon is the same in all nuclei—that associated with an interaction with a few neighbors. This tells us that the total binding energy of the nucleus is proportional to A and therefore proportional to the nuclear volume. The contribution to the binding energy of the entire nucleus is $C_1 A$, where C_1 is an adjustable constant that we can determine by fitting the prediction of the model to experimental results.

- **The surface effect.** Because nucleons on the surface of the drop have fewer neighbors than those in the interior, surface nucleons reduce the binding energy by an amount proportional to their number. Because the number of surface nucleons is proportional to the surface area $4\pi r^2$ of the nucleus, and because $r^2 \propto A^{2/3}$ (Eq. 44.1), the surface term can be expressed as $-C_2 A^{2/3}$, where C_2 is a second adjustable constant.

- **The Coulomb repulsion effect.** Each proton repels every other proton in the nucleus. The corresponding potential energy per pair of interacting protons is $k_e e^2/r$, where k_e is the Coulomb constant. The total electric potential energy is equivalent to the work required to assemble Z protons, initially infinitely far apart, into a sphere of volume V. This energy is proportional to the number of proton pairs $Z(Z-1)/2$ and inversely proportional to the nuclear radius. Consequently, the reduction in binding energy that results from the Coulomb effect is $-C_3 Z(Z-1)/A^{1/3}$, where C_3 is yet another adjustable constant.

- **The symmetry effect.** Another effect that lowers the binding energy is related to the symmetry of the nucleus in terms of values of N and Z. For small values of A, stable nuclei tend to have $N \approx Z$. Any large asymmetry between N and Z for light nuclei reduces the binding energy and makes the nucleus less stable. For larger A, the value of N for stable nuclei is naturally larger than Z. This effect can be described by a binding-energy term of the form $-C_4(N-Z)^2/A$, where C_4 is another adjustable constant.[2] For small A, any large asymmetry between values of N and Z makes this term relatively large and reduces the binding energy. For large A, the A in the denominator of this term reduces its value so that it has little effect on the overall binding energy.

Adding these contributions, we arrive at an expression for the total binding energy

$$E_b = C_1 A - C_2 A^{2/3} - C_3 \frac{Z(Z-1)}{A^{1/3}} - C_4 \frac{(N-Z)^2}{A} \qquad (44.3)$$

Semiempirical binding-energy formula

This equation, often referred to as the **semiempirical binding-energy formula,** contains four constants that are adjusted to fit the theoretical expression to experimental data. For nuclei having $A \geq 15$, the constants have the values

$$C_1 = 15.7 \text{ MeV} \qquad C_2 = 17.8 \text{ MeV}$$
$$C_3 = 0.71 \text{ MeV} \qquad C_4 = 23.6 \text{ MeV}$$

---

[2]   The liquid-drop model *describes* the fact that heavy nuclei have $N > Z$. The shell model, as we shall see shortly, *explains* why this is true with a physical argument.

Equation 44.3, together with these constants, fits the known nuclear mass values very well. However, the liquid-drop model does not account for some finer details of nuclear structure, such as stability rules and angular momentum. On the other hand, it does provide a qualitative description of nuclear fission, as we shall discuss in Chapter 45.

Example 44.3   Applying the Semiempirical Binding-Energy Formula

The nucleus ^{64}Zn has a tabulated binding energy of 559.09 MeV. Use the semiempirical binding-energy formula to generate a theoretical estimate of the binding energy for this nucleus.

Solution For this nucleus, we have $Z = 30$, $N = 34$, and $A = 64$. The four terms of the semiempirical binding-energy formula have the following values:

$$C_1 A = (15.7 \text{ MeV})(64) = 1\,005 \text{ MeV}$$

$$C_2 A^{2/3} = (17.8 \text{ MeV})(64)^{2/3} = 285 \text{ MeV}$$

$$C_3 \frac{Z(Z-1)}{A^{1/3}} = (0.71 \text{ MeV}) \frac{(30)(29)}{(64)^{1/3}} = 154 \text{ MeV}$$

$$C_4 \frac{(N-Z)^2}{A} = (23.6 \text{ MeV}) \frac{(34-30)^2}{64} = 5.90 \text{ MeV}$$

Substituting these values into Equation 44.3, we find

$$E_b = 1\,005 \text{ MeV} - 285 \text{ MeV} - 154 \text{ MeV} - 5.90 \text{ MeV}$$

$$= \boxed{560 \text{ MeV}}$$

This value differs from the tabulated value by less than 0.2%. Notice how the sizes of the terms decrease from the first to the fourth term. The fourth term is particularly small, especially for this nucleus, which does not have an excessive number of neutrons.

The Shell Model

The liquid-drop model describes the general behavior of nuclear binding energies relatively well. However, when the binding energies are studied more closely, we find the following features:

- Most stable nuclei have an even value of A. Furthermore, only eight stable nuclei have odd values for both Z and N.

- Figure 44.6 shows a graph of the difference between the binding energy per nucleon calculated by Equation 44.3 and the measured binding energy. There is evidence for regularly spaced peaks in the data that are not described by the semiempirical binding-energy formula. The peaks occur at values of N or Z that have become known as **magic numbers:**

Magic numbers
$$Z \text{ or } N = 2, 8, 20, 28, 50, 82 \tag{44.4}$$

Figure 44.6 The difference between measured binding energies and those calculated from the liquid-drop model as a function of A. The appearance of regular peaks in the experimental data suggests behavior that is not predicted in the liquid-drop model. (Adapted from R. A. Dunlap, *The Physics of Nuclei and Particles*, Brooks/Cole, Belmont CA, 2004.)

- High-precision studies of nuclear radii show deviations from the simple expression in Equation 44.1. Graphs of experimental data show peaks in the curve of radius versus N at values of N equal to the magic numbers.

- We defined isotopes in Section 44.1. A group of *isotones* is a collection of nuclei having the same value of N and varying values of Z. When the number of stable isotones is graphed as function of N, there are peaks in the graph—again at the magic numbers in Equation 44.4.

- Several other nuclear measurements show anomalous behavior at the magic numbers.[3]

These peaks in graphs of experimental data are reminiscent of the peaks in Figure 42.20 for the ionization energy of atoms, which arose because of the shell structure of the atom. The **shell model** of the nucleus, also called the **independent-particle model,** was developed independently in 1949 and 1950 by Maria Goeppert-Mayer and Hans Jensen, both German scientists, who shared the 1963 Nobel Prize in physics. In this model, each nucleon is assumed to exist in a shell, similar to an atomic shell for an electron. The nucleons exist in quantized energy states, and there are few collisions between nucleons. Obviously, the assumptions of this model differ greatly from those made in the liquid-drop model.

The quantized states occupied by the nucleons can be described by a set of quantum numbers. Because both the proton and the neutron have spin $\frac{1}{2}$, the exclusion principle can be applied to describe the allowed states (as we did for electrons in Chapter 42). That is, each state can contain only two protons (or two neutrons) having *opposite* spins (Fig. 44.7). The protons have a set of allowed states, and these states differ from those of the neutrons because the two species move in different potential wells. The proton energy levels are farther apart than the neutron levels because the protons experience a superposition of the Coulomb force and the nuclear force, whereas the neutrons experience only the nuclear force.

One factor influencing the observed characteristics of nuclear ground states is *nuclear spin–orbit* effects. The spin–orbit interaction between the spin of an electron and its orbital motion in an atom gives rise to the sodium doublet discussed in Section 42.6 and is magnetic in origin. In contrast, the spin–orbit effect for nucleons in a nucleus is due to the nuclear force. It is much stronger than in the atomic case, and it has opposite sign. When these effects are taken into account, the shell model is able to account for the observed magic numbers.

The shell model helps us understand why nuclei containing an even number of protons and neutrons are more stable than other nuclei. (There are 160 stable even–even isotopes.) Any particular state is filled when it contains two protons (or two neutrons) having opposite spins. An extra proton or neutron can be added to the nucleus only at

Maria Goeppert-Mayer
German Scientist (1906–1972)

Goeppert-Mayer was born and educated in Germany. She is best known for her development of the shell model (independent-particle model) of the nucleus, published in 1950. A similar model was simultaneously developed by Hans Jensen, another German scientist. Goeppert-Mayer and Jensen were awarded the Nobel Prize in physics in 1963 for their extraordinary work in understanding the structure of the nucleus. (*Courtesy of Louise Barker/AIP Niels Bohr Library*)

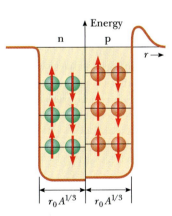

Figure 44.7 A square potential well containing 12 nucleons. The orange circles represent protons, and the green circles represent neutrons. The energy levels for the protons are slightly higher than those for the neutrons because of the electric potential energy associated with the system of protons. The difference in the levels increases as Z increases. Note that only two nucleons having opposite spins can occupy a given level, as required by the exclusion principle.

---

[3]   For further details, see Chapter 5 of R. A. Dunlap, *The Physics of Nuclei and Particles*, Brooks/Cole, 2004.

the expense of increasing the energy of the nucleus. This increase in energy leads to a nucleus that is less stable than the original nucleus. A careful inspection of the stable nuclei shows that the majority have a special stability when their nucleons combine in pairs, which results in a total angular momentum of zero.

The shell model also helps us to understand why nuclei tend to have more neutrons than protons. As in Figure 44.7, the proton energy levels are higher than those for neutrons, due to the extra energy associated with Coulomb repulsion. This effect becomes more pronounced as Z increases. As a result, as Z increases and higher states are filled, a proton level for a given quantum number will be much higher in energy than the neutron level for the same quantum number. In fact, it will be even higher in energy than neutron levels for higher quantum numbers. As a result, it is more energetically favorable for the nucleus to form with neutrons in the lower energy levels rather than protons in the higher energy levels, so that the number of neutrons is greater than the number of protons.

More sophisticated models of the nucleus have been developed and continue to be developed. For example, the *collective model* combines features of the liquid-drop and shell models. The development of theoretical models of the nucleus continues to be an active area of research.

44.4 Radioactivity

In 1896, Henri Becquerel accidentally discovered that uranyl potassium sulfate crystals emit an invisible radiation that can darken a photographic plate when the plate is covered to exclude light. After a series of experiments, he concluded that the radiation emitted by the crystals was of a new type, one that requires no external stimulation and was so penetrating that it could darken protected photographic plates and ionize gases. This process of spontaneous emission of radiation by uranium was soon to be called **radioactivity.**

Subsequent experiments by other scientists showed that other substances were more powerfully radioactive. The most significant investigations of this type were conducted by Marie and Pierre Curie. After several years of careful and laborious chemical separation processes on tons of pitchblende, a radioactive ore, the Curies reported the discovery of two previously unknown elements, both radioactive. These were named polonium and radium. Additional experiments, including Rutherford's famous work on alpha-particle scattering, suggested that radioactivity is the result of the *decay*, or disintegration, of unstable nuclei.

Three types of radioactive decay occur in radioactive substances: alpha (α) decay, in which the emitted particles are ^{4}He nuclei; beta (β) decay, in which the emitted particles are either electrons or positrons; and gamma (γ) decay, in which the emitted "rays" are high-energy photons. A **positron** is a particle like the electron in all respects except that the positron has a charge of $+e$. (The positron is the *antimatter twin* of the electron; see Section 46.2.) The symbol e^{-} is used to designate an electron, and e^{+} designates a positron.

It is possible to distinguish among these three forms of radiation by using the scheme described in Figure 44.8. The radiation from a variety of radioactive samples is directed into a region in which there is a magnetic field. The radiation beam splits into three components, two bending in opposite directions and the third experiencing no change in direction. From this simple observation, we can conclude that the radiation of the undeflected beam carries no charge (the gamma ray), the component deflected upward corresponds to positively charged particles (alpha particles), and the component deflected downward corresponds to negatively charged particles (e^{-}). If the beam includes a positron (e^{+}), it is deflected upward like the alpha particle but follows a different trajectory due to its smaller mass.

The three types of radiation have quite different penetrating powers. Alpha particles barely penetrate a sheet of paper, beta particles (electrons and positrons) can penetrate a few millimeters of aluminum, and gamma rays can penetrate several centimeters of lead.

Marie Curie
Polish Scientist (1867–1934)

In 1903, Marie Curie shared the Nobel Prize in physics with her husband Pierre and with Becquerel for their studies of radioactive substances. In 1911 she was awarded a Nobel Prize in chemistry for the discovery of radium and polonium. She died of leukemia caused by years of exposure to radioactive substances. "I persist in believing that the ideas that then guided us are the only ones which can lead to true social progress. We cannot hope to build a better world without improving the individual. Toward this end, each of us must work toward his own highest development, accepting at the same time his share of responsibility in the general life of humanity." (*FPG International*)

▲ **PITFALL PREVENTION**

44.3 Rays or Particles?

Early in the history of nuclear physics, the term *radiation* was used to describe the emanations from radioactive nuclei. We now know that alpha radiation and beta radiation involve the emission of particles with nonzero rest energy. Even though these are not examples of electromagnetic radiation, the use of the term *radiation* for all three types of emission is deeply entrenched in our language and in the physics community.

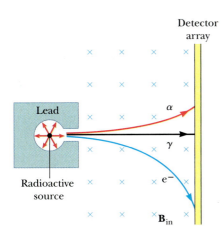

Figure 44.8 The radiation from radioactive sources can be separated into three components by using a magnetic field to deflect the charged particles. The detector array at the right records the events. The gamma ray is not deflected by the magnetic field.

The decay process is probabilistic in nature and can be described with statistical calculations for a radioactive substance of macroscopic size, so that the number of radioactive nuclei is large. For large numbers of radioactive nuclei, the rate at which a particular decay process occurs in a sample is proportional to the number of radioactive nuclei present (that is, the number of nuclei that have not yet decayed). If N is the number of undecayed radioactive nuclei present at some instant, the rate of change of N is

$$\frac{dN}{dt} = -\lambda N \qquad (44.5)$$

where λ, called the **decay constant,** is the probability of decay per nucleus per second. The negative sign indicates that dN/dt is negative; that is, N decreases in time.

Equation 44.5 can be written in the form

$$\frac{dN}{N} = -\lambda\, dt$$

which, upon integration, gives

$$N = N_0 e^{-\lambda t} \qquad (44.6)$$

where the constant N_0 represents the number of undecayed radioactive nuclei at $t = 0$. Equation 44.6 shows that the number of undecayed radioactive nuclei in a sample decreases exponentially with time. The plot of N versus t shown in Figure 44.9 illustrates the exponential nature of the decay.

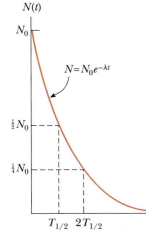

Active Figure 44.9 Plot of the exponential decay law for radioactive nuclei. The vertical axis represents the number of undecayed radioactive nuclei present at any time t, and the horizontal axis is time. The time $T_{1/2}$ is the half-life of the sample.

▲ **PITFALL PREVENTION**

44.4 Notation Warning

In Section 44.1, we introduced the symbol N as an integer representing the number of neutrons in a nucleus. In the present discussion, the symbol N represents the number of undecayed nuclei in a radioactive sample remaining after some time interval. As you read further, be sure to consider the context to determine the appropriate meaning for the symbol N.

Exponential behavior of the number of undecayed nuclei

At the Active Figures link at http://www.pse6.com, observe the decay curves for nuclei with varying half-lives.

The **decay rate** R, which is the number of decays per second, can be obtained by combining Equations 44.5 and 44.6:

Exponential behavior of the decay rate

$$R = \left| \frac{dN}{dt} \right| = \lambda N = N_0 \lambda e^{-\lambda t} = R_0 e^{-\lambda t} \tag{44.7}$$

where $R_0 = N_0 \lambda$ is the decay rate at $t = 0$. The decay rate R of a sample is often referred to as its **activity.** Note that both N and R decrease exponentially with time.

Another parameter useful in characterizing nuclear decay is the **half-life** $T_{1/2}$:

⚠ **PITFALL PREVENTION**

44.5   Half-Life

It is *not* true that all of the original nuclei have decayed after two half-lives! In one half-life, *half of those nuclei that are left* will decay.

The **half-life** of a radioactive substance is the time interval during which half of a given number of radioactive nuclei decay.

To find an expression for the half-life, we first set $N = N_0/2$ and $t = T_{1/2}$ in Equation 44.6 to give

$$\frac{N_0}{2} = N_0 e^{-\lambda T_{1/2}}$$

Canceling the N_0 factors and then taking the reciprocal of both sides, we obtain $e^{\lambda T_{1/2}} = 2$. Taking the natural logarithm of both sides gives

Half-life

$$T_{1/2} = \frac{\ln 2}{\lambda} = \frac{0.693}{\lambda} \tag{44.8}$$

This is a convenient expression relating the half-life $T_{1/2}$ to the decay constant λ. After a time interval equal to one half-life, there are $N_0/2$ radioactive nuclei remaining (by definition); after two half-lives, half of these remaining nuclei have decayed and $N_0/4$ radioactive nuclei are left; after three half-lives, $N_0/8$ are left, and so on. In general, after n half-lives, the number of undecayed radioactive nuclei remaining is $N_0/2^n$.

A frequently used unit of activity is the **curie** (Ci), defined as

The curie

$$1 \text{ Ci} \equiv 3.7 \times 10^{10} \text{ decays/s}$$

This value was originally selected because it is the approximate activity of 1 g of radium. The SI unit of activity is the **becquerel** (Bq):

The becquerel

$$1 \text{ Bq} \equiv 1 \text{ decay/s}$$

Therefore, $1 \text{ Ci} = 3.7 \times 10^{10}$ Bq. The curie is a rather large unit, and the more frequently used activity units are the millicurie and the microcurie.

Quick Quiz 44.4 On your birthday, you measure the activity of a sample of ^{210}Bi, which has a half-life of 5.01 days. The activity you measure is $1.000 \ \mu$Ci. What is the activity of this sample on your next birthday? (a) $1.000 \ \mu$Ci (b) 0 (c) $\sim 0.2 \ \mu$Ci (d) $\sim 0.01 \ \mu$Ci (e) $\sim 10^{-22} \ \mu$Ci

Quick Quiz 44.5 Suppose you have a pure radioactive material with a half-life of $T_{1/2}$. You begin with N_0 undecayed nuclei of the material at $t = 0$. At $t = \frac{1}{2}T_{1/2}$, how many of the nuclei *have decayed?* (a) $\frac{1}{4}N_0$ (b) $\frac{1}{2}N_0$ (c) $\frac{3}{4}N_0$ (d) $0.707N_0$ (e) $0.293N_0$

Example 44.4   How Many Nuclei Are Left?

The isotope carbon-14, $^{14}_{6}C$, is radioactive and has a half-life of 5 730 years. If you start with a sample of 1 000 carbon-14 nuclei, how many will still be undecayed in 22 920 years?

Solution In 5 730 years, half the sample will have decayed, leaving 500 carbon-14 nuclei remaining. In another 5 730 years (for a total time interval of 11 460 years), the number will be reduced to 250 nuclei. After another 5 730 years (total time interval 17 190 years), 125 remain. Finally, after four half-lives (22 920 years), only about 62 remain.

What If? What if you were able to perform this experiment? How would the results compare to our predictions?

Answer The numbers found in the example represent ideal circumstances. As we have mentioned, radioactive decay is actually a probabilistic statistical process and accurate statistical predictions are only possible with a very large number of atoms. The original sample in this example contained only 1 000 nuclei, certainly not a very large number. Thus, if you counted the number remaining, for example, after one half-life for this small sample, it probably would not be exactly 500.

Example 44.5   The Activity of Radium

Interactive

The half-life of the radioactive nucleus radium-226, $^{226}_{88}Ra$, is 1.6×10^3 yr.

(A) What is the decay constant λ of this nucleus?

Solution We can calculate λ using Equation 44.8 and the fact that

$$T_{1/2} = 1.6 \times 10^3 \text{ yr} \left(\frac{3.16 \times 10^7 \text{ s}}{1 \text{ yr}} \right)$$

$$= 5.0 \times 10^{10} \text{ s}$$

Therefore,

$$\lambda = \frac{0.693}{T_{1/2}} = \frac{0.693}{5.0 \times 10^{10} \text{ s}} = \boxed{1.4 \times 10^{-11} \text{ s}^{-1}}$$

Note that this result is also the probability that any single $^{226}_{88}Ra$ nucleus will decay in a time interval of one second.

(B) If a sample contains 3.0×10^{16} $^{226}_{88}Ra$ nuclei at $t = 0$, determine its activity in curies at this time.

Solution By definition (Eq. 44.7) R_0, the activity at $t = 0$, is λN_0, where N_0 is the number of radioactive nuclei present at $t = 0$. With $N_0 = 3.0 \times 10^{16}$, we have

$$R_0 = \lambda N_0 = (1.4 \times 10^{-11} \text{ s}^{-1})(3.0 \times 10^{16})$$

$$= (4.2 \times 10^5 \text{ Bq}) \left(\frac{1 \text{ Ci}}{3.7 \times 10^{10} \text{ Bq}} \right)$$

$$= \boxed{11 \ \mu\text{Ci}}$$

(C) What is the activity after the sample is 2.0×10^3 yr old?

Solution We use Equation 44.7 and the fact that the age of the sample is 2.0×10^3 yr $= 6.3 \times 10^{10}$ s:

$$R = R_0 e^{-\lambda t}$$

$$= (4.2 \times 10^5 \text{ Bq}) e^{-(1.4 \times 10^{-11} \text{ s}^{-1})(6.3 \times 10^{10} \text{ s})}$$

$$= 1.7 \times 10^5 \text{ Bq} \left(\frac{1 \text{ Ci}}{3.7 \times 10^{10} \text{ Bq}} \right)$$

$$= \boxed{4.7 \ \mu\text{Ci}}$$

At the Interactive Worked Example link at **http://www.pse6.com,** *practice evaluating the parameters for radioactive decay of various isotopes of radium.*

Example 44.6   The Activity of Carbon

At time $t = 0$, a radioactive sample contains 3.50 μg of pure $^{11}_{6}C$, which has a half-life of 20.4 min.

(A) Determine the number N_0 of nuclei in the sample at $t = 0$.

Solution The molar mass of $^{11}_{6}C$ is approximately 11.0 g/mol, and so 11.0 g contains Avogadro's number (6.02×10^{23}) of nuclei. Therefore, 3.50 μg contains N_0 nuclei, where

$$\frac{N_0}{6.02 \times 10^{23} \text{ nuclei/mol}} = \frac{3.50 \times 10^{-6} \text{ g}}{11.0 \text{ g/mol}}$$

$$N_0 = \boxed{1.92 \times 10^{17} \text{ nuclei}}$$

(B) What is the activity of the sample initially and after 8.00 h?

Solution With $T_{1/2} = 20.4$ min $= 1\,224$ s, the decay constant is

$$\lambda = \frac{0.693}{T_{1/2}} = \frac{0.693}{1\,224 \text{ s}} = 5.66 \times 10^{-4} \text{ s}^{-1}$$

Therefore, the initial activity of the sample is

$$R_0 = \lambda N_0 = (5.66 \times 10^{-4} \text{ s}^{-1})(1.92 \times 10^{17})$$

$$= \boxed{1.09 \times 10^{14} \text{ Bq}}$$

We use Equation 44.7 to find the activity at $t = 8.00$ h $= 2.88 \times 10^4$ s:

$$R = R_0 e^{-\lambda t} = (1.09 \times 10^{14} \text{ Bq}) e^{-(5.66 \times 10^{-4} \text{ s}^{-1})(2.88 \times 10^4 \text{ s})}$$

$$= \boxed{9.08 \times 10^6 \text{ Bq}}$$

A listing of activity versus time for this situation is given in Table 44.2.

Table 44.2

Activity Versus Time for the Sample Described in Example 44.6	
t (h)	R (Bq)
0	1.09×10^{14}
1	1.42×10^{13}
2	1.85×10^{12}
3	2.41×10^{11}
4	3.15×10^{10}
5	4.10×10^9
6	5.34×10^8
7	6.97×10^7
8	9.08×10^6

Example 44.7   A Radioactive Isotope of Iodine

A sample of the isotope ^{131}I, which has a half-life of 8.04 days, has an activity of 5.0 mCi at the time of shipment. Upon receipt in a medical laboratory, the activity is 4.2 mCi. How much time has elapsed between the two measurements?

Solution To conceptualize this problem, consider that the sample is continuously decaying as it is in transit. The decrease in the activity is 16% during the time interval between shipment and receipt, so we expect the elapsed time to be less than the half-life of 8.04 d. The stated activity corresponds to many decays per second, so that N is large and we can categorize this problem as one in which we can use our statistical analysis of radioactivity. To analyze the problem we use Equation 44.7 in the form

$$\frac{R}{R_0} = e^{-\lambda t}$$

where the sample is shipped at $t = 0$, at which time the activity is R_0. Taking the natural logarithm of each side, we have

$$\ln\left(\frac{R}{R_0}\right) = -\lambda t$$

$$(1) \qquad t = -\frac{1}{\lambda} \ln\left(\frac{R}{R_0}\right)$$

To find λ, we use Equation 44.8:

$$(2) \qquad \lambda = \frac{0.693}{T_{1/2}} = \frac{0.693}{8.04 \text{ d}} = 8.62 \times 10^{-2} \text{ d}^{-1}$$

Substituting Equation (2) into Equation (1) gives

$$t = -\left(\frac{1}{8.62 \times 10^{-2} \text{ d}^{-1}}\right) \ln\left(\frac{4.2 \text{ mCi}}{5.0 \text{ mCi}}\right) = \boxed{2.0 \text{ d}}$$

To finalize this problem, note that this is indeed less than the half-life, as we expected. This example demonstrates the difficulty in shipping radioactive samples with short half-lives. If the shipment were to be delayed by several days, only a small fraction of the sample might remain upon receipt. This difficulty can be addressed by shipping a combination of isotopes in which the desired isotope is the product of a decay occurring within the sample. It is possible for the desired isotope to be in *equilibrium*, in which case it is created at the same rate as it decays. Therefore, the amount of the desired isotope remains constant during the shipping process. Upon receipt, the desired isotope can be separated from the rest of the sample and its decay from the initial activity begins upon receipt rather than upon shipment.

44.5   The Decay Processes

As we stated in the preceding section, a radioactive nucleus spontaneously decays by one of three processes: alpha decay, beta decay, or gamma decay. Figure 44.10 shows a close-up view of a portion of Figure 44.4 from $Z = 65$ to $Z = 80$. The blue circles are the stable nuclei seen in Figure 44.4. In addition, unstable nuclei above and below the line of stability for each value of Z are shown. Above the line of stability, the red circles show unstable nuclei that are neutron-rich and undergo a beta decay process in which an electron is emitted. Below the blue circles are green circles corresponding to proton-rich unstable nuclei that primarily undergo a beta-decay process in which a positron is emitted or a competing process called electron capture. Beta decay and electron capture are described in more detail below. Further below the line of stability (with a few exceptions) are yellow circles that represent very proton-rich nuclei for which the primary decay mechanism is alpha decay, which we will discuss first.

Alpha Decay

A nucleus emitting an alpha particle ($_2^4$He) loses two protons and two neutrons. Therefore, the atomic number Z decreases by 2, the mass number A decreases by 4, and the neutron number decreases by 2. The decay can be written

$$_Z^A X \longrightarrow _{Z-2}^{A-4} Y + _2^4 He \tag{44.9}$$

where X is called the **parent nucleus** and Y the **daughter nucleus.** As a general rule in any decay expression such as this, (1) the sum of the mass numbers A must be the same on both sides of the decay and (2) the sum of the atomic numbers Z must be the same on both sides of the decay. As examples, ^{238}U and ^{226}Ra are both alpha emitters and decay according to the schemes

$$_{92}^{238} U \longrightarrow _{90}^{234} Th + _2^4 He \tag{44.10}$$

$$_{88}^{226} Ra \longrightarrow _{86}^{222} Rn + _2^4 He \tag{44.11}$$

The decay of ^{226}Ra is shown in Figure 44.11.

When the nucleus of one element changes into the nucleus of another, as happens in alpha decay, the process is called **spontaneous decay.** In any spontaneous decay, relativistic energy and momentum of the isolated parent nucleus must be conserved. If we call M_X the mass of the parent nucleus, M_Y the mass of the daughter nucleus, and M_α the mass of the alpha particle, we can define the **disintegration energy** Q of the system as

$$Q = (M_X - M_Y - M_\alpha)c^2 \tag{44.12}$$

The energy Q is in joules when the masses are in kilograms and c is the speed of light, 3.00×10^8 m/s. However, when the masses are expressed in the more convenient unit u, Q can be calculated in MeV using the expression

$$Q = (M_X - M_Y - M_\alpha) \times 931.494 \text{ MeV/u} \tag{44.13}$$

The disintegration energy Q appears in the form of kinetic energy in the daughter nucleus and the alpha particle, and is sometimes referred to as the Q value of the nuclear decay. In the case of the ^{226}Ra decay described in Figure 44.11, if the parent nucleus is at rest before the decay, the total kinetic energy of the products is 4.87 MeV. Most of this kinetic energy is associated with the alpha particle because this particle is much less massive than the daughter nucleus ^{222}Rn. That is, because momentum must be conserved, the lighter alpha particle recoils with a much higher speed than does the daughter nucleus. Generally, less massive particles carry off most of the energy in nuclear decays.

Experimental observations of alpha-particle energies show a number of discrete energies rather than a single energy. This is due to the fact that the daughter nucleus may

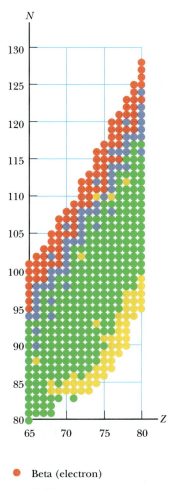

- 🔴 Beta (electron)
- 🔵 Stable
- 🟢 Beta (positron) or electron capture
- 🟡 Alpha

Active Figure 44.10 A close-up view of the line of stability in Figure 44.4 from $Z = 65$ to $Z = 80$. The blue dots represent stable nuclei as in Figure 44.4. The other colored dots represent unstable isotopes above and below the line of stability, with the color of the dot indicating the primary means of decay.

Study the decay modes and decay energies by clicking on any of the colored dots at the Active Figures link at http://www.pse6.com.

Active Figure 44.11 The alpha decay of radium-226. The radium nucleus is initially at rest. After the decay, the radon nucleus has kinetic energy K_{Rn} and momentum $\mathbf{p}_{Rn}$, and the alpha particle has kinetic energy K_α and momentum $\mathbf{p}_\alpha$.

At the Active Figures link at http://www.pse6.com, observe the decay of radium-226. For a large number of decays, observe the development of the graph in Figure 44.14b.

⚠ **PITFALL PREVENTION**

44.6 Another Q

We have seen the symbol Q before, but this is a brand new meaning for this symbol—the disintegration energy. In this context, it is neither heat nor charge, for which we have used Q before.

be left in an excited quantum state after the decay. As a result, not all of the disintegration energy is available as kinetic energy of the alpha particle and daughter nucleus. The emission of an alpha particle is followed by one or more gamma-ray photons (see below) as the excited nucleus decays to the ground state. The observed discrete alpha-particle energies represent evidence of the quantized nature of the nucleus and allow a determination of the energies of the quantum states.

Finally, it is interesting to note that if one assumed that ^{238}U (or any other alpha emitter) decayed by emitting either a proton or a neutron, the mass of the decay products would exceed that of the parent nucleus, corresponding to a negative Q value. A negative Q value indicates that such a proposed decay does not occur spontaneously.

Quick Quiz 44.6 Which of the following is the correct daughter nucleus associated with the alpha decay of $^{157}_{72}$Hf? (a) $^{153}_{72}$Hf (b) $^{153}_{70}$Yb (c) $^{157}_{70}$Yb

---

Example 44.8 The Energy Liberated when Radium Decays

The ^{226}Ra nucleus undergoes alpha decay according to Equation 44.11. Calculate the Q value for this process. Take the masses to be 226.025 403 u for ^{226}Ra, 222.017 570 u for ^{222}Rn, and 4.002 603 u for $^4_2$He, as found in Table A.3.

Solution We may add 88 electrons to both sides of the decay expression in Equation 44.11. The differences in electron binding energies are negligible when compared with the Q value for the nuclear decay process. Then, we may use the masses of neutral atoms in Equation 44.13 to see that

$$Q = (M_X - M_Y - M_\alpha) \times 931.494 \text{ MeV/u}$$
$$= (226.025\,403 \text{ u} - 222.017\,570 \text{ u} - 4.002\,603 \text{ u})$$
$$\times 931.494 \text{ MeV/u}$$
$$= (0.005\,230 \text{ u}) \times 931.494 \text{ MeV/u} = \boxed{4.87 \text{ MeV}}$$

What If? Suppose we measured the kinetic energy of the alpha particle from this decay. Would we measure 4.87 MeV?

Answer The value of Q that we have calculated is the disintegration energy, not the kinetic energy of the alpha particle. The energy of 4.87 MeV includes the kinetic energy of both the alpha particle and the daughter nucleus after the decay. Thus, the kinetic energy of the alpha particle must be *less* than 4.87 MeV.

Let us determine this kinetic energy mathematically. The parent nucleus is an isolated system that decays into an alpha particle and a daughter nucleus. Thus, momentum must be conserved for the system. The initial momentum is

zero, so

$$(1) \qquad 0 = M_Y v_Y - M_\alpha v_\alpha$$

where we have explicitly indicated with the minus sign that the particles move in opposite directions, so we can interpret v_Y and v_α as speeds rather than velocity components.

The disintegration energy is equal to the sum of the kinetic energies of the alpha particle and the daughter nucleus (assuming that the daughter nucleus is left in the ground state):

$$(2) \qquad Q = \tfrac{1}{2} M_\alpha v_\alpha^2 + \tfrac{1}{2} M_Y v_Y^2$$

If we solve Equation (1) for v_Y and substitute into Equation (2), we find

$$Q = \tfrac{1}{2} M_\alpha v_\alpha^2 + \tfrac{1}{2} M_Y \left(\frac{M_\alpha v_\alpha}{M_Y}\right)^2 = \tfrac{1}{2} M_\alpha v_\alpha^2 \left(1 + \frac{M_\alpha}{M_Y}\right)$$
$$= K_\alpha \left(\frac{M_Y + M_\alpha}{M_Y}\right)$$

Thus, in general, the kinetic energy of the alpha particle is related to the disintegration energy according to

$$K_\alpha = Q \left(\frac{M_Y}{M_Y + M_\alpha}\right)$$

For the specific decay of ^{226}Ra that we are exploring in this example, we have

$$K_\alpha = (4.87 \text{ MeV}) \left(\frac{222}{222 + 4}\right) = 4.78 \text{ MeV}$$

To understand the mechanism of alpha decay, let us model the parent nucleus as a system consisting of (1) the alpha particle, already formed as an entity within the nucleus, and (2) the daughter nucleus that will result when the alpha particle is emitted. Figure 44.12 shows a plot of potential energy versus separation distance r between the alpha particle and the daughter nucleus, where the distance marked R is the range of the nuclear force. The curve represents the combined effects of (1) the repulsive Coulomb force, which gives the positive part of the curve for $r > R$, and (2) the attractive nuclear force, which causes the curve to be negative for $r < R$. As we saw in Example 44.8, a typical disintegration energy Q is about 5 MeV, which is the approximate kinetic energy of the alpha particle, represented by the lower dashed line in Figure 44.12.

According to classical physics, the alpha particle is trapped in a potential well. How, then, does it ever escape from the nucleus? The answer to this question was first provided by George Gamow (1904–1968) in 1928 and independently by R. W. Gurney and E. U. Condon in 1929, using quantum mechanics. In Section 41.6, we discussed the view of quantum mechanics that there is always some probability that a particle can tunnel through a barrier. This is exactly how we can describe alpha decay—the alpha particle tunnels through the barrier in Figure 44.12, escaping the nucleus. Furthermore, this model agrees with the observation that higher-energy alpha particles come from nuclei with shorter half-lives. For higher-energy alpha particles in Figure 44.12, the barrier is narrower and the probability is higher that tunneling will occur. The higher probability translates to a shorter half-life.

As an example, consider the decays of ^{238}U and ^{226}Ra in Equations 44.10 and 44.11, along with the corresponding half-lives and alpha-particle energies:

$$^{238}U: \quad T_{1/2} = 4.47 \times 10^9 \, yr \quad K_\alpha = 4.20 \, MeV$$

$$^{226}Ra: \quad T_{1/2} = 1.60 \times 10^3 \, yr \quad K_\alpha = 4.78 \, MeV$$

Notice that a relatively small difference in alpha-particle energy is associated with a tremendous difference of six orders of magnitude in the half-life. The origin of this effect can be seen as follows. First, in Figure 44.12, notice that the curve below an alpha-particle energy of 5 MeV has a slope with a relatively small magnitude. Thus, a small difference in energy on the vertical axis has a relatively large effect on the width of the potential barrier. Second, recall Equation 41.20, which describes the exponential dependence of the probability of transmission on the barrier width. These two factors combine to give the very sensitive relationship between half-life and alpha-particle energy that the data above suggest.

The Smoke Detector. A life-saving application of alpha decay is in the household smoke detector, shown in Figure 44.13. Most of the common ones use a radioactive material. The detector consists of an ionization chamber, a sensitive current detector, and an alarm. A weak radioactive source (usually $^{241}_{95}Am$) ionizes the air in the chamber of the

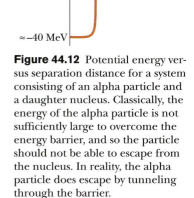

Figure 44.12 Potential energy versus separation distance for a system consisting of an alpha particle and a daughter nucleus. Classically, the energy of the alpha particle is not sufficiently large to overcome the energy barrier, and so the particle should not be able to escape from the nucleus. In reality, the alpha particle does escape by tunneling through the barrier.

(a)   (b)

Figure 44.13 (a) A smoke detector uses alpha decay to determine whether smoke is in the air. The alpha source is in the black cylinder at the right. (b) Smoke entering the chamber reduces the detected current, causing the alarm to sound. © *1998 by the AAAS.*

detector, creating charged particles. A voltage is maintained between the plates inside the chamber, setting up a small but detectable current in the external circuit due to the ions acting as charge carriers between the plates. As long as the current is maintained, the alarm is deactivated. However, if smoke drifts into the chamber, the ions become attached to the smoke particles. These heavier particles do not drift as readily as do the lighter ions, which causes a decrease in the detector current. The external circuit senses this decrease in current and sets off the alarm.

Beta Decay

When a radioactive nucleus undergoes beta decay, the daughter nucleus contains the same number of nucleons as the parent nucleus but the atomic number is changed by 1, which means that the number of protons changes:

$$\begin{array}{ccc}
{}^{A}_{Z}X & \longrightarrow & {}^{A}_{Z+1}Y + e^{-} \qquad \text{(incomplete expression)} & \text{(44.14)}
\end{array}$$

$$\begin{array}{ccc}
{}^{A}_{Z}X & \longrightarrow & {}^{A}_{Z-1}Y + e^{+} \qquad \text{(incomplete expression)} & \text{(44.15)}
\end{array}$$

where, as we mentioned in Section 44.4, e^{-} is used to designate an electron and e^{+} designates a positron, with *beta particle* being the general term referring to either. *Beta decay is not described completely by these expressions.* We shall give reasons for this shortly.

As with alpha decay, the nucleon number and total charge are both conserved in beta decays. From the fact that A does not change but Z does, we conclude that in beta decay, either a neutron changes to a proton (Eq. 44.14) or a proton changes to a neutron (Eq. 44.15). Note that the electron or positron emitted in these decays is not present beforehand in the nucleus; it is created in the process of the decay from the rest energy of the decaying nucleus. Two typical beta-decay processes are

$$\begin{array}{ccc}
{}^{14}_{6}C & \longrightarrow & {}^{14}_{7}N + e^{-} \qquad \text{(incomplete expression)} & \text{(44.16)}
\end{array}$$

$$\begin{array}{ccc}
{}^{12}_{7}N & \longrightarrow & {}^{12}_{6}C + e^{+} \qquad \text{(incomplete expression)} & \text{(44.17)}
\end{array}$$

Let us consider the energy of the system undergoing beta decay before and after the decay. As with alpha decay, energy of the isolated system must be conserved. Experimentally, it is found that beta particles from a single type of nucleus are emitted over a continuous range of energies (Fig. 44.14a), as opposed to alpha decay, in which the alpha particles are emitted with discrete energies (Fig. 44.14b). The kinetic energy of the system after the decay is equal to the decrease in mass-energy of the system, that is, the Q value. However, because all decaying nuclei in the sample have the same initial mass, *the Q value must be the same for each decay.* In view of this, why do the emitted particles have the range of kinetic energies shown in Figure 44.14a? The law of conservation of energy seems to be violated! And it becomes worse: further analysis of the decay processes

At the Active Figures link at http://www.pse6.com, observe the development of these graphs for the decays in Figures 44.11 and 44.15.

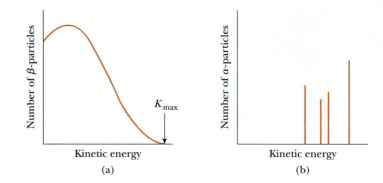

Active Figure 44.14 (a) Distribution of beta particle energies in a typical beta decay. All energies are observed up to a maximum value. (b) In contrast, the energies of alpha particles from an alpha decay are discrete.

described by Equations 44.14 and 44.15 shows that the laws of conservation of both angular momentum (spin) and linear momentum are also violated!

After a great deal of experimental and theoretical study, Pauli in 1930 proposed that a third particle must be present in the decay products to carry away the "missing" energy and momentum. Fermi later named this particle the **neutrino** (little neutral one) because it had to be electrically neutral and have little or no mass. Although it eluded detection for many years, the neutrino (symbol ν, Greek nu) was finally detected experimentally in 1956 by Frederick Reines, who received the Nobel Prize for this work in 1995. It has the following properties:

- It has zero electric charge.

- Its mass is either zero (in which case it travels at the speed of light) or very small; there is much recent persuasive experimental evidence that suggests that the neutrino mass is not zero. Current experiments place the upper bound of the mass of the neutrino at about 7 eV/c^2.

- It has a spin of $\frac{1}{2}$, which allows the law of conservation of angular momentum to be satisfied in beta decay.

- It interacts very weakly with matter and is therefore very difficult to detect.

Properties of the neutrino

We can now write the beta-decay processes (Eqs. 44.14 and 44.15) in their correct and complete form:

$$^{A}_{Z}X \longrightarrow ^{A}_{Z+1}Y + e^- + \bar{\nu} \quad \text{(complete expression)} \quad (44.18)$$

$$^{A}_{Z}X \longrightarrow ^{A}_{Z-1}Y + e^+ + \nu \quad \text{(complete expression)} \quad (44.19)$$

Beta decay processes

as well as those for carbon-14 and nitrogen-12 (Eqs. 44.16 and 44.17):

$$^{14}_{6}C \longrightarrow ^{14}_{7}N + e^- + \bar{\nu} \quad \text{(complete expression)} \quad (44.20)$$

$$^{12}_{7}N \longrightarrow ^{12}_{6}C + e^+ + \nu \quad \text{(complete expression)} \quad (44.21)$$

where the symbol $\bar{\nu}$ represents the **antineutrino,** the antiparticle to the neutrino. We shall discuss antiparticles further in Chapter 46. For now, it suffices to say that **a neutrino is emitted in positron decay and an antineutrino is emitted in electron decay.** As with alpha decay, the decays listed above are analyzed by applying conservation laws, but relativistic expressions must be used for beta particles because their kinetic energy is large (typically 1 MeV) compared with their rest energy of 0.511 MeV. Figure 44.15 shows a pictorial representation of the decays described by Equations 44.20 and 44.21.

Active Figure 44.15 (a) The beta decay of carbon-14. The final products of the decay are the nitrogen-14 nucleus, an electron, and an antineutrino. (b) The beta decay of nitrogen-12. The final products of the decay are the carbon-12 nucleus, a positron, and a neutrino.

At the Active Figures link at http://www.pse6.com, *observe the decay of carbon-14. For a large number of decays, observe the development of the graph in Figure 44.14a.*

In Equation 44.18, the number of protons has increased by one and the number of neutrons has decreased by one. We can write the fundamental process of e^- decay in terms of a neutron changing into a proton as follows:

$$\text{n} \longrightarrow \text{p} + e^- + \bar{\nu} \tag{44.22}$$

The electron and the antineutrino are ejected from the nucleus, with the net result that there is one more proton and one fewer neutron, consistent with the changes in Z and $A - Z$. A similar process occurs in e^+ decay, with a proton changing into a neutron, a positron, and a neutrino. This latter process can only occur within the nucleus, with the result that the nuclear mass decreases. It cannot occur for an isolated proton because its mass is less than that of the neutron.

A process that competes with e^+ decay is **electron capture.** This occurs when a parent nucleus captures one of its own orbital electrons and emits a neutrino. The final product after decay is a nucleus whose charge is $Z - 1$:

$$^A_Z\text{X} + \,_{-1}^{\ 0}e \longrightarrow \,_{Z-1}^{\ \ A}\text{Y} + \nu \tag{44.23}$$

In most cases, it is a K-shell electron that is captured, and for this reason the process is referred to as **K capture.** One example is the capture of an electron by ^7_4Be:

$$^7_4\text{Be} + \,_{-1}^{\ 0}e \longrightarrow \,^7_3\text{Li} + \nu$$

> ⚠ **PITFALL PREVENTION**
>
> **44.7 Mass Number of the Electron**
>
> An alternative notation for an electron is the symbol $_{-1}^{\ 0}e$. This does not imply that the electron has zero rest energy. The mass of the electron is so much smaller than that of the lightest nucleon, however, that we approximate it as zero in the context of nuclear decays and reactions.

Because the neutrino is very difficult to detect, electron capture is usually observed by the x-rays given off as higher-shell electrons cascade downward to fill the vacancy created in the K shell.

Finally, we specify Q values for the beta-decay processes. The Q values for e^- decay and electron capture are given by $Q = (M_X - M_Y)c^2$, where M_X and M_Y are the masses of neutral atoms. The Q values for e^+ decay are given by $Q = (M_X - M_Y - 2m_e)c^2$. The extra term $-2m_ec^2$ in this expression is due to the fact that the atomic number of the parent decreases by one when the daughter is formed. To form a neutral atom, the daughter atom sheds one electron. Thus, the final products are the daughter atom, the shed electron, and the ejected positron. In e^- decay, the parent nucleus experiences an increase in atomic number and must add an electron in order for the atom to become neutral. If we imagine the neutral parent atom and an electron that combine with the daughter as the initial system and the final system as the neutral daughter atom and the beta-ejected electron, there is a free electron in the system both before and after the decay. Thus, in subtracting the initial and final mass-energies of the system, this electron mass cancels.

These relationships are useful in determining whether or not a process is energetically possible. For example, the expression for proposed e^+ decay for a particular parent nucleus may turn out to be negative. In this case, this decay will not occur. The expression for electron capture for this parent nucleus, however, may give a positive number, so that electron capture can occur even though e^+ decay is not possible. This is the case for the decay of ^7_4Be shown above.

> **Quick Quiz 44.7** Which of the following is the correct daughter nucleus associated with the beta decay of $^{184}_{72}\text{Hf}$? (a) $^{183}_{72}\text{Hf}$ (b) $^{183}_{73}\text{Ta}$ (c) $^{184}_{73}\text{Ta}$

Carbon Dating

The beta decay of ^{14}C (Eq. 44.20) is commonly used to date organic samples. Cosmic rays in the upper atmosphere cause nuclear reactions (Section 44.7) that create ^{14}C. The ratio of ^{14}C to ^{12}C in the carbon dioxide molecules of our atmosphere has a constant value of approximately 1.3×10^{-12}. The carbon atoms in all living organisms have this same $^{14}\text{C}/^{12}\text{C}$ ratio because the organisms continuously exchange carbon dioxide with their surroundings. When an organism dies, however, it no longer absorbs ^{14}C from the atmosphere, and so the $^{14}\text{C}/^{12}\text{C}$ ratio decreases as the ^{14}C decays with a half-life of 5 730 yr. It is therefore possible to measure the age of a material by measuring its ^{14}C activity. Using this technique, scientists have been able to identify samples of wood, charcoal, bone, and shell as having lived from 1 000 to 25 000 years ago. This knowledge has helped us reconstruct the history of living organisms—including humans—during this time span.

A particularly interesting example is the dating of the Dead Sea Scrolls. This group of manuscripts was discovered by a shepherd in 1947. Translation showed them to be religious documents, including most of the books of the Old Testament. Because of their historical and religious significance, scholars wanted to know their age. Carbon dating applied to the material in which they were wrapped established their age at approximately 1 950 yr.

Conceptual Example 44.9   The Age of Ice Man

In 1991, a German tourist discovered the well-preserved remains of a man, now called the "Ice Man," trapped in a glacier in the Italian Alps. (See the chapter opening photograph.) Radioactive dating with ^{14}C revealed that this person was alive about 5 300 years ago. Why did scientists date a sample of the Ice Man using ^{14}C rather than ^{11}C, which is a beta emitter having a half-life of 20.4 min?

Solution Because ^{14}C has a half-life of 5 730 yr, the fraction of ^{14}C nuclei remaining after one half-life is high enough to allow accurate measurements of changes in the sample's activity. Because ^{11}C has a very short half-life, it is not useful—its activity decreases to a vanishingly small value over the age of the sample, making it impossible to detect.

An isotope used to date a sample must be present in a known amount in the sample when it is formed. As a general rule, the isotope chosen to date a sample should also have a half-life that is on the same order of magnitude as the age of the sample. If the half-life is much less than the age of the sample, there won't be enough activity left to measure because almost all of the original radioactive nuclei will have decayed. If the half-life is much greater than the age of the sample, the amount of decay that has taken place since the sample died will be too small to measure. For example, if you have a specimen estimated to have died 50 years ago, neither ^{14}C (5 730 yr) nor ^{11}C (20 min) is suitable. If you know your sample contains hydrogen, however, you can measure the activity of ^{3}H (tritium), a beta emitter that has a half-life of 12.3 yr.

Example 44.10   Radioactive Dating   **Interactive**

A piece of charcoal containing 25.0 g of carbon is found in some ruins of an ancient city. The sample shows a ^{14}C activity R of 250 decays/min. How long has the tree from which this charcoal came been dead?

Solution First, let us calculate the decay constant for ^{14}C, which has a half-life of 5 730 yr:

$$\lambda = \frac{0.693}{T_{1/2}} = \frac{0.693}{(5\ 730\ \text{yr})(3.16 \times 10^7\ \text{s/yr})}$$

$$= 3.83 \times 10^{-12}\ \text{s}^{-1}$$

The number of ^{14}C nuclei can be calculated in two steps. First, the number of ^{12}C nuclei in 25.0 g of carbon is

$$N(^{12}\text{C}) = \frac{6.02 \times 10^{23}\ \text{nuclei/mol}}{12.0\ \text{g/mol}}(25.0\ \text{g})$$

$$= 1.25 \times 10^{24}\ \text{nuclei}$$

Knowing that the ratio of ^{14}C to ^{12}C in the live sample was 1.3×10^{-12}, we see that the number of ^{14}C nuclei in 25.0 g *before* decay was

$$N_0(^{14}\text{C}) = (1.3 \times 10^{-12})(1.25 \times 10^{24}) = 1.6 \times 10^{12}\ \text{nuclei}$$

Hence, the initial activity of the sample was

$$R_0 = N_0\lambda = (1.6 \times 10^{12}\ \text{nuclei})(3.83 \times 10^{-12}\ \text{s}^{-1})$$

$$= 6.13\ \text{decays/s} = 368\ \text{decays/min}$$

We now use Equation 44.7, which relates the activity R at any time t to the initial activity R_0:

$$R = R_0 e^{-\lambda t}$$

$$e^{-\lambda t} = \frac{R}{R_0}$$

Using $R = 250$ decays/min and $R_0 = 368$ decays/min, we calculate t by taking the natural logarithm of both sides of this expression:

$$-\lambda t = \ln\left(\frac{R}{R_0}\right) = \ln\left(\frac{250}{368}\right) = -0.39$$

$$t = \frac{0.39}{\lambda} = \frac{0.39}{3.83 \times 10^{-12}\ \text{s}^{-1}}$$

$$= 1.0 \times 10^{11}\ \text{s} = \boxed{3.2 \times 10^3\ \text{yr}}$$

Practice using carbon dating on samples at the Interactive Worked Example link at **http://www.pse6.com.**

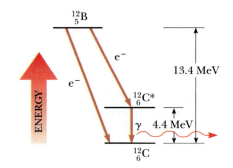

Figure 44.16 An energy-level diagram showing the initial nuclear state of a ^{12}B nucleus and two possible lower-energy states of the ^{12}C nucleus. The beta decay of the ^{12}B nucleus can result in either of two situations: the ^{12}C nucleus is in the ground state or in the excited state, in which case the nucleus is denoted as ^{12}C*. In the latter case, the beta decay to ^{12}C* is followed by a gamma decay to ^{12}C as the excited nucleus makes a transition to the ground state.

Gamma Decay

Very often, a nucleus that undergoes radioactive decay is left in an excited energy state. The nucleus can then undergo a second decay to a lower energy state, perhaps to the ground state, by emitting a high-energy photon:

Gamma decay

$$\,_{Z}^{A}X^* \longrightarrow \,_{Z}^{A}X + \gamma \qquad (44.24)$$

where X^* indicates a nucleus in an excited state. The typical half-life of an excited nuclear state is 10^{-10} s. Photons emitted in such a de-excitation process are called gamma rays. Such photons have very high energy (1 MeV–1 GeV) relative to the energy of visible light (about 1 eV). Recall from Section 42.3 that the energy of a photon emitted or absorbed by an atom equals the difference in energy between the two electronic states involved in the transition. Similarly, a gamma-ray photon has an energy hf that equals the energy difference ΔE between two nuclear energy levels. When a nucleus decays by emitting a gamma ray, the only change in the nucleus is that it ends up in a lower energy state. There are no changes in Z, N, or A.

A nucleus may reach an excited state as the result of a violent collision with another particle. However, it is more common for a nucleus to be in an excited state after it has undergone alpha or beta decay. The following sequence of events represents a typical situation in which gamma decay occurs:

$$\,_{5}^{12}B \longrightarrow \,_{6}^{12}C^* + e^- + \bar{\nu} \qquad (44.25)$$

$$\,_{6}^{12}C^* \longrightarrow \,_{6}^{12}C + \gamma \qquad (44.26)$$

Figure 44.16 shows the decay scheme for ^{12}B, which undergoes beta decay to either of two levels of ^{12}C. It can either (1) decay directly to the ground state of ^{12}C by emitting a 13.4-MeV electron or (2) undergo beta decay to an excited state of ^{12}C* followed by gamma decay to the ground state. The latter process results in the emission of a 9.0-MeV electron and a 4.4-MeV photon.

The various pathways by which a radioactive nucleus can undergo decay are summarized in Table 44.3.

Table 44.3

Various Decay Pathways	
Alpha decay	$\,_{Z}^{A}X \rightarrow \,_{Z-2}^{A-4}Y + \,_{2}^{4}He$
Beta decay (e^-)	$\,_{Z}^{A}X \rightarrow \,_{Z+1}^{A}Y + e^- + \bar{\nu}$
Beta decay (e^+)	$\,_{Z}^{A}X \rightarrow \,_{Z-1}^{A}Y + e^+ + \nu$
Electron capture	$\,_{Z}^{A}X + e^- \rightarrow \,_{Z-1}^{A}Y + \nu$
Gamma decay	$\,_{Z}^{A}X^* \rightarrow \,_{Z}^{A}X + \gamma$

Table 44.4

The Four Radioactive Series			
Series	**Starting Isotope**	**Half-Life (years)**	**Stable End Product**
Uranium	$^{238}_{92}U$	4.47×10^9	$^{206}_{82}Pb$
Actinium } Natural	$^{235}_{92}U$	7.04×10^8	$^{207}_{82}Pb$
Thorium	$^{232}_{90}Th$	1.41×10^{10}	$^{208}_{82}Pb$
Neptunium	$^{237}_{93}Np$	2.14×10^6	$^{209}_{83}Bi$

44.6 Natural Radioactivity

Radioactive nuclei are generally classified into two groups: (1) unstable nuclei found in nature, which give rise to **natural radioactivity,** and (2) unstable nuclei produced in the laboratory through nuclear reactions, which exhibit **artificial radioactivity.**

As Table 44.4 shows, there are three series of naturally occurring radioactive nuclei. Each series starts with a specific long-lived radioactive isotope whose half-life exceeds that of any of its unstable descendants. The three natural series begin with the isotopes ^{238}U, ^{235}U, and ^{232}Th, and the corresponding stable end products are three isotopes of lead: ^{206}Pb, ^{207}Pb, and ^{208}Pb. The fourth series in Table 44.4 begins with ^{237}Np and has as its stable end product ^{209}Bi. The element ^{237}Np is a *transuranic* element (one having an atomic number greater than that of uranium) not found in nature. This element has a half-life of "only" 2.14×10^6 years.

Figure 44.17 shows the successive decays for the ^{232}Th series. Note that ^{232}Th first undergoes alpha decay to ^{228}Ra. Next, ^{228}Ra undergoes two successive beta decays to ^{228}Th. The series continues and finally branches when it reaches ^{212}Bi. At this point, there are two decay possibilities. The end of the decay series is the stable isotope ^{208}Pb. The sequence shown in Figure 44.17 is characterized by a mass-number decrease of either 4 (for alpha decays) or 0 (for beta or gamma decays). The two uranium series are more complex than the ^{232}Th series. Also, there are several naturally occurring radioactive isotopes, such as ^{14}C and ^{40}K, which are not part of any decay series.

Because of these radioactive series, our environment is constantly replenished with radioactive elements that would otherwise have disappeared long ago. For example, because the Solar System is approximately 5×10^9 years old, the supply of ^{226}Ra (whose half-life is only 1 600 years) would have been depleted by radioactive decay long ago if it were not for the radioactive series starting with ^{238}U.

Figure 44.17 Successive decays for the ^{232}Th series.

44.7 Nuclear Reactions

We have studied radioactivity, which is a spontaneous process in which the structure of a nucleus changes. It is also possible to stimulate changes in the structure of nuclei by bombarding them with energetic particles. Such collisions, which change the identity of the target nuclei, are called **nuclear reactions.** Rutherford was the first to observe them, in 1919, using naturally occurring radioactive sources for the bombarding particles. Since then, thousands of nuclear reactions have been observed following the development of charged-particle accelerators in the 1930s. With today's advanced technology in particle accelerators and particle detectors, it is possible to achieve particle energies of at least 1 000 GeV = 1 TeV. These high-energy particles are used to create new particles whose properties are helping to solve the mysteries of the nucleus.

Consider a reaction in which a target nucleus X is bombarded by a particle a, resulting in a daughter nucleus Y and an outgoing particle b:

$$a + X \longrightarrow Y + b$$

(44.27)   **Nuclear reaction**

Sometimes this reaction is written in the more compact form

$$X(a, b)Y$$

In Section 44.5, the Q value, or disintegration energy, of a radioactive decay was defined as the mass-energy transformed to kinetic energy as a result of the decay process. Likewise, we define the **reaction energy** Q associated with a nuclear reaction as *the total change in mass-energy resulting from the reaction*:

Reaction energy Q

$$Q = (M_a + M_X - M_Y - M_b)\,c^2 \qquad (44.28)$$

As an example, consider the reaction ^{7}Li (p, α)^{4}He. The notation p indicates a proton, which is a hydrogen nucleus. Thus, we can write this reaction in the expanded form

$$^1_1\text{H} + {}^7_3\text{Li} \longrightarrow {}^4_2\text{He} + {}^4_2\text{He}$$

The Q value for this reaction is 17.3 MeV. A reaction such as this, for which Q is positive, is called **exothermic.** A reaction for which Q is negative is called **endothermic.** In order to satisfy conservation of momentum, an endothermic reaction does not occur unless the bombarding particle has a kinetic energy greater than Q. (See problem 56.) The minimum energy necessary for such a reaction to occur is called the **threshold energy.**

Nuclear reactions must obey the law of conservation of linear momentum. Generally the only force acting on the interacting particles is their mutual force of interaction; that is, there are no external accelerating electric fields present near the colliding particles. Thus, the system is considered to be isolated for purposes of momentum conservation.

If particles a and b in a nuclear reaction are identical, so that X and Y are also necessarily identical, the reaction is called a **scattering event.** If the kinetic energy of the system (a and X) before the event is the same as that of the system (b and Y) after the event, it is classified as *elastic scattering.* If the kinetic energies of the system before and after the event are not the same, the reaction is described as *inelastic scattering.* In this case, the difference in energy is accounted for by the fact that the target nucleus has been raised to an excited state by the event. The final system now consists of b and an excited nucleus Y*, and eventually will become b, Y, and γ, where γ is the gamma-ray photon that is emitted when the system returns to the ground state. This elastic and inelastic terminology is identical to that used in describing collisions between macroscopic objects (Section 9.3).

Measured Q values for a number of nuclear reactions involving light nuclei are given in Table 44.5.

In addition to energy and momentum, the total charge and total number of nucleons must be conserved in any nuclear reaction. For example, consider the reaction ^{19}F(p, α)^{16}O, which has a Q value of 8.11 MeV. We can show this reaction more completely as

$$^1_1\text{H} + {}^{19}_9\text{F} \longrightarrow {}^{16}_8\text{O} + {}^4_2\text{He} \qquad (44.29)$$

The total number of nucleons before the reaction (1 + 19 = 20) is equal to the total number after the reaction (16 + 4 = 20). Furthermore, the total charge is the same before (1 + 9) and after (8 + 2) the reaction.

Example 44.11   Bombarding Fluorine

Verify that the Q value for the reaction in Equation 44.29 is 8.11 MeV.

Solution Using Table A.3 in the Appendix, we can evaluate the atomic masses in the expression for the Q value:

$$Q = (M_a + M_X - M_Y - M_b) \times 931.494 \text{ MeV/u}$$
$$= (1.007\,825 + 18.998\,403 \text{ u} - 15.994\,915 \text{ u}$$
$$- 4.002\,603 \text{ u}) \times 931.494 \text{ MeV/u}$$
$$= (0.008\,710 \text{ u}) \times 931.494 \text{ MeV/u} = \boxed{8.11 \text{ MeV}}$$

Table 44.5

Q Values for Nuclear Reactions Involving Light Nuclei	
Reaction[a]	**Q Value (MeV)**
$^2\text{H}(\text{n}, \gamma)\,^3\text{H}$	6.257
$^2\text{H}(\text{d}, \text{p})\,^3\text{H}$	4.033
$^6\text{Li}(\text{p}, \alpha)\,^3\text{He}$	4.019
$^6\text{Li}(\text{d}, \text{p})\,^7\text{Li}$	5.025
$^7\text{Li}(\text{p}, \text{n})\,^7\text{Be}$	-1.644
$^7\text{Li}(\text{p}, \alpha)\,^4\text{He}$	17.347
$^9\text{Be}(\text{n}, \gamma)\,^{10}\text{Be}$	6.812
$^9\text{Be}(\gamma, \text{n})\,^8\text{Be}$	-1.665
$^9\text{Be}(\text{d}, \text{p})\,^{10}\text{Be}$	4.588
$^9\text{Be}(\text{p}, \alpha)\,^6\text{Li}$	2.125
$^{10}\text{B}(\text{n}, \alpha)\,^7\text{Li}$	2.789
$^{10}\text{B}(\text{p}, \alpha)\,^7\text{Be}$	1.145
$^{12}\text{C}(\text{n}, \gamma)\,^{13}\text{C}$	4.946
$^{13}\text{C}(\text{p}, \text{n})\,^{13}\text{N}$	-3.003
$^{14}\text{N}(\text{n}, \text{p})\,^{14}\text{C}$	0.626
$^{14}\text{N}(\text{n}, \gamma)\,^{15}\text{N}$	10.833
$^{18}\text{O}(\text{p}, \text{n})\,^{18}\text{F}$	-2.438
$^{19}\text{F}(\text{p}, \alpha)\,^{16}\text{O}$	8.114

a The symbols n, p, d, α, and γ denote the neutron, proton, deuteron, alpha particle, and photon, respectively.

44.8 Nuclear Magnetic Resonance and Magnetic Resonance Imaging

In this section, we shall describe an important application of nuclear physics in medicine called magnetic resonance imaging. In order to understand this application, we must first discuss the spin angular momentum of the nucleus. This discussion will have parallels with the discussion of spin for atomic electrons.

In Chapter 42, we discussed the fact that the electron has an intrinsic angular momentum, which we called spin. Nuclei also have spin because their component particles—neutrons and protons—each have spin $\frac{1}{2}$, as well as orbital angular momentum within the nucleus. All types of angular momentum obey the quantum rules that we outlined for orbital and spin angular momentum in Chapter 42. In particular, there are two quantum numbers associated with the angular momentum that determine the allowed values of the magnitude of the angular momentum vector and its direction in space. The magnitude of the nuclear angular momentum is $\sqrt{I(I+1)}\,\hbar$, where I is called the **nuclear spin quantum number** and may be an integer or a half-integer, depending on how the individual proton and neutron spins combine. The maximum value of the z component of the spin angular momentum vector is $I\hbar$. Figure 44.18 is a vector model (see Section 42.6) illustrating the possible orientations of the nuclear spin vector and its projections along the z axis for the case in which $I = \frac{3}{2}$.

Nuclear spin has an associated, corresponding nuclear magnetic moment, similar to that of the electron. The spin magnetic moment of a nucleus is measured in terms of the **nuclear magneton** μ_n, a unit of moment defined as

$$\mu_n \equiv \frac{e\hbar}{2m_p} = 5.05 \times 10^{-27}\,\text{J/T} \qquad (44.30)$$

Nuclear magneton

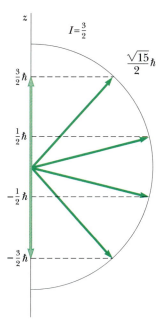

Figure 44.18 A vector model showing possible orientations of the nuclear spin angular momentum vector and its projections along the z axis for the case $I = \frac{3}{2}$.

where m_p is the mass of the proton. This definition is analogous to that of the Bohr magneton μ_B, which corresponds to the spin magnetic moment of a free electron (see Section 42.6). Note that μ_n is smaller than μ_B ($= 9.274 \times 10^{-24}$ J/T) by a factor of 1 836 because of the large difference between the proton mass and the electron mass.

The magnetic moment of a free proton is $2.792\,8\mu_n$. Unfortunately, there is no general theory of nuclear magnetism that explains this value. The neutron also has a magnetic moment, which has a value of $-1.913\,5\mu_n$. The negative sign indicates that this moment is opposite the spin angular momentum of the neutron. The existence of a magnetic moment for the neutron is surprising in view of the fact that the neutron is uncharged. This suggests that the neutron is not a fundamental particle but rather has an underlying structure consisting of charged constituents. We shall explore this structure in Chapter 46.

Quick Quiz 44.8 Which of the following do you expect *not* to vary substantially among different isotopes of an element? (a) atomic mass number, (b) nuclear spin magnetic moment, (c) chemical properties?

Nuclear magnetic moments, as well as electronic magnetic moments, precess when placed in an external magnetic field. This is due to the fact that a magnetic torque is applied to the spin angular momentum, so that precession occurs just as it does for a rotating top in a gravitational field, as discussed in Section 11.5. The frequency at which a magnetic moment precesses, called the **Larmor precessional frequency** ω_p, is directly proportional to the magnitude of the magnetic field. This is described schematically in Figure 44.19a, in which the external magnetic field is along the z axis. For example, the Larmor frequency of a proton in a 1-T magnetic field is 42.577 MHz. The potential energy of a magnetic dipole moment $\boldsymbol{\mu}$ in an external magnetic field $\mathbf{B}$ is given by $-\boldsymbol{\mu} \cdot \mathbf{B}$ (Eq. 29.12). When the magnetic moment $\boldsymbol{\mu}$ is lined up with the field as closely as quantum physics allows, the potential energy of the dipole moment in the field has its minimum value E_{min}. When $\boldsymbol{\mu}$ is as antiparallel to the field as possible, the potential energy has its maximum value E_{max}. In general, there are other energy states between these values corresponding to the quantized directions of the magnetic moment with respect to the field. For a nucleus with spin $\frac{1}{2}$, there are only two allowed states, with energies E_{min} and E_{max}. These two energy states are shown in Figure 44.19b.

It is possible to observe transitions between these two spin states using a technique called **NMR,** for **nuclear magnetic resonance.** A constant magnetic field (**B** in

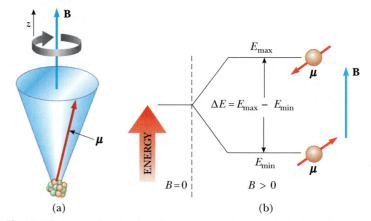

Figure 44.19 (a) When a nucleus is placed in an external magnetic field **B**, the nuclear spin magnetic moment precesses about the magnetic field with a frequency proportional to the magnitude of the field. (b) A nucleus with spin $\frac{1}{2}$ can occupy one of two energy states when placed in an external magnetic field. The lower energy state E_{min} corresponds to the case where the spin is aligned with the field as much as possible according to quantum mechanics, and the higher energy state E_{max} corresponds to the case where the spin is opposite the field as much as possible.

Figure 44.20 Experimental arrangement for nuclear magnetic resonance. The radio-frequency magnetic field created by the coil surrounding the sample and provided by the variable-frequency oscillator is perpendicular to the constant magnetic field created by the electromagnet. When the nuclei in the sample meet the resonance condition, the nuclei absorb energy from the radio-frequency field of the coil, and this absorption changes the characteristics of the circuit in which the coil is included. Most modern NMR spectrometers use superconducting magnets at fixed field strengths and operate at frequencies of approximately 200 MHz.

Fig. 44.19a) is introduced to define a z axis. A second, weaker, oscillating magnetic field is then applied perpendicular to **B**. When the frequency of the oscillating field is adjusted to match the Larmor precessional frequency of the nuclei in the sample, a torque that acts on the precessing magnetic moments causes them to "flip" between the two spin states shown in Figure 44.19b. These transitions result in a net absorption of energy by the nuclei, an absorption that can be detected electronically.

A diagram of the apparatus used in nuclear magnetic resonance is illustrated in Figure 44.20. The energy absorbed by the nuclei is supplied by the generator producing the oscillating magnetic field. Nuclear magnetic resonance and a related technique called *electron spin resonance* are extremely important methods for studying nuclear and atomic systems and the ways in which these systems interact with their surroundings.

A widely used medical diagnostic technique called **MRI,** for **magnetic resonance imaging,** is based on nuclear magnetic resonance. Because nearly two thirds of the atoms in the human body are hydrogen (which gives a strong signal), MRI works exceptionally well for viewing internal tissues. The patient is placed inside a large solenoid that supplies a magnetic field that is constant in time but whose magnitude varies spatially across the body. Because of the variation in the field, protons in different parts of the body precess at different frequencies, so the resonance signal can be used to provide information about the positions of the protons. A computer is used to analyze the position information to provide data for constructing a final image. An MRI scan showing incredible detail in internal body structure is shown in Figure 44.21.

The main advantage of MRI over other imaging techniques is that it causes minimal cellular damage. The photons associated with the radio-frequency signals used in MRI

SBHA/Getty Images

Figure 44.21 A color-enhanced MRI scan of a human brain.

have energies of only about 10^{-7} eV. Because molecular bond strengths are much larger (approximately 1 eV), the radio-frequency radiation causes little cellular damage. In comparison, x-rays have energies ranging from 10^4 to 10^6 eV and can cause considerable cellular damage. Thus, despite some individuals' fears of the word *nuclear* associated with MRI, the radio-frequency radiation involved is overwhelmingly safer than the x-rays that these individuals might accept more readily! A disadvantage of MRI is that the equipment required to conduct the procedure is very expensive so MRI images are costly.

The magnetic field produced by the solenoid is sufficient to lift a car, and the radio signal is about the same magnitude as that from a small commercial broadcasting station! While MRI is inherently safe in normal use, the strong magnetic field of the solenoid requires diligent care to make sure that no ferromagnetic materials are located in the room near the MRI apparatus. Several accidents have occurred, such as a 2000 incident in which a gun was pulled from a police officer's hand and discharged upon striking the machine.

SUMMARY

Take a practice test for this chapter by clicking on the Practice Test link at http://www.pse6.com.

A nucleus is represented by the symbol $^A_Z X$, where A is the **mass number** (the total number of nucleons), and Z is the **atomic number** (the total number of protons). The total number of neutrons in a nucleus is the **neutron number** N, where $A = N + Z$. Nuclei having the same Z value but different A and N values are **isotopes** of each other. Assuming that nuclei are spherical, their radius is given by

$$r = r_0 A^{1/3} \qquad (44.1)$$

where $r_0 = 1.2$ fm.

Nuclei are stable because of the **nuclear force** between nucleons. This short-range force dominates the Coulomb repulsive force at distances of less than about 2 fm and is independent of charge. Light stable nuclei have equal numbers of protons and neutrons. Heavy stable nuclei have more neutrons than protons. The most stable nuclei have Z and N values that are both even.

The difference between the sum of the masses of a group of separate nucleons and the mass of the compound nucleus containing these nucleons, when multiplied by c^2, gives the **binding energy** E_b of the nucleus. We can calculate the binding energy of the nucleus of mass M_A using the expression

$$E_b = (Z m_p + N m_n - M_A) \times 931.494 \text{ MeV/u} \qquad (44.2)$$

where m_p is the mass of the proton and m_n is the mass of the neutron.

The **liquid-drop model** of nuclear structure treats the nucleons as molecules in a drop of liquid. The four main contributions influencing binding energy are the volume effect, the surface effect, the Coulomb repulsion effect, and the symmetry effect. Summing such contributions results in the **semiempirical binding-energy formula:**

$$E_b = C_1 A - C_2 A^{2/3} - C_3 \frac{Z(Z-1)}{A^{1/3}} - C_4 \frac{(N-Z)^2}{A} \qquad (44.3)$$

The **shell model,** or **independent-particle model,** assumes that each nucleon exists in a shell and can only have discrete energy values. The stability of certain nuclei can be explained with this model.

A radioactive substance decays by **alpha decay, beta decay,** or **gamma decay.** An alpha particle is the ^{4}He nucleus; a beta particle is either an electron (e^-) or a positron (e^+); a gamma particle is a high-energy photon.

If a radioactive material contains N_0 radioactive nuclei at $t = 0$, the number N of nuclei remaining after a time t has elapsed is

$$N = N_0 e^{-\lambda t} \qquad (44.6)$$

where λ is the **decay constant,** a number equal to the probability per second that a nucleus will decay. The **decay rate,** or **activity,** of a radioactive substance is

$$R = \left| \frac{dN}{dt} \right| = R_0 e^{-\lambda t} \tag{44.7}$$

where $R_0 = N_0 \lambda$ is the activity at $t = 0$. The **half-life** $T_{1/2}$ is defined as the time interval required for half of a given number of radioactive nuclei to decay, where

$$T_{1/2} = \frac{0.693}{\lambda} \tag{44.8}$$

In alpha decay, a helium nucleus is ejected from the parent nucleus with a discrete set of kinetic energies. A nucleus undergoing beta decay emits either an electron (e^-) and an antineutrino ($\bar{\nu}$) or a positron (e^+) and a neutrino (ν). The electron or positron is ejected with a range of energies. In **electron capture,** the nucleus of an atom absorbs one of its own electrons and emits a neutrino. In gamma decay, a nucleus in an excited state decays to its ground state and emits a gamma ray.

Nuclear reactions can occur when a target nucleus X is bombarded by a particle a, resulting in a daughter nucleus Y and an outgoing particle b:

$$a + X \longrightarrow Y + b \tag{44.27}$$

The mass-energy conversion in such a reaction, called the **reaction energy** Q, is

$$Q = (M_a + M_X - M_Y - M_b)c^2 \tag{44.28}$$

Nuclei have an intrinsic spin angular momentum of magnitude $\sqrt{I(I+1)}\,\hbar$, where I is the **nuclear spin quantum number.** The magnetic moment of a nucleus is measured in terms of the **nuclear magneton** μ_n, where

$$\mu_n \equiv \frac{e\hbar}{2m_p} = 5.05 \times 10^{-27} \, \text{J/T} \tag{44.30}$$

When a nuclear spin magnetic moment is placed in an external magnetic field, it precesses about the field with a frequency (the **Larmor precessional frequency**) that is proportional to the magnitude of the field.

QUESTIONS

1. In Rutherford's experiment, assume that an alpha particle is headed directly toward the nucleus of an atom. Why doesn't the alpha particle make physical contact with the nucleus?

2. How many protons are there in the nucleus $^{222}_{86}\text{Rn}$? How many neutrons? How many orbiting electrons are there in the neutral atom?

3. Element X has several isotopes. What do these isotopes have in common? How do they differ?

4. Why are very heavy nuclei unstable?

5. Explain why nuclei that are well off the line of stability in Figure 44.4 tend to be unstable.

6. Consider two heavy nuclei X and Y having similar mass numbers. If X has the higher binding energy, which nucleus tends to be more unstable?

7. Why do nearly all the naturally occurring isotopes lie above the $N = Z$ line in Figure 44.4?

8. Discuss the differences between the liquid-drop model and the shell model of the nucleus.

9. Would the liquid-drop or shell model be more appropriate to predict the behavior of a nucleus in a fission reaction? Which would be more successful in predicting the magnetic moment of a given nucleus? Which could better explain the gamma-ray spectrum of an excited nucleus?

10. From Table A.3, identify the four stable nuclei that have magic numbers in both Z and N.

11. If a nucleus has a half-life of 1 year, does this mean it will be completely decayed after 2 years? Explain.

12. "If no more people were to be born, the law of population growth would strongly resemble the radioactive decay law." Discuss this statement.

13. Two samples of the same radioactive nuclide are prepared. Sample A has twice the initial activity of sample B. How does the half-life of A compare with the half-life of B? After each has passed through five half-lives, what is the ratio of their activities?

14. What fraction of a radioactive sample has decayed after two half-lives have elapsed?

15. Explain why the half-lives for radioactive nuclei are essentially independent of temperature.

16. The radioactive nucleus $^{226}_{88}$Ra has a half-life of approximately 1.6×10^3 years. Being that the Solar System is about 5 billion years old, why do we still find this nucleus in nature (Figure Q44.16)?

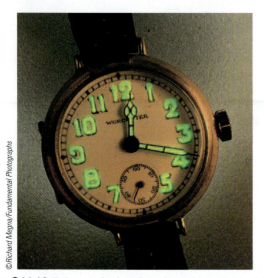

Figure Q44.16 Paint on the hands and numbers of this older watch contains minute amounts of natural radium mixed with a phosphorescent material. The radioactive decay of radium causes the phosphor to glow continuously.

17. A free neutron undergoes beta decay with a half-life of about 10 min. Can a free proton undergo a similar decay?

18. What is the difference between a neutrino and a photon?

19. Pick any beta-decay process and show that the neutrino must have zero charge.

20. Use Equations 44.20 to 44.22 to explain why the neutrino must have a spin of $\frac{1}{2}$.

21. If a nucleus such as ^{226}Ra initially at rest undergoes alpha decay, which has more kinetic energy after the decay, the alpha particle or the daughter nucleus?

22. Use the analogy of a bullet and a rifle to explain why the recoiling nucleus carries off only a very small fraction of the disintegration energy in the alpha decay of a nucleus.

23. Can a nucleus emit alpha particles that have different energies? Explain.

24. Explain why many heavy nuclei undergo alpha decay but do not spontaneously emit neutrons or protons.

25. If an alpha particle and an electron have the same kinetic energy, which undergoes the greater deflection when passed through a magnetic field?

26. If film is kept in a wooden box, alpha particles from a radioactive source outside the box cannot expose the film but beta particles can. Explain.

27. Does the reaction energy of a nuclear reaction represent the quantity (final mass − initial mass)c^2, or does it represent the quantity (initial mass − final mass)c^2?

28. Explain how you can carbon-date the age of a sample.

29. Suppose it could be shown that the cosmic-ray intensity at the Earth's surface was much greater 10 000 years ago. How would this difference affect what we accept as valid carbon-dated values of the age of ancient samples of once-living matter?

30. Why is carbon dating unable to provide accurate measurements of the ages of very old material?

31. Explain the main differences between alpha, beta, and gamma rays.

32. How many values of I_z are possible for $I = 5/2$? for $I = 3$?

33. In nuclear magnetic resonance, how does increasing the value of the constant magnetic field change the frequency of the radio-frequency field that excites a particular transition?

34. Do all natural events have causes? Is the Universe intelligible? Give reasons for your answers. *Note:* You may wish to consider again Question 21 in Chapter 6, on whether the future is determinate.

PROBLEMS

1, 2, 3 = straightforward, intermediate, challenging   ☐ = full solution available in the *Student Solutions Manual and Study Guide*

🌐 = coached solution with hints available at http://www.pse6.com   🖥 = computer useful in solving problem

▭ = paired numerical and symbolic problems

Note: Atomic masses are listed in Table A.3 in Appendix A.

Section 44.1 Some Properties of Nuclei

1. What is the order of magnitude of the number of protons in your body? of the number of neutrons? of the number of electrons?

2. **Review problem.** Singly ionized carbon is accelerated through 1 000 V and passed into a mass spectrometer to determine the isotopes present (see Chapter 29). The

magnitude of the magnetic field in the spectrometer is 0.200 T. (a) Determine the orbit radii for the ^{12}C and the ^{13}C isotopes as they pass through the field. (b) Show that the ratio of radii may be written in the form

$$\frac{r_1}{r_2} = \sqrt{\frac{m_1}{m_2}}$$

and verify that your radii in part (a) agree with this.

3. **Review problem.** An alpha particle ($Z = 2$, mass 6.64×10^{-27} kg) approaches to within 1.00×10^{-14} m of a carbon nucleus ($Z = 6$). What are (a) the maximum Coulomb

force on the alpha particle, (b) the acceleration of the alpha particle at this point, and (c) the potential energy of the alpha particle–nucleus system when the alpha particle is at this point?

4. In a Rutherford scattering experiment, alpha particles having kinetic energy of 7.70 MeV are fired toward a gold nucleus. (a) Use energy conservation to determine the distance of closest approach between the alpha particle and gold nucleus. Assume the nucleus remains at rest. (b) **What If?** Calculate the de Broglie wavelength for the 7.70-MeV alpha particle and compare it to the distance obtained in part (a). (c) Based on this comparison, why is it proper to treat the alpha particle as a particle and not as a wave in the Rutherford scattering experiment?

5. (a) Use energy methods to calculate the distance of closest approach for a head-on collision between an alpha particle having an initial energy of 0.500 MeV and a gold nucleus (^{197}Au) at rest. (Assume the gold nucleus remains at rest during the collision.) (b) What minimum initial speed must the alpha particle have to get as close as 300 fm?

6. How much energy (in MeV units) must an alpha particle have to reach the surface of a gold nucleus ($Z = 79$, $A = 197$)? Assume the gold nucleus remains stationary.

7. Find the radius of (a) a nucleus of $^4_2$He and (b) a nucleus of $^{238}_{92}$U.

8. Find the nucleus that has a radius approximately equal to half the radius of uranium $^{238}_{92}$U.

9. A star ending its life with a mass of two times the mass of the Sun is expected to collapse, combining its protons and electrons to form a neutron star. Such a star could be thought of as a gigantic atomic nucleus. If a star of mass $2 \times 1.99 \times 10^{30}$ kg collapsed into neutrons ($m_n = 1.67 \times 10^{-27}$ kg), what would its radius be? (Assume that $r = r_0 A^{1/3}$.)

10. **Review problem.** What would be the gravitational force between two golf balls (each with a 4.30-cm diameter), 1.00 m apart, if they were made of nuclear matter?

11. From Table A.3, identify the stable nuclei that correspond to the magic numbers given by Equation 44.4.

12. Consider the stable isotopes in Table A.3 with abundances over 25.0%. That is, select the nuclei that are not radioactive and that occur in more than 25.0% of the atoms of each element in nature. Count the number of these nuclei that are (a) even Z, even N; (b) even Z, odd N; (c) odd Z, even N; and (d) odd Z, odd N.

13. Nucleus 1 has eight times as many protons as nucleus 2, five times as many neutrons, and six times as many nucleons as nucleus 2. Nucleus 1 has four more neutrons than protons. (a) What are the two nuclei? (b) Is each nucleus stable? If not, what is the minimum number of neutrons that must be added to, or removed from, each unstable nucleus to make it stable?

Section 44.2 Nuclear Binding Energy

14. Calculate the binding energy per nucleon for (a) ^{2}H, (b) ^{4}He, (c) ^{56}Fe, and (d) ^{238}U.

15. The iron isotope ^{56}Fe is near the peak of the stability curve. This is why iron is generally prominent in the spectrum of the Sun and stars. Show that ^{56}Fe has a higher binding energy per nucleon than its neighbors ^{55}Mn and ^{59}Co. Compare your results with Figure 44.5.

16. Two nuclei having the same mass number are known as *isobars*. Calculate the difference in binding energy per nucleon for the isobars $^{23}_{11}$Na and $^{23}_{12}$Mg. How do you account for the difference?

17. Nuclei having the same mass numbers are called *isobars*. The isotope $^{139}_{57}$La is stable. A radioactive isobar, $^{139}_{59}$Pr, is located below the line of stable nuclei in Figure 44.4 and decays by e^+ emission. Another radioactive isobar of $^{139}_{57}$La, $^{139}_{55}$Cs, decays by e^- emission and is located above the line of stable nuclei in Figure 44.4. (a) Which of these three isobars has the highest neutron-to-proton ratio? (b) Which has the greatest binding energy per nucleon? (c) Which do you expect to be heavier, $^{139}_{59}$Pr or $^{139}_{55}$Cs?

18. The energy required to construct a uniformly charged sphere of total charge Q and radius R is $U = 3k_e Q^2/5R$, where k_e is the Coulomb constant (see Problem 71). Assume that a ^{40}Ca nucleus contains 20 protons uniformly distributed in a spherical volume. (a) How much energy is required to counter their electrical repulsion according to the above equation? (*Suggestion:* First calculate the radius of a ^{40}Ca nucleus.) (b) Calculate the binding energy of ^{40}Ca. (c) Explain what you can conclude from comparing the result of part (b) and that of part (a).

19. A pair of nuclei for which $Z_1 = N_2$ and $Z_2 = N_1$ are called *mirror isobars* (the atomic and neutron numbers are interchanged). Binding-energy measurements on these nuclei can be used to obtain evidence of the charge independence of nuclear forces (that is, proton–proton, proton–neutron, and neutron–neutron nuclear forces are equal). Calculate the difference in binding energy for the two mirror isobars $^{15}_8$O and $^{15}_7$N. The electric repulsion among eight protons rather than seven accounts for the difference.

20. Calculate the minimum energy required to remove a neutron from the $^{43}_{20}$Ca nucleus.

Section 44.3 Nuclear Models

21. Using the graph in Figure 44.5, estimate how much energy is released when a nucleus of mass number 200 fissions into two nuclei each of mass number 100.

22. (a) In the liquid-drop model of nuclear structure, why does the surface-effect term $-C_2 A^{2/3}$ have a negative sign? (b) **What If?** The binding energy of the nucleus increases as the volume-to-surface ratio increases. Calculate this ratio for both spherical and cubical shapes, and explain which is more plausible for nuclei.

23. (a) Use the semiempirical binding-energy formula to compute the binding energy for $^{56}_{26}$Fe. (b) What percentage is contributed to the binding energy by each of the four terms?

Section 44.4 Radioactivity

24. The half-life of ^{131}I is 8.04 days. On a certain day, the activity of an iodine-131 sample is 6.40 mCi. What is its activity 40.2 days later?

25. A sample of radioactive material contains 1.00×10^{15} atoms and has an activity of 6.00×10^{11} Bq. What is its half-life?

26. Determine the activity of 1.00 g of ^{60}Co. The half-life of ^{60}Co is 5.27 yr.

27. A freshly prepared sample of a certain radioactive isotope has an activity of 10.0 mCi. After 4.00 h, its activity is 8.00 mCi. (a) Find the decay constant and half-life. (b) How many atoms of the isotope were contained in the freshly prepared sample? (c) What is the sample's activity 30.0 h after it is prepared?

28. How much time elapses before 90.0% of the radioactivity of a sample of $^{72}_{33}$As disappears, as measured by its activity? The half-life of $^{72}_{33}$As is 26 h.

29. The radioactive isotope ^{198}Au has a half-life of 64.8 h. A sample containing this isotope has an initial activity ($t = 0$) of 40.0 μCi. Calculate the number of nuclei that decay in the time interval between $t_1 = 10.0$ h and $t_2 = 12.0$ h.

30. A radioactive nucleus has half-life $T_{1/2}$. A sample containing these nuclei has initial activity R_0. Calculate the number of nuclei that decay during the interval between the times t_1 and t_2.

31. In an experiment on the transport of nutrients in the root structure of a plant, two radioactive nuclides X and Y are used. Initially 2.50 times more nuclei of type X are present than of type Y. Just three days later there are 4.20 times more nuclei of type X than of type Y. Isotope Y has a half-life of 1.60 d. What is the half-life of isotope X?

32. (a) The daughter nucleus formed in radioactive decay is often radioactive. Let N_{10} represent the number of parent nuclei at time $t = 0$, $N_1(t)$ the number of parent nuclei at time t, and λ_1 the decay constant of the parent. Suppose the number of daughter nuclei at time $t = 0$ is zero, let $N_2(t)$ be the number of daughter nuclei at time t, and let λ_2 be the decay constant of the daughter. Show that $N_2(t)$ satisfies the differential equation

$$\frac{dN_2}{dt} = \lambda_1 N_1 - \lambda_2 N_2$$

(b) Verify by substitution that this differential equation has the solution

$$N_2(t) = \frac{N_{10}\lambda_1}{\lambda_1 - \lambda_2}(e^{-\lambda_2 t} - e^{-\lambda_1 t})$$

This equation is the law of successive radioactive decays. (c) ^{218}Po decays into ^{214}Pb with a half-life of 3.10 min, and ^{214}Pb decays into ^{214}Bi with a half-life of 26.8 min. On the same axes, plot graphs of $N_1(t)$ for ^{218}Po and $N_2(t)$ for ^{214}Pb. Let $N_{10} = 1\,000$ nuclei, and choose values of t from 0 to 36 min in 2-min intervals. The curve for ^{214}Pb at first rises to a maximum and then starts to decay. At what instant t_m is the number of ^{214}Pb nuclei a maximum? (d) By applying the condition for a maximum $dN_2/dt = 0$, derive a symbolic equation for t_m in terms of λ_1 and λ_2. Does the value obtained in (c) agree with this equation?

Section 44.5   The Decay Processes

33. Find the energy released in the alpha decay

$$^{238}_{92}\text{U} \longrightarrow {}^{234}_{90}\text{Th} + {}^{4}_{2}\text{He}$$

You will find Table A.3 useful.

34. Identify the missing nuclide or particle (X):
(a) X $\rightarrow$ $^{65}_{28}$Ni + γ
(b) $^{215}_{84}$Po $\rightarrow$ X + α
(c) X $\rightarrow$ $^{55}_{26}$Fe + e$^+$ + ν
(d) $^{109}_{48}$Cd + X $\rightarrow$ $^{109}_{47}$Ag + ν
(e) $^{14}_{7}$N + $^{4}_{2}$He $\rightarrow$ X + $^{17}_{8}$O

35. A living specimen in equilibrium with the atmosphere contains one atom of ^{14}C (half-life = 5 730 yr) for every 7.7×10^{11} stable carbon atoms. An archeological sample of wood (cellulose, $C_{12}H_{22}O_{11}$) contains 21.0 mg of carbon. When the sample is placed inside a shielded beta counter with 88.0% counting efficiency, 837 counts are accumulated in one week. Assuming that the cosmic-ray flux and the Earth's atmosphere have not changed appreciably since the sample was formed, find the age of the sample.

36. A certain African artifact is found to have a carbon-14 activity of (0.12 ± 0.01) Bq per gram of carbon. Assume the uncertainty is negligible in the half-life of ^{14}C (5 730 yr) and in the activity of atmospheric carbon (0.25 Bq per gram). The age of the object lies within what range?

37. A ^{3}H nucleus beta decays into ^{3}He by creating an electron and an antineutrino according to the reaction

$$^{3}_{1}\text{H} \longrightarrow {}^{3}_{2}\text{He} + \text{e}^- + \bar{\nu}$$

The symbols in this reaction refer to nuclei. Write the reaction referring to neutral atoms by adding one electron to both sides. Then use Table A.3 to determine the total energy released in this reaction.

38. Determine which decays can occur spontaneously:
(a) $^{40}_{20}$Ca $\rightarrow$ e$^+$ + $^{40}_{19}$K
(b) $^{98}_{44}$Ru $\rightarrow$ $^{4}_{2}$He + $^{94}_{42}$Mo
(c) $^{144}_{60}$Nd $\rightarrow$ $^{4}_{2}$He + $^{140}_{58}$Ce

39. The nucleus $^{15}_{8}$O decays by electron capture. The nuclear reaction is written

$$^{15}_{8}\text{O} + \text{e}^- \longrightarrow {}^{15}_{7}\text{N} + \nu$$

(a) Write the process going on for a single particle within the nucleus. (b) Write the decay process referring to neutral atoms. (c) Determine the energy of the neutrino. Disregard the daughter's recoil.

Section 44.6   Natural Radioactivity

40. A rock sample contains traces of ^{238}U, ^{235}U, ^{232}Th, ^{208}Pb, ^{207}Pb, and ^{206}Pb. Careful analysis shows that the ratio of the amount of ^{238}U to ^{206}Pb is 1.164. (a) Assume that the rock originally contained no lead, and determine the age of the rock. (b) What should be the ratios of ^{235}U to ^{207}Pb and of ^{232}Th to ^{208}Pb so that they would yield the same age for the rock? Ignore the minute amounts of the intermediate decay products in the decay chains. Note that this form of multiple dating gives reliable geological dates.

41. Enter the correct isotope symbol in each open square in Figure P44.41, which shows the sequences of decays in the natural radioactive series starting with the long-lived isotope uranium-235 and ending with the stable nucleus lead-207.

Figure P44.41

42. *Indoor air pollution.* Uranium is naturally present in rock and soil. At one step in its series of radioactive decays, ^{238}U produces the chemically inert gas radon-222, with a half-life of 3.82 days. The radon seeps out of the ground to mix into the atmosphere, typically making open air radioactive with activity 0.3 pCi/L. In homes ^{222}Rn can be a serious pollutant, accumulating to reach much higher activities in enclosed spaces. If the radon radioactivity exceeds 4 pCi/L, the Environmental Protection Agency suggests taking action to reduce it, such as by reducing infiltration of air from the ground. (a) Convert the activity 4 pCi/L to units of becquerel per cubic meter. (b) How many ^{222}Rn atoms are in one cubic meter of air displaying this activity? (c) What fraction of the mass of the air does the radon constitute?

43. Two radioactive samples consist of nuclei X in one sample, and nuclei Y in the other. The two samples have the same initial activity. After 0.685 h, the activity of the sample containing nuclei X is 1.04 times the activity of the other sample. The half-life of nucleus X exceeds the half-life of nucleus Y by 77.2 h. (a) Find the half-life of each nucleus. (b) Use Table A.3 to identify the two nuclei. (c) Nucleus X decays to nucleus Y by an alpha-decay chain. How many alpha-decay events are there in the chain?

44. The most common isotope of radon is ^{222}Rn, which has half-life 3.82 days. (a) What fraction of the nuclei that were on Earth one week ago are now undecayed? (b) What fraction of those that existed one year ago? (c) In view of these results, explain why radon remains a problem, contributing significantly to our background radiation exposure.

Section 44.7   Nuclear Reactions

45. The reaction $^{27}_{13}$Al$(\alpha, n)^{30}_{15}$P, achieved in 1934, was the first known in which the product nucleus is radioactive. Calculate the Q value of this reaction.

46. Identify the unknown nuclei and particles X and X' in the following nuclear reactions:
    (a) $X + {}^{4}_{2}He \rightarrow {}^{24}_{12}Mg + {}^{1}_{0}n$
    (b) ${}^{235}_{92}U + {}^{1}_{0}n \rightarrow {}^{90}_{38}Sr + X + 2{}^{1}_{0}n$
    (c) $2{}^{1}_{1}H \rightarrow {}^{2}_{1}H + X + X'$

47. Natural gold has only one isotope, ${}^{197}_{79}$Au. If natural gold is irradiated by a flux of slow neutrons, electrons are emitted. (a) Write the reaction equation. (b) Calculate the maximum energy of the emitted electrons.

48. A beam of 6.61-MeV protons is incident on a target of ${}^{27}_{13}$Al. Those that collide produce the reaction

$$p + {}^{27}_{13}Al \longrightarrow {}^{27}_{14}Si + n$$

($^{27}_{14}$Si has mass 26.986 705 u.) Ignoring any recoil of the product nucleus, determine the kinetic energy of the emerging neutrons.

49. Using the mass of ^{9}Be from Table A.3 and Q values of appropriate reactions from Table 44.5, calculate the masses of ^{8}Be and ^{10}Be in atomic mass units to four decimal places.

50. (a) Suppose ${}^{10}_{5}$B is struck by an alpha particle, releasing a proton and a product nucleus in the reaction. What is the product nucleus? (b) An alpha particle and a product nucleus are produced when ${}^{13}_{6}$C is struck by a proton. What is the product nucleus?

51. Determine the Q value associated with the spontaneous fission of ^{236}U into the fragments ^{90}Rb and ^{143}Cs, which have mass 89.914 809 u and 142.927 330 u, respectively. The masses of the other particles involved in the reaction are given in Table A.3.

Section 44.8   Nuclear Magnetic Resonance and Magnetic Resonance Imaging

52. Construct a diagram like that of Figure 44.18 for the cases when I equals (a) 5/2 and (b) 4.

53. The radio frequency at which a nucleus displays resonance absorption between spin states is called the Larmor precessional frequency and is given by

$$f = \frac{\Delta E}{h} = \frac{2\mu B}{h}$$

Calculate the Larmor frequency for (a) free neutrons in a magnetic field of 1.00 T, (b) free protons in a magnetic field of 1.00 T, and (c) free protons in the earth's magnetic field at a location where the magnitude of the field is 50.0 μT.

Additional Problems

54. As part of his discovery of the neutron in 1932, James Chadwick determined the mass of the newly identified particle by firing a beam of fast neutrons, all having the same speed, at two different targets and measuring the maximum recoil speeds of the target nuclei. The maximum speeds arise when an elastic head-on collision occurs between a neutron and a stationary target nucleus. (a) Represent the masses and final speeds of the two target nuclei as m_1, v_1, m_2, and v_2, and assume Newtonian mechanics

applies. Show that the neutron mass can be calculated from the equation

$$m_n = \frac{m_1 v_1 - m_2 v_2}{v_2 - v_1}$$

(b) Chadwick directed a beam of neutrons (produced from a nuclear reaction) on paraffin, which contains hydrogen. The maximum speed of the protons ejected was found to be 3.3×10^7 m/s. Because the velocity of the neutrons could not be determined directly, a second experiment was performed using neutrons from the same source and nitrogen nuclei as the target. The maximum recoil speed of the nitrogen nuclei was found to be 4.7×10^6 m/s. The masses of a proton and a nitrogen nucleus were taken as 1 u and 14 u, respectively. What was Chadwick's value for the neutron mass?

55. (a) One method of producing neutrons for experimental use is bombardment of light nuclei with alpha particles. In the method used by Chadwick in 1932, alpha particles emitted by polonium are incident on beryllium nuclei:

$$^4_2\text{He} + ^9_4\text{Be} \longrightarrow ^{12}_6\text{C} + ^1_0\text{n}$$

What is the Q value? (b) Neutrons are also often produced by small-particle accelerators. In one design, deuterons accelerated in a Van de Graaff generator bombard other deuterium nuclei:

$$^2_1\text{H} + ^2_1\text{H} \longrightarrow ^3_2\text{He} + ^1_0\text{n}$$

Is this reaction exothermic or endothermic? Calculate its Q value.

56. When the nuclear reaction represented by Equation 44.27 is endothermic, the reaction energy Q is negative. For the reaction to proceed, the incoming particle must have a minimum energy called the threshold energy, E_{th}. Some fraction of the energy of the incident particle is transferred to the compound nucleus to conserve momentum. Therefore, E_{th} must be greater than Q. (a) Show that

$$E_{th} = -Q\left(1 + \frac{M_a}{M_X}\right)$$

(b) Calculate the threshold energy of the incident alpha particle in the reaction

$$^4_2\text{He} + ^{14}_7\text{N} \longrightarrow ^{17}_8\text{O} + ^1_1\text{H}$$

57. One method of producing neutrons for experimental use is to bombard ^7_3Li with protons. The neutrons are emitted according to the reaction

$$^1_1\text{H} + ^7_3\text{Li} \longrightarrow ^7_4\text{Be} + ^1_0\text{n}$$

What is the minimum kinetic energy the incident proton must have if this reaction is to occur? You may use the result of Problem 56.

58. A byproduct of some fission reactors is the isotope $^{239}_{94}\text{Pu}$, an alpha emitter having a half-life of 24 120 yr:

$$^{239}_{94}\text{Pu} \longrightarrow ^{235}_{92}\text{U} + \alpha$$

Consider a sample of 1.00 kg of pure $^{239}_{94}\text{Pu}$ at $t = 0$. Calculate (a) the number of $^{239}_{94}\text{Pu}$ nuclei present at $t = 0$ and

(b) the initial activity in the sample. (c) **What If?** How long does the sample have to be stored if a "safe" activity level is 0.100 Bq?

59. The atomic mass of ^{57}Co is 56.936 296 u. (a) Can ^{57}Co decay by e^+ emission? Explain. (b) **What If?** Can ^{14}C decay by e^- emission? Explain. (c) If either answer is yes, what is the range of kinetic energies available for the beta particle?

60. (a) Find the radius of the $^{12}_6\text{C}$ nucleus. (b) Find the force of repulsion between a proton at the surface of a $^{12}_6\text{C}$ nucleus and the remaining five protons. (c) How much work (in MeV) has to be done to overcome this electric repulsion to put the last proton into the nucleus? (d) Repeat (a), (b), and (c) for $^{238}_{92}\text{U}$.

61. (a) Why is the beta decay $p \rightarrow n + e^+ + \nu$ forbidden for a free proton? (b) **What If?** Why is the same reaction possible if the proton is bound in a nucleus? For example, the following reaction occurs:

$$^{13}_7\text{N} \longrightarrow ^{13}_6\text{C} + e^+ + \nu$$

(c) How much energy is released in the reaction given in (b)? *Suggestion:* Add seven electrons to both sides of the reaction to write it for neutral atoms. You may use the masses $m(e^+) = 0.000\,549$ u, $M(^{13}\text{C}) = 13.003\,355$ u, $M(^{13}\text{N}) = 13.005\,739$ u.

62. 🖥 The activity of a radioactive sample was measured over 12 h, with the net count rates shown in the table.

Time(h)	Counting Rate (counts/min)
1.00	3 100
2.00	2 450
4.00	1 480
6.00	910
8.00	545
10.0	330
12.0	200

(a) Plot the logarithm of counting rate as a function of time. (b) Determine the decay constant and half-life of the radioactive nuclei in the sample. (c) What counting rate would you expect for the sample at $t = 0$? (d) Assuming the efficiency of the counting instrument to be 10.0%, calculate the number of radioactive atoms in the sample at $t = 0$.

63. The ^{145}Pm nucleus decays by alpha emission. (a) Determine the daughter nucleus. (b) Using the values given in Table A.3, determine the energy released in this decay. (c) What fraction of this energy is carried away by the alpha particle when the recoil of the daughter is taken into account?

64. When, after a reaction or disturbance of any kind, a nucleus is left in an excited state, it can return to its normal (ground) state by emission of a gamma-ray photon (or several photons). This process is illustrated by Equation 44.24. The emitting nucleus must recoil to conserve both energy and momentum. (a) Show that the recoil energy of

the nucleus is

$$E_r = \frac{(\Delta E)^2}{2Mc^2}$$

where ΔE is the difference in energy between the excited and ground states of a nucleus of mass M. (b) Calculate the recoil energy of the ^{57}Fe nucleus when it decays by gamma emission from the 14.4-keV excited state. For this calculation, take the mass to be 57 u. (*Suggestions:* When writing the equation for conservation of energy, use $(Mv)^2/2M$ for the kinetic energy of the recoiling nucleus. Also, assume that $hf \ll Mc^2$ and use the binomial expansion.)

65. After the sudden release of radioactivity from the Chernobyl nuclear reactor accident in 1986, the radioactivity of milk in Poland rose to 2 000 Bq/L due to iodine-131 present in the grass eaten by dairy cattle. Radioactive iodine, with half-life 8.04 days, is particularly hazardous, because the thyroid gland concentrates iodine. The Chernobyl accident caused a measurable increase in thyroid cancers among children in Belarus. (a) For comparison, find the activity of milk due to potassium. Assume that one liter of milk contains 2.00 g of potassium, of which 0.011 7% is the isotope ^{40}K with a half-life 1.28×10^9 yr. (b) After what elapsed time would the activity due to iodine fall below that due to potassium?

66. Europeans named a certain direction in the sky as between the horns of Taurus the Bull. On the day they named as July 4, 1054 A.D., a brilliant light appeared there. Europeans left no surviving record of the supernova, which could be seen in daylight for some days. As it faded it remained visible for years, dimming for a time with the 77.1-day half-life of the radioactive cobalt-56 that had been created in the explosion. (a) The remains of the star now form the Crab nebula (see page 1066). In it, the cobalt-56 has now decreased to what fraction of its original activity? (b) Suppose that an American, of the people called the Anasazi, made a charcoal drawing of the supernova. The carbon-14 in the charcoal has now decayed to what fraction of its original activity?

67. A theory of nuclear astrophysics proposes that all the elements heavier than iron are formed in supernova explosions ending the lives of massive stars. Assume that at the time of the explosion the amounts of ^{235}U and ^{238}U were equal. How long ago did the star(s) explode that released the elements that formed our Earth? The present ^{235}U/^{238}U ratio is 0.007 25. The half-lives of ^{235}U and ^{238}U are 0.704×10^9 yr and 4.47×10^9 yr.

68. After determining that the Sun has existed for hundreds of millions of years, but before the discovery of nuclear physics, scientists could not explain why the Sun has continued to burn for such a long time. For example, if it were a coal fire, it would have burned up in about 3 000 yr. Assume that the Sun, whose mass is 1.99×10^{30} kg, originally consisted entirely of hydrogen and that its total power output is 3.77×10^{26} W. (a) If the energy-generating mechanism of the Sun is the fusion of hydrogen into helium via the net reaction

$$4(^1_1\text{H}) + 2(\text{e}^-) \longrightarrow {}^4_2\text{He} + 2\nu + \gamma$$

calculate the energy (in joules) given off by this reaction. (b) Determine how many hydrogen atoms constitute the Sun. Take the mass of one hydrogen atom to be 1.67×10^{-27} kg. (c) If the total power output remains constant, after what time interval will all the hydrogen be converted into helium, making the Sun die? The actual projected lifetime of the Sun is about 10 billion years, because only the hydrogen in a relatively small core is available as a fuel. Only in the core are temperatures and densities high enough for the fusion reaction to be self-sustaining.

69. **Review problem.** Consider the Bohr model of the hydrogen atom, with the electron in the ground state. The magnetic field at the nucleus produced by the orbiting electron has a value of 12.5 T. (See Chapter 30, Problem 1.) The proton can have its magnetic moment aligned in either of two directions perpendicular to the plane of the electron's orbit. Because of the interaction of the proton's magnetic moment with the electron's magnetic field, there will be a difference in energy between the states with the two different orientations of the proton's magnetic moment. Find that energy difference in eV.

70. Many radioisotopes have important industrial, medical, and research applications. One of these is ^{60}Co, which has a half-life of 5.27 yr and decays by the emission of a beta particle (energy 0.31 MeV) and two gamma photons (energies 1.17 MeV and 1.33 MeV). A scientist wishes to prepare a ^{60}Co sealed source that will have an activity of 10.0 Ci after 30.0 months of use. (a) What is the initial mass of ^{60}Co required? (b) At what rate will the source emit energy after 30.0 months?

71. **Review problem.** Consider a model of the nucleus in which the positive charge (Ze) is uniformly distributed throughout a sphere of radius R. By integrating the energy density $\frac{1}{2}\epsilon_0 E^2$ over all space, show that the electric potential energy may be written

$$U = \frac{3Z^2e^2}{20\pi\epsilon_0 R} = \frac{3k_e Z^2 e^2}{5R}$$

72. The ground state of $^{93}_{43}$Tc (molar mass 92.910 2 g/mol) decays by electron capture and e^+ emission to energy levels of the daughter (molar mass 92.906 8 g/mol in ground state) at 2.44 MeV, 2.03 MeV, 1.48 MeV, and 1.35 MeV. (a) For which of these levels are electron capture and e^+ decay allowed? (b) Identify the daughter and sketch the decay scheme, assuming all excited states de-excite by direct γ decay to the ground state.

73. Free neutrons have a characteristic half-life of 10.4 min. What fraction of a group of free neutrons with kinetic energy 0.040 0 eV will decay before traveling a distance of 10.0 km?

74. In a piece of rock from the Moon, the ^{87}Rb content is assayed to be 1.82×10^{10} atoms per gram of material, and the ^{87}Sr content is found to be 1.07×10^9 atoms per gram. (a) Calculate the age of the rock. (b) **What If?** Could the material in the rock actually be much older? What assumption is implicit in using the radioactive dating method? (The relevant decay is ^{87}Rb $\rightarrow$ ^{87}Sr $+ \text{e}^- + \overline{\nu}$. The half-life of the decay is 4.75×10^{10} yr.)

75. *Student determination of the half-life of* ^{137}Ba. The radioactive barium isotope ^{137}Ba has a relatively short half-life and can be easily extracted from a solution containing its parent cesium (^{137}Cs). This barium isotope is commonly used in an undergraduate laboratory exercise for demonstrating the radioactive decay law. Undergraduate students using modest experimental equipment took the data presented in Figure P44.75. Determine the half-life for the decay of ^{137}Ba using their data.

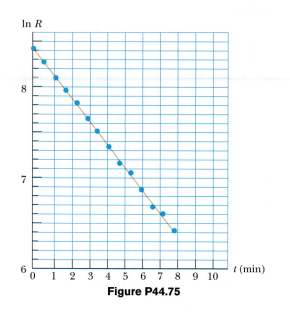

Figure P44.75

Answers to Quick Quizzes

44.1 (b). The value of $N = A - Z$ is the same for all three nuclei.

44.2 (a). The value of Z is the same for all three nuclei because they are all nuclei of nitrogen.

44.3 (c). The value of A is the same for all three nuclei, as seen by the unchanging pre-superscript.

44.4 (e). A year of 365 days is equivalent to 365 d/5.01 d $\approx$ 73 half-lives. Thus, the activity will be reduced after one year to approximately $(1/2)^{73}$ (1.000 μCi) $\sim 10^{-22}$ μCi.

44.5 (e). The time we are interested in is half of a half-life. Thus, the number of *remaining* nuclei is $(\frac{1}{2})^{1/2}N_0 = N_0/\sqrt{2} = 0.707N_0$. The number of nuclei that *have decayed* is $N_0 - 0.707N_0 = 0.293N_0$.

44.6 (b). In alpha decay, the atomic number decreases by two and the atomic mass number decreases by four.

44.7 (c). In e^- decay, the atomic number increases by one and the atomic mass number stays fixed. None of the choices is consistent with e^+ decay, so we assume that the decay must be by e^-.

44.8 (c). Isotopes of a given element correspond to nuclei with different numbers of neutrons. This results in different atomic mass numbers of the nucleus—and different magnetic moments, because the neutron, despite being uncharged, has a magnetic moment. The chemical behavior, however, is governed by the electrons. All isotopes of a given element have the same number of electrons and, therefore, the same chemical behavior.

Chapter 45

Applications of Nuclear Physics

▲ *The San Onofre Nuclear Generating Station, south of San Clemente, California, is one of dozens of nuclear power plants around the world that provide energy from uranium. These plants operate via a nuclear process called* fission, *while plants based on a second process, fusion,* are years in the future. (Tony Freeman/Index Stock Imagery)

In this chapter, we study two means for deriving energy from nuclear reactions: fission, in which a large nucleus splits into two smaller nuclei, and fusion, in which two small nuclei fuse to form a larger one. In both cases, the released energy can be used either constructively (as in the production of electric power) or destructively (as in nuclear weapons). We also examine the ways in which radiation interacts with matter and look at several devices used to detect radiation. The chapter concludes with a discussion of some industrial and biological applications of radiation.

45.1 Interactions Involving Neutrons

Nuclear fission is the process that occurs in a nuclear reactor and ultimately results in energy supplied to a community by electrical transmission. Nuclear fusion is an area of active research but has not yet been commercially developed for the supply of energy. We will discuss fission first and then explore fusion in Section 45.4.

To understand nuclear fission and the physics of nuclear reactors, we must first understand how neutrons interact with nuclei. Because of their charge neutrality, neutrons are not subject to Coulomb forces and as a result do not interact electrically with electrons or the nucleus. Therefore, neutrons can easily penetrate deep into an atom and collide with the nucleus.

A fast neutron (energy greater than about 1 MeV) traveling through matter undergoes many collisions with nuclei. In each collision, the neutron gives up some of its kinetic energy to a nucleus. For some materials and for fast neutrons, elastic collisions dominate. Materials for which this occurs are called **moderators** because they slow down (or moderate) the originally energetic neutrons very effectively. Moderator nuclei should be of low mass so that more kinetic energy is transferred to them in elastic collisions. For this reason, materials that are abundant in hydrogen, such as paraffin and water, are good moderators for neutrons.

Sooner or later, most neutrons bombarding a moderator become **thermal neutrons,** which means they are in thermal equilibrium with the moderator material. Their average kinetic energy at room temperature is, from Equation 21.4,

$$K_{av} = \tfrac{3}{2} k_B T \approx \tfrac{3}{2} k_B (300 \text{ K}) \approx 0.04 \text{ eV}$$

which corresponds to a neutron root-mean-square speed of about 2 800 m/s. Thermal neutrons have a distribution of speeds, just as the molecules in a container of gas do (see Chapter 21). High-energy neutrons, those with energy of several MeV, *thermalize* (that is, their average energy reaches K_{av}) in less than 1 ms when they are incident on a moderator.

Once the neutrons have thermalized and the energy of a particular neutron is sufficiently low, there is a high probability that the neutron will be captured by a nucleus, an event that is accompanied by the emission of a gamma ray. This **neutron capture** reaction can be written

Neutron capture reaction

$$^{1}_{0}\text{n} + ^{A}_{Z}\text{X} \longrightarrow ^{A+1}_{Z}\text{X}^{*} \longrightarrow ^{A+1}_{Z}\text{X} + \gamma \qquad (45.1)$$

Once the neutron is captured, the nucleus $^{A+1}_{\ Z}X^*$ is in an excited state for a very short time before it undergoes gamma decay. The product nucleus $^{A+1}_{\ Z}X$ is usually radioactive and decays by beta emission.

The neutron-capture rate for neutrons passing through any sample depends on the type of atoms in the sample and on the energy of the incident neutrons. The interaction of neutrons with matter increases with decreasing neutron energy because a slow neutron spends more time in the vicinity of target nuclei.

45.2 Nuclear Fission

As we mentioned in Section 44.2, nuclear **fission** occurs when a heavy nucleus, such as ^{235}U, splits into two smaller nuclei. Fission is initiated when a heavy nucleus captures a thermal neutron, as described by the first step in Equation 45.1. The absorption of the neutron creates a nucleus that is unstable and can change to a lower-energy configuration by splitting into two smaller nuclei. In such a reaction, the combined mass of the daughter nuclei is less than the mass of the parent nucleus, and the difference in mass is called the **mass defect.** Multiplying the mass defect by c^2 gives the numerical value of the released energy. Energy is released because the binding energy per nucleon of the daughter nuclei is about 1 MeV greater than that of the parent nucleus (see Fig. 44.5).

Nuclear fission was first observed in 1938 by Otto Hahn (1879–1968) and Fritz Strassman (1902–1980) following some basic studies by Fermi. After bombarding uranium with neutrons, Hahn and Strassman discovered among the reaction products two medium-mass elements, barium and lanthanum. Shortly thereafter, Lise Meitner (1878–1968) and her nephew Otto Frisch (1904–1979) explained what had happened. The uranium nucleus had split into two nearly equal fragments plus several neutrons after absorbing a neutron. Such an occurrence was of considerable interest to physicists attempting to understand the nucleus, but it was to have even more far-reaching consequences. Measurements showed that about 200 MeV of energy was released in each fission event, and this fact was to affect the course of history.

The fission of ^{235}U by thermal neutrons can be represented by the reaction

$$\,^1_0n + \,^{235}_{92}U \longrightarrow \,^{236}_{92}U^* \longrightarrow X + Y + \text{neutrons} \qquad (45.2)$$

General fission reaction for ^{235}U

where $^{236}U^*$ is an intermediate excited state that lasts only for about 10^{-12} s before splitting into medium-mass nuclei X and Y, which are called **fission fragments.** In any fission reaction, there are many combinations of X and Y that satisfy the requirements of conservation of energy and charge. In the case of uranium, for example, there are about 90 daughter nuclei that can be formed.

Fission also results in the production of several neutrons, typically two or three. On the average, about 2.5 neutrons are released per event. A typical fission reaction for uranium is

$$\,^1_0n + \,^{235}_{92}U \longrightarrow \,^{141}_{56}Ba + \,^{92}_{36}Kr + 3\,(^1_0n) \qquad (45.3)$$

Figure 45.1 is a graph of the distribution of fission products versus mass number A. The most probable products have mass numbers $A \approx 140$ and $A \approx 95$. Suppose these products are $^{140}_{53}I$ (with 87 neutrons) and $^{95}_{39}Y$ (with 56 neutrons). If these nuclei are located on the graph of Figure 44.4, it is seen that both are well above the line of stability. These fragments, because they are very unstable owing to their unusually high number of neutrons, almost instantaneously release two or three neutrons.

The breakup of the uranium nucleus can be compared to what happens to a drop of water when excess energy is added to it. (Recall the liquid-drop model of the nucleus described in Section 44.3.) Initially, all the atoms in the drop have some energy, but this is not enough to break up the drop. Imagine adding energy to a drop

▲ **PITFALL PREVENTION**

45.1 Binding Energy Reminder

Remember from Chapter 44 that binding energy is the absolute value of the system energy and is related to the system mass. Therefore, when considering Figure 44.5, imagine flipping it upside down for a curve representing system mass. In a fission reaction, the system mass decreases. This decrease in mass appears in the system as kinetic energy of the fission products.

Figure 45.1 Distribution of fission products versus mass number for the fission of ^{235}U bombarded with thermal neutrons. Note that the vertical axis is logarithmic.

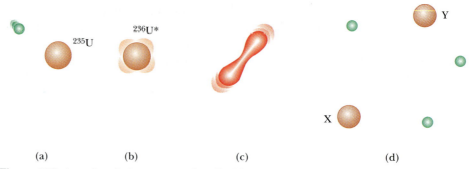

(a)　　　　　　　(b)　　　　　　　(c)　　　　　　　(d)

Figure 45.2 A nuclear fission event as described by the liquid-drop model of the nucleus. (a) A slow neutron approaches a ^{235}U nucleus. (b) The neutron is absorbed by the ^{235}U nucleus, changing it to ^{236}U*, that is the ^{236}U nucleus in an excited state. (c) The nucleus deforms and oscillates like a liquid drop. (d) The nucleus undergoes fission, resulting in two lighter nuclei X, and Y, and several neutrons.

of water floating in empty space so that it begins to elongate and compress (as in Fig. 45.2c). If enough energy is added, the drop vibrates until the amplitude of vibration becomes large enough to cause it to break. In the uranium nucleus, a similar process occurs, as described in the four steps of Figure 45.2.

Quick Quiz 45.1 Which of the following is a possible set of products for a $^{235}_{92}$U fission reaction in which two neutrons are released? (a) $^{141}_{56}$Ba, $^{93}_{36}$Kr (b) $^{142}_{55}$Cs, $^{93}_{37}$Rb (c) $^{139}_{54}$Xe, $^{96}_{38}$Sr (d) $^{140}_{53}$I, $^{93}_{38}$Sr

Let us estimate the disintegration energy Q released in a typical fission process. From Figure 44.5 we see that the binding energy per nucleon is about 7.2 MeV for heavy nuclei ($A \approx 240$) and about 8.2 MeV for nuclei of intermediate mass. This means that the nuclei in fission fragments are more tightly bound and therefore have less combined mass than the parent nucleus. This decrease in mass appears as released energy when fission occurs. The amount of energy released is approximately 8.2 MeV − 7.2 MeV = 1 MeV per nucleon. Because there are a total of 235 nucleons in $^{235}_{92}$U, we estimate that the energy released per fission event is

$$Q \approx (235 \text{ nucleons})(8.2 \text{ MeV/nucleon} - 7.2 \text{ MeV/nucleon}) = 235 \text{ MeV}$$

This is a large amount of energy relative to the amount released in chemical processes. For example, the energy released in the combustion of one molecule of octane used in gasoline engines is about one millionth of the energy released in a single fission event!

Quick Quiz 45.2 When a nucleus undergoes fission, the two daughter nuclei are generally radioactive. By which process are they most likely to decay? (a) alpha decay (b) beta decay (e^-) (c) beta decay (e^+).

Quick Quiz 45.3 Which of the following are possible fission reactions?

(a) $^1_0\text{n} + ^{235}_{92}\text{U} \longrightarrow ^{140}_{54}\text{Xe} + ^{94}_{38}\text{Sr} + 2\,(^1_0\text{n})$

(b) $^1_0\text{n} + ^{235}_{92}\text{U} \longrightarrow ^{132}_{50}\text{Sn} + ^{101}_{42}\text{Mo} + 3\,(^1_0\text{n})$

(c) $^1_0\text{n} + ^{239}_{94}\text{Pu} \longrightarrow ^{137}_{53}\text{I} + ^{97}_{41}\text{Nb} + 3\,(^1_0\text{n})$

Example 45.1   The Energy Released in the Fission of ^{235}U

Calculate the energy released when 1.00 kg of ^{235}U fissions, taking the disintegration energy per event to be $Q = 208$ MeV.

Solution We need to know the number of nuclei in 1.00 kg of uranium. Because $A = 235$ for uranium, we know that the molar mass of this isotope is 235 g/mol. Therefore, the number of nuclei in our sample is

$$N = \left(\frac{6.02 \times 10^{23} \text{ nuclei/mol}}{235 \text{ g/mol}} \right)(1.00 \times 10^3 \text{ g})$$

$$= 2.56 \times 10^{24} \text{ nuclei}$$

Hence, the total disintegration energy is

$$E = NQ = (2.56 \times 10^{24} \text{ nuclei})(208 \text{ MeV/nucleus})$$

$$= \boxed{5.32 \times 10^{26} \text{ MeV}}$$

Let us convert this energy to kWh:

$$E = (5.32 \times 10^{26} \text{ MeV}) \left(\frac{1.60 \times 10^{-13} \text{ J}}{\text{MeV}} \right) \left(\frac{1 \text{ kWh}}{3.60 \times 10^6 \text{ J}} \right)$$

$$= 2.36 \times 10^7 \text{ kWh}$$

This is enough energy to keep a 100-W lightbulb burning for 30 000 years! If the available fission energy in 1 kg of ^{235}U were suddenly released, it would be equivalent to detonating about 20 000 tons of TNT.

45.3   Nuclear Reactors

In the preceding section, we learned that, when ^{235}U fissions, an average of 2.5 neutrons are emitted per event. These neutrons can trigger other nuclei to fission, with the possibility of a chain reaction (Fig. 45.3). Calculations show that if the chain reaction is not controlled (that is, if it does not proceed slowly), it can result in a violent explosion, with the

Active Figure 45.3 A nuclear chain reaction initiated by the capture of a neutron.

Observe the chain reaction at the Active Figures link at http://www.pse6.com.

Courtesy of Chicago Historical Society

Figure 45.4 Artist's rendition of the world's first nuclear reactor. Because of wartime secrecy, there are few photographs of the completed reactor, which was composed of layers of moderating graphite interspersed with uranium. A self-sustained chain reaction was first achieved on December 2, 1942. Word of the success was telephoned immediately to Washington with this message: "The Italian navigator has landed in the New World and found the natives very friendly." The historic event took place in an improvised laboratory in the racquet court under the stands of the University of Chicago's Stagg Field, and the Italian navigator was Enrico Fermi.

Enrico Fermi
Italian Physicist (1901–1954)

Fermi was awarded the Nobel Prize in physics in 1938 for producing transuranic elements by neutron irradiation and for his discovery of nuclear reactions brought about by thermal neutrons. He made many other outstanding contributions to physics, including his theory of beta decay, the free-electron theory of metals, and the development of the world's first fission reactor in 1942. Fermi was truly a gifted theoretical and experimental physicist. He was also well known for his ability to present physics in a clear and exciting manner. "Whatever Nature has in store for mankind, unpleasant as it may be, men must accept, for ignorance is never better than knowledge." (*National Accelerator Laboratory*)

release of an enormous amount of energy. When the reaction is controlled, however, the energy released can be put to constructive use. In the United States, for example, nearly 20% of the electricity generated each year comes from nuclear power plants, and nuclear power is used extensively in many countries, including France, Japan, and Germany.

A nuclear reactor is a system designed to maintain what is called a **self-sustained chain reaction.** This important process was first achieved in 1942 by Enrico Fermi and his team at the University of Chicago, with naturally occurring uranium as the fuel.[1] In the first nuclear reactor (Fig. 45.4), Fermi placed bricks of graphite (carbon) between the fuel elements. Carbon nuclei are about 12 times more massive than neutrons, but after several collisions with carbon nuclei, a neutron is slowed sufficiently to increase its likelihood of fission with ^{235}U. In this design, carbon is the moderator; most modern reactors use water as the moderator.

Most reactors in operation today also use uranium as fuel. However, naturally occurring uranium contains only about 0.7% of the ^{235}U isotope, with the remaining 99.3% being ^{238}U. This fact is important to the operation of a reactor because ^{238}U almost never fissions. Instead, it tends to absorb neutrons without a subsequent fission event, producing neptunium and plutonium. For this reason, reactor fuels must be artificially *enriched* to contain at least a few percent ^{235}U.

In the process of slowing down, neutrons may be captured by nuclei that do not fission. The most common event of this type is neutron capture by ^{238}U, which constitutes more than 90% of the uranium in the enriched fuel elements. The probability of neutron capture by ^{238}U is very high when the neutrons have high kinetic energies and very low when they have low kinetic energies. Thus, the slowing down of the neutrons by the moderator serves the secondary purpose of making them available for reaction with ^{235}U and decreasing their chances of being captured by ^{238}U.

To achieve a self-sustained chain reaction, an average of one neutron emitted in each ^{235}U fission must be captured by another ^{235}U nucleus and cause that nucleus to undergo fission. A useful parameter for describing the level of reactor operation is the **reproduction constant** K, defined as **the average number of neutrons from each**

---

[1]   Although Fermi's reactor was the first manufactured nuclear reactor, there is evidence that a natural fission reaction may have sustained itself for perhaps hundreds of thousands of years in a deposit of uranium in Gabon, West Africa. See G. Cowan, "A Natural Fission Reactor," *Scientific American* 235(5):36, 1976.

Figure 45.5 Main components of a pressurized-water nuclear reactor.

fission event that cause another fission event. As we have seen, K has an average value of 2.5 in the uncontrolled fission of uranium.

A self-sustained chain reaction is achieved when $K = 1$. Under this condition, the reactor is said to be **critical.** When $K < 1$, the reactor is subcritical and the reaction dies out. When $K > 1$, the reactor is supercritical and a runaway reaction occurs. In a nuclear reactor used to furnish power to a utility company, it is necessary to maintain a value of K close to 1. If K rises above this value, the internal energy produced in the reaction could melt the reactor.

Several types of reactor systems allow the kinetic energy of fission fragments to be transformed to other types of energy and eventually transferred out of the reactor plant by electrical transmission. The most common reactor in use in the United States is the pressurized-water reactor (Fig. 45.5). We shall examine this type because its main parts are common to all reactor designs. Fission events in the uranium **fuel elements** in the reactor core raise the temperature of the water contained in the primary (closed) loop, which is maintained at high pressure to keep the water from boiling. (This water also serves as the moderator to slow down the neutrons released in the fission events with energy of about 2 MeV.) The hot water is pumped through a heat exchanger, where the internal energy of the water is transferred by conduction to the water contained in the secondary loop. The hot water in the secondary loop is converted to steam, which does work to drive a turbine–generator system to create electric power. The water in the secondary loop is isolated from the water in the primary loop to avoid contamination of the secondary water and the steam by radioactive nuclei from the reactor core.

In any reactor, a fraction of the neutrons produced in fission leak out of the uranium fuel elements before inducing other fission events. If the fraction leaking out is too large, the reactor will not operate. The percentage lost is large if the fuel elements are very small because leakage is a function of the ratio of surface area to volume. Therefore, a critical feature of the reactor design is an optimal surface area–to–volume ratio of the fuel elements.

Control of Power Level

Safety is of critical importance in the operation of a nuclear reactor. The reproduction constant K must not be allowed to rise above 1, lest a runaway reaction occur. Consequently, reactor design must include a means of controlling the value of K.

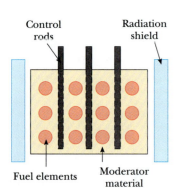

Figure 45.6 Cross section of a reactor core showing the control rods, fuel elements containing enriched fuel, and moderating material, all surrounded by a radiation shield.

The basic design of a nuclear reactor core is shown in Figure 45.6. The fuel elements consist of uranium that has been enriched in the ^{235}U isotope. To control the power level, **control rods** are inserted into the reactor core. These rods are made of materials, such as cadmium, that are very efficient in absorbing neutrons. By adjusting the number and position of the control rods in the reactor core, the K value can be varied and any power level within the design range of the reactor can be achieved.

Quick Quiz 45.4 Which of the following has the function of slowing down fast neutrons? (a) control rods (b) moderator (c) fuel elements.

Quick Quiz 45.5 To reduce the value of the reproduction constant K, do you (a) push the control rods deeper into the core or (b) pull the control rods farther out of the core?

Safety and Waste Disposal

The 1979 near-disaster at a nuclear power plant at Three Mile Island in Pennsylvania and the 1986 accident at the Chernobyl reactor in Ukraine rightfully focused attention on reactor safety. The Three Mile Island accident was the result of inadequate control-room instrumentation and poor emergency-response training. There were no injuries or detectable health impacts from the event, even though more than one third of the fuel melted.

This unfortunately was not the case at Chernobyl, where the activity of the materials released immediately after the accident totaled approximately 1.2×10^{19} Bq and resulted in the evacuation of 135 000 people. Thirty individuals died during the accident or shortly thereafter, and data from the Ukraine Radiological Institute suggest that more than 2 500 deaths could be attributed to the Chernobyl accident. In the period 1986–1997 there was a tenfold increase in the number of children contracting thyroid cancer from the ingestion of radioactive iodine in milk from cows that ate contaminated grass. One conclusion of an international conference studying the Ukraine accident was that the main causes of the Chernobyl accident were the coincidence of severe deficiencies in the reactor physical design and a violation of safety procedures. Most of these deficiencies have been addressed at plants of similar design in Russia and neighboring countries of the former Soviet Union.

Commercial reactors achieve safety through careful design and rigid operating protocol, and it is only when these variables are compromised that reactors pose a danger. Radiation exposure and the potential health risks associated with such exposure are controlled by three layers of containment. The fuel and radioactive fission products are contained inside the reactor vessel. Should this vessel rupture, the reactor building acts as a second containment structure to prevent radioactive material from contaminating the environment. Finally, the reactor facilities must be in a remote location to protect the general public from exposure should radiation escape the reactor building.

A continuing concern about nuclear fission reactors is the safe disposal of radioactive material when the reactor core is replaced. This waste material contains long-lived, highly radioactive isotopes and must be stored over long time intervals in such a way that there is no chance of environmental contamination. At present, sealing radioactive wastes in waterproof containers and burying them in deep geologic repositories seems to be the most promising solution.

Transport of reactor fuel and reactor wastes poses additional safety risks. Accidents during transport of nuclear fuel could expose the public to harmful levels of radiation. The Department of Energy requires stringent crash tests of all containers used to transport nuclear materials. Container manufacturers must demonstrate that their containers will not rupture even in high-speed collisions.

Despite these risks, there are advantages to the use of nuclear power to be weighed against the risks. For example, nuclear power plants do not produce air pollution and greenhouse gases as do fossil fuel plants, and the supply of uranium on the Earth is predicted to last longer than the supply of fossil fuels. For each source of energy, whether nuclear, hydroelectric, fossil fuel, wind, or solar, the risks must be weighed against the benefits and the availability of the energy source.

45.4   Nuclear Fusion

In Chapter 44 we found that the binding energy for light nuclei ($A < 20$) is much smaller than the binding energy for heavier nuclei. This suggests a process that is the reverse of fission. As we mentioned in Section 39.9, when two light nuclei combine to form a heavier nucleus, the process is called nuclear **fusion.** Because the mass of the final nucleus is less than the combined masses of the original nuclei, there is a loss of mass accompanied by a release of energy.

Two examples of such energy-liberating fusion reactions are as follows:

$$^1_1\text{H} + {}^1_1\text{H} \longrightarrow {}^2_1\text{H} + e^+ + \nu$$

$$^1_1\text{H} + {}^2_1\text{H} \longrightarrow {}^3_2\text{He} + \gamma$$

These reactions occur in the core of a star and are responsible for the outpouring of energy from the star. This second reaction is followed by either hydrogen–helium fusion or helium–helium fusion:

$$^1_1\text{H} + {}^3_2\text{He} \longrightarrow {}^4_2\text{He} + e^+ + \nu$$

$$^3_2\text{He} + {}^3_2\text{He} \longrightarrow {}^4_2\text{He} + {}^1_1\text{H} + {}^1_1\text{H}$$

These are the basic reactions in the **proton–proton cycle,** believed to be one of the basic cycles by which energy is generated in the Sun and other stars that contain an abundance of hydrogen. Most of the energy production takes place in the Sun's interior, where the temperature is approximately 1.5×10^7 K. Because such high temperatures are required to drive these reactions, they are called **thermonuclear fusion reactions.** All of the reactions in the proton–proton cycle are exothermic. An overview of the cycle is that four protons combine to form an alpha particle and two positrons.

▲ **PITFALL PREVENTION**

45.2   Fission and Fusion

The words *fission* and *fusion* sound similar, but they correspond to different processes. Consider the binding-energy curve in Figure 44.5. There are two directions from which you can approach the peak of the curve so that energy is released: combining two light nuclei—fusion—or separating a heavy nucleus into two lighter nuclei—fission.

Quick Quiz 45.6 In the core of a star, hydrogen nuclei combine in fusion reactions. Once the hydrogen has been exhausted, fusion of helium nuclei can occur. Once the helium is used up, if the star is sufficiently massive, fusion of heavier and heavier nuclei can occur. Consider fusion reactions involving two nuclei with the same value of A. For these types of reactions, which of the following values of A are impossible? (a) 12 (b) 20 (c) 28 (d) 64

Example 45.2   Energy Released in Fusion

Find the total energy released in the fusion reactions in the proton–proton cycle.

Solution As mentioned earlier, the net result of the proton–proton cycle is to fuse four protons to form an alpha particle. The initial mass of four protons is, using atomic masses from Table A.3,

$$4(1.007\,825\text{ u}) = 4.031\,300\text{ u}$$

The change in mass of the system is this value minus the mass of the resultant alpha particle:

$$4.031\,300\text{ u} - 4.002\,603\text{ u} = 0.028\,697\text{ u}$$

Now we convert this mass change into energy units:

$$E = 0.028\,697\text{ u} \times 931.494\text{ MeV/u} = \boxed{26.7\text{ MeV}}$$

Terrestrial Fusion Reactions

The enormous amount of energy released in fusion reactions suggests the possibility of harnessing this energy for useful purposes. A great deal of effort is currently under way to develop a sustained and controllable thermonuclear reactor—a fusion power reactor. Controlled fusion is often called the ultimate energy source because of the availability of its fuel source: water. For example, if deuterium were used as the fuel, 0.12 g of it could be extracted from 1 gal of water at a cost of about four cents. This amount of deuterium would release about 10^{10} J if all nuclei underwent fusion. By comparison, a gallon of gasoline releases about 10^8 J upon burning, and costs far more than four cents.

An additional advantage of fusion reactors is that comparatively few radioactive byproducts are formed. For the proton–proton cycle, for instance, the end product is safe, nonradioactive helium. Unfortunately, a thermonuclear reactor that can deliver a net power output spread out over a reasonable time interval is not yet a reality, and many difficulties must be resolved before a successful device is constructed.

The Sun's energy is based, in part, upon a set of reactions in which hydrogen is converted to helium. However, the proton–proton interaction is not suitable for use in a fusion reactor because the event requires very high temperatures and densities. The process works in the Sun only because of the extremely high density of protons in the Sun's interior.

The reactions that appear most promising for a fusion power reactor involve deuterium ($^2_1$H) and tritium ($^3_1$H):

$$^2_1\text{H} + {}^2_1\text{H} \longrightarrow {}^3_2\text{He} + {}^1_0\text{n} \qquad Q = 3.27 \text{ MeV}$$

$$^2_1\text{H} + {}^2_1\text{H} \longrightarrow {}^3_1\text{H} + {}^1_1\text{H} \qquad Q = 4.03 \text{ MeV} \qquad (45.4)$$

$$^2_1\text{H} + {}^3_1\text{H} \longrightarrow {}^4_2\text{He} + {}^1_0\text{n} \qquad Q = 17.59 \text{ MeV}$$

As noted earlier, deuterium is available in almost unlimited quantities from our lakes and oceans and is very inexpensive to extract. Tritium, however, is radioactive ($T_{1/2} = 12.3$ yr) and undergoes beta decay to ^{3}He. For this reason, tritium does not occur naturally to any great extent and must be artificially produced.

One of the major problems in obtaining energy from nuclear fusion is that the Coulomb repulsive force between two nuclei, which carry positive charges, must be overcome before they can fuse. Potential energy as a function of the separation distance between two deuterons (deuterium nuclei, each having charge $+e$) is shown in Figure 45.7. The potential energy is positive in the region $r > R$, where the Coulomb repulsive force dominates ($R \approx 1$ fm), and negative in the region $r < R$, where the nuclear force dominates. The fundamental problem then is to give the two nuclei enough kinetic energy to overcome this repulsive force. This can be accomplished by raising the fuel to extremely high temperatures (to about 10^8 K, far greater than the interior temperature of the Sun). At these high temperatures, the atoms are ionized and the system consists of a collection of electrons and nuclei, commonly referred to as a *plasma*.

Figure 45.7 Potential energy as a function of separation distance between two deuterons. The Coulomb repulsive force is dominant at long range, and the nuclear force is dominant at short range, where R is on the order of 1 fm. If we neglect tunneling, to undergo fusion the two deuterons require an energy E greater than the height of the barrier.

Example 45.3   The Fusion of Two Deuterons   **Interactive**

The separation distance between two deuterons must be about 1.0×10^{-14} m in order for the nuclear force to overcome the repulsive Coulomb force.

(A) Calculate the height of the potential barrier due to the repulsive force.

Solution Conceptualize this problem by imagining moving two deuterons toward each other. As they move closer together, the Coulomb repulsion force becomes stronger.

Work must be done to push against this force, and this work appears in the system of two deuterons as electric potential energy. We categorize this problem as one involving the electric potential energy of a system of two charged particles. To analyze the problem, recall that the potential energy associated with two charges separated by a distance r is, from Equation 25.13,

$$U = k_e \frac{q_1 q_2}{r}$$

where k_e is the Coulomb constant. For the case of two deuterons, $q_1 = q_2 = +e$, so that

$$U = k_e \frac{e^2}{r} = (8.99 \times 10^9 \text{ N} \cdot \text{m}^2/\text{C}^2) \frac{(1.60 \times 10^{-19} \text{ C})^2}{1.0 \times 10^{-14} \text{ m}}$$

$$= 2.3 \times 10^{-14} \text{ J} = \boxed{0.14 \text{ MeV}}$$

(B) Estimate the temperature required for a deuteron to overcome the potential barrier, assuming an energy of $\frac{3}{2}k_B T$ per deuteron (where k_B is Boltzmann's constant).

Solution Because the total Coulomb energy of the pair is 0.14 MeV, the Coulomb energy per deuteron is 0.07 MeV = 1.1×10^{-14} J. Setting this energy equal to the average energy per deuteron gives

$$\tfrac{3}{2} k_B T = 1.1 \times 10^{-14} \text{ J}$$

Solving for T gives

$$T = \frac{2(1.1 \times 10^{-14} \text{ J})}{3(1.38 \times 10^{-23} \text{ J/K})} = \boxed{5.3 \times 10^8 \text{ K}}$$

To finalize this problem, note that this calculated temperature is too high because the particles in the plasma have a Maxwellian speed distribution (Section 21.6), and therefore some fusion reactions are caused by particles in the high-energy tail of this distribution. Furthermore, even those particles that do not have enough energy to overcome the barrier have some probability of tunneling through. When these effects are taken into account, a temperature of "only" 4×10^8 K appears adequate to fuse two deuterons in a plasma.

(C) Find the energy released in the deuterium–deuterium reaction

$$^2_1\text{H} + {}^2_1\text{H} \longrightarrow {}^3_1\text{H} + {}^1_1\text{H}$$

Solution The mass of a single deuterium atom is equal to 2.014 102 u. Thus, the total mass before the reaction is 4.028 204 u. After the reaction, the sum of the masses is 3.016 049 u + 1.007 825 u = 4.023 874 u. The excess mass is 0.004 33 u, equivalent to an energy of

$$0.004\ 33 \text{ u} \times 931.494 \text{ MeV/u} = \boxed{4.03 \text{ MeV}}$$

Notice that this energy value is consistent with that already given in Equation 45.4.

What If? Suppose that the tritium resulting from the reaction in part (C) reacts with another deuterium in the reaction

$$^2_1\text{H} + {}^3_1\text{H} \longrightarrow {}^4_2\text{He} + {}^1_0\text{n}$$

How much energy is released in the sequence of two reactions?

Answer The overall effect of the sequence of two reactions is that three deuterium nuclei have combined to form a helium nucleus, a hydrogen nucleus, and a neutron. The initial mass is 3(2.014 102 u) = 6.042 306 u. After the reaction, the sum of the masses is 4.002 603 u + 1.007 825 u + 1.008 665 = 6.019 093 u. The excess mass is 0.023 213 u, equivalent to an energy of 21.6 MeV. Notice that this value is the sum of the Q values for the second and third reactions in Equation 45.4.

> At the *Interactive Worked Example* link at **http://www.pse6.com**, *practice evaluating parameters for a number of different fusion reactions.*

The temperature at which the power generation rate in any fusion reaction exceeds the loss rate is called the **critical ignition temperature** T_{ignit}. This temperature for the deuterium–deuterium (D–D) reaction is 4×10^8 K. From the relationship $E \approx \frac{3}{2}k_B T$, the ignition temperature is equivalent to approximately 52 keV. The critical ignition temperature for the deuterium-tritium (D–T) reaction is about 4.5×10^7 K, or only 6 keV. A plot of the power $\mathscr{P}_{\text{gen}}$ generated by fusion versus temperature for the two reactions is shown in Figure 45.8. The green straight line represents the power $\mathscr{P}_{\text{lost}}$ lost via the radiation mechanism known as bremsstrahlung (Section 42.8). In this principal mechanism of energy loss, radiation (primarily x-rays) is emitted as the result of electron–ion collisions within the plasma. The intersections of the $\mathscr{P}_{\text{lost}}$ line with the $\mathscr{P}_{\text{gen}}$ curves give the critical ignition temperatures.

Figure 45.8 Power generated versus temperature for deuterium–deuterium (D–D) and deuterium–tritium (D–T) fusion. The green line represents power lost as a function of temperature. When the generation rate exceeds the loss rate, ignition takes place.

In addition to the high-temperature requirements, there are two other critical parameters that determine whether or not a thermonuclear reactor is successful: the **ion density** n and **confinement time** τ, which is the time interval during which energy injected into the plasma remains within the plasma. The British physicist J. D. Lawson has shown that the ion density and confinement time must both be large enough to ensure that more fusion energy is released than the amount required to raise the temperature of the plasma. A graph of the value of $n\tau$ necessary to achieve a net energy output for the D–T and D–D reactions at different temperatures is shown in Figure 45.9. The product $n\tau$ is referred to as the **Lawson number** of a reaction. In particular, **Lawson's criterion** states that a net energy output is possible for values of $n\tau$ that meet the following conditions:

Lawson's criterion

$$n\tau \geq 10^{14} \text{ s/cm}^3 \quad \text{(D–T)}$$
$$n\tau > 10^{16} \text{ s/cm}^3 \quad \text{(D–D)}$$

(45.5)

These values represent the minima of the curves in Figure 45.9.

Lawson's criterion was arrived at by comparing the energy required to raise the temperature of a given plasma with the energy generated by the fusion process.[2] The energy E_{in} required to raise the temperature of the plasma is proportional to the ion density n, which we can express as $E_{\text{in}} = C_1 n$, where C_1 is some constant. The energy generated by the fusion process is proportional to $n^2\tau$, or $E_{\text{gen}} = C_2 n^2 \tau$. This may be understood by realizing that the fusion energy released is proportional to both the rate at which interacting ions collide ($\propto n^2$) and the confinement time τ. Net energy is produced when $E_{\text{gen}} > E_{\text{in}}$. When the constants C_1 and C_2 are calculated for different reactions, the condition that $E_{\text{gen}} \geq E_{\text{in}}$ leads to Lawson's criterion.

In summary, the two basic requirements of a successful thermonuclear power reactor are

- The plasma temperature must be very high—about 4.5×10^7 K for the D–T reaction and 4×10^8 K for the D–D reaction.

- To meet Lawson's criterion, the product $n\tau$ must be large. For a given value of n, the probability of fusion between two particles increases as τ increases. For a given value of τ, the collision rate between nuclei increases as n increases.

Current efforts are aimed at meeting Lawson's criterion at temperatures exceeding T_{ignit}. Although the minimum required plasma densities have been achieved, the problem of confinement time is more difficult. How can a plasma be confined at 10^8 K and achieve a Lawson number of 10^{14} s/cm³? The two basic techniques under investigation are magnetic confinement and inertial confinement.

Figure 45.9 The Lawson number $n\tau$ at which net energy output is possible versus temperature for the D–T and D–D fusion reactions. The regions above the colored curves represent favorable conditions for fusion.

Magnetic Confinement

Many fusion-related plasma experiments use **magnetic confinement** to contain the plasma. A toroidal device called a **tokamak,** first developed in Russia, is shown in Figure 45.10a. A combination of two magnetic fields is used to confine and stabilize the plasma: (1) a strong toroidal field produced by the current in the toroidal windings surrounding a donut-shaped vacuum chamber and (2) a weaker "poloidal" field produced by the toroidal current. In addition to confining the plasma, the toroidal current is used to raise its temperature. The resultant helical magnetic field lines spiral around the plasma and keep it from touching the walls of the vacuum chamber. (If the plasma touches the walls, its temperature is reduced and heavy impurities sputtered from the walls "poison" it and lead to large power losses.)

---

[2] Lawson's criterion neglects the energy needed to set up the strong magnetic field used to confine the hot plasma in a magnetic confinement approach. This energy is expected to be about 20 times greater than the energy required to raise the temperature of the plasma. For this reason, it is necessary either to have a magnetic energy recovery system or to use superconducting magnets.

Figure 45.10 (a) Diagram of a tokamak used in the magnetic confinement scheme. (b) Interior view of the closed Tokamak Fusion Test Reactor (TFTR) vacuum vessel at the Princeton Plasma Physics Laboratory. (c) The National Spherical Torus Experiment (NSTX) that began operation in March 1999.

One of the major breakthroughs in magnetic confinement in the 1980s was in the area of auxiliary energy input to reach ignition temperatures. Experiments have shown that injecting a beam of energetic neutral particles into the plasma is a very efficient method of raising it to ignition temperatures. Radio-frequency energy input will probably be needed for reactor-size plasmas.

When it was in operation, the Tokamak Fusion Test Reactor (TFTR, Fig. 45.10b) at Princeton reported central ion temperatures of 510 million degrees Celsius, more than 30 times hotter than the center of the Sun. The $n\tau$ values in the TFTR for the D–T reaction were well above 10^{13} s/cm^3 and close to the value required by Lawson's criterion. In 1991, reaction rates of 6×10^{17} D–T fusions per second were reached in the JET tokamak at Abington, England.

One of the new generation of fusion experiments is the National Spherical Torus Experiment (NSTX) shown in Figure 45.10c. Rather than the donut-shaped plasma of a tokamak, the NSTX produces a spherical plasma that has a hole through its center. The major advantage of the spherical configuration is its ability to confine the plasma at a higher pressure in a given magnetic field. This approach could lead to development of smaller, more economical fusion reactors.

An international collaborative effort involving Canada, Europe, Japan, and Russia is currently under way to build a fusion reactor called ITER (International Thermonuclear Experimental Reactor). China and the United States began to participate in program activities in early 2003. This facility will address the remaining technological and scientific issues concerning the feasibility of fusion power. The design is completed (Fig. 45.11), and site and construction negotiations are under way. As of the printing of this text, four possible sites are proposed—Cadarache, France; Clarington, Canada; Rokkasho-mura, Japan; and Vandellòs, Spain. If the planned device works as expected, the Lawson number for ITER will be about six times greater than the current record holder, the JT-60U tokamak in Japan. ITER will produce 1.5 GW of power, and the

Courtesy of ITER

Figure 45.11 Cutaway diagram of the ITER (International Thermonuclear Experimental Reactor). Note the size of the reactor relative to that of a person *(arrow)*.

energy content of the alpha particles inside the reactor will be so intense that they will sustain the fusion reaction, allowing the auxiliary energy sources to be turned off once the reaction is initiated.

Example 45.4 Inside a Fusion Reactor

In 1998, the JT-60U tokamak in Japan was operated with a D–T plasma density of 4.8×10^{13} cm^{-3} at a temperature (in energy units) of 24.1 keV. It was able to confine this plasma inside a magnetic field for 1.1 s.

(A) Does this meet Lawson's criterion?

Solution Equation 45.5 says that for a D–T plasma, the Lawson number $n\tau$ must be greater than 10^{14} s/cm^3. For the JT-60U,

$$n\tau = (4.8 \times 10^{13} \text{ cm}^{-3})(1.1 \text{ s}) = 5.3 \times 10^{13} \text{ s/cm}^3$$

which is close to meeting Lawson's criterion. In fact, scientists recorded a power gain of 1.25, indicating the reactor was operating slightly past the break-even point and was producing more energy than required to maintain the plasma.

(B) How does the plasma density compare with the density of atoms in an ideal gas when the gas is at room temperature and pressure?

Solution The density of atoms in a sample of ideal gas is given by N_A / V_{mol}, where N_A is Avogadro's number and V_{mol} is the molar volume of an ideal gas under standard conditions, 2.24×10^{-2} m^3/mol. Thus, the density of the gas is

$$\frac{N_A}{V_{\text{mol}}} = \frac{6.02 \times 10^{23} \text{ atoms/mol}}{2.24 \times 10^{-2} \text{ m}^3/\text{mol}} = 2.7 \times 10^{25} \text{ atoms/m}^3$$

This is more than 500 000 times greater than the plasma density in the reactor.

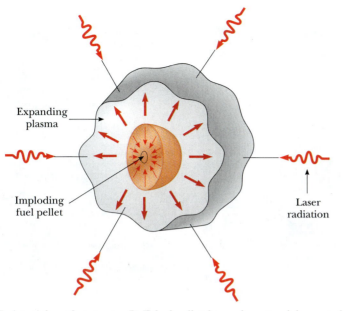

Figure 45.12 In inertial confinement, a D–T fuel pellet fuses when struck by several high-intensity laser beams simultaneously.

Inertial Confinement

The second technique for confining a plasma is called **inertial confinement** and makes use of a D–T target that has a very high particle density. In this scheme, the confinement time is very short (typically 10^{-11} to 10^{-9} s), so that, because of their own inertia, the particles do not have a chance to move appreciably from their initial positions. Thus, Lawson's criterion can be satisfied by combining a high particle density with a short confinement time.

Laser fusion is the most common form of inertial confinement. A small D–T pellet, about 1 mm in diameter, is struck simultaneously by several focused, high-intensity laser beams, resulting in a large pulse of input energy that causes the surface of the fuel pellet to evaporate (Fig. 45.12). The escaping particles exert a third-law reaction force on the core of the pellet, resulting in a strong, inwardly moving compressive shock wave. This shock wave increases the pressure and density of the core and produces a corresponding increase in temperature. When the temperature of the core reaches ignition temperature, fusion reactions occur.

Two of the leading laser fusion laboratories in the United States are the Omega facility at the University of Rochester in New York and the Nova facility at Lawrence Livermore National Laboratory in California. The Omega facility focuses 24 laser beams on the target, and the Nova facility employs 10 beams. Figure 45.13a shows the target chamber at Nova, and Figure 45.13b shows the tiny, spherical D–T pellets used. Nova is capable of injecting a power of 2×10^{14} W into a 0.5-mm D–T pellet and has achieved values of $n\tau \approx 5 \times 10^{14}$ s/cm^3 and ion temperatures of 5.0 keV. These values are close to those required for D–T ignition. This steady progress has led the U.S. Department of Energy and other groups to plan a national facility that will involve a laser fusion device with an input energy in the 5- to 10-MJ range.

Fusion Reactor Design

In the D–T fusion reaction

$$\,^2_1\text{H} + \,^3_1\text{H} \longrightarrow \,^4_2\text{He} + \,^1_0\text{n} \qquad Q = 17.59 \text{ MeV}$$

the alpha particle carries 20% of the energy and the neutron carries 80%, or about 14 MeV. The alpha particles, because they are charged, are primarily absorbed by the

(a)   (b)

Figure 45.13 (a) The target chamber of the Nova Laser Facility at Lawrence Livermore Laboratory. (b) Spherical plastic target shells used to contain the D–T fuel, shown clustered on a quarter. The shells have very smooth surfaces and are about 100 nm thick.

plasma, causing the plasma's temperature to increase. In contrast, the 14-MeV neutrons, being electrically neutral, pass through the plasma and are absorbed by a surrounding blanket material, where their large kinetic energy is extracted and used to generate electric power. A diagram of the deuterium—tritium fusion reaction is shown in Figure 45.14.

One scheme is to use molten lithium metal as the neutron-absorbing material and to circulate the lithium in a closed heat-exchange loop to produce steam and drive turbines as in a conventional power plant. Figure 45.15 shows a diagram of such a reactor. It is estimated that a blanket of lithium about 1 m thick will capture nearly 100% of the neutrons from the fusion of a small D–T pellet.

The capture of neutrons by lithium is described by the reaction

$$\,^1_0n + \,^6_3Li \longrightarrow \,^3_1H + \,^4_2He$$

where the kinetic energies of the charged tritium $\,^3_1H$ and alpha particle are converted to internal energy in the molten lithium. An extra advantage of using lithium as the energy-transfer medium is that the tritium produced can be separated from the lithium and returned as fuel to the reactor.

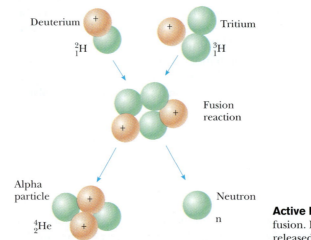

At the Active Figures link at http://www.pse6.com, choose to observe several fusion reactions and measure the energy released.

Active Figure 45.14 Deuterium–tritium fusion. Eighty percent of the energy released is in the 14-MeV neutron.

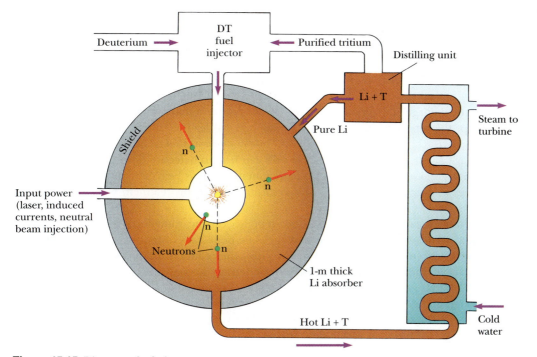

Figure 45.15 Diagram of a fusion reactor.

Advantages and Problems of Fusion

If fusion power can ever be harnessed, it will offer several advantages over fission-generated power: (1) low cost and abundance of fuel (deuterium), (2) impossibility of runaway accidents, and (3) decreased radiation hazard. Some of the anticipated problems and disadvantages include (1) scarcity of lithium, (2) limited supply of helium, which is needed for cooling the superconducting magnets used to produce strong confining fields, and (3) structural damage and induced radioactivity caused by neutron bombardment. If such problems and the engineering design factors can be resolved, nuclear fusion may become a feasible source of energy by the middle of the twenty-first century.

45.5 Radiation Damage

In Chapter 34 we learned that electromagnetic radiation is all around us in the form of radio waves, microwaves, light waves, and so on. In this section, we turn to forms of radiation that can cause severe damage as they pass through matter. These include radiation resulting from radioactive processes and radiation in the form of energetic particles such as neutrons and protons. It is these forms that we refer to here and in the following two sections when we use the word *radiation*.

The degree and type of damage depend upon several factors, including the type and energy of the radiation and the properties of the matter. The metals used in nuclear reactor structures can be severely weakened by high fluxes of energetic neutrons because these high fluxes often lead to metal fatigue. The damage in such situations is in the form of atomic displacements, often resulting in major alterations in the properties of the material.

Radiation damage in biological organisms is primarily due to ionization effects in cells. The normal operation of a cell may be disrupted when highly reactive ions are formed as the result of ionizing radiation. For example, hydrogen and the hydroxyl radical OH^- produced from water molecules can induce chemical reactions that may break bonds in proteins and other vital molecules. Furthermore, the ionizing radiation may affect vital molecules directly by removing electrons from their structure. Large doses of radiation are especially dangerous because damage to a great number of molecules in a cell may cause the cell to die. Although the death of a single cell is usually not a problem, the death of many cells may result in irreversible damage to the organism. Cells that divide rapidly, such as those of the digestive tract, reproductive organs, and hair follicles, are especially susceptible. Also, cells that survive the radiation may become defective. These defective cells can produce more defective cells and lead to cancer.

In biological systems, it is common to separate radiation damage into two categories: somatic damage and genetic damage. *Somatic damage* is that associated with any body cell except the reproductive cells. Somatic damage can lead to cancer or seriously alter the characteristics of specific organisms. *Genetic damage* affects only reproductive cells. Damage to the genes in reproductive cells can lead to defective offspring. It is important to be aware of the effect of diagnostic treatments, such as x-rays and other forms of radiation exposure, and to balance the significant benefits of treatment with the damaging effects.

Damage caused by radiation also depends on the radiation's penetrating power. Alpha particles cause extensive damage, but penetrate only to a shallow depth in a material due to the strong interaction with other charged particles. Neutrons do not interact via the electric force and hence penetrate deeper, causing significant damage. Gamma rays are high-energy photons that can cause severe damage, but often pass through matter without interaction.

Several units have been used historically to quantify the amount, or dose, of any radiation that interacts with a substance.

> The **roentgen** (R) is that amount of ionizing radiation that produces an electric charge of 3.33×10^{-10} C in 1 cm^3 of air under standard conditions.

Equivalently, the roentgen is that amount of radiation that increases the energy of 1 kg of air by 8.76×10^{-3} J.

For most applications, the roentgen has been replaced by the rad (an acronym for *radiation absorbed dose*):

> One **rad** is that amount of radiation that increases the energy of 1 kg of absorbing material by 1×10^{-2} J.

Although the rad is a perfectly good physical unit, it is not the best unit for measuring the degree of biological damage produced by radiation because damage depends not only on the dose but also on the type of the radiation. For example, a given dose of alpha particles causes about ten times more biological damage than an equal dose of x-rays. The **RBE** (relative biological effectiveness) factor for a given type of radiation is **the number of rads of x-radiation or gamma radiation that produces the same biological damage as 1 rad of the radiation being used.** The RBE factors for different types of radiation are given in Table 45.1. The values are only approximate because

Table 45.1

RBE[a] Factors for Several Types of Radiation	
Radiation	**RBE Factor**
X-rays and gamma rays	1.0
Beta particles	1.0–1.7
Alpha particles	10–20
Thermal neutrons	4–5
Fast neutrons and protons	10
Heavy ions	20

[a] RBE = relative biological effectiveness.

Table 45.2

| Units for Radiation Dosage | | | | | |
Quantity	SI Unit	Symbol	Relation to Other SI units	Older Unit	Conversion
Absorbed dose	gray	Gy	$= 1\,J/kg$	rad	1 Gy = 100 rad
Dose equivalent	sievert	Sv	$= 1\,J/kg$	rem	1 Sv = 100 rem

they vary with particle energy and with the form of the damage. The RBE factor should be considered only a first-approximation guide to the actual effects of radiation.

Finally, the **rem** (radiation equivalent in man) is the product of the dose in rad and the RBE factor:

$$\text{Dose in rem} \equiv \text{dose in rad} \times \text{RBE} \qquad (45.6)$$

Radiation dose in rem

According to this definition, 1 rem of any two types of radiation produces the same amount of biological damage. From Table 45.1, we see that a dose of 1 rad of fast neutrons represents an effective dose of 10 rem, but 1 rad of gamma radiation is equivalent to a dose of only 1 rem.

Low-level radiation from natural sources, such as cosmic rays and radioactive rocks and soil, delivers to each of us a dose of about 0.13 rem/yr; this radiation is called *background radiation*. Background radiation varies with geography, with the main factors being altitude (exposure to cosmic rays) and geology (radon gas released by some rock formations, deposits of naturally radioactive minerals).

The upper limit of radiation dose rate recommended by the U.S. government (apart from background radiation) is about 0.5 rem/yr. Many occupations involve much higher radiation exposures, and so an upper limit of 5 rem/yr has been set for combined whole-body exposure. Higher upper limits are permissible for certain parts of the body, such as the hands and the forearms. A dose of 400 to 500 rem results in a mortality rate of about 50% (which means that half the people exposed to this radiation level die). The most dangerous form of exposure for most people is either ingestion or inhalation of radioactive isotopes, especially isotopes of those elements the body retains and concentrates, such as ^{90}Sr.

This discussion has focused on measurements of radiation dosage in units such as rads and rems because these units are still widely used. These units, however, have been formally replaced with new SI units. The rad has been replaced with the *gray* (Gy), equal to 100 rad. The rem has been replaced with the *sievert* (Sv), equal to 100 rem. Table 45.2 summarizes the older and the current SI units of radiation dosage.

45.6 Radiation Detectors

Particles passing through matter interact with the matter in several ways. The particle can ionize atoms, scatter from atoms, be absorbed by atoms, etc. Radiation detectors exploit these interactions to allow a measurement of the particle's energy, momentum, or charge, and sometimes the very existence of the particle if it is otherwise difficult to detect. Various devices have been developed for detecting radiation. These devices are used for a variety of purposes, including medical diagnoses, radioactive dating measurements, measuring background radiation, and measuring the mass, energy, and momentum of particles created in high-energy nuclear reactions.

In the early part of the twentieth century, detectors were much simpler than those used today. We will discuss three of these early detectors first. A **photographic emulsion** is the simplest example of a detector. A charged particle ionizes the atoms in an

(a)   (b)

Figure 45.16 (a) Artificially colored bubble-chamber photograph showing tracks of particles that have passed through the chamber. (b) This research scientist is studying a photograph of particle tracks made in a bubble chamber at Fermilab. The curved tracks are produced by charged particles moving through the chamber in the presence of an applied magnetic field. Negatively charged particles deflect in one direction, while positively charged particles deflect in the opposite direction.

emulsion layer. The path of the particle corresponds to a family of points at which chemical changes have occurred in the emulsion. When the emulsion is developed, the particle's track becomes visible. A **cloud chamber** contains a gas that has been supercooled to just below its usual condensation point. An energetic particle passing through ionizes the gas along the particle's path. The ions serve as centers for condensation of the supercooled gas. The track of the particle can be seen with the naked eye and can be photographed. A magnetic field can be applied to determine the charges of the particles, as well as their momentum and energy. A device called a **bubble chamber,** invented in 1952 by D. Glaser, uses a liquid (usually liquid hydrogen) maintained near its boiling point. Ions produced by incoming charged particles leave bubble tracks, which can be photographed (Fig. 45.16). Because the density of the detecting medium in a bubble chamber is much higher than the density of the gas in a cloud chamber, the bubble chamber has a much higher sensitivity.

More contemporary detectors involve more sophisticated processes. In an **ion chamber** (Fig. 45.17), electron–ion pairs are generated as radiation passes through a gas and produces an electrical signal. Two plates in the chamber are connected to a voltage supply and thereby maintained at different electric potentials. The positive plate attracts the electrons, and the negative plate attracts positive ions, causing a current pulse that is proportional to the number of electron–ion pairs produced when a particle passes through the chamber. When an ion chamber is used both to detect the presence of a particle and to measure its energy, it is called a **proportional counter.**

The **Geiger counter** (Fig. 45.18) is perhaps the most common form of ion chamber used to detect radioactivity. It can be considered the prototype of all counters that use the ionization of a medium as the basic detection process. It consists of a thin wire electrode aligned along the central axis of a cylindrical metallic tube filled with a gas at low pressure. The wire is maintained at a high positive electric potential (about 10^3 V) relative to the tube. When a high-energy particle resulting, for example, from a radioactive decay enters the tube through a thin window at one end, some of the gas atoms are ionized. The electrons removed from these atoms are attracted toward the

Figure 45.17 Simplified diagram of an ion chamber. The radioactive source creates electrons and positive ions that are collected by the charged plates. The current set up in the external circuit is proportional to a radioactive particle's kinetic energy if the particle stops in the chamber.

wire electrode, and in the process they ionize other atoms in their path. This sequential ionization results in an *avalanche* of electrons that produces a current pulse. After the pulse has been amplified, it can either be used to trigger an electronic counter or be delivered to a loudspeaker that clicks each time a particle is detected. Although a Geiger counter easily detects the presence of a particle, the energy lost by the particle in the counter is *not* proportional to the current pulse produced. Thus, a Geiger counter cannot be used to measure the energy of a particle.

A **semiconductor-diode detector** is essentially a reverse-bias p–n junction. Recall from Section 43.7 that a p–n junction passes current readily when forward-biased and

Figure 45.18 (a) Diagram of a Geiger counter. The voltage between the wire electrode and the metallic tube is usually about 1 000 V. (b) A scientist uses a Geiger counter to make a measurement.

prohibits a current when reverse-biased. As an energetic particle passes through the junction, electrons are excited into the conduction band and holes are formed in the valence band. The internal electric field sweeps the electrons toward the positive (*n*) side of the junction and the holes toward the negative (*p*) side. This movement of electrons and holes creates a pulse of current that is measured with an electronic counter. In a typical device, the duration of the pulse is 10^{-8} s.

A **scintillation counter** usually uses a solid or liquid material whose atoms are easily excited by radiation. The excited atoms then emit photons when they return to their ground state. Common materials used as scintillators are transparent crystals of sodium iodide and certain plastics. If the scintillator material is attached to a photomultiplier (PM) tube (Section 40.2), the photons emitted by the scintillator can be detected and an electrical signal produced.

Both the scintillator and the semiconductor-diode detector are much more sensitive than a Geiger counter, mainly because of the higher density of the detecting medium. Both measure the total energy deposited in the detector, which can be very useful in particle identification. In addition, if the particle stops in the detector, they can both be used to measure the total particle energy.

Track detectors are devices used to view the tracks of charged particles directly. High-energy particles produced in particle accelerators may have energies ranging from 10^9 to 10^{12} eV. Therefore, they often cannot be stopped and cannot have their energy measured with the detectors already mentioned. Instead, the energy and momentum of these energetic particles are found from the curvature of their path in a magnetic field of known magnitude and direction.

A **spark chamber** is a counting device that consists of an array of conducting parallel plates and is capable of recording a three-dimensional track record. Even-numbered plates are grounded, and odd-numbered plates are maintained at a high electric potential (about 10 kV). The spaces between the plates contain an inert gas at atmospheric pressure. When a charged particle passes through the chamber, gas atoms are ionized, resulting in a current surge and visible sparks along the particle path. These sparks may be photographed or electronically detected and the data sent to a computer for path reconstruction and determination of particle mass, momentum, and energy.

Newer versions of the spark chamber have been developed. A **drift chamber** has thousands of high-voltage wires arrayed through the space of the detector, which is filled with gas. The result is an array of thousands of Geiger counters. When a charged particle passes through the detector, it ionizes gas molecules, and the ejected electrons drift toward the high-voltage wires, creating an electrical signal upon arrival. A computer detects the signals and reconstructs the path through the detector. A large-volume, sophisticated drift chamber that has provided significant results in studying particles formed in collisions of atoms is the Solenoidal Tracker at RHIC (STAR). (The acronym RHIC stands for Relativistic Heavy Ion Collider, a facility at Brookhaven National Laboratory that began operation in 2000.) This type of drift chamber is called a **time projection chamber.** A photograph of the STAR detector is shown in Figure 45.19.

Figure 45.19 The STAR detector at the Relativistic Heavy Ion Collider at Brookhaven National Laboratory.

45.7 Uses of Radiation

Nuclear physics applications are extremely widespread in manufacturing, medicine, and biology. Even a brief discussion of all the possibilities would fill an entire book, and to keep such a book up to date would require frequent revisions. In this section, we present a few of these applications and the underlying theories supporting them.

Tracing

Radioactive tracers are used to track chemicals participating in various reactions. One of the most valuable uses of radioactive tracers is in medicine. For example, iodine, a nutrient needed by the human body, is obtained largely through the intake of iodized

① ^{24}NaCl solution injected

② Blood carries ^{24}NaCl to both legs

Site of constriction

③ High reading– good circulation

④ Low reading– poor circulation

Figure 45.20 A tracer technique for determining the condition of the human circulatory system.

salt and seafood. To evaluate the performance of the thyroid, the patient drinks a very small amount of radioactive sodium iodide containing ^{131}I, an artificially produced isotope of iodine (the natural, nonradioactive isotope is ^{127}I). The amount of iodine in the thyroid gland is determined as a function of time by measuring the radiation intensity at the neck area. How much of the isotope ^{131}I remains in the thyroid is a measure of how well that gland is functioning.

A second medical application is indicated in Figure 45.20. A solution containing radioactive sodium is injected into a vein in the leg, and the time at which the radioisotope arrives at another part of the body is detected with a radiation counter. The elapsed time is a good indication of the presence or absence of constrictions in the circulatory system.

Tracers are also useful in agricultural research. Suppose the best method of fertilizing a plant is to be determined. A certain element in a fertilizer, such as nitrogen, can be *tagged* (identified) with one of its radioactive isotopes. The fertilizer is then sprayed on one group of plants, sprinkled on the ground for a second group, and raked into the soil for a third. A Geiger counter is then used to track the nitrogen through the three groups.

Tracing techniques are as wide ranging as human ingenuity can devise. Present applications range from checking how teeth absorb fluoride to monitoring how cleansers contaminate food-processing equipment to studying deterioration inside an automobile engine. In the latter case, a radioactive material is used in the manufacture of the piston rings, and the oil is checked for radioactivity to determine the amount of wear on the rings.

Materials Analysis

For centuries, a standard method of identifying the elements in a sample of material has been chemical analysis, which involves determining how the material reacts with various chemicals. A second method is spectral analysis, which uses the fact that, when excited, each element emits its own characteristic set of electromagnetic wavelengths. These methods are now supplemented by a third technique, **neutron activation analysis.** Both chemical and spectral methods have the disadvantage that a fairly large sample of the material must be destroyed for the analysis. In addition, extremely small quantities of an element may go undetected by either method. Neutron activation analysis has an advantage over the other two methods in both respects.

When a material is irradiated with neutrons, nuclei in the material absorb the neutrons and are changed to different isotopes, most of which are radioactive. For

AP/Wide World Photos

Figure 45.21 A bomb detector will irradiate this package with neutrons. If there are hidden explosives inside, chemicals within the explosive materials become radioactive and can be easily detected. The half-life of the resulting radiation is so short that there is no danger to the security personnel removing the package for further inspection.

example, ^{65}Cu absorbs a neutron to become ^{66}Cu, which undergoes beta decay:

$$^1_0\text{n} + \,^{65}_{29}\text{Cu} \longrightarrow \,^{66}_{29}\text{Cu} \longrightarrow \,^{66}_{30}\text{Zn} + \text{e}^- + \overline{\nu}$$

The presence of the copper can be deduced because it is known that ^{66}Cu has a half-life of 5.1 min and decays with the emission of beta particles having maximum energies of 2.63 and 1.59 MeV. Also emitted in the decay of ^{66}Cu is a 1.04-MeV gamma ray. By examining the radiation emitted by a substance after it has been exposed to neutron irradiation, one can detect extremely small amounts of an element in that substance.

Neutron activation analysis is used routinely in a number of industries—for example, in commercial aviation for the checking of airline luggage for hidden explosives (Fig. 45.21). The following nonroutine use is of interest. Napoleon died on the island of St. Helena in 1821, supposedly of natural causes. Over the years, suspicion has existed that his death was not all that natural. After his death, his head was shaved and locks of his hair were sold as souvenirs. In 1961, the amount of arsenic in a sample of this hair was measured by neutron activation analysis, and an unusually large quantity of arsenic was found. (Activation analysis is so sensitive that very small pieces of a single hair could be analyzed.) Results showed that the arsenic was fed to him irregularly. In fact, the arsenic concentration pattern corresponded to the fluctuations in the severity of Napoleon's illness as determined from historical records.

Art historians use neutron activation analysis to detect forgeries. The pigments used in paints have changed throughout history, and old and new pigments react differently to neutron activation. The method can even reveal hidden works of art behind existing paintings because an older, hidden layer of paint reacts differently than the surface layer to neutron activation.

Radiation Therapy

Martin Dohrn/Science Photo Library/ Photo Researchers

Figure 45.22 This large machine is being set to deliver a dose of radiation from ^{60}Co in an effort to destroy a cancerous tumor. Cancer cells are especially susceptible to this type of therapy because they tend to divide more often than cells of healthy tissue nearby.

Radiation causes the most damage to rapidly dividing cells. Therefore, it is useful in cancer treatment because tumor cells divide extremely rapidly. Several mechanisms can be used to deliver radiation to a tumor. In some cases, a narrow beam of x-rays or radiation from a source such as ^{60}Co is used, as shown in Figure 45.22. In other situations, thin radioactive needles called *seeds* are implanted in the cancerous tissue. The radioactive isotope ^{131}I is used to treat cancer of the thyroid.

Council for Agricultural Science & Technology

NON - IRRADIATED -

IRRADIATED - (0.2 M RAD)

Figure 45.23 The strawberries on the left are untreated and have become moldy. The unspoiled strawberries on the right have been irradiated. The radiation has killed or incapacitated the mold spores that have spoiled the strawberries on the left.

Food Preservation

Radiation is finding increasing use as a means of preserving food because exposure to high levels of radiation can destroy or incapacitate bacteria and mold spores (Fig. 45.23). Techniques include exposing foods to gamma rays, high-energy electron beams, or x-rays. Food preserved this way can be placed in a sealed container (to keep out new spoiling agents) and stored for long periods of time. There is no evidence of adverse effect on the taste or nutritional value of food from irradiation. The safety of irradiated foods has been endorsed by the World Health Organization (WHO), the Centers for Disease Control and Prevention (CDC), the U.S. Department of Agriculture (USDA), and the Food and Drug Administration (FDA).

SUMMARY

Take a practice test for this chapter by clicking on the Practice Test link at http://www.pse6.com.

The probability that neutrons are captured as they move through matter generally increases with decreasing neutron energy. A **thermal neutron** is a slow-moving neutron that has a high probability of being captured by a nucleus in a **neutron capture event:**

$$_{0}^{1}n + {}_{Z}^{A}X \longrightarrow {}_{Z}^{A+1}X^{*} \longrightarrow {}_{Z}^{A+1}X + \gamma \tag{45.1}$$

where ${}_{Z}^{A+1}X^{*}$ is an excited intermediate nucleus that rapidly emits a photon.

Nuclear fission occurs when a very heavy nucleus, such as ${}^{235}U$, splits into two smaller **fission fragments.** Thermal neutrons can create fission in ${}^{235}U$:

$$_{0}^{1}n + {}_{92}^{235}U \longrightarrow {}_{92}^{236}U^{*} \longrightarrow X + Y + \text{neutrons} \tag{45.2}$$

where X and Y are the fission fragments and ${}^{236}U^{*}$ is an intermediate excited state. On the average, 2.5 neutrons are released per fission event. The fragments then undergo a series of beta and gamma decays to various stable isotopes. The energy released per fission event is about 200 MeV.

The **reproduction constant** K is the average number of neutrons released from each fission event that cause another event. In a fission reactor, it is necessary to maintain $K \approx 1$. The value of K is affected by such factors as reactor geometry, mean neutron energy, and probability of neutron capture. Neutron energies are reduced with a moderator material that slows down energetic neutrons and therefore increases the probability of neutron capture by other ${}^{235}U$ nuclei. The power level of the reactor is adjusted with control rods made of a material that is very efficient in absorbing neutrons.

In **nuclear fusion,** two light nuclei fuse to form a heavier nucleus and release energy. The major obstacle in obtaining useful energy from fusion is the large Coulomb repulsive force between the charged nuclei at small separation distances. Sufficient energy must be supplied to the particles to overcome this Coulomb barrier. The temperature required to produce fusion is on the order of 10^8 K, and at this temperature all matter occurs as a plasma.

In a fusion reactor, the plasma temperature must reach the **critical ignition temperature,** the temperature at which the power generated by the fusion reactions exceeds the power lost in the system. The most promising fusion reaction is the D–T reaction, which has a critical ignition temperature of approximately 4.5×10^7 K. Two critical parameters in fusion reactor design are **ion density** n and **confinement time** τ, the time interval during which the interacting particles must be maintained at $T > T_{\text{ignit}}$. **Lawson's criterion** states that for the D–T reaction, $n\tau \geq 10^{14}$ s/cm^3.

Energy absorbed in a substance due to radiation is measured with a unit called the **rad,** where one rad is that amount of radiation that increases the energy of 1 kg of absorbing material by 1×10^{-2} J. The radiation dose that is related to the amount of biological damage in an organism is measured by a unit called the **rem** (radiation equivalent in man) and is the product of the dose in rad and the RBE factor:

$$\text{Dose in rem} \equiv \text{dose in rad} \times \text{RBE} \qquad (45.6)$$

where the **RBE** (relative biological effectiveness) factor for a given type of radiation is the number of rads of x-radiation or gamma radiation that produces the same biological damage as 1 rad of the radiation being used.

The rad and the rem have been formally replaced with new SI units. The rad has been replaced with the *gray* (Gy), equal to 100 rad. The rem has been replaced with the *sievert* (Sv), equal to 100 rem.

QUESTIONS

1. Explain the function of a moderator in a fission reactor.

2. Why is water a better shield against neutrons than lead or steel?

3. If a nucleus captures a slow-moving neutron, the product is left in a highly excited state, with an energy approximately 8 MeV above the ground state. Explain the source of the excitation energy.

4. Discuss the advantages and disadvantages of fission reactors from the point of view of safety, pollution, and resources. Make a comparison with power generated from the burning of fossil fuels.

5. Why would a fusion reactor produce less radioactive waste than a fission reactor?

6. Lawson's criterion states that the product of ion density and confinement time must exceed a certain number before a break-even fusion reaction can occur. Why should these two parameters determine the outcome?

7. Why is the temperature required for the D–T fusion less than that needed for the D–D fusion? Estimate the relative importance of Coulomb repulsion and nuclear attraction in each case.

8. What factors make a fusion reaction difficult to achieve?

9. Discuss the similarities and differences between fusion and fission.

10. Discuss the advantages and disadvantages of fusion power from the viewpoint of safety, pollution, and resources.

11. Discuss three major problems associated with the development of a controlled fusion reactor.

12. Describe two techniques being pursued in an effort to obtain power from nuclear fusion.

13. If two radioactive samples have the same activity measured in curies, will they necessarily create the same damage to a medium? Explain.

14. Why should a radiologist be extremely cautious about x-ray doses when treating pregnant women?

15. The design of a PM tube (Figure 40.12) might suggest that any number of dynodes may be used to amplify a weak signal. What factors do you suppose would limit the amplification in this device?

16. *And swift, and swift past comprehension*
    Turn round Earth's beauty and her might.
    The heavens blaze in alternation
    With deep and chill and rainy night.
    In mighty currents foams the ocean
    Up from the rocks' abyssal base,
    With rock and sea torn into motion
    In ever-swift celestial race.
    And tempests bluster in a contest
    From sea to land, from land to sea.
    In rage they forge a chain around us
    Of deepest meaning, energy.
    There flames a lightning disaster
    Before the thunder, in its way.
    But all Your servants honor, Master,
    The gentle order of Your day.
    Johann Wolfgang von Goethe wrote this song of the archangels in *Faust* half a century before the law of conservation of energy was recognized. Students often find it useful to think of a list of several "forms of energy," from kinetic to nuclear. Argue for or against the view that these lines of poetry make an obvious or oblique reference to every form of energy and energy transfer.

PROBLEMS

1, **2**, **3** = straightforward, intermediate, challenging   ☐ = full solution available in the *Student Solutions Manual and Study Guide*

🌐 = coached solution with hints available at http://www.pse6.com   💻 = computer useful in solving problem

▨ = paired numerical and symbolic problems

Section 45.2   Nuclear Fission

> *Note:* Problem 53 in Chapter 25 and Problems 21, 51, and 68 in Chapter 44 can be assigned with this section.

1. Burning one metric ton (1 000 kg) of coal can yield an energy of 3.30×10^{10} J. Fission of one nucleus of uranium-235 yields an average of about 208 MeV. What mass of uranium produces the same energy as a ton of coal?

2. Find the energy released in the fission reaction

$$^{1}_{0}n + ^{235}_{92}U \longrightarrow ^{98}_{40}Zr + ^{135}_{52}Te + 3(^{1}_{0}n)$$

The atomic masses of the fission products are: $^{98}_{40}Zr$, 97.912 7 u; $^{135}_{52}Te$, 134.916 5 u.

3. Strontium-90 is a particularly dangerous fission product of ^{235}U because it is radioactive and it substitutes for calcium in bones. What other direct fission products would accompany it in the neutron-induced fission of ^{235}U? (*Note:* This reaction may release two, three, or four free neutrons.)

4. List the nuclear reactions required to produce ^{239}Pu from ^{238}U under fast neutron bombardment.

5. 🌐 List the nuclear reactions required to produce ^{233}U from ^{232}Th under fast neutron bombardment.

6. (a) The following fission reaction is typical of those occurring in a nuclear electric generating station:

$$^{1}_{0}n + ^{235}_{92}U \longrightarrow ^{141}_{56}Ba + ^{92}_{36}Kr + 3(^{1}_{0}n)$$

Find the energy released. The required masses are

$$M(^{1}_{0}n) = 1.008\ 665\ u$$

$$M(^{235}_{92}U) = 235.043\ 923\ u$$

$$M(^{141}_{56}Ba) = 140.914\ 4\ u$$

$$M(^{92}_{36}Kr) = 91.926\ 2\ u$$

(b) What fraction of the initial mass of the system is transformed?

7. A reaction that has been considered as a source of energy is the absorption of a proton by a boron-11 nucleus to produce three alpha particles:

$$^{1}_{1}H + ^{11}_{5}B \longrightarrow 3(^{4}_{2}He)$$

This is an attractive possibility because boron is easily obtained from the Earth's crust. A disadvantage is that the protons and boron nuclei must have large kinetic energies in order for the reaction to take place. This is in contrast to the initiation of uranium fission by slow neutrons. (a) How much energy is released in each reaction? (b) Why must the reactant particles have high kinetic energies?

8. A typical nuclear fission power plant produces about 1.00 GW of electrical power. Assume that the plant has an overall efficiency of 40.0% and that each fission produces 200 MeV of energy. Calculate the mass of ^{235}U consumed each day.

9. **Review problem.** Suppose enriched uranium containing 3.40% of the fissionable isotope $^{235}_{92}U$ is used as fuel for a ship. The water exerts an average friction force of magnitude 1.00×10^5 N on the ship. How far can the ship travel per kilogram of fuel? Assume that the energy released per fission event is 208 MeV and that the ship's engine has an efficiency of 20.0%.

Section 45.3   Nuclear Reactors

10. To minimize neutron leakage from a reactor, the surface area-to-volume ratio should be a minimum. For a given volume V, calculate this ratio for (a) a sphere, (b) a cube, and (c) a parallelepiped of dimensions $a \times a \times 2a$. (d) Which of these shapes would have minimum leakage? Which would have maximum leakage?

11. 🌐 It has been estimated that on the order of 10^9 tons of natural uranium is available at concentrations exceeding 100 parts per million, of which 0.7% is the fissionable isotope ^{235}U. Assume that all the world's energy use (7×10^{12} J/s) were supplied by ^{235}U fission in conventional nuclear reactors, releasing 208 MeV for each reaction. How long would the supply last? The estimate of uranium supply is taken from K. S. Deffeyes and I. D. MacGregor, "World Uranium Resources," *Scientific American* **242**(1):66, 1980.

12. If the reproduction constant is 1.000 25 for a chain reaction in a fission reactor and the average time interval between successive fissions is 1.20 ms, by what factor will the reaction rate increase in one minute?

13. A large nuclear power reactor produces about 3 000 MW of power in its core. Three months after a reactor is shut down, the core power from radioactive byproducts is 10.0 MW. Assuming that each emission delivers 1.00 MeV of energy to the power, find the activity in becquerels three months after the reactor is shut down.

Section 45.4   Nuclear Fusion

14. (a) Consider a fusion generator built to create 3.00 GW of power. Determine the rate of fuel burning in grams per hour if the D–T reaction is used. (b) Do the same for the D–D reaction assuming that the reaction products are split evenly between (n, ^{3}He) and (p, ^{3}H).

15. Two nuclei having atomic numbers Z_1 and Z_2 approach each other with a total energy E. (a) Suppose they will

spontaneously fuse if they approach within a distance of 1.00×10^{-14} m. Find the minimum value of E required to produce fusion, in terms of Z_1 and Z_2. (b) Evaluate the minimum energy for fusion for the D–D and D–T reactions (the first and third reactions in Eq. 45.4).

16. **Review problem.** Consider the deuterium–tritium fusion reaction with the tritium nucleus at rest:

$$^2_1H + {}^3_1H \longrightarrow {}^4_2He + {}^1_0n$$

(a) Suppose that the reactant nuclei will spontaneously fuse if their surfaces touch. From Equation 44.1, determine the required distance of closest approach between their centers. (b) What is the electric potential energy (in eV) at this distance? (c) Suppose the deuteron is fired straight at an originally stationary tritium nucleus with just enough energy to reach the required distance of closest approach. What is the common speed of the deuterium and tritium nuclei as they touch, in terms of the initial deuteron speed v_i? (*Suggestion:* At this point, the two nuclei have a common velocity equal to the center-of-mass velocity.) (d) Use energy methods to find the minimum initial deuteron energy required to achieve fusion. (e) Why does the fusion reaction actually occur at much lower deuteron energies than that calculated in (d)?

17. To understand why plasma containment is necessary, consider the rate at which an unconfined plasma would be lost. (a) Estimate the rms speed of deuterons in a plasma at 4.00×10^8 K. (b) **What If?** Estimate the order of magnitude of the time interval during which such a plasma would remain in a 10-cm cube if no steps were taken to contain it.

18. Of all the hydrogen in the oceans, 0.030 0% of the mass is deuterium. The oceans have a volume of 317 million mi^3. (a) If nuclear fusion were controlled and all the deuterium in the oceans were fused to $^4_2$He, how many joules of energy would be released? (b) **What If?** World power consumption is about 7.00×10^{12} W. If consumption were 100 times greater, how many years would the energy calculated in part (a) last?

19. It has been suggested that fusion reactors are safe from explosion because there is never enough energy in the plasma to do much damage. (a) In 1992, the TFTR reactor achieved an ion temperature of 4.0×10^8 K, an ion density of 2.0×10^{13} cm^{-3}, and a confinement time of 1.4 s. Calculate the amount of energy stored in the plasma of the TFTR reactor. (b) How many kilograms of water could be boiled away by this much energy? (The plasma volume of the TFTR reactor is about 50 m^3.)

20. **Review problem.** To confine a stable plasma, the magnetic energy density in the magnetic field (Eq. 32.14) must exceed the pressure $2nk_BT$ of the plasma by a factor of at least 10. In the following, assume a confinement time $\tau = 1.00$ s. (a) Using Lawson's criterion, determine the ion density required for the D–T reaction. (b) From the ignition-temperature criterion, determine the required plasma pressure. (c) Determine the magnitude of the magnetic field required to contain the plasma.

21. Find the number of ^{6}Li and the number of ^{7}Li nuclei present in 2.00 kg of lithium. (The natural abundance of ^{6}Li is 7.5%; the remainder is ^{7}Li.)

22. One old prediction for the future was to have a fusion reactor supply energy to dissociate the molecules in garbage into separate atoms and then to ionize the atoms. This material could be put through a giant mass spectrometer, so that trash would be a new source of isotopically pure elements—the mine of the future. Assuming an average atomic mass of 56 and an average charge of 26 (a high estimate, considering all the organic materials), at a beam current of 1.00 MA, how long would it take to process 1.00 metric ton of trash?

Section 45.5 Radiation Damage

23. A building has become accidentally contaminated with radioactivity. The longest-lived material in the building is strontium-90. ($^{90}_{38}$Sr has an atomic mass 89.907 7 u, and its half-life is 29.1 yr. It is particularly dangerous because it substitutes for calcium in bones.) Assume that the building initially contained 5.00 kg of this substance uniformly distributed throughout the building (a very unlikely situation) and that the safe level is defined as less than 10.0 decays/min (to be small in comparison to background radiation). How long will the building be unsafe?

24. **Review problem.** A particular radioactive source produces 100 mrad of 2-MeV gamma rays per hour at a distance of 1.00 m. (a) How long could a person stand at this distance before accumulating an intolerable dose of 1 rem? (b) **What If?** Assuming the radioactive source is a point source, at what distance would a person receive a dose of 10.0 mrad/h?

25. Assume that an x-ray technician takes an average of eight x-rays per day and receives a dose of 5 rem/yr as a result. (a) Estimate the dose in rem per photograph taken. (b) How does the technician's exposure compare with low-level background radiation?

26. When gamma rays are incident on matter, the intensity of the gamma rays passing through the material varies with depth x as $I(x) = I_0e^{-\mu x}$, where μ is the absorption coefficient and I_0 is the intensity of the radiation at the surface of the material. For 0.400-MeV gamma rays in lead, the absorption coefficient is 1.59 cm^{-1}. (a) Determine the "half-thickness" for lead—that is, the thickness of lead that would absorb half the incident gamma rays. (b) What thickness will reduce the radiation by a factor of 10^4?

27. A "clever" technician decides to warm some water for his coffee with an x-ray machine. If the machine produces 10.0 rad/s, how long will it take to raise the temperature of a cup of water by 50.0°C?

28. **Review problem.** The danger to the body from a high dose of gamma rays is not due to the amount of energy absorbed but occurs because of the ionizing nature of the radiation. To illustrate this, calculate the rise in body temperature that would result if a "lethal" dose of 1 000 rad were absorbed strictly as internal energy. Take the specific heat of living tissue as 4 186 J/kg · °C.

29. Technetium-99 is used in certain medical diagnostic procedures. Assume 1.00×10^{-8} g of ^{99}Tc is injected into a 60.0-kg patient and half of the 0.140-MeV gamma rays are absorbed in the body. Determine the total radiation dose received by the patient.

30. Strontium-90 from the testing of nuclear bombs can still be found in the atmosphere. Each decay of ^{90}Sr releases 1.1 MeV of energy into the bones of a person who has had strontium replace the calcium. Assume a 70.0-kg person receives 1.00 μg of ^{90}Sr from contaminated milk. Calculate the absorbed dose rate (in J/kg) in one year. Take the half-life of ^{90}Sr to be 29.1 yr.

Section 45.6   Radiation Detectors

31. In a Geiger tube, the voltage between the electrodes is typically 1.00 kV and the current pulse discharges a 5.00-pF capacitor. (a) What is the energy amplification of this device for a 0.500-MeV electron? (b) How many electrons participate in the avalanche caused by the single initial electron?

32. Assume a photomultiplier tube (Figure 40.12) has seven dynodes with potentials of 100, 200, 300, . . . , 700 V. The average energy required to free an electron from the dynode surface is 10.0 eV. Assume that just one electron is incident and that the tube functions with 100% efficiency. (a) How many electrons are freed at the first dynode? (b) How many electrons are collected at the last dynode? (c) What is the energy available to the counter for each electron?

33. (a) Your grandmother recounts to you how, as young children, your father, aunts, and uncles made the screen door slam continually as they ran between the house and the back yard. The time interval between one slam and the next varied randomly, but the average slamming rate stayed constant at 38.0/h from dawn to dusk every summer day. If the slamming rate suddenly dropped to zero, the children would have found a nest of baby field mice or gotten into some other mischief requiring adult intervention. How long after the last screen-door slam would a prudent and attentive parent wait before leaving her or his work to see about the children? Explain your reasoning. (b) A student wishes to measure the half-life of a radioactive substance, using a small sample. Consecutive clicks of her Geiger counter are randomly spaced in time. The counter registers 372 counts during one 5.00-min interval, and 337 counts during the next 5.00 min. The average background rate is 15 counts per minute. Find the most probable value for the half-life. (c) Estimate the uncertainty in the half-life determination. Explain your reasoning.

Section 45.7   Uses of Radiation

34. During the manufacture of a steel engine component, radioactive iron (^{59}Fe) is included in the total mass of 0.200 kg. The component is placed in a test engine when the activity due to this isotope is 20.0 μCi. After a 1 000-h test period, some of the lubricating oil is removed from the engine and found to contain enough ^{59}Fe to produce 800 disintegrations/min/L of oil. The total volume of oil in the engine is 6.50 L. Calculate the total mass worn from the engine component per hour of operation. (The half-life of ^{59}Fe is 45.1 d.)

35. At some time in your past or future, you may find yourself in a hospital to have a PET scan. The acronym stands for *positron-emission tomography*. In the procedure, a radioactive element that undergoes e^+ decay is introduced into your body. The equipment detects the gamma rays that result from pair annihilation when the emitted positron encounters an electron in your body's tissue. Suppose you receive an injection of glucose that contains on the order of 10^{10} atoms of ^{14}O. Assume that the oxygen is uniformly distributed through 2 L of blood after 5 min. What will be the order of magnitude of the activity of the oxygen atoms in 1 cm^3 of the blood?

36. You want to find out how many atoms of the isotope ^{65}Cu are in a small sample of material. You bombard the sample with neutrons to ensure that on the order of 1% of these copper nuclei absorb a neutron. After activation you turn off the neutron flux, and then use a highly efficient detector to monitor the gamma radiation that comes out of the sample. Assume that half of the ^{66}Cu nuclei emit a 1.04-MeV gamma ray in their decay. (The other half of the activated nuclei decay directly to the ground state of ^{66}Ni.) If after 10 min (two half-lives) you have detected 10^4 MeV of photon energy at 1.04 MeV, (a) about how many ^{65}Cu atoms are in the sample? (b) Assume the sample contains natural copper. Refer to the isotopic abundances listed in Table A.3 and estimate the total mass of copper in the sample.

37. *Neutron activation analysis* is a method for chemical analysis at the level of isotopes. When a sample is irradiated by neutrons, radioactive atoms are produced continuously and then decay according to their characteristic half-lives. (a) Assume that one species of radioactive nuclei is produced at a constant rate R and that its decay is described by the conventional radioactive decay law. If irradiation begins at time $t = 0$, show that the number of radioactive atoms accumulated at time t is

$$N = \frac{R}{\lambda} (1 - e^{-\lambda t})$$

(b) What is the maximum number of radioactive atoms that can be produced?

Additional Problems

38. The nuclear bomb dropped on Hiroshima on August 6, 1945, released 5×10^{13} J of energy, equivalent to that from 12 000 tons of TNT. The fission of one $^{235}_{92}$U nucleus releases an average of 208 MeV. Estimate (a) the number of nuclei fissioned, and (b) the mass of this $^{235}_{92}$U.

39. Carbon detonations are powerful nuclear reactions that temporarily tear apart the cores inside massive stars late in their lives. These blasts are produced by carbon fusion, which requires a temperature of about 6×10^8 K to overcome the strong Coulomb repulsion between carbon nuclei. (a) Estimate the repulsive energy barrier to fusion, using the temperature required for carbon fusion. (In other words, what is the average kinetic energy of a carbon nucleus at 6×10^8 K?) (b) Calculate the energy (in MeV) released in each of these "carbon-burning" reactions:

$$^{12}\text{C} + ^{12}\text{C} \longrightarrow ^{20}\text{Ne} + ^{4}\text{He}$$
$$^{12}\text{C} + ^{12}\text{C} \longrightarrow ^{24}\text{Mg} + \gamma$$

(c) Calculate the energy (in kWh) given off when 2.00 kg of carbon completely fuses according to the first reaction.

40. **Review problem.** Consider a nucleus at rest, which then spontaneously splits into two fragments of masses m_1 and m_2. Show that the fraction of the total kinetic energy that is carried by fragment m_1 is

$$\frac{K_1}{K_{tot}} = \frac{m_2}{m_1 + m_2}$$

and the fraction carried by m_2 is

$$\frac{K_2}{K_{tot}} = \frac{m_1}{m_1 + m_2}$$

assuming relativistic corrections can be ignored. (*Note:* If the parent nucleus was moving before the decay, then the fission products still divide the kinetic energy as shown, as long as all velocities are measured in the center-of-mass frame of reference, in which the total momentum of the system is zero.)

41. A stationary $^{236}_{92}$U nucleus fissions spontaneously into two primary fragments, $^{87}_{35}$Br and $^{149}_{57}$La. (a) Calculate the disintegration energy. The required atomic masses are 86.920 711 u for $^{87}_{35}$Br, 148.934 370 u for $^{149}_{57}$La, and 236.045 562 u for $^{236}_{92}$U. (b) How is the disintegration energy split between the two primary fragments? You may use the result of Problem 40. (c) Calculate the speed of each fragment immediately after the fission.

42. Compare the fractional energy loss in a typical ^{235}U fission reaction with the fractional energy loss in D–T fusion.

43. The half-life of tritium is 12.3 yr. If the TFTR fusion reactor contained 50.0 m^3 of tritium at a density equal to 2.00×10^{14} ions/cm^3, how many curies of tritium were in the plasma? Compare this value with a fission inventory (the estimated supply of fissionable material) of 4×10^{10} Ci.

44. **Review problem.** A very slow neutron (with speed approximately equal to zero) can initiate the reaction

$$^1_0n + {}^{10}_5B \longrightarrow {}^7_3Li + {}^4_2He$$

The alpha particle moves away with speed 9.25×10^6 m/s. Calculate the kinetic energy of the lithium nucleus. Use nonrelativistic equations.

45. **Review problem.** A nuclear power plant operates by using the energy released in nuclear fission to convert 20°C water into 400°C steam. How much water could theoretically be converted to steam by the complete fissioning of 1.00 g of ^{235}U at 200 MeV/fission?

46. **Review problem.** A nuclear power plant operates by using the energy released in nuclear fission to convert liquid water at T_c into steam at T_h. How much water could theoretically be converted to steam by the complete fissioning of a mass m of ^{235}U at 200 MeV/fission?

47. About 1 of every 3 300 water molecules contains one deuterium atom. (a) If all the deuterium nuclei in 1 L of water are fused in pairs according to the D–D fusion reaction ^{2}H + ^{2}H → ^{3}He + n + 3.27 MeV, how much energy in joules is liberated? (b) **What If?** Burning gasoline produces about 3.40×10^7 J/L. Compare the energy obtainable from the fusion of the deuterium in a liter of water with the energy liberated from the burning of a liter of gasoline.

48. **Review problem.** The first nuclear bomb was a fissioning mass of plutonium-239 exploded in the Trinity test, before dawn on July 16, 1945, at Alamogordo, New Mexico. Enrico Fermi was 14 km away, lying on the ground facing away from the bomb. After the whole sky had flashed with unbelievable brightness, Fermi stood up and began dropping bits of paper to the ground. They first fell at his feet in the calm and silent air. As the shock wave passed, about 40 s after the explosion, the paper then in flight jumped about 5 cm away from ground zero. (a) Assume that the shock wave in air propagated equally in all directions without absorption. Find the change in volume of a sphere of radius 14 km as it expands by 5 cm. (b) Find the work $P\Delta V$ done by the air in this sphere on the next layer of air farther from the center. (c) Assume the shock wave carried on the order of one tenth of the energy of the explosion. Make an order-of-magnitude estimate of the bomb yield. (d) One ton of exploding trinitrotoluene (TNT) releases 4.2 GJ of energy. What was the order of magnitude of the energy of the Trinity test in equivalent tons of TNT? The dawn revealed the mushroom cloud. Fermi's immediate knowledge of the bomb yield agreed with that determined days later by analysis of elaborate measurements.

49. A certain nuclear plant generates internal energy at a rate of 3.065 GW and transfers energy out of the plant by electrical transmission at a rate of 1.000 GW. Of the wasted energy, 3.0% is ejected to the atmosphere and the remainder is passed into a river. A state law requires that the river water be warmed by no more than 3.50°C when it is returned to the river. (a) Determine the amount of cooling water necessary (in kg/h and m^3/h) to cool the plant. (b) Assume fission generates 7.80×10^{10} J/g of ^{235}U. Determine the rate of fuel burning (in kg/h) of ^{235}U.

50. The alpha-emitter polonium-210 ($^{210}_{84}$Po) is used in a nuclear energy source on a spacecraft (Fig. P45.50). Determine the initial power output of the source. Assume that it

Figure P45.50 The *Pioneer 10* spacecraft leaves the Solar System. It carries radioactive power supplies at the ends of two booms. Solar panels would not work in this region far from the Sun.

Courtesy of NASA Ames

contains 0.155 kg of ^{210}Po and that the efficiency for conversion of radioactive decay energy to energy transferred by electrical transmission is 1.00%.

51. Natural uranium must be processed to produce uranium enriched in ^{235}U for bombs and power plants. The processing yields a large quantity of nearly pure ^{238}U as a byproduct, called "depleted uranium." Because of its high mass density, it is used in armor-piercing artillery shells. (a) Find the edge dimension of a 70.0-kg cube of ^{238}U. (Refer to Table 1.5.) (b) The isotope ^{238}U has a long half-life of 4.47×10^9 yr. As soon as one nucleus decays, it begins a relatively rapid series of 14 steps that together constitute the net reaction

$$^{238}_{92}\text{U} \longrightarrow 8(^{4}_{2}\text{He}) + 6(^{0}_{-1}\text{e}) + ^{206}_{82}\text{Pb} + 6\bar{\nu} + Q_{net}$$

Find the net decay energy. (Refer to Table A.3.) (c) Argue that a radioactive sample with decay rate R and decay energy Q has power output $\mathcal{P} = QR$. (d) Consider an artillery shell with a jacket of 70.0 kg of ^{238}U. Find its power output due to the radioactivity of the uranium and its daughters. Assume the shell is old enough that the daughters have reached steady-state amounts. Express the power in joules per year. (e) **What If?** A 17-year-old soldier of mass 70.0 kg works in an arsenal where many such artillery shells are stored. Assume his radiation exposure is limited to 5.00 rem per year. Find the rate at which he can absorb energy of radiation, in joules per year. Assume an average RBE factor of 1.10.

52. A 2.0-MeV neutron is emitted in a fission reactor. If it loses half its kinetic energy in each collision with a moderator atom, how many collisions must it undergo in order to become a thermal neutron, with energy 0.039 eV?

53. Assuming that a deuteron and a triton are at rest when they fuse according to the reaction $^2_1\text{H} + ^3_1\text{H} \rightarrow ^4_2\text{He} + ^1_0\text{n} + 17.6$ MeV, determine the kinetic energy acquired by the neutron.

54. A sealed capsule containing the radiopharmaceutical phosphorus-32 ($^{32}_{15}$P), an e^- emitter, is implanted into a patient's tumor. The average kinetic energy of the beta particles is 700 keV. The initial activity is 5.22 MBq. Determine the absorbed dose during a 10.0-day period. Assume the beta particles are completely absorbed in 100 g of tissue. (*Suggestion:* Find the number of beta particles emitted.)

55. (a) Calculate the energy (in kilowatt-hours) released if 1.00 kg of ^{239}Pu undergoes complete fission and the energy released per fission event is 200 MeV. (b) Calculate the energy (in electron volts) released in the deuterium–tritium fusion reaction

$$^2_1\text{H} + ^3_1\text{H} \longrightarrow ^4_2\text{He} + ^1_0\text{n}$$

(c) Calculate the energy (in kilowatt-hours) released if 1.00 kg of deuterium undergoes fusion according to this reaction. (d) **What If?** Calculate the energy (in kilowatt-hours) released by the combustion of 1.00 kg of coal if each $C + O_2 \rightarrow CO_2$ reaction yields 4.20 eV. (e) List advantages and disadvantages of each of these methods of energy generation.

56. The Sun radiates energy at the rate of 3.77×10^{26} W. Suppose that the net reaction

$$4(^1_1\text{H}) + 2(^0_{-1}\text{e}) \longrightarrow ^4_2\text{He} + 2\nu + \gamma$$

accounts for all the energy released. Calculate the number of protons fused per second.

57. Consider the two nuclear reactions

$$(\text{I}) \qquad \text{A} + \text{B} \longrightarrow \text{C} + \text{E}$$
$$(\text{II}) \qquad \text{C} + \text{D} \longrightarrow \text{F} + \text{G}$$

(a) Show that the net disintegration energy for these two reactions ($Q_{net} = Q_\text{I} + Q_\text{II}$) is identical to the disintegration energy for the net reaction

$$\text{A} + \text{B} + \text{D} \longrightarrow \text{E} + \text{F} + \text{G}$$

(b) One chain of reactions in the proton–proton cycle in the Sun's core is

$$^1_1\text{H} + ^1_1\text{H} \longrightarrow ^2_1\text{H} + ^0_1\text{e} + \nu$$
$$^0_1\text{e} + ^0_{-1}\text{e} \longrightarrow 2\gamma$$
$$^1_1\text{H} + ^2_1\text{H} \longrightarrow ^3_2\text{He} + \gamma$$
$$^1_1\text{H} + ^3_2\text{He} \longrightarrow ^4_2\text{He} + ^0_1\text{e} + \nu$$
$$^0_1\text{e} + ^0_{-1}\text{e} \longrightarrow 2\gamma$$

Based on part (a), what is Q_{net} for this sequence?

58. Suppose the target in a laser fusion reactor is a sphere of solid hydrogen that has a diameter of 1.50×10^{-4} m and a density of 0.200 g/cm^3. Also assume that half of the nuclei are ^{2}H and half are ^{3}H. (a) If 1.00% of a 200-kJ laser pulse is delivered to this sphere, what temperature does the sphere reach? (b) If all of the hydrogen "burns" according to the D–T reaction, how many joules of energy are released?

59. In addition to the proton–proton cycle described in the chapter text, the carbon cycle, first proposed by Hans Bethe in 1939, is another cycle by which energy is released in stars as hydrogen is converted to helium. The carbon cycle requires higher temperatures than the proton–proton cycle. The series of reactions is

$$^{12}\text{C} + ^1\text{H} \longrightarrow ^{13}\text{N} + \gamma$$
$$^{13}\text{N} \longrightarrow ^{13}\text{C} + e^+ + \nu$$
$$e^+ + e^- \longrightarrow 2\gamma$$
$$^{13}\text{C} + ^1\text{H} \longrightarrow ^{14}\text{N} + \gamma$$
$$^{14}\text{N} + ^1\text{H} \longrightarrow ^{15}\text{O} + \gamma$$
$$^{15}\text{O} \longrightarrow ^{15}\text{N} + e^+ + \nu$$
$$e^+ + e^- \longrightarrow 2\gamma$$
$$^{15}\text{N} + ^1\text{H} \longrightarrow ^{12}\text{C} + ^4\text{He}$$

(a) If the proton–proton cycle requires a temperature of 1.5×10^7 K, estimate by proportion the temperature required for the carbon cycle. (b) Calculate the Q value for each step in the carbon cycle and the overall energy released. (c) Do you think the energy carried off by the neutrinos is deposited in the star? Explain.

60. When photons pass through matter, the intensity I of the beam (measured in watts per square meter) decreases exponentially according to

$$I = I_0 e^{-\mu x}$$

where I_0 is the intensity of the incident beam and I is the intensity of the beam that just passed through a thickness

x of material. The constant μ is known as the *linear absorption coefficient*, and its value depends on the absorbing material and the wavelength of the photon beam. This wavelength (or energy) dependence allows us to filter out unwanted wavelengths from a broad-spectrum x-ray beam. (a) Two x-ray beams of wavelengths λ_1 and λ_2 and equal incident intensities pass through the same metal plate. Show that the ratio of the emergent beam intensities is

$$\frac{I_2}{I_1} = e^{-(\mu_2 - \mu_1)x}$$

(b) Compute the ratio of intensities emerging from an aluminum plate 1.00 mm thick if the incident beam contains equal intensities of 50 pm and 100 pm x-rays. The values of μ for aluminum at these two wavelengths are $\mu_1 = 5.4 \text{ cm}^{-1}$ at 50 pm and $\mu_2 = 41.0 \text{ cm}^{-1}$ at 100 pm. (c) Repeat for an aluminum plate 10.0 mm thick.

61. *To build a bomb.* (a) At time $t = 0$ a sample of uranium is exposed to a neutron source that causes N_0 nuclei to undergo fission. The sample is in a supercritical state, with a reproduction constant $K > 1$. A chain reaction occurs which proliferates fission throughout the mass of uranium. The chain reaction can be thought of as a succession of *generations*. The N_0 fissions produced initially are the zeroth generation of fissions. From this generation, N_0K neutrons go off to produce fission of new uranium nuclei. The N_0K fissions that occur subsequently are the first generation of fissions, and from this generation, N_0K^2 neutrons go in search of uranium nuclei in which to cause fission. The subsequent N_0K^2 fissions are the second generation of fissions. This process can continue until all the uranium nuclei have fissioned. Show that the cumulative total of fissions N that have occurred up to and including the nth generation after the zeroth generation, is given by

$$N = N_0 \left(\frac{K^{n+1} - 1}{K - 1} \right)$$

(b) Consider a hypothetical uranium bomb made from 5.50 kg of isotopically pure ^{235}U. The chain reaction has a reproduction constant of 1.10, and starts with a zeroth generation of 1.00×10^{20} fissions. The average time interval between one fission generation and the next is 10.0 ns. How long after the zeroth generation does it take the uranium in this bomb to fission completely? (c) Assume that the bulk modulus of uranium is 150 GPa. Find the speed of sound in uranium. You may ignore the density difference between ^{235}U and natural uranium. (d) Find the time interval required for a compressional wave to cross the radius of a 5.50-kg sphere of uranium. This time interval indicates how quickly the motion of explosion begins. (e) Fission must occur in a time interval that is short compared to that in part (d), for otherwise most of the uranium will disperse in small chunks without having fissioned. Can the bomb considered in part (b) release the explosive energy of all of its uranium? If so, how much energy does it release, in equivalent tons of TNT? Assume that one ton of TNT releases 4.20 GJ and that each uranium fission releases 200 MeV of energy.

Answers to Quick Quizzes

45.1 (a). This is the only set of products of the four that satisfies the conditions that the atomic numbers add to 92 and the atomic mass numbers add to 234 (allowing for two neutrons to make up the total of 236).

45.2 (b). According to Figure 44.4, the ratio N/Z increases with increasing Z. As a result, when a heavy nucleus fissions to two lighter nuclei, the lighter nuclei tend to have too many neutrons for the nucleus to be stable. Beta decay in which electrons are ejected decreases the number of neutrons and increases the number of protons in order to stabilize the nucleus.

45.3 (a) and (b). In both these cases the Z and A values balance on the two sides of the equations. In reaction (c), $Z_{\text{left}} = Z_{\text{right}}$ but $A_{\text{left}} \neq A_{\text{right}}$.

45.4 (b). Collisions of the neutrons with moderator nuclei slow them down.

45.5 (a). To reduce the value of K, more neutrons need to be absorbed, so a larger volume of the control rods must be inside the reactor core.

45.6 (d). Figure 44.5 shows that the curve representing the binding energy per nucleon peaks at $A \approx 60$. Consequently, combining two nuclei with equal values of $A > 60$ results in an increase in mass, so that a fusion reaction will not occur.

Chapter 46

Particle Physics and Cosmology

▲ *A shower of particle tracks from a head-on collision of gold nuclei, each moving with energy 100 GeV. This collision occurred at the Relativistic Heavy Ion Collider (RHIC) at Brookhaven National Laboratory and was recorded with the STAR (Solenoidal Tracker at RHIC) detector. The tracks represent many fundamental particles arising from the energy of the collision. (Courtesy of Brookhaven National Laboratory/RHIC-STAR)*

The word *atom* comes from the Greek *atomos*, which means "indivisible." The early Greeks believed that atoms were the indivisible constituents of matter; that is, they regarded them as elementary particles. After 1932 physicists viewed all matter as consisting of three constituent particles: electrons, protons, and neutrons. Beginning in the 1940s, many "new" particles were discovered in experiments involving high-energy collisions between known particles. The new particles are characteristically very unstable and have very short half-lives, ranging between 10^{-6} s and 10^{-23} s. So far, more than 300 of them have been catalogued.

Until the 1960s, physicists were bewildered by the great number and variety of subatomic particles that were being discovered. They wondered whether the particles had no systematic relationship connecting them, or whether a pattern was emerging that would provide a better understanding of the elaborate structure in the subatomic world. The fact that the neutron has a magnetic moment despite having zero electric charge (Section 44.8) suggests an underlying structure to the neutron. The fact that the periodic table explains how over a hundred elements can be formed from three types of particles (electrons, protons, and neutrons) suggests that perhaps there is a means of forming over 300 subatomic particles from a small number of basic building blocks.

Recall Figure 1.2, in which we illustrated the various levels of structure in matter. We studied the atomic structure of matter in Chapter 42. In Chapter 44, we delved deeper and investigated the substructure of the atom by describing the structure of the nucleus. As mentioned in Section 1.2, the protons and neutrons in the nucleus, and a host of other exotic particles, are now known to be composed of six different varieties of particles called *quarks*. In this concluding chapter, we examine the current theory of elementary particles, in which all matter is constructed from only two families of particles, quarks and leptons. We also discuss how clarifications of such models might help scientists understand the birth and evolution of the Universe.

46.1 The Fundamental Forces in Nature

As noted in Section 5.1, all natural phenomena can be described by four fundamental forces acting between particles. In order of decreasing strength, they are the nuclear force, the electromagnetic force, the weak force, and the gravitational force.

The nuclear force (Chapter 44) is an attractive force between nucleons. It is very short-range and is negligible for separation distances between nucleons greater than about 10^{-15} m (about the size of the nucleus). The electromagnetic force, which binds atoms and molecules together to form ordinary matter, has about 10^{-2} times the strength of the nuclear force. It is a long-range force that decreases in magnitude as the inverse square of the separation between interacting particles. The weak force is a short-range force that tends to produce instability in certain nuclei. It is responsible for decay processes, and its strength is only about 10^{-5} times that of the nuclear force. Finally, the gravitational force is a long-range force that has a strength of only about

Table 46.1

Particle Interactions				
Interaction	Relative Strength	Range of Force	Mediating Field Particle	Mass of Field Particle (GeV/c^2)
Nuclear	1	Short (≈ 1 fm)	Gluon	0
Electromagnetic	10^{-2}	∞	Photon	0
Weak	10^{-5}	Short ($\approx 10^{-3}$ fm)	$W^{\pm}$, Z^0 bosons	80.4, 80.4, 91.2
Gravitational	10^{-39}	∞	Graviton	0

10^{-39} times that of the nuclear force. Although this familiar interaction is the force that holds the planets, stars, and galaxies together, its effect on elementary particles is negligible.

In Section 13.5, we discussed the difficulty that early scientists had with the notion of the gravitational force acting at a distance, with no physical contact between the interacting objects. To resolve this difficulty, the concept of the gravitational field was introduced. Similarly, in Chapter 23, we introduced the electric field to describe the electric force acting between charged objects, followed by a discussion of the magnetic field in Chapter 29. In modern physics, the nature of the interaction between particles is carried a step further. These interactions are described in terms of the exchange of entities called **field particles** or **exchange particles.** Field particles are also called **gauge bosons.**[1] The interacting particles continuously emit and absorb field particles. The emission of a field particle by one particle and its absorption by another manifests as a force between the two interacting particles. In the case of the electromagnetic interaction, for instance, the field particles are photons. In the language of modern physics, the electromagnetic force is said to be *mediated* by photons, and photons are the field particles of the electromagnetic field. Likewise, the nuclear force is mediated by field particles called *gluons* (so called because they act as a "glue" to bond the nucleons together). The weak force is mediated by field particles called W and Z *bosons,* and the gravitational force is proposed to be mediated by field particles called *gravitons.* These interactions, their ranges, and their relative strengths are summarized in Table 46.1.

46.2 Positrons and Other Antiparticles

In the 1920s, Paul Dirac developed a relativistic quantum-mechanical description of the electron that successfully explained the origin of the electron's spin and its magnetic moment. His theory had one major problem, however: its relativistic wave equation required solutions corresponding to negative energy states, and if negative energy states existed, an electron in a state of positive energy would be expected to make a rapid transition to one of these states, emitting a photon in the process.

Dirac circumvented this difficulty by postulating that all negative energy states are filled. The electrons occupying these negative energy states are collectively called the *Dirac sea.* Electrons in the Dirac sea are not directly observable because the Pauli exclusion principle does not allow them to react to external forces—there are no available states to which an electron can make a transition in response to an external force. Therefore, an electron in such a state acts as an isolated system, unless an interaction

Paul Adrien Maurice Dirac

British Physicist (1902–1984)

Dirac was instrumental in the understanding of antimatter and the unification of quantum mechanics and relativity. He made many contributions to the development of quantum physics and cosmology. Dirac won the Nobel Prize for physics in 1933. *(Courtesy of AIP Emilio Segrè Visual Archives)*

---

[1] The word *bosons* suggests that the field particles have integral spin, as discussed in Section 43.8. The word *gauge* comes from *gauge theory,* which is a sophisticated mathematical analysis that is beyond the scope of this book.

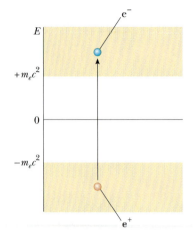

Figure 46.1 Dirac's model for the existence of antielectrons (positrons). The states lower in energy than $-m_ec^2$ are filled with electrons (the Dirac sea). One of these electrons can make a transition out of its state only if it is provided with energy equal to or larger than $2m_ec^2$. This leaves a vacancy in the Dirac sea, which can behave as a particle identical to the electron except for its positive charge.

⚠ **PITFALL PREVENTION**

46.1 Antiparticles

An antiparticle is not identified solely on the basis of opposite charge; even neutral particles have antiparticles, which are defined in terms of other properties, such as spin.

with the environment is strong enough to excite the electron to a positive energy state. Such an excitation causes one of the negative energy states to be vacant, as in Figure 46.1, leaving a hole in the sea of filled states. *The hole can react to external forces and is observable.* The hole reacts in a way similar to that of the electron, except that it has a positive charge—it is the *antiparticle* to the electron.

This theory strongly suggested that *for every particle an antiparticle exists,* not only for fermions such as electrons but also for bosons. This has subsequently been verified for *all* particles known today. The antiparticle for a charged particle has the same mass as the particle but opposite charge. For example, the electron's antiparticle (the *positron* mentioned in Section 44.4) has a rest energy of 0.511 MeV and a positive charge of $+1.60 \times 10^{-19}$ C.

Carl Anderson (1905–1991) observed the positron experimentally in 1932, and in 1936 he was awarded a Nobel Prize for his achievement. Anderson discovered the positron while examining tracks created in a cloud chamber by electron-like particles of positive charge. (These early experiments used cosmic rays—mostly energetic protons passing through interstellar space—to initiate high-energy reactions on the order of several GeV.) To discriminate between positive and negative charges, Anderson placed the cloud chamber in a magnetic field, causing moving charges to follow curved paths. He noted that some of the electron-like tracks deflected in a direction corresponding to a positively charged particle.

Since Anderson's discovery, positrons have been observed in a number of experiments. A common source of positrons is **pair production.** In this process, a gamma-ray photon with sufficiently high energy interacts with a nucleus, and an electron–positron pair is created from the photon. (The presence of the nucleus allows the principle of conservation of momentum to be satisfied.) Because the total rest energy of the electron–positron pair is $2m_ec^2 = 1.02$ MeV (where m_e is the mass of the electron), the photon must have at least this much energy to create an electron–positron pair. Therefore, the energy of a photon is converted to rest energy of the electron and positron in accordance with Einstein's relationship $E_R = mc^2$. If the gamma-ray photon has energy in excess of the rest energy of the electron–positron pair, the excess appears as kinetic energy of the two particles. Figure 46.2 shows early observations of tracks of electron–positron pairs in a bubble chamber created by 300-MeV gamma rays striking a lead sheet.

Quick Quiz 46.1 Given the identification of the particles in Figure 46.2b, what is the direction of the external magnetic field in Figure 46.2a? (a) into the page (b) out of the page (c) impossible to determine.

Figure 46.2 (a) Bubble-chamber tracks of electron–positron pairs produced by 300-MeV gamma rays striking a lead sheet. (b) The pertinent pair-production events. The positrons deflect upward and the electrons downward in an applied magnetic field.

The reverse process can also occur. Under the proper conditions, an electron and a positron can annihilate each other to produce two gamma-ray photons that have a combined energy of at least 1.02 MeV:

$$e^- + e^+ \longrightarrow 2\gamma$$

Because the initial momentum of the electron–positron system is approximately zero, the two gamma rays travel in opposite directions after the annihilation, satisfying the principle of conservation of momentum for the system.

Practically every known elementary particle has a distinct antiparticle. Among the exceptions are the photon and the neutral pion (π^0—see Section 46.3). Following the construction of high-energy accelerators in the 1950s, many other antiparticles were revealed. These included the antiproton, discovered by Emilio Segré (1905–1989) and Owen Chamberlain (b. 1920) in 1955, and the antineutron,[2] discovered shortly thereafter.

Electron–positron annihilation is used in the medical diagnostic technique called *positron emission tomography* (PET). The patient is injected with a glucose solution containing a radioactive substance that decays by positron emission, and the material is carried by the blood throughout the body. A positron emitted during a decay event in one of the radioactive nuclei in the glucose solution annihilates with an electron in the surrounding tissue, resulting in two gamma-ray photons emitted in opposite directions. A gamma detector surrounding the patient pinpoints the source of the photons and, with the assistance of a computer, displays an image of the sites at which the glucose accumulates. (Glucose is metabolized rapidly in cancerous tumors and accumulates at those sites, providing a strong signal for a PET detector system.) The images from a PET scan can indicate a wide variety of disorders in the brain, including Alzheimer's disease (Fig. 46.3). In addition, because glucose metabolizes more rapidly in active areas of the brain, a PET scan can indicate which areas of the brain are involved in the activities in which the patient is engaging at the time of the scan, such as language use, music, and vision.

National Institutes of Health

Figure 46.3 PET scans of the brain of a healthy older person *(left)* and that of a patient suffering from Alzheimer's disease *(right)*. Lighter regions contain higher concentrations of radioactive glucose, indicating higher metabolism rates and therefore increased brain activity.

---

[2] Antiparticles of charged particles have the opposite charge. Antiparticles for uncharged particles, such as the neutron, are a little more difficult to describe. One basic process that can detect the existence of an antiparticle is pair annihilation. For example, a neutron and an antineutron can annihilate to form two gamma rays. Because the photon and the neutral pion do not have distinct antiparticles, we do not observe pair annihilation with either of these particles.

46.3 Mesons and the Beginning of Particle Physics

Physicists in the mid-1930s had a fairly simple view of the structure of matter. The building blocks were the proton, the electron, and the neutron. Three other particles were either known or postulated at the time: the photon, the neutrino, and the positron. Together these six particles were considered the fundamental constituents of matter. With this simple picture, however, no one was able to answer the following important question: in view of the fact that the protons in any nucleus should strongly repel one another due to their charges of the same sign, what is the nature of the force that holds the nucleus together? Scientists recognized that this mysterious force must be much stronger than anything encountered in nature up to that time. This is the nuclear force discussed in Section 44.1 and examined in historical perspective in the following paragraphs.

The first theory to explain the nature of the nuclear force was proposed in 1935 by the Japanese physicist Hideki Yukawa—an effort that earned him a Nobel Prize in physics in 1949. To understand Yukawa's theory, recall the introduction of field particles in Section 46.1, which stated that each fundamental force is mediated by a field particle exchanged between the interacting particles. Yukawa used this idea to explain the nuclear force, proposing the existence of a new particle whose exchange between nucleons in the nucleus causes the nuclear force. He established that the range of the force is inversely proportional to the mass of this particle and predicted the mass to be about 200 times the mass of the electron. (Yukawa's predicted particle is *not* the gluon mentioned in Section 46.1, which is massless and is today considered to be the field particle for the nuclear force.) Because the new particle would have a mass between that of the electron and that of the proton, it was called a **meson** (from the Greek *meso,* "middle").

In efforts to substantiate Yukawa's predictions, physicists began experimental searches for the meson by studying cosmic rays entering the Earth's atmosphere. In 1937, Carl Anderson and his collaborators discovered a particle of mass $106 \text{ MeV}/c^2$, about 207 times the mass of the electron. This was thought to be Yukawa's meson. However, subsequent experiments showed that the particle interacted very weakly with matter and hence could not be the field particle for the nuclear force. That puzzling situation inspired several theoreticians to propose two mesons having slightly different masses equal to about 200 times that of the electron—one having been discovered by Anderson and the other, still undiscovered, predicted by Yukawa. This idea was confirmed in 1947 with the discovery of the **pi meson** (π), or simply **pion.** The particle discovered by Anderson in 1937, the one initially thought to be Yukawa's meson, is not really a meson. (We shall discuss the characteristics of mesons in Section 46.4.) Instead, it takes part in the weak and electromagnetic interactions only and is now called the **muon** (μ).

The pion comes in three varieties, corresponding to three charge states: π^+, π^-, and π^0. The π^+ and π^- particles (π^- is the antiparticle of π^+) each have a mass of $139.6 \text{ MeV}/c^2$, and the π^0 mass is $135.0 \text{ MeV}/c^2$. Two muons exist: μ^- and its antiparticle μ^+.

Pions and muons are very unstable particles. For example, the π^-, which has a mean lifetime of 2.6×10^{-8} s, decays to a muon and an antineutrino.[3] The muon, which has a mean lifetime of $2.2 \ \mu$s, then decays to an electron, a neutrino, and an antineutrino:

$$\pi^- \longrightarrow \mu^- + \overline{\nu}$$

$$\mu^- \longrightarrow e^- + \nu + \overline{\nu} \tag{46.1}$$

Hideki Yukawa

Japanese Physicist
(1907–1981)

Yukawa was awarded the Nobel Prize in 1949 for predicting the existence of mesons. This photograph of him at work was taken in 1950 in his office at Columbia University. Yukawa came to Columbia in 1949 after spending the early part of his career in Japan. *(UPI/Corbis-Bettman)*

---

[3] The antineutrino is another zero-charge particle for which the identification of the antiparticle is more difficult than that for a charged particle. Although the details are beyond the scope of this text, the neutrino and antineutrino can be differentiated by means of the relationship between the linear momentum and the spin angular momentum of the particles.

Note that for chargeless particles (as well as some charged particles, such as the proton), a bar over the symbol indicates an antiparticle, as in beta decay (see Section 44.5). Other antiparticles, such as e^+ and μ^+, use a different notation.

The interaction between two particles can be represented in a simple diagram called a **Feynman diagram,** developed by the American physicist Richard P. Feynman. Figure 46.4 is such a diagram for the electromagnetic interaction between two electrons. A Feynman diagram is a qualitative graph of time on the vertical axis versus space on the horizontal axis. It is qualitative in the sense that the actual values of time and space are not important, but the overall appearance of the graph provides a representation of the process.

In the simple case of the electron–electron interaction in Figure 46.4, a photon (the field particle) mediates the electromagnetic force between the electrons. Imagining time as the vertical axis, notice that the entire interaction is represented in the diagram as occurring at a single point in time. Therefore, the paths of the electrons appear to undergo a discontinuous change in direction at the moment of interaction. This is different from the *actual* paths, which would be curved due to the continuous exchange of large numbers of field particles. This is another aspect of the qualitative nature of Feynman diagrams.

In the electron–electron interaction, the photon, which transfers energy and momentum from one electron to the other, is called a *virtual photon* because it vanishes during the interaction without having been detected. In Chapter 40, we discussed the fact that a photon has energy $E = hf$, where f is its frequency. Consequently, for a system of two electrons initially at rest, the system has energy $2m_e c^2$ before a virtual photon is released and energy $2m_e c^2 + hf$ after the virtual photon is released (plus any kinetic energy of the electron resulting from the emission of the photon). Is this a violation of the law of conservation of energy for an isolated system? No; this process does *not* violate the law of conservation of energy because the virtual photon has a very short lifetime Δt that makes the uncertainty in the energy $\Delta E \approx \hbar/2\Delta t$ of the system consisting of two electrons and the photon greater than the photon energy. Therefore, within the constraints of the uncertainty principle, the energy of the system is conserved.

Now consider a pion mediating the nuclear force between a proton and a neutron, as in Yukawa's model (Fig. 46.5a). The rest energy E_R of a pion of mass m_π is given by Einstein's equation $E_R = m_\pi c^2$. As with the photon in Figure 46.4, in order to conserve energy, the uncertainty in the system energy must be greater than the rest energy of the pion: $\Delta E > E_R$. The existence of the pion would violate the law of conservation of energy if the particle existed for a time interval greater than $\Delta t \approx \hbar/2E_R$ (from the uncertainty principle), where E_R is the rest energy of the pion and Δt is the time interval required for the pion to transfer from one nucleon to the other. Therefore,

$$\Delta t \approx \frac{\hbar}{2E_R} = \frac{\hbar}{2m_\pi c^2} \qquad (46.2)$$

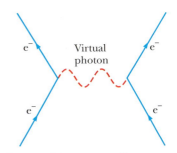

Figure 46.4 Feynman diagram representing a photon mediating the electromagnetic force between two electrons.

Richard Feynman
American Physicist
(1918–1988)

Inspired by Dirac, Feynman developed quantum electrodynamics, the theory of the interaction of light and matter on a relativistic and quantum basis. Feynman won the Nobel Prize for physics in 1965. The prize was shared by Feynman, Julian Schwinger, and Sin Itiro Tomonaga. Early in his career, he was a leading member of the team developing the first nuclear weapon in the Manhattan Project. Toward the end of his career, he worked on the commission investigating the 1986 *Challenger* tragedy and demonstrated the effects of cold temperatures on the rubber O-rings used in the space shuttle. *(© Shelly Gazin/CORBIS)*

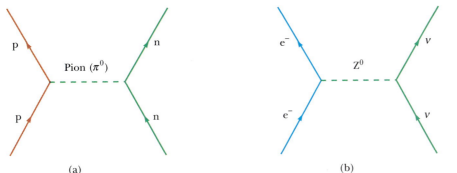

(a)   (b)

Figure 46.5 (a) Feynman diagram representing a proton and a neutron interacting via the nuclear force with a neutral pion mediating the force. (This is *not* the current model for nucleon interaction.) (b) Feynman diagram for an electron and a neutrino interacting via the weak force, with a Z^0 boson mediating the force.

Because the pion cannot travel faster than the speed of light, the maximum distance d it can travel in a time interval Δt is $c\Delta t$. Therefore,

$$d = c\Delta t \approx \frac{\hbar}{2m_\pi c} \tag{46.3}$$

From Table 46.1, we know that the range of the nuclear force is approximately 1×10^{-15} m. Using this value for d in Equation 46.3, we estimate the rest energy of the pion to be

$$m_\pi c^2 \approx \frac{\hbar c}{2d} = \frac{(1.05 \times 10^{-34}\,\text{J}\cdot\text{s})(3.00 \times 10^8\,\text{m/s})}{2(1 \times 10^{-15}\,\text{m})}$$
$$= 1.6 \times 10^{-11}\,\text{J} \approx 100\,\text{MeV}$$

Because this is the same order of magnitude as the observed rest energies of the pions, we have some confidence in the field-particle model.

The concept we have just described is quite revolutionary. In effect, it says that a system of two nucleons can change into two nucleons plus a pion, as long as it returns to its original state in a very short time interval. (Remember that this is the older historical model, which assumes that the pion is the field particle for the nuclear force; keep in mind that the gluon is the actual field particle in current models.) Physicists often say that a nucleon undergoes *fluctuations* as it emits and absorbs field particles. These fluctuations are a consequence of a combination of quantum mechanics (through the uncertainty principle) and special relativity (through Einstein's energy–mass relationship $E_R = mc^2$).

This section has dealt with the field particles that were originally proposed to mediate the nuclear force (pions) and those that mediate the electromagnetic force (photons). The graviton, the field particle for the gravitational force, has yet to be observed. $W^\pm$ and Z^0 particles, which mediate the weak force, were discovered in 1983 by the Italian physicist Carlo Rubbia (b. 1934) and his associates, using a proton–antiproton collider. Rubbia and Simon van der Meer (b. 1925), both at CERN,[4] shared the 1984 Nobel Prize in physics for the discovery of the $W^\pm$ and Z^0 particles and the development of the proton–antiproton collider. Figure 46.5b shows a Feynman diagram for a weak interaction mediated by a Z^0 boson.

46.2 The Nuclear Force and the Strong Force

The nuclear force that we discussed in Chapter 44 was originally called the strong force. Once the quark theory (Section 46.9) was established, however, the phrase *strong force* was reserved for the force between quarks. We shall follow this convention—the strong force is between quarks and the nuclear force is between nucleons. The nuclear force is a secondary result of the strong force, as we discuss in Section 46.10. Because of this historical development of the names for these forces, other books sometimes refer to the nuclear force as the strong force.

46.4 Classification of Particles

All particles other than field particles can be classified into two broad categories, *hadrons* and *leptons*. The criterion for separating these particles into categories is whether or not they interact via the strong force. The nuclear force between nucleons in a nucleus is a particular manifestation of the strong force, but we will use the term *strong force* in general to refer to any interaction between particles made up of quarks. (For more detail on quarks and the strong force, see Section 46.9.) Table 46.2 provides a summary of the properties of hadrons and leptons.

Hadrons

Particles that interact through the strong force (as well as through the other fundamental forces) are called **hadrons.** The two classes of hadrons, *mesons* and *baryons*, are distinguished by their masses and spins.

---

[4] CERN was originally the Conseil Européen pour la Recherche Nucléaire—European Organization for Nuclear Research; the name has been altered to European Laboratory for Particle Physics, but the CERN acronym has been retained.

Table 46.2

Some Particles and Their Properties											
Category	Particle Name	Symbol	Anti-particle	Mass (MeV/c^2)	B	L_e	L_μ	L_τ	S	Lifetime(s)	Principal Decay Modes[a]
Leptons	Electron	e^-	e^+	0.511	0	+1	0	0	0	Stable	
	Electron–neutrino	ν_e	$\bar{\nu}_e$	$<7\ eV/c^2$	0	+1	0	0	0	Stable	
	Muon	μ^-	μ^+	105.7	0	0	+1	0	0	2.20×10^{-6}	$e^-\bar{\nu}_e\nu_\mu$
	Muon–neutrino	ν_μ	$\bar{\nu}_\mu$	<0.3	0	0	+1	0	0	Stable	
	Tau	τ^-	τ^+	1 784	0	0	0	+1	0	$<4 \times 10^{-13}$	$\mu^-\bar{\nu}_\mu\nu_\tau, e^-\bar{\nu}_e\nu_\tau$
	Tau–neutrino	ν_τ	$\bar{\nu}_\tau$	<30	0	0	0	+1	0	Stable	
Hadrons											
Mesons	Pion	π^+	π^-	139.6	0	0	0	0	0	2.60×10^{-8}	$\mu^+\nu_\mu$
		π^0	Self	135.0	0	0	0	0	0	0.83×10^{-16}	2γ
	Kaon	K^+	K^-	493.7	0	0	0	0	+1	1.24×10^{-8}	$\mu^+\nu_\mu, \pi^+\pi^0$
		K_s^0	$\bar{K}_s^0$	497.7	0	0	0	0	+1	0.89×10^{-10}	$\pi^+\pi^-, 2\pi^0$
		K_L^0	$\bar{K}_L^0$	497.7	0	0	0	0	+1	5.2×10^{-8}	$\pi^\pm e^\mp\bar{\nu}_e, 3\pi^0$ $\pi^\pm\mu^\mp\bar{\nu}_\mu$
	Eta	η	Self	548.8	0	0	0	0	0	$<10^{-18}$	$2\gamma, 3\pi^0$
		η'	Self	958	0	0	0	0	0	2.2×10^{-21}	$\eta\pi^+ \pi^-$
Baryons	Proton	p	$\bar{p}$	938.3	+1	0	0	0	0	Stable	
	Neutron	n	$\bar{n}$	939.6	+1	0	0	0	0	614	$pe^-\bar{\nu}_e$
	Lambda	Λ^0	$\bar{\Lambda}^0$	1 115.6	+1	0	0	0	−1	2.6×10^{-10}	$p\pi^-, n\pi^0$
	Sigma	Σ^+	$\bar{\Sigma}^-$	1 189.4	+1	0	0	0	−1	0.80×10^{-10}	$p\pi^0, n\pi^+$
		Σ^0	$\bar{\Sigma}^0$	1 192.5	+1	0	0	0	−1	6×10^{-20}	$\Lambda^0\gamma$
		Σ^-	$\bar{\Sigma}^+$	1 197.3	+1	0	0	0	−1	1.5×10^{-10}	$n\pi^-$
	Delta	Δ^{++}	$\bar{\Delta}^{--}$	1 230	+1	0	0	0	0	6×10^{-24}	$p\pi^+$
		Δ^+	$\bar{\Delta}^-$	1 231	+1	0	0	0	0	6×10^{-24}	$p\pi^0, n\pi^+$
		Δ^0	$\bar{\Delta}^0$	1 232	+1	0	0	0	0	6×10^{-24}	$n\pi^0, p\pi^-$
		Δ^-	$\bar{\Delta}^+$	1 234	+1	0	0	0	0	6×10^{-24}	$n\pi^-$
	Xi	Ξ^0	$\bar{\Xi}^0$	1 315	+1	0	0	0	−2	2.9×10^{-10}	$\Lambda^0\pi^0$
		Ξ^-	$\bar{\Xi}^+$	1 321	+1	0	0	0	−2	1.64×10^{-10}	$\Lambda^0\pi^-$
	Omega	Ω^-	Ω^+	1 672	+1	0	0	0	−3	0.82×10^{-10}	$\Xi^-\pi^0, \Xi^0\pi^-$, Λ^0K^-

a   Notations in this column such as $p\pi^-$, $n\pi^0$ mean two possible decay modes. In this case, the two possible decays are $\Lambda^0 \rightarrow p + \pi^-$ and $\Lambda^0 \rightarrow n + \pi^0$.

Mesons all have zero or integer spin (0 or 1). As indicated in Section 46.3, the name comes from the expectation that Yukawa's proposed meson mass would lie between the masses of the electron and the proton. Several meson masses do lie in this range, although mesons having masses greater than that of the proton have been found to exist.

All mesons are known to decay finally into electrons, positrons, neutrinos, and photons. The pions are the lightest known mesons; they have masses of about 1.4×10^2 MeV/c^2 and all three pions—π^+, π^-, and π^0—have a spin of 0. (This indicates that the particle discovered by Anderson in 1937, the muon, is not a meson; the muon has spin $\frac{1}{2}$. It belongs in the *lepton* classification, described on page 1520.)

Baryons, the second class of hadrons, have masses equal to or greater than the proton mass (the name *baryon* means "heavy" in Greek), and their spin is always a half-integer value ($\frac{1}{2}$ or $\frac{3}{2}$). Protons and neutrons are baryons, as are many other particles. With the exception of the proton, all baryons decay, in such a way that the end products include a proton. For example, the baryon called the Ξ hyperon (Greek capital xi) decays to the Λ^0 baryon (Greek capital lambda) in about 10^{-10} s. The Λ^0 then decays to a proton and a π^- in approximately 3×10^{-10} s.

Today it is believed that hadrons are not elementary particles but are composed of more elementary units called quarks, per Section 46.9.

Leptons

Leptons (from the Greek *leptos*, meaning "small" or "light") are a group of particles that do not interact by means of the strong force. All leptons have spin $\frac{1}{2}$. Unlike hadrons, which have size and structure, leptons appear to be truly elementary, meaning that they have no structure and are point-like.

Quite unlike the case with hadrons, the number of known leptons is small. Currently, scientists believe that only six leptons exist—the electron, the muon, the tau, and a neutrino associated with each: e^-, μ^-, τ^-, ν_e, ν_μ, ν_τ. The tau lepton, discovered in 1975, has a mass about twice that of the proton. Direct experimental evidence for the neutrino associated with the tau was announced by the Fermi National Accelerator Laboratory (Fermilab) in July, 2000. Each of the six leptons has an antiparticle.

Current studies indicate that neutrinos have a small but nonzero mass. If they do have mass, then they cannot travel at the speed of light. Also, because so many neutrinos exist, their combined mass may be sufficient to cause all the matter in the Universe to eventually collapse into a single point, which might then explode and create a completely new Universe! We shall discuss this possibility in more detail in Section 46.12.

46.5 Conservation Laws

In general, the laws of conservation of energy, linear momentum, angular momentum, and electric charge provide us with a set of rules that all processes must follow. In Chapter 44 we learned that conservation laws are important for understanding why certain radioactive decays and nuclear reactions occur and others do not. In the study of elementary particles, a number of additional conservation laws are important. Although the two described here have no theoretical foundation, they are supported by abundant empirical evidence.

Baryon Number

Experimental results tell us that whenever a baryon is created in a decay or nuclear reaction, an antibaryon is also created. This scheme can be quantified by assigning every particle a quantum number, the **baryon number,** as follows: $B = +1$ for all baryons, $B = -1$ for all antibaryons, and $B = 0$ for all other particles. (See Table 46.2.) Therefore, the **law of conservation of baryon number** states that **whenever a nuclear reaction or decay occurs, the sum of the baryon numbers before the process must equal the sum of the baryon numbers after the process.**

Conservation of baryon number

If baryon number is absolutely conserved, the proton must be absolutely stable. For example, a decay of the proton to a positron and a neutral pion would satisfy conservation of energy, momentum, and electric charge. However, such a decay has never been observed. The law of conservation of baryon number would be consistent with the absence of this decay, as the proposed decay would involve the loss of a baryon. At the present, all we can say is that protons have a half-life of at least 10^{33} years (the estimated age of the Universe is only 10^{10} years), from experimental observations as pointed out in Example 46.2. Some recent theories, however, predict that the proton is unstable. According to this theory, baryon number is not absolutely conserved.

Quick Quiz 46.2 Consider the following decay: $n \rightarrow \pi^+ + \pi^- + \mu^+ + \mu^-$. What conservation laws are violated by this decay? (a) energy (b) electric charge (c) baryon number (d) angular momentum (e) no conservation laws.

Quick Quiz 46.3 Consider the following decay: $n \rightarrow p + \pi^-$. What conservation laws are violated by this decay? (a) energy (b) electric charge (c) baryon number (d) angular momentum (e) no conservation laws.

Example 46.1 Checking Baryon Numbers

Use the law of conservation of baryon number to determine whether the following reactions can occur:

(A) $p + n \rightarrow p + p + n + \overline{p}$

(B) $p + n \rightarrow p + p + \overline{p}$

Solution (A) The left side of the equation gives a total baryon number of $1 + 1 = 2$. The right side gives a total baryon number of $1 + 1 + 1 + (-1) = 2$. Therefore, baryon number is conserved and the reaction can occur (provided the initial particles have sufficient kinetic energy that energy conservation is satisfied). (B) The left side of the equation gives a total baryon number of $1 + 1 = 2$. However, the right side gives $1 + 1 + (-1) = 1$. Because baryon number is not conserved, the reaction cannot occur.

Example 46.2 Detecting Proton Decay
Interactive

Measurements taken at the Super Kamiokande neutrino detection facility (Fig. 46.6) indicate that the half-life of protons is at least 10^{33} years.

(A) Estimate how long we would have to watch, on average, to see a proton in a glass of water decay.

Solution To conceptualize the problem, imagine the number of protons in a glass of water. Although this number is huge, we know that the probability of a single proton undergoing decay is small, so we would expect to wait a long time before observing a decay. Because a half-life is provided in the problem, we categorize this problem as one in which we can apply our statistical analysis techniques from Section 44.4. To analyze the problem, let us estimate that a glass contains about 250 g of water. The number of molecules of water is

$$\frac{(250 \text{ g})(6.02 \times 10^{23} \text{ molecules/mol})}{18 \text{ g/mol}}$$

$$= 8.4 \times 10^{24} \text{ molecules}$$

Each water molecule contains one proton in each of its two hydrogen atoms plus eight protons in its oxygen atom, for a total of ten. Therefore, there are 8.4×10^{25} protons in the glass of water. The decay constant (Section 44.4) is given by Equation 44.8:

$$\lambda = \frac{0.693}{T_{1/2}} = \frac{0.693}{10^{33} \text{ yr}} = 6.9 \times 10^{-34} \text{ yr}^{-1}$$

This is the probability that *one* proton will decay in one year. The probability that *any* proton in our glass of water will decay in the one-year interval is (Eqs. 44.5 and 44.7)

$$R = (8.4 \times 10^{25})(6.9 \times 10^{-34} \text{ yr}^{-1}) = 5.8 \times 10^{-8} \text{ yr}^{-1}$$

To finalize this part of the problem, note that we have to watch our glass of water for $1/R \approx$ 17 million years! This indeed is a long time, as we suspected.

(B) The Super Kamiokande neutrino facility contains 50 000 metric tons of water. Estimate the average time interval between detected proton decays in this much water if the half-life of a proton is 10^{33} yr.

Courtesy of KBR [Institute for Cosmic Ray Research], University of Tokyo

Figure 46.6 (Example 46.2) This detector at the Super Kamiokande neutrino facility in Japan is used to study photons and neutrinos. It holds 50 000 metric tons of highly purified water and 13 000 photomultipliers. The photograph was taken while the detector was being filled. Technicians use a raft to clean the photodetectors before they are submerged.

Solution We find the ratio of the number of molecules in 50 000 metric tons of water to that in the glass of water in part (A), which will be the same as the ratio of masses:

$$\frac{N_{\text{Kamiokande}}}{N_{\text{glass}}} = \frac{m_{\text{Kamiokande}}}{m_{\text{glass}}}$$

$$= \frac{50\,000\,\text{metric ton}}{250\,\text{g}}\left(\frac{1\,000\,\text{kg}}{1\,\text{metric ton}}\right)\left(\frac{1\,000\,\text{g}}{1\,\text{kg}}\right)$$

$$= 2.0 \times 10^{8}$$

$$N_{\text{Kamiokande}} = (2.0 \times 10^{8})N_{\text{glass}}$$

$$= (2.0 \times 10^{8})(8.4 \times 10^{24}\,\text{molecules})$$

$$= 1.7 \times 10^{33}\,\text{molecules}$$

Each of these molecules contains ten protons. The probability that one of these protons will decay in one year is

$$R = (10)(1.7 \times 10^{33})(6.9 \times 10^{-34}\,\text{yr}^{-1}) \approx 12\,\text{yr}^{-1}$$

Note that the average time interval between decays is about one twelfth of a year, or approximately one month.

This is much shorter than the time interval in part (A), due to the tremendous amount of water in the detector facility.

Practice the statistics of proton decay at the Interactive Worked Example link at **http://www.pse6.com.**

Lepton Number

We have three conservation laws involving lepton numbers, one for each variety of lepton. The **law of conservation of electron lepton number** states that whenever a nuclear reaction or decay occurs, **the sum of the electron lepton numbers before the process must equal the sum of the electron lepton numbers after the process.**

Conservation of lepton number

The electron and the electron neutrino are assigned an electron lepton number $L_e = +1$ and the antileptons e^+ and $\bar{\nu}_e$ are assigned an electron lepton number $L_e = -1$. All other particles have $L_e = 0$. For example, consider the decay of the neutron:

$$\text{n} \longrightarrow \text{p} + \text{e}^- + \bar{\nu}_e$$

Before the decay, the electron lepton number is $L_e = 0$; after the decay, it is $0 + 1 + (-1) = 0$. Therefore, electron lepton number is conserved. (Baryon number must also be conserved, of course, and it is: before the decay, $B = +1$, and after the decay $B = +1 + 0 + 0 = +1$.)

Similarly, when a decay involves muons, the muon lepton number L_μ is conserved. The μ^- and the ν_μ are assigned a muon lepton number $L_\mu = +1$ and the antimuons μ^+ and $\bar{\nu}_\mu$ are assigned a muon lepton number $L_\mu = -1$. All other particles have $L_\mu = 0$.

Finally, tau lepton number L_τ is conserved with similar assignments made for the tau lepton, its neutrino, and their two antiparticles.

Quick Quiz 46.4 Consider the following decay: $\pi^0 \rightarrow \mu^- + e^+ + \nu_\mu$. What conservation laws are violated by this decay? (a) energy (b) angular momentum (c) electric charge (d) baryon number (e) electron lepton number (f) muon lepton number (g) tau lepton number (h) no conservation laws.

Quick Quiz 46.5 Suppose a claim is made that the decay of the neutron is given by $\text{n} \rightarrow \text{p} + \text{e}^-$. What conservation laws are violated by this decay? (a) energy (b) angular momentum (c) electric charge (d) baryon number (e) electron lepton number (f) muon lepton number (g) tau lepton number (h) no conservation laws.

Quick Quiz 46.6 A student claims to have observed a decay of an electron into two electron neutrinos, traveling in opposite directions. What conservation laws would be violated by this decay? (a) energy (b) angular momentum (c) electric charge (d) baryon number (e) electron lepton number (f) muon lepton number (g) tau lepton number (h) no conservation laws.

Example 46.3   **Checking Lepton Numbers**

Use the law of conservation of lepton numbers to determine which of the following decay schemes can occur:

(A) $\mu^- \rightarrow e^- + \bar{\nu}_e + \nu_\mu$

(B) $\pi^+ \rightarrow \mu^+ + \nu_\mu + \nu_e$

Solution (A) Because this decay involves a muon and an electron, L_μ and L_e must both be conserved. Before

the decay, $L_\mu = +1$ and $L_e = 0$. After the decay, $L_\mu = 0 + 0 + 1 = +1$ and $L_e = +1 + (-1) + 0 = 0$. Therefore, both numbers are conserved, and on this basis the decay is possible.

(B) Before the decay, $L_\mu = 0$ and $L_e = 0$. After the decay, $L_\mu = -1 + 1 + 0 = 0$, but $L_e = 0 + 0 + 1 = 1$. Therefore, the decay is not possible because electron lepton number is not conserved.

46.6   Strange Particles and Strangeness

Many particles discovered in the 1950s were produced by the interaction of pions with protons and neutrons in the atmosphere. A group of these—the kaon (K), lambda (Λ), and sigma (Σ) particles—exhibited unusual properties both as they were created and as they decayed and hence were called *strange particles.*

One unusual property of strange particles is that they are always produced in pairs. For example, when a pion collides with a proton, a highly probable result is the production of two neutral strange particles (Fig. 46.7):

$$\pi^- + p \longrightarrow K^0 + \Lambda^0$$

However, the reaction $\pi^- + p \rightarrow K^0 + n^0$, where only one of the final particles is strange, never occurs, even though no known conservation laws would be violated and even though the energy of the pion is sufficient to initiate the reaction.

Courtesy Lawrence Berkeley Laboratory, University of California, Photographic Services

Figure 46.7 This bubble-chamber photograph shows many events, and the inset is a drawing of identified tracks. The strange particles Λ^0 and K^0 are formed at the bottom as a π^- particle interacts with a proton in the reaction $\pi^- + p \rightarrow \Lambda^0 + K^0$. (Note that the neutral particles leave no tracks, as indicated by the dashed lines in the inset.) The Λ^0 then decays in the reaction $\Lambda^0 \rightarrow \pi^- + p$ and the K^0 in the reaction $K^0 \rightarrow \pi^+ + \mu^- + \bar{\nu}_\mu$.

The second peculiar feature of strange particles is that, although they are produced in reactions involving the strong interaction at a high rate, they do not decay into particles that interact via the strong force at a high rate. Instead, they decay very slowly, which is characteristic of the weak interaction. Their half-lives are in the range 10^{-10} s to 10^{-8} s, whereas most other particles that interact via the strong force have much shorter lifetimes on the order of 10^{-23} s.

To explain these unusual properties of strange particles, a new quantum number S, called **strangeness,** was introduced, together with a conservation law. The strangeness numbers for some particles are given in Table 46.2. The production of strange particles in pairs is explained by assigning $S = +1$ to one of the particles, $S = -1$ to the other, and $S = 0$ to all nonstrange particles. The **law of conservation of strangeness** states that **in a nuclear reaction or decay that occurs via the strong force, strangeness is conserved, that is, the sum of the strangeness numbers before the process must equal the sum of the strangeness numbers after the process. In processes that occur via the weak interaction, strangeness may not be conserved.**

Conservation of strangeness number

The low decay rate of strange particles can be explained by assuming that the strong and electromagnetic interactions obey the law of conservation of strangeness but the weak interaction does not. Because the decay of a strange particle involves the loss of one strange particle, it violates strangeness conservation and hence proceeds slowly via the weak interaction.

Example 46.4   Is Strangeness Conserved?

(A) Use the law of strangeness conservation to determine whether the reaction $\pi^0 + n \rightarrow K^+ + \Sigma^-$ occurs.

Solution From Table 46.2, we see that the initial strangeness is $S = 0 + 0 = 0$. Because the strangeness of the K^+ is $S = +1$ and the strangeness of the Σ^- is $S = -1$, the strangeness of the final products is $+1 - 1 = 0$. Therefore, strangeness is conserved, and the reaction is allowed.

(B) Show that the reaction $\pi^- + p \rightarrow \pi^- + \Sigma^+$ does not conserve strangeness.

Solution Before: $S = 0 + 0 = 0$; after: $S = 0 + (-1) = -1$. Therefore, strangeness is not conserved.

46.7   Making Particles and Measuring Their Properties

The bewildering array of entries in Table 46.2 leaves one yearning for firm ground. It is natural to wonder about an entry, for example, which shows a particle (Σ^0) which exists for 10^{-20} s and has a mass of 1 192.5 MeV/c^2. How is it possible to detect a particle that exists for only 10^{-20} s? In this section we answer such questions and explain how elementary particles are produced and their properties measured.

Most elementary particles are unstable and are created in nature only rarely, in cosmic ray showers. In the laboratory, however, great numbers of these particles are created in controlled collisions between high-energy particles and a suitable target. The incident particles must have very high energy, and it takes a relatively long time interval for electromagnetic fields to accelerate particles to high energies. Therefore, stable charged particles such as electrons or protons generally make up the incident beam. In addition, targets must be simple and stable, and the simplest target, hydrogen, serves nicely as both a target (the proton) and a detector.

Figure 46.7 documents a typical event in which a bubble chamber served as both target source and detector. Many parallel tracks of negative pions are visible entering the photograph from the bottom. As the labels in the inset drawing show, one of the pions has hit a stationary proton in the hydrogen and produced two strange particles, Λ^0 and K^0, according to the reaction

$$\pi^- + p \longrightarrow \Lambda^0 + K^0$$

Neither neutral strange particle leaves a track, but their subsequent decay into charged particles can be seen in Figure 46.7. A magnetic field directed into the plane of the page causes the track of each charged particle to curve, and from the measured curvature we can determine the particle's charge and linear momentum. If the mass and momentum of the incident particle are known, we can then usually calculate the product particle's mass, kinetic energy, and speed from the laws of conservation of momentum and energy. Finally, combining a product particle's speed with the length of the track it leaves, we can calculate the particle's lifetime. Figure 46.7 shows that sometimes we can use this lifetime technique even for a neutral particle, which leaves no track. As long as the beginning and end of the missing track are known, as well as the particle speed, we can infer the missing track length and so determine the lifetime of the neutral particle.

Resonance Particles

Using drift chambers or other modern detectors, one can measure decay track lengths as short as 10^{-6} m. This means that lifetimes as short as 10^{-16} s can be measured for high-energy particles traveling at about the speed of light. We arrive at this result by assuming that a decaying particle travels 1 μm at a speed of $0.99c$ in the reference frame of the laboratory, yielding a lifetime of $\Delta t_{lab} = 1 \times 10^{-6}$ m$/0.99c \approx 3.4 \times 10^{-15}$ s. This is not our final result, however, because we must account for the relativistic effects of time dilation. Because the proper lifetime Δt_p measured in the decaying particle's reference frame is shorter than the laboratory-frame value Δt_{lab} by a factor of $\sqrt{1 - (v^2/c^2)}$ (see Eq. 39.7), we can calculate the proper lifetime as follows:

$$\Delta t_p = \Delta t_{lab} \sqrt{1 - \frac{v^2}{c^2}} = (3.4 \times 10^{-15} \text{ s}) \sqrt{1 - \frac{(0.99c)^2}{c^2}} = 4.8 \times 10^{-16} \text{ s}$$

Unfortunately, even with Einstein's help, the best answer we can obtain with the track-length method is several orders of magnitude away from lifetimes of 10^{-20} s. How, then, can we detect the presence of particles that exist for time intervals such as 10^{-20} s? For such short-lived particles, known as **resonance particles,** all we can do is infer their masses, their lifetimes, and their very existence from data on their decay products.

Let us consider this detection process in detail by examining the case of the resonance particle called the delta plus (Δ^+), which has a mass of 1 231 MeV$/c^2$ and a lifetime of about 6×10^{-24} s. This particle is produced in the reaction

$$e^- + p \longrightarrow e^- + \Delta^+ \tag{46.4}$$

followed in 6×10^{-24} s by the decay

$$\Delta^+ \longrightarrow \pi^+ + n \tag{46.5}$$

Because the Δ^+ lifetime is so short, the particle leaves no measurable track in a drift chamber. It might therefore seem impossible to distinguish the reactions given in Equations 46.4 and 46.5 from the reaction

$$e^- + p \longrightarrow e^- + \pi^+ + n \tag{46.6}$$

in which the reactants of Equation 46.4 decay directly to e^-, π^+, and n with no intermediate step in which a Δ^+ is produced. Distinguishing between these two possibilities is not impossible, however. If a Δ^+ particle exists, it has a distinct rest energy, which must come from the kinetic energy of the incoming particles. If we imagine firing electrons with increasing kinetic energy at protons, eventually we will provide enough energy to the system to create the Δ^+ particle. This is very similar to firing photons of increasing energy at an atom until you provide them with enough energy to excite the atom to a higher quantum state. In fact, the Δ^+ particle is an excited state of the proton, which we can understand via the quark theory discussed in Section 46.9. After the Δ^+ particle is formed, its rest energy becomes the energies of the outgoing pion and neutron. Equation 39.27 can be solved for the rest energy of the Δ^+ particle in terms of its kinetic energy and linear momentum:

$$(m_{\Delta^+}c^2)^2 = E_{\Delta^+}^2 - p_{\Delta^+}^2 c^2 = E_{\Delta^+}^2 - (\mathbf{p}_{\Delta^+})^2 c^2$$

When the Δ^+ particle decays into a pion and a neutron, conservation of energy and momentum require that

$$E_{\Delta^+} = E_{\pi^+} + E_n \qquad \mathbf{p}_{\Delta^+} = \mathbf{p}_{\pi^+} + \mathbf{p}_n$$

Therefore, the rest energy of the Δ^+ particle can be expressed in terms of the energies and momenta of the outgoing particles, which can all be measured in the drift-chamber computer reconstruction:

$$(m_{\Delta^+}c^2)^2 = (E_{\pi^+} + E_n)^2 - (\mathbf{p}_{\pi^+} + \mathbf{p}_n)^2 c^2$$

Any pions and neutrons that come from the decay of a Δ^+ particle must have energies and momenta that combine in this equation to give the rest energy of the Δ^+ particle. Pions and neutrons coming from the reaction of Equation 46.6 will have a variety of energies and momenta with no particular pattern because the energy of the reactants can divide up in many ways among the three outgoing particles in this reaction. At the energy at which the rest energy of the Δ^+ particle can be created, many reactions occur, as evidenced by the proper combinations of energy and momentum already described.

To show the existence of the Δ^+ particle, we analyze a large number of events in which a π^+ and a neutron are produced. Then the number of events in a given energy range is plotted versus energy. Following this procedure, we obtain a slowly varying curve that has a sharp peak superimposed on it. The peak represents the incident electron energy at which the rest energy of the Δ^+ particle was created, revealing the existence of the particle.

Figure 46.8 is an experimental plot for the Δ^+ particle. The dashed broad curve was produced by direct events in which no Δ^+ was created (see Eq. 46.6). The sharp peak near 1 230 MeV was produced by all the events in which a Δ^+ was formed and decayed to a pion and a neutron. Therefore, the rest energy of the Δ^+ particle is near 1 230 MeV. Peaks corresponding to two resonance particles with masses greater than that of the Δ^+ particle can also be seen in Figure 46.8.

Graphs such as Figure 46.8 can tell us not only the mass of a short-lived particle but also its lifetime. The width of the resonance peak and the uncertainty relation $\Delta E \, \Delta t \approx \hbar/2$ are used to infer the lifetime Δt of the particle. The measured width of

Figure 46.8 Experimental evidence for the existence of the Δ^+ particle. The sharp peak near 1 230 MeV was produced by events in which a Δ^+ formed and promptly decayed to a π^+ and a neutron.

115 MeV in Figure 46.8 leads to a lifetime of 6×10^{-24} s for the Δ^+ particle. In this incredibly short lifetime, a Δ^+ particle moving at $0.99c$ travels only 10^{-14} m, which is about ten nuclear diameters.

46.8 Finding Patterns in the Particles

One of the tools scientists use is the detection of patterns in data, patterns that contribute to our understanding of nature. One of the best examples of the use of this tool is the development of the periodic table, which provides a fundamental understanding of the chemical behavior of the elements. As mentioned in the introduction, the periodic table explains how more than 100 elements can be formed from three particles—the electron, the proton, and the neutron. The table of nuclides, part of which is shown in Table A.3, contains hundreds of nuclides, but all can be built from protons and neutrons.

The number of observed particles and resonances observed by particle physicists is also in the hundreds. Is it possible that a small number of entities exist from which all of these can be built? Taking a hint from the success of the periodic table and the table of nuclides, let us explore the historical search for patterns among the particles.

Many classification schemes have been proposed for grouping particles into families. Consider, for instance, the baryons listed in Table 46.2 that have spins of $\frac{1}{2}$: p, n, Λ^0, Σ^+, Σ^0, Σ^-, Ξ^0, and Ξ^-. If we plot strangeness versus charge for these baryons using a sloping coordinate system, as in Figure 46.9a, we observe a fascinating pattern: six of the baryons form a hexagon, and the remaining two are at the hexagon's center.

As a second example, consider the following nine spin-zero mesons listed in Table 46.2: π^+, π^0, π^-, K^+, K^0, K^-, η, η', and the antiparticle $\overline{K}^0$. Figure 46.9b is a plot of strangeness versus charge for this family. Again, a hexagonal pattern emerges. In this case, each particle on the perimeter of the hexagon lies opposite its antiparticle, and the remaining three (which form their own antiparticles) are at the center of the hexagon. These and related symmetric patterns were developed independently in 1961 by Murray Gell-Mann and Yuval Ne'eman (b. 1925). Gell-Mann called the patterns the **eightfold way,** after the eightfold path to nirvana in Buddhism.

Groups of baryons and mesons can be displayed in many other symmetric patterns within the framework of the eightfold way. For example, the family of spin-$\frac{3}{2}$ baryons known in 1961 contains nine particles arranged in a pattern like that of the pins in a bowling alley, as in Figure 46.10. (The particles Σ^{*+}, Σ^{*0}, Σ^{*-}, Ξ^{*0}, and Ξ^{*-} are excited states of the particles Σ^+, Σ^0, Σ^-, Ξ^0, and Ξ^-. In these higher-energy states, the spins

Murray Gell-Mann
American Physicist (b. 1929)

Murray Gell-Mann was awarded the Nobel Prize in 1969 for his theoretical studies dealing with subatomic particles. *(Courtesy of Michael R. Dressler)*

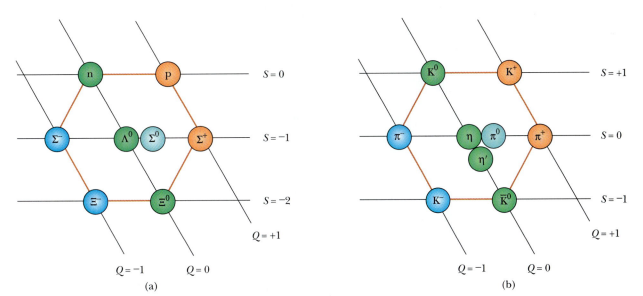

Figure 46.9 (a) The hexagonal eightfold-way pattern for the eight spin-$\frac{1}{2}$ baryons. This strangeness-versus-charge plot uses a sloping axis for charge number Q and a horizontal axis for strangeness S. (b) The eightfold-way pattern for the nine spin-zero mesons.

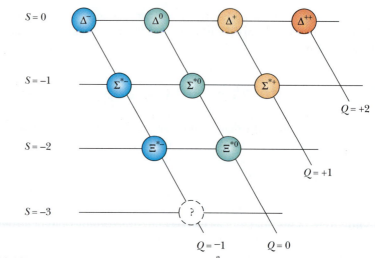

Figure 46.10 The pattern for the higher-mass, spin-$\frac{3}{2}$ baryons known at the time the pattern was proposed. The three Σ^* and two Ξ^* particles are excited states of the corresponding spin-$\frac{1}{2}$ particles in Figure 46.9. These excited states have higher mass and spin $\frac{3}{2}$. The absence of a particle in the bottom position was evidence of a new particle yet to be discovered, the Ω^-.

of the three quarks (see Section 46.9) making up the particle are aligned so that the total spin of the particle is $\frac{3}{2}$.) When this pattern was proposed, an empty spot occurred in it (at the bottom position), corresponding to a particle that had never been observed. Gell-Mann predicted that the missing particle, which he called the omega minus (Ω^-), should have spin $\frac{3}{2}$, charge -1, strangeness -3, and rest energy of approximately 1 680 MeV. Shortly thereafter, in 1964, scientists at the Brookhaven National Laboratory found the missing particle through careful analyses of bubble-chamber photographs (Fig. 46.11) and confirmed all its predicted properties.

The prediction of the missing particle in the eightfold way has much in common with the prediction of missing elements in the periodic table. Whenever a vacancy occurs in an organized pattern of information, experimentalists have a guide for their investigations.

Courtesy of Brookhaven National Laboratory

Figure 46.11 Discovery of the Ω^- particle. The photograph on the left shows the original bubble-chamber tracks. The drawing on the right isolates the tracks of the important events. The K^- particle at the bottom collides with a proton to produce the first detected Ω^- particle plus a K^0 and a K^+.

46.9 Quarks

As we have noted, leptons appear to be truly elementary particles because there are only a few types of them, and experiments indicate that they have no measurable size or internal structure. Hadrons, on the other hand, are complex particles having size and structure. The existence of the strangeness–charge patterns of the eightfold way suggests that hadrons have substructure. Furthermore, we know that hundreds of types of hadrons exist and that many of them decay into other hadrons.

The Original Quark Model

In 1963 Gell-Mann and George Zweig (b. 1937) independently proposed a model for the substructure of hadrons. According to their model, all hadrons are composed of two or three elementary constituents called **quarks.** (Gell-Mann borrowed the word *quark* from the passage "Three quarks for Muster Mark" in James Joyce's *Finnegans Wake.* In Zweig's model, he called the constituents "aces.") The model has three types of quarks, designated by the symbols u, d, and s. These are given the arbitrary names **up, down,** and **strange.** The various types of quarks are called **flavors.** Figure 46.12 is a pictorial representation of the quark compositions of several hadrons.

An unusual property of quarks is that they carry a fractional electronic charge. The u, d, and s quarks have charges of $+2e/3$, $-e/3$, and $-e/3$, respectively, where e is the elementary charge 1.60×10^{-19} C. These and other properties of quarks and antiquarks are given in Table 46.3. Notice that quarks have spin $\frac{1}{2}$, which means that all quarks are fermions, defined as any particle having half-integral spin, as pointed out in Section 43.8. As Table 46.3 shows, associated with each quark is an antiquark of opposite charge, baryon number, and strangeness.

The compositions of all hadrons known when Gell-Mann and Zweig presented their model can be completely specified by three simple rules:

- A meson consists of one quark and one antiquark, giving it a baryon number of 0, as required.
- A baryon consists of three quarks.
- An antibaryon consists of three antiquarks.

The theory put forth by Gell-Mann and Zweig is referred to as the *original quark model.*

> **Quick Quiz 46.7** Using a coordinate system like that in Figure 46.9, draw an eightfold-way diagram for the three quarks in the original quark model.

Mesons
π^+

K$^-$

Baryons
p

n

Active Figure 46.12 Quark composition of two mesons and two baryons.

At the Active Figures link at http://www.pse6.com, observe the quark composition for the mesons and baryons in Tables 46.4 and 46.5.

Table 46.3

Properties of Quarks and Antiquarks

Quarks

Name	Symbol	Spin	Charge	Baryon Number	Strangeness	Charm	Bottomness	Topness
Up	u	$\frac{1}{2}$	$+\frac{2}{3}e$	$\frac{1}{3}$	0	0	0	0
Down	d	$\frac{1}{2}$	$-\frac{1}{3}e$	$\frac{1}{3}$	0	0	0	0
Strange	s	$\frac{1}{2}$	$-\frac{1}{3}e$	$\frac{1}{3}$	−1	0	0	0
Charmed	c	$\frac{1}{2}$	$+\frac{2}{3}e$	$\frac{1}{3}$	0	+1	0	0
Bottom	b	$\frac{1}{2}$	$-\frac{1}{3}e$	$\frac{1}{3}$	0	0	+1	0
Top	t	$\frac{1}{2}$	$+\frac{2}{3}e$	$\frac{1}{3}$	0	0	0	+1

Antiquarks

Name	Symbol	Spin	Charge	Baryon Number	Strangeness	Charm	Bottomness	Topness
Anti-up	$\bar{u}$	$\frac{1}{2}$	$-\frac{2}{3}e$	$-\frac{1}{3}$	0	0	0	0
Anti-down	$\bar{d}$	$\frac{1}{2}$	$+\frac{1}{3}e$	$-\frac{1}{3}$	0	0	0	0
Anti-strange	$\bar{s}$	$\frac{1}{2}$	$+\frac{1}{3}e$	$-\frac{1}{3}$	+1	0	0	0
Anti-charmed	$\bar{c}$	$\frac{1}{2}$	$-\frac{2}{3}e$	$-\frac{1}{3}$	0	−1	0	0
Anti-bottom	$\bar{b}$	$\frac{1}{2}$	$+\frac{1}{3}e$	$-\frac{1}{3}$	0	0	−1	0
Anti-top	$\bar{t}$	$\frac{1}{2}$	$-\frac{2}{3}e$	$-\frac{1}{3}$	0	0	0	−1

Charm and Other Developments

Although the original quark model was highly successful in classifying particles into families, some discrepancies occurred between its predictions and certain experimental decay rates. Consequently, several physicists proposed a fourth quark flavor in 1967. They argued that if four types of leptons exist (as was thought at the time), then there should also be four flavors of quarks because of an underlying symmetry in nature. The fourth quark, designated c, was assigned a property called **charm.** A *charmed* quark has charge $+2e/3$, just as the up quark does, but its charm distinguishes it from the other three quarks. This introduces a new quantum number C, representing charm. The new quark has charm $C = +1$, its antiquark has charm of $C = -1$, and all other quarks have $C = 0$. Charm, like strangeness, is conserved in strong and electromagnetic interactions but not in weak interactions.

Evidence that the charmed quark exists began to accumulate in 1974, when a heavy meson called the J/Ψ particle (or simply Ψ, uppercase Greek psi) was discovered independently by two groups, one led by Burton Richter (b. 1931) at the Stanford Linear Accelerator (SLAC), and the other led by Samuel Ting (b. 1936) at the Brookhaven National Laboratory. In 1976 Richter and Ting were awarded a Nobel Prize in physics for this work. The J/Ψ particle does not fit into the three-quark model; instead, it has properties of a combination of the proposed charmed quark and its antiquark ($c\bar{c}$). It is much more massive than the other known mesons ($\sim 3\,100$ MeV/c^2), and its lifetime is much longer than the lifetimes of particles that interact via the strong force. Soon, related mesons were discovered, corresponding to such quark combinations as $\bar{c}d$ and $c\bar{d}$, all of which have great masses and long lifetimes. The existence of these new mesons provided firm evidence for the fourth quark flavor.

In 1975, researchers at Stanford University reported strong evidence for the tau (τ) lepton, mass 1 784 MeV/c^2. This was the fifth type of lepton, which led physicists to propose that more flavors of quarks might exist, on the basis of symmetry arguments similar to those leading to the proposal of the charmed quark. These proposals led to more elaborate quark models and the prediction of two new quarks, **top** (t) and

Table 46.4

Quark[a] Composition of Mesons										
		Antiquarks								
		$\bar{b}$		$\bar{c}$		$\bar{s}$		$\bar{d}$		$\bar{u}$
Quarks	b	Y	$(\bar{b}b)$	B_c^-	$(\bar{c}b)$	$\bar{B}_s^{\,0}$	$(\bar{s}b)$	$\bar{B}_d^{\,0}$	$(\bar{d}b)$	B^-	$(\bar{u}b)$
	c	B_c^+	$(\bar{b}c)$	J/Ψ	$(\bar{c}c)$	D_s^+	$(\bar{s}c)$	D^+	$(\bar{d}c)$	D^0	$(\bar{u}c)$
	s	B_s^0	$(\bar{b}s)$	D_s^-	$(\bar{c}s)$	η, η'	$(\bar{s}s)$	$\bar{K}^0$	$(\bar{d}s)$	K^-	$(\bar{u}s)$
	d	B_d^0	$(\bar{b}d)$	D^-	$(\bar{c}d)$	K^0	$(\bar{s}d)$	π^0, η, η'	$(\bar{d}d)$	π^-	$(\bar{u}d)$
	u	B^+	$(\bar{b}u)$	$\bar{D}^0$	$(\bar{c}u)$	K^+	$(\bar{s}u)$	π^+	$(\bar{d}u)$	π^0, η, η'	$(\bar{u}u)$

[a] The top quark does not form mesons because it decays too quickly.

bottom (b). (Some physicists prefer *truth* and *beauty*.) To distinguish these quarks from the others, quantum numbers called *topness* and *bottomness* (with allowed values + 1, 0, − 1) were assigned to all quarks and antiquarks (see Table 46.3). In 1977, researchers at the Fermi National Laboratory, under the direction of Leon Lederman (b. 1922), reported the discovery of a very massive new meson Y^- (Greek capital upsilon), whose composition is considered to be $b\bar{b}$, providing evidence for the bottom quark. In March 1995, researchers at Fermilab announced the discovery of the top quark (supposedly the last of the quarks to be found), which has a mass of 173 GeV/c^2.

Table 46.4 lists the quark compositions of mesons formed from the up, down, strange, charmed, and bottom quarks. Table 46.5 shows the quark combinations for the baryons listed in Table 46.2. Note that only two flavors of quarks, u and d, are contained in all hadrons encountered in ordinary matter (protons and neutrons).

You are probably wondering whether the discoveries of elementary particles will ever end. How many "building blocks" of matter really exist? At the present, physicists believe that the elementary particles in nature are six quarks and six leptons, together with their antiparticles, and the four field particles listed in Table 46.1. Table 46.6 lists the rest energies and charges of the quarks and leptons.

Table 46.5

Quark Composition of Several Baryons[a]	
Particle	**Quark Compositon**
p	uud
n	udd
Λ^0	uds
Σ^+	uus
Σ^0	uds
Σ^-	dds
Δ^{++}	uuu
Δ^+	uud
Δ^0	udd
Δ^-	ddd
Ξ^0	uss
Ξ^-	dss
Ω^-	sss

[a] Some baryons have the same quark composition, such as the p and the Δ^+ and the n and the Δ^0. In these cases, the Δ particles are considered to be excited states of the proton and neutron.

Table 46.6

The Elementary Particles and Their Rest Energies and Charges		
Particle	**Rest Energy**	**Charge**
Quarks		
u	360 MeV	$+\frac{2}{3}e$
d	360 MeV	$-\frac{1}{3}e$
s	540 MeV	$-\frac{1}{3}e$
c	1 500 MeV	$+\frac{2}{3}e$
b	5 GeV	$-\frac{1}{3}e$
t	173 GeV	$+\frac{2}{3}e$
Leptons		
e^-	511 keV	$-e$
μ^-	105.7 MeV	$-e$
τ^-	1 784 MeV	$-e$
ν_e	<7 eV	0
ν_μ	<0.3 MeV	0
ν_τ	<30 MeV	0

Despite extensive experimental effort, no isolated quark has ever been observed. Physicists now believe that at ordinary temperatures quarks are permanently confined inside ordinary particles because of an exceptionally strong force that prevents them from escaping, called (appropriately) the **strong force**[5] (which we introduced at the beginning of Section 46.4 and we discuss further in Section 46.10). This force increases with separation distance, similar to the force exerted by a stretched spring. Current efforts are underway to form a **quark–gluon plasma,** a state of matter in which the quarks are freed from neutrons and protons. In 2000, scientists at CERN announced evidence for a quark–gluon plasma formed by colliding lead nuclei. Experiments continue at CERN as well as at the Relativistic Heavy Ion Collider (RHIC) at Brookhaven to verify the production of a quark–gluon plasma.

Quick Quiz 46.8 Doubly charged baryons, such as the Δ^{++}, are known to exist. True or false: doubly charged mesons also exist.

Quick Quiz 46.9 Imagine hypothetical super-heavy baryons with the following quark combinations: (i) ccc (ii) bbb (iii) ttt. Rank these baryons according to (a) mass, from smallest to largest (b) spin (c) electric charge and (d) baryon number.

46.10 Multicolored Quarks

Shortly after the concept of quarks was proposed, scientists recognized that certain particles had quark compositions that violated the exclusion principle. In Section 42.7, we applied the exclusion principle to electrons in atoms. The principle is more general, however, and applies to all particles with half-integral spin ($\frac{1}{2}, \frac{3}{2}$, etc.), which we collectively call fermions. Because all quarks are fermions having spin $\frac{1}{2}$, they are expected to follow the exclusion principle. One example of a particle that appears to violate the exclusion principle is the Ω^- (sss) baryon, which contains three strange quarks having parallel spins, giving it a total spin of $\frac{3}{2}$. All three quarks have the same spin quantum number, in violation of the exclusion principle. Other examples of baryons made up of identical quarks having parallel spins are the Δ^{++} (uuu) and the Δ^- (ddd).

To resolve this problem, it was suggested that quarks possess an additional property called **color charge.** This property is similar in many respects to electric charge except that it occurs in six varieties rather than two. The colors assigned to quarks are red, green, and blue, and antiquarks have the colors antired, antigreen, and antiblue. Therefore, the colors red, green, and blue serve as the "quantum numbers" for the color of the quark. To satisfy the exclusion principle, the three quarks in any baryon must all have different colors. The three colors "neutralize" to white. A quark and an antiquark in a meson must be of a color and the corresponding anticolor and will consequently neutralize to white, similar to the way electric charges + and − neutralize to zero net charge. The result is that baryons and mesons are always colorless (or white). Therefore, the apparent violation of the exclusion principle in the Ω^- baryon is removed because the three quarks in the particle have different colors.

Note that the new property of color increases the number of quarks by a factor of three, since each of the six quarks comes in three colors. Although the concept of

> ⚠ **PITFALL PREVENTION**
>
> **46.3 Color Charge Is Not Really Color**
>
> The description of color for a quark has nothing to do with visual sensation from light. It is simply a convenient name for a property that is analogous to electric charge.

---

[5] As a reminder, the original meaning of the term *strong force* was the short-range attractive force between nucleons, which we have called the *nuclear force*. The nuclear force between nucleons is a secondary effect of the strong force between quarks.

color in the quark model was originally conceived to satisfy the exclusion principle, it also provided a better theory for explaining certain experimental results. For example, the modified theory correctly predicts the lifetime of the π^0 meson.

The theory of how quarks interact with each other is called **quantum chromo-dynamics,** or QCD, to parallel the name *quantum electrodynamics* (the theory of the electrical interaction between light and matter). In QCD, each quark is said to carry a color charge, in analogy to electric charge. The strong force between quarks is often called the **color force.** Therefore, the terms *strong force* and *color force* are used interchangeably.

In Section 46.1 we stated that the nuclear interaction between hadrons is mediated by massless field particles called **gluons.** As we have mentioned, the nuclear force is actually a secondary effect of the strong force between quarks. The gluons are the mediators of the strong force. When a quark emits or absorbs a gluon, the quark's color may change. For example, a blue quark that emits a gluon may become a red quark, and a red quark that absorbs this gluon becomes a blue quark.

The color force between quarks is analogous to the electric force between charges: particles with the same color repel, and those with opposite colors attract. Therefore, two green quarks repel each other, but a green quark is attracted to an antigreen quark. The attraction between quarks of opposite color to form a meson ($q\bar{q}$) is indicated in Figure 46.13a. Differently colored quarks also attract one another, although with less intensity than the oppositely colored quark and antiquark. For example, a cluster of red, blue, and green quarks all attract one another to form a baryon, as in Figure 46.13b. Therefore, every baryon contains three quarks of three different colors.

Although the nuclear force between two colorless hadrons is negligible at large separations, the net strong force between their constituent quarks is not exactly zero at small separations. This residual strong force is the nuclear force that binds protons and neutrons to form nuclei. It is similar to the force between two electric dipoles. Each dipole is electrically neutral. An electric field surrounds the dipoles, however, because of the separation of the positive and negative charges (see Section 23.6). As a result, an electric interaction occurs between the dipoles that is weaker than the force between single charges. In Section 43.1, we explored how this interaction results in the Van der Waals force between neutral molecules.

According to QCD, a more basic explanation of the nuclear force can be given in terms of quarks and gluons. Figure 46.14a shows the nuclear interaction between a neutron and a proton by means of Yukawa's pion, in this case a π^-. This drawing differs from Figure 46.5a, in which the field particle is a π^0—there is no transfer of

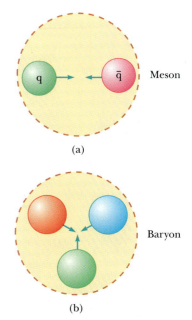

Figure 46.13 (a) A green quark is attracted to an antigreen quark. This forms a meson whose quark structure is ($q\bar{q}$). (b) Three quarks of different colors attract each other to form a baryon.

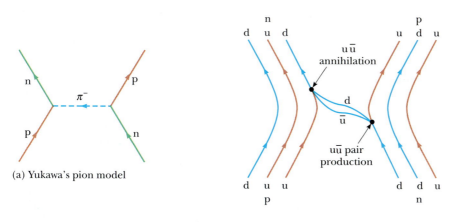

Figure 46.14 (a) A nuclear interaction between a proton and a neutron explained in terms of Yukawa's pion-exchange model. Because the pion carries charge, the proton and neutron switch identities. (b) The same interaction, explained in terms of quarks and gluons. Note that the exchanged $\bar{u}d$ quark pair makes up a π^- meson.

charge from one nucleon to the other in Figure 46.5a. In Figure 46.14a, the charged pion carries charge from one nucleon to the other, so the nucleons change identities—the proton becomes a neutron and the neutron becomes a proton.

Let us look at the same interaction from the viewpoint of the quark model, shown in Figure 46.14b. In this Feynman diagram, the proton and neutron are represented by their quark constituents. Each quark in the neutron and proton is continuously emitting and absorbing gluons. The energy of a gluon can result in the creation of quark–antiquark pairs. This is similar to the creation of electron–positron pairs in pair production, which we investigated in Section 46.2. When the neutron and proton approach to within 1 fm of each other, these gluons and quarks can be exchanged between the two nucleons, and such exchanges produce the nuclear force. Figure 46.14b depicts one possibility for the process shown in Figure 46.14a. A down quark in the neutron on the right emits a gluon. The energy of the gluon is then transformed to create a $u\bar{u}$ pair. The u quark stays within the nucleon (which has now changed to a proton), and the recoiling d quark and the $\bar{u}$ antiquark are transmitted to the proton on the left side of the diagram. Here the $\bar{u}$ annihilates a u quark within the proton (with the creation of a gluon), and the d is captured. The net effect is to change a u quark to a d quark, and the proton on the left has changed to a neutron.

As the d quark and $\bar{u}$ antiquark in Figure 46.14b transfer between the nucleons, the d and $\bar{u}$ exchange gluons with each other and can be considered to be bound to each other by means of the strong force. If we look back at Table 46.4, we see that this combination is a π^-—Yukawa's field particle! Therefore, the quark model of interactions between nucleons is consistent with the pion-exchange model.

46.11   The Standard Model

Scientists now believe that there are three classifications of truly elementary particles: leptons, quarks, and field particles. These three particles are further classified as either fermions or bosons. Note that quarks and leptons have spin $\frac{1}{2}$ and hence are fermions, while the field particles have integral spin of 1 or higher and are bosons.

Recall from Section 46.1 that the weak force is believed to be mediated by the W^+, W^-, and Z^0 bosons. These particles are said to have *weak charge*, just as quarks have color charge. Therefore, each elementary particle can have mass, electric charge, color charge, and weak charge. Of course, one or more of these could be zero.

In 1979, Sheldon Glashow (b. 1932), Abdus Salam (1926–1996), and Steven Weinberg (b. 1933) won a Nobel Prize in physics for developing a theory that unifies the electromagnetic and weak interactions. This **electroweak theory** postulates that the weak and electromagnetic interactions have the same strength when the particles involved have very high energies. The two interactions are viewed as different manifestations of a single unifying electroweak interaction. The theory makes many concrete predictions, but perhaps the most spectacular is the prediction of the masses of the W and Z particles at about 82 GeV/c^2 and 93 GeV/c^2, respectively. These predictions are close to the masses in Table 46.1 determined by experiment.

The combination of the electroweak theory and QCD for the strong interaction is referred to in high-energy physics as the **Standard Model.** Although the details of the Standard Model are complex, its essential ingredients can be summarized with the help of Figure 46.15. (The Standard Model does not include the gravitational force at present; however, we include gravity in Figure 46.15 because physicists hope to eventually incorporate this force into a unified theory.) This diagram shows that quarks participate in all of the fundamental forces and that leptons participate in all except the strong force.

The Standard Model does not answer all questions. A major question that is still unanswered is why, of the two mediators of the electroweak interaction, the photon has no mass but the W and Z bosons do. Because of this mass difference, the elec-

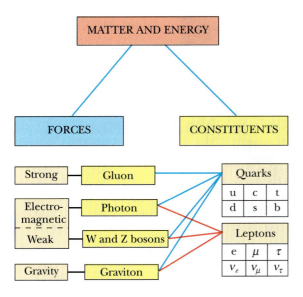

Figure 46.15 The Standard Model of particle physics.

tromagnetic and weak forces are quite distinct at low energies but become similar at very high energies—when the rest energy is negligible relative to the total energy. The behavior as one goes from high to low energies is called *symmetry breaking* because the forces are similar, or symmetric, at high energies but are very different at low energies. The nonzero rest energies of the W and Z bosons raise the question of the origin of particle masses. To resolve this problem, a hypothetical particle called the **Higgs boson,** which provides a mechanism for breaking the electroweak symmetry, has been proposed. The Standard Model modified to include the Higgs mechanism provides a logically consistent explanation of the massive nature of the W and Z bosons. Unfortunately, the Higgs boson has not yet been found, but physicists know that its rest energy should be less than 1 TeV. In order to determine whether the Higgs boson exists, two quarks each having at least 1 TeV of energy must collide. However, calculations show that such a collision requires injecting 40 TeV of energy within the volume of a proton.

Because of the limited energy available in conventional accelerators using fixed targets, it is necessary to employ colliding-beam accelerators called **colliders.** The concept of colliders is straightforward. Particles that have equal masses and equal kinetic energies, traveling in opposite directions in an accelerator ring, collide head-on to produce the required reaction and form new particles. Because the total momentum of the interacting particles is zero, all of their kinetic energy is available for the reaction. The Large Electron–Positron (LEP) Collider at CERN (Fig. 46.16) and the Stanford Linear Collider collide both electrons and positrons. The Super Proton Synchrotron at CERN accelerates protons and antiprotons to energies of 270 GeV. The world's highest-energy proton accelerator, the Tevatron at the Fermi National Laboratory in Illinois, produces protons at almost 1 000 GeV (1 TeV). The Relativistic Heavy Ion Collider (RHIC) at Brookhaven National Laboratory will collide heavy ions to search for the quark–gluon plasma, as discussed earlier. CERN expects a 2007 completion date for the Large Hadron Collider (LHC), a proton–proton collider that will provide a center-of-mass energy of 14 TeV and enable exploration of Higgs-boson physics. The accelerator will be constructed in the same 27-km circumference tunnel now housing the LEP Collider, and many countries will participate in the project.

In addition to increasing energies in modern accelerators, detection techniques have become increasingly sophisticated. We saw simple bubble-chamber photographs earlier in this chapter that required hours of analysis by hand. Figure 46.17 shows a modern detection display of particle tracks after a reaction; the tracks are analyzed rapidly by computer. The chapter opening photograph shows a complex set of tracks from a collision of gold nuclei.

Courtesy of CERN

Figure 46.16 A view from inside the Large Electron–Positron (LEP) Collider tunnel, which is 27 km in circumference.

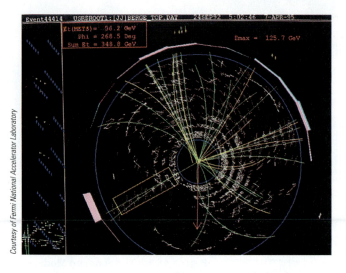

Figure 46.17 Computers at Fermilab create a pictorial representation such as this of the paths of particles after a collision.

46.12 The Cosmic Connection

In this section we describe one of the most fascinating theories in all of science—the Big Bang theory of the creation of the Universe—and the experimental evidence that supports it. This theory of cosmology states that the Universe had a beginning and, furthermore, that the beginning was so cataclysmic that it is impossible to look back beyond it. According to this theory, the Universe erupted from an infinitely dense singularity about 10 billion years ago.[6] The first few minutes after the Big Bang saw such extremely high energy that it is believed that all four interactions of physics were unified and all matter was contained in a quark–gluon plasma.

The evolution of the four fundamental forces from the Big Bang to the present is shown in Figure 46.18. During the first 10^{-43} s (the ultrahot epoch, $T \sim 10^{32}$ K), it is

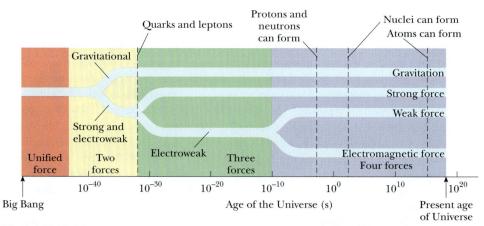

Figure 46.18 A brief history of the Universe from the Big Bang to the present. The four forces became distinguishable during the first nanosecond. Following this, all the quarks combined to form particles that interact via the nuclear force. However, the leptons remained separate and to this day exist as individual, observable particles.

---

[6] An updated estimate for the age of the Universe of between 11.2 billion and 20 billion years was published shortly before this book went to press: L. M. Krauss and B. Chaboyer, "Age Estimates of Globular Clusters in the Milky Way: Constraints on Cosmology," *Science,* January 3, 2003; 299: pp. 65–69. Data released from the Wilkinson Microwave Anisotropy Probe in February 2003 pinpoints the age as 13.7 ± 0.2 billion years ("WMAP Spacecraft Maps the Entire Cosmic Microwave Sky with Unprecedented Precision", *Physics Today,* April 2003; 56(4), pp. 21–24.)

presumed that the strong, electroweak, and gravitational forces were joined to form a completely unified force. In the first 10^{-35} s following the Big Bang (the hot epoch, $T \sim 10^{29}$ K), symmetry breaking occurred for gravity while the strong and electroweak forces remained unified. This was a period when particle energies were so great ($> 10^{16}$ GeV) that very massive particles as well as quarks, leptons, and their antiparticles existed. Then, after 10^{-35} s, the Universe rapidly expanded and cooled (the warm epoch, $T \sim 10^{29}$ to 10^{15} K) and the strong and electroweak forces parted company. As the Universe continued to cool, the electroweak force split into the weak force and the electromagnetic force about 10^{-10} s after the Big Bang.

After a few minutes, protons and neutrons condensed out of the plasma. For half an hour the Universe underwent thermonuclear detonation, exploding as a hydrogen bomb and producing most of the helium nuclei that now exist. The Universe continued to expand, and its temperature dropped. Until about 700 000 years after the Big Bang, the Universe was dominated by radiation. Energetic radiation prevented matter from forming single hydrogen atoms because collisions would instantly ionize any atoms that happened to form. Photons experienced continuous Compton scattering from the vast numbers of free electrons, resulting in a Universe that was opaque to radiation. By the time the Universe was about 700 000 years old, it had expanded and cooled to about 3 000 K, and protons could bind to electrons to form neutral hydrogen atoms. Because of the quantized energies of the atoms, far more wavelengths of radiation were not absorbed by atoms than were absorbed, and the Universe suddenly became transparent to photons. Radiation no longer dominated the Universe, and clumps of neutral matter steadily grew—first atoms, then molecules, gas clouds, stars, and finally galaxies.

Observation of Radiation from the Primordial Fireball

In 1965, Arno A. Penzias (b. 1933) and Robert W. Wilson (b. 1936) of Bell Laboratories were testing a sensitive microwave receiver and made an amazing discovery. A pesky signal producing a faint background hiss was interfering with their satellite communications experiments. In spite of their valiant efforts, the signal remained. Ultimately, it became clear that they were detecting microwave background radiation (at a wavelength of 7.35 cm), which represented the leftover "glow" from the Big Bang.

The microwave horn that served as their receiving antenna is shown in Figure 46.19. The intensity of the detected signal remained unchanged as the antenna was

AT&T Bell Laboratories

Figure 46.19 Robert W. Wilson *(left)* and Arno A. Penzias with the Bell Telephone Laboratories horn-reflector antenna.

pointed in different directions. The fact that the radiation had equal strengths in all directions suggested that the entire Universe was the source of this radiation. Evicting a flock of pigeons from the 20-ft horn and cooling the microwave detector both failed to remove the signal. Through a casual conversation, Penzias and Wilson discovered that a group at Princeton had predicted the residual radiation from the Big Bang and were planning an experiment to attempt to confirm the theory. The excitement in the scientific community was high when Penzias and Wilson announced that they had already observed an excess microwave background compatible with a 3-K blackbody source, which was consistent with the predicted temperature of the Universe at this time after the Big Bang.

Because Penzias and Wilson made their measurements at a single wavelength, they did not completely confirm the radiation as 3-K blackbody radiation. Subsequent experiments by other groups added intensity data at different wavelengths, as shown in Figure 46.20. The results confirm that the radiation is that of a black body at 2.7 K. This figure is perhaps the most clear-cut evidence for the Big Bang theory. The 1978 Nobel Prize in physics was awarded to Penzias and Wilson for this most important discovery.

The discovery of the cosmic background radiation brought with it a problem, however—the radiation was too uniform. Scientists believed that slight fluctuations in this background had to occur to act as nucleation sites for the formation of the galaxies and other objects we now see in the sky. In 1989, NASA launched a satellite called COBE (KOH-bee), for Cosmic Background Explorer, to study this radiation in greater detail. In 1992, George Smoot (b. 1945) at the Lawrence Berkeley Laboratory found, on the basis of the data collected, that the background was not perfectly uniform but instead contained irregularities that corresponded to temperature variations of 0.000 3 K. In 2000, the BOOMERANG experiment was carried out by an international team of scientists from Italy, the U.K., the U.S., Canada, and France. This team used instruments flying in a balloon over Antarctica to generate data with over 25 times the angular resolution of the COBE data. These data clearly show a peak in fluctuation intensity 300 000 years after the Big Bang. The Wilkinson Microwave Anisotropy Probe, launched in June 2001, exhibited data in early 2003 that allowed observation of temperature differences in the cosmos in the microkelvin range.

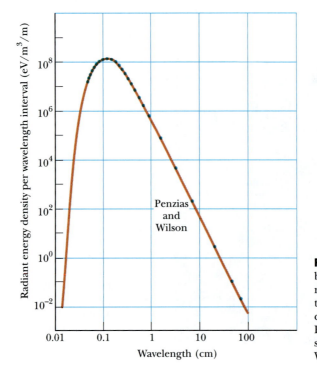

Figure 46.20 Theoretical blackbody (brown curve) and measured radiation spectra (blue points) of the Big Bang. Most of the data were collected from the Cosmic Background Explorer (COBE) satellite. The datum of Penzias and Wilson is indicated.

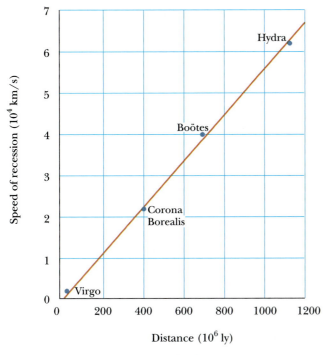

Figure 46.21 Hubble's law: a plot of speed of recession versus distance for four galaxies.

Other Evidence for an Expanding Universe

The Big Bang theory of cosmology predicts that the Universe is expanding. Most of the key discoveries supporting the theory of an expanding Universe were made in the twentieth century. Vesto Melvin Slipher (1875–1969), an American astronomer, reported in 1912 that most nebulae are receding from the Earth at speeds up to several million miles per hour. Slipher was one of the first scientists to use Doppler shifts (see Section 17.4) in spectral lines to measure velocities.

In the late 1920s, Edwin P. Hubble (1889–1953) made the bold assertion that the whole Universe is expanding. From 1928 to 1936, until they reached the limits of the 100-inch telescope, Hubble and Milton Humason (1891–1972) worked at Mount Wilson in California to prove this assertion. The results of that work and of its continuation with the use of a 200-inch telescope in the 1940s showed that the speeds at which galaxies are receding from the Earth increase in direct proportion to their distance R from us (Fig. 46.21). This linear relationship, known as **Hubble's law,** may be written

$$v = HR \qquad (46.7) \qquad \text{\color{blue}\textbf{Hubble's law}}$$

where H, called the **Hubble constant,** has the approximate value

$$H \approx 17 \times 10^{-3} \text{ m/s} \cdot \text{ly}$$

Example 46.5   Recession of a Quasar

A quasar is an object that appears similar to a star and is very distant from the Earth. Its speed can be determined from Doppler-shift measurements in the light it emits. A certain quasar recedes from the Earth at a speed of $0.55c$. How far away is it?

Solution We can find the distance through Hubble's law:

$$R = \frac{v}{H} = \frac{(0.55)(3.00 \times 10^8 \text{ m/s})}{17 \times 10^{-3} \text{ m/s} \cdot \text{ly}}$$

$$= \boxed{9.7 \times 10^9 \text{ ly}}$$

What If? Suppose we assume that the quasar has moved at this speed ever since the Big Bang. With this assumption, estimate the age of the Universe.

Answer We approximate the distance from Earth to the quasar as the distance that the quasar has moved from the singularity since the Big Bang. We can then find the time interval from a calculation as performed in Chapter 2: $\Delta t = \Delta x/v = R/v = 1/H \approx 18$ billion years, which is in approximate agreement with other calculations.

Figure 46.22 Red shift, or speed of recession, versus magnitude (which is related to brightness) of 18 faint galaxy clusters. Significant scatter of the data occurs, so the extrapolation of the curve to the upper right is uncertain. Curve *A* is the trend suggested by the six faintest clusters. Curve *C* corresponds to a Universe having a constant rate of expansion. If more data are taken and the complete set of data indicates a curve that falls between *B* and *C*, the expansion will slow but never stop. If the data fall to the left of *B*, expansion will eventually stop and the Universe will begin to contract.

Will the Universe Expand Forever?

In the 1950s and 1960s, Allan R. Sandage (b. 1926) used the 200-inch telescope at Mount Palomar to measure the speeds of galaxies at distances of up to 6 billion lightyears away from the Earth. These measurements showed that these very distant galaxies were moving about 10 000 km/s faster than Hubble's law predicted. According to this result, the Universe must have been expanding more rapidly 1 billion years ago, and consequently we conclude from these data that the expansion rate is slowing[7] (Fig. 46.22). Today, astronomers and physicists are trying to determine the rate of expansion. If the average mass density of the Universe is less than some critical value, the galaxies will slow in their outward rush but still escape to infinity. If the average density exceeds the critical value, the expansion will eventually stop and contraction will begin, possibly leading to a superdense state followed by another expansion. In this scenario, we have an oscillating Universe.

Example 46.6   The Critical Density of the Universe

(A) Starting from energy conservation, derive an expression for the critical mass density of the Universe ρ_c in terms of the Hubble constant H and the universal gravitational constant G.

Solution Figure 46.23 shows a large section of the Universe, contained within a sphere of radius R. The total mass of the galaxies in this volume is M. A galaxy of mass $m \ll M$ that has a speed v at a distance R from the center of the sphere will escape to infinity (at which its speed will approach zero) if the sum of its kinetic energy and the gravitational potential energy

of the system—the galaxy plus the rest of the Universe—is zero at any time. The Universe may be infinite in spatial extent, but Gauss's law for gravitation (Problem 73 in Chapter 24) implies that only the mass M inside the sphere contributes to the gravitational potential energy of the galaxy:

$$E_{\text{total}} = 0 = K + U = \tfrac{1}{2}mv^2 - \frac{GmM}{R}$$

We substitute for the mass M contained within the sphere the product of the critical density and the volume of the

---

[7]   The data at large distances have large observational uncertainties and may be systematically in error from effects such as abnormal brightness in the most distant visible clusters.

sphere:

$$\tfrac{1}{2}mv^2 = \frac{Gm\left(\tfrac{4}{3}\pi R^3 \rho_c\right)}{R}$$

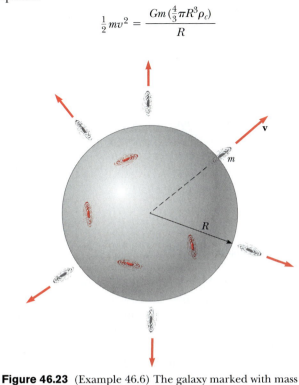

Figure 46.23 (Example 46.6) The galaxy marked with mass m is escaping from a large cluster of galaxies contained within a spherical volume of radius R. Only the mass within R slows the galaxy.

Solving for the critical density gives

$$\rho_c = \frac{3v^2}{8\pi GR^2}$$

From Hubble's law, the ratio of v to R is $v/R = H$, so this expression becomes

$$\rho_c = \frac{3H^2}{8\pi G}$$

(B) Estimate a numerical value for the critical density in grams per cubic centimeter.

Solution Using $H = 17 \times 10^{-3}$ m/s·ly where 1 ly = 9.46 $\times 10^{15}$ m, we find for the critical density

$$\rho_c = \frac{3H^2}{8\pi G} = \frac{3(17 \times 10^{-3}\,\mathrm{m/s \cdot ly})^2}{8\pi(6.67 \times 10^{-11}\,\mathrm{N \cdot m^2/kg^2})}$$

$$\times \left(\frac{1\,\mathrm{ly}}{9.46 \times 10^{15}\,\mathrm{m}}\right)^2$$

$$= 6 \times 10^{-27}\,\mathrm{kg/m^3} = 6 \times 10^{-30}\,\mathrm{g/cm^3}$$

Because the mass of a hydrogen atom is 1.67×10^{-24} g, this value of ρ_c corresponds to 3×10^{-6} hydrogen atoms per cubic centimeter or 3 atoms per cubic meter.

Missing Mass in the Universe?

The luminous matter in galaxies averages out to a Universe density of 5×10^{-33} g/cm³. The radiation in the Universe has a mass equivalent of approximately 2% of the luminous matter. The total mass of all nonluminous matter (such as interstellar gas and black holes) may be estimated from the speeds of galaxies orbiting each other in a cluster. The higher the galaxy speeds, the more mass in the cluster. Measurements on the Coma cluster of galaxies indicate, surprisingly, that the amount of nonluminous matter is 20 to 30 times the amount of luminous matter present in stars and luminous gas clouds. Yet even this large invisible component of *dark matter*, if extrapolated to the Universe as a whole, leaves the observed mass density a factor of 10 less than ρ_c. The deficit, called *missing mass,* has been the subject of intense theoretical and experimental work, with exotic particles such as axions, photinos, and superstring particles suggested as candidates for the missing mass. Some researchers have made the more mundane proposal that the missing mass is present in neutrinos. In fact, neutrinos are so abundant that a tiny neutrino rest energy on the order of only 20 eV would furnish the missing mass and "close" the Universe. Current experiments designed to measure the rest energy of the neutrino will have an impact on predictions for the future of the Universe.

Mysterious Energy in the Universe?

A surprising twist in the story of the Universe arose in 1998 with the observation of a class of supernovae that have a fixed absolute brightness. By combining the apparent brightness and the redshift of light from these explosions, their distance and speed of recession of the Earth can be determined. These observations led to the conclusion that the expansion of the Universe is not slowing down, but it is accelerating! Observations by other groups also led to the same interpretation.

In order to explain this acceleration, physicists have proposed *dark energy*, which is energy possessed by the vacuum of space. In the early life of the Universe, gravity dominated over the dark energy. As the Universe expanded and the gravitational force between galaxies became smaller because of the great distances between them, the dark energy became more important. The dark energy results in an effective repulsive force that causes the expansion rate to increase.[8]

Although we have some degree of certainty about the beginning of the Universe, we are uncertain about how the story will end. Will the Universe keep on expanding forever, or will it someday collapse and then expand again, perhaps in an endless series of oscillations? Results and answers to these questions remain inconclusive, and the exciting controversy continues.

46.13 Problems and Perspectives

While particle physicists have been exploring the realm of the very small, cosmologists have been exploring cosmic history back to the first microsecond of the Big Bang. Observation of the events that occur when two particles collide in an accelerator is essential for reconstructing the early moments in cosmic history. For this reason, perhaps the key to understanding the early Universe is to first understand the world of elementary particles. Cosmologists and physicists now find that they have many common goals and are joining hands in an attempt to understand the physical world at its most fundamental level.

Our understanding of physics at short distances is far from complete. Particle physics is faced with many questions. Why does so little antimatter exist in the Universe? Is it possible to unify the strong and electroweak theories in a logical and consistent manner? Why do quarks and leptons form three similar but distinct families? Are muons the same as electrons apart from their difference in mass, or do they have other subtle differences that have not been detected? Why are some particles charged and others neutral? Why do quarks carry a fractional charge? What determines the masses of the elementary constituents of matter? Can isolated quarks exist?

An important and obvious question that remains is whether leptons and quarks have an underlying structure. If they do, we can envision an infinite number of deeper structure levels. However, if leptons and quarks are indeed the ultimate constituents of matter, as physicists today tend to believe, we should be able to construct a final theory of the structure of matter, just as Einstein dreamed of doing. This theory, whimsically called the Theory of Everything, is a combination of the Standard Model and a quantum theory of gravity.

String Theory—a New Perspective

Let us briefly discuss one current effort at answering some of these questions by proposing a new perspective on particles. While reading this book, you may recall starting off with the particle model and doing quite a bit of physics with it. In Part 2, we introduced the wave model, and there was more physics to be investigated via the properties of waves. We used a wave model for light in Part 5. Early in Part 6, however, we saw the need to return to the particle model for light. Furthermore, we found that

---

[8] For an update on dark energy, see S. Perlmutter, "Supernovae, Dark Energy, and the Accelerating Universe", *Physics Today*, April 2003; 56(4), pp. 53–60.

(a)   (b)

Figure 46.24 (a) A piece of paper is cut into a rectangular shape. As a rectangle, the shape has two dimensions. (b) The paper is rolled up into a soda straw. From far away, it appears to be one-dimensional. The curled-up second dimension is not visible when viewed from a distance large compared to the diameter of the straw.

material particles had wave-like characteristics. The quantum particle model discussed in Chapter 40 allowed us to build particles out of waves, suggesting that a wave is the fundamental entity. In the current chapter, however, we introduced elementary particles as the fundamental entities. It seems as if we cannot make up our mind! In this final section, we shall discuss a current research effort to build particles out of waves and vibrations.

String theory is an effort to unify the four fundamental forces by modeling all particles as various vibrational modes of a single entity—an incredibly small string. The typical length of such a string is on the order of 10^{-35} m, called the **Planck length**. We have seen quantized modes before—the frequencies of vibrating guitar strings in Chapter 18 and the quantized energy levels of atoms in Chapter 42. In string theory, each quantized mode of vibration of the string corresponds to a different elementary particle in the Standard Model.

One of the complicating factors in string theory is that it requires space–time to have ten dimensions. Despite the theoretical and conceptual difficulties in dealing with ten dimensions, string theory holds promise in incorporating gravity with the other forces. Four of the ten dimensions are visible to us—three space dimensions and one time dimension—and the other six are said to be *compactified*. That is, the six dimensions are curled up so tightly that they are not visible in the macroscopic world.

As an analogy, consider a soda straw. We can build a soda straw by cutting a rectangular piece of paper (Fig. 46.24a), which clearly has two dimensions, and rolling it up into a small tube (Fig. 46.24b). From far away, the soda straw looks like a one-dimensional straight line. The second dimension has been curled up and is not visible. String theory claims that six space–time dimensions are curled up in an analogous way, with the curling being on the size of the Planck length—impossible to see from our viewpoint.

Another complicating factor with string theory is that it is difficult for string theorists to guide experimentalists as to what to look for in an experiment. The Planck length is so small that direct experimentation on strings is impossible. Until the theory has been further developed, string theorists are restricted to applying the theory to known results and testing for consistency.

One of the predictions of string theory, called **supersymmetry** (SUSY), suggests that every elementary particle has a superpartner that has not yet been observed. It is believed that supersymmetry is a broken symmetry (like the broken electroweak symmetry at low energies) and the masses of the superpartners are above our current capabilities of detection by accelerators. Some theorists claim that the mass of superpartners is the missing mass discussed in Section 46.12. Keeping with the whimsical trend in naming particles and their properties, superpartners are given names such as the *squark* (the superpartner to a quark), the *selectron* (electron), and the *gluinos* (gluon).

Other theorists are working on **M-theory**, which is an eleven-dimensional theory based on membranes rather than strings. In a way reminiscent of the correspondence principle, M-theory is claimed to reduce to string theory if one compactifies from eleven dimensions to ten dimensions.

The questions that we listed at the beginning of this section go on and on. Because of the rapid advances and new discoveries in the field of particle physics, many of these questions may be resolved in the next decade while other new questions may emerge.

SUMMARY

Take a practice test for this chapter by clicking on the Practice Test link at http://www.pse6.com.

Before quark theory was developed, the four fundamental forces in nature were identified as nuclear, electromagnetic, weak, and gravitational. All the interactions in which these forces take part are mediated by **field particles.** The electromagnetic interaction is mediated by the photon; the weak interaction is mediated by the $W^{\pm}$ and Z^0 bosons; the gravitational interaction is mediated by gravitons; the nuclear interaction is mediated by gluons.

A charged particle and its **antiparticle** have the same mass but opposite charge, and other properties will have opposite values, such as lepton number and baryon number. It is possible to produce particle–antiparticle pairs in nuclear reactions if the available energy is greater than $2mc^2$, where m is the mass of the particle (or antiparticle).

Particles other than field particles are classified as hadrons or leptons. **Hadrons** interact via all four fundamental forces. They have size and structure, and are not elementary particles. There are two types—**baryons** and **mesons.** Baryons, which generally are the most massive particles, have nonzero **baryon number** and a spin of $\frac{1}{2}$ or $\frac{3}{2}$. Mesons have baryon number zero and either zero or integral spin.

Leptons have no structure or size and are considered truly elementary. They interact only via the weak, gravitational, and electromagnetic forces. Six types of leptons exist: the electron e^-, the muon μ^-, the tau τ^-; and their neutrinos ν_e, ν_μ, and ν_τ.

In all reactions and decays, quantities such as energy, linear momentum, angular momentum, electric charge, baryon number, and lepton number are strictly conserved. Certain particles have properties called **strangeness** and **charm.** These unusual properties are conserved in all decays and nuclear reactions except those that occur via the weak force.

Theorists in elementary particle physics have postulated that all hadrons are composed of smaller units known as **quarks,** and experimental evidence agrees with this model. Quarks have fractional electric charge and come in six **flavors**: up (u), down (d), strange (s), charmed (c), top (t), and bottom (b). Each baryon contains three quarks, and each meson contains one quark and one antiquark.

According to the theory of **quantum chromodynamics,** quarks have a property called **color,** and the force between quarks is referred to as the **strong force** or the **color force.** The strong force is now considered to be a fundamental force. The nuclear force, which was originally considered to be fundamental, is now understood to be a secondary effect of the strong force, due to gluon exchanges between hadrons.

The electromagnetic and weak forces are now considered to be manifestations of a single force called the **electroweak force.** The combination of quantum chromodynamics and the electroweak theory is called the **Standard Model.**

The background microwave radiation discovered by Penzias and Wilson strongly suggests that the Universe started with a Big Bang 11 to 20 billion years ago. The background radiation is equivalent to that of a black body at 3 K. Various astronomical measurements strongly suggest that the Universe is expanding. According to **Hubble's law,** distant galaxies are receding from the Earth at a speed $v = HR$, where R is the distance from the Earth to the galaxy and H is the **Hubble constant,** $H \approx 17 \times 10^{-3}$ m/s·ly.

QUESTIONS

1. Name the four fundamental interactions and the field particle that mediates each.

2. When an electron and a positron meet at low speeds in free space, why are *two* 0.511-MeV gamma rays produced, rather than one gamma ray with an energy of 1.02 MeV?

3. Describe the quark model of hadrons, including the properties of quarks.

4. What are the differences between hadrons and leptons?

5. Describe the properties of baryons and mesons and the important differences between them.

6. Particles known as resonances have very short lifetimes, on the order of 10^{-23} s. From this information, would you guess that they are hadrons or leptons? Explain.

7. Kaons all decay into final states that contain no protons or neutrons. What is the baryon number of kaons?

8. The Ξ^0 particle decays by the weak interaction according to the decay mode $\Xi^0 \rightarrow \Lambda^0 + \pi^0$. Would you expect this decay to be fast or slow? Explain.

9. Identify the particle decays in Table 46.2 that occur by the weak interaction. Justify your answers.

10. Identify the particle decays in Table 46.2 that occur by the electromagnetic interaction. Justify your answers.

11. Two protons in a nucleus interact via the nuclear interaction. Are they also subject to the weak interaction?

12. Discuss the following conservation laws: energy, linear momentum, angular momentum, electric charge, baryon number, lepton number, and strangeness. Are all of these laws based on fundamental properties of nature? Explain.

13. An antibaryon interacts with a meson. Can a baryon be produced in such an interaction? Explain.

14. Describe the essential features of the Standard Model of particle physics.

15. How many quarks are in each of the following: (a) a baryon, (b) an antibaryon, (c) a meson, (d) an antimeson? How do you account for the fact that baryons have half-integral spins while mesons have spins of 0 or 1? (*Note:* Quarks have spin $\frac{1}{2}$.)

16. In the theory of quantum chromodynamics, quarks come in three colors. How would you justify the statement that "all baryons and mesons are colorless"?

17. Which baryon did Murray Gell-Mann predict in 1961? What is the quark composition of this particle?

18. What is the quark composition of the Ξ^- particle? (See Table 46.5.)

19. The W and Z bosons were first produced at CERN in 1983 by causing a beam of protons and a beam of antiprotons to meet at high energy. Why was this an important discovery?

20. How did Edwin Hubble determine in 1928 that the Universe is expanding?

21. Neutral atoms did not exist until hundreds of thousands of years after the Big Bang. Why?

22. **Review question.** A girl and her grandmother grind corn while the woman tells the girl stories about what is most important. A boy keeps crows away from ripening corn while his grandfather sits in the shade and explains to him the Universe and his place in it. What the children do not understand this year they will better understand next year. Now you must take the part of the adults. State the most general, most fundamental, most universal truths that you know. If you need to repeat someone else's ideas, get the best version of those ideas you can, and state your source. If there is something you do not understand, make a plan to understand it better within the next year.

PROBLEMS

1, 2, 3 = straightforward, intermediate, challenging   □ = full solution available in the *Student Solutions Manual and Study Guide*

= coached solution with hints available at http://www.pse6.com   💻 = computer useful in solving problem

= paired numerical and symbolic problems

Section 46.1 The Fundamental Forces in Nature
Section 46.2 Positrons and Other Antiparticles

1. A photon produces a proton–antiproton pair according to the reaction $\gamma \rightarrow p + \bar{p}$. What is the minimum possible frequency of the photon? What is its wavelength?

2. Two photons are produced when a proton and antiproton annihilate each other. In the reference frame in which the center of mass of the proton–antiproton system is stationary, what are the minimum frequency and corresponding wavelength of each photon?

3. A photon with an energy $E_\gamma = 2.09$ GeV creates a proton–antiproton pair in which the proton has a kinetic energy of 95.0 MeV. What is the kinetic energy of the antiproton? ($m_p c^2 = 938.3$ MeV.)

Section 46.3 Mesons and the Beginning of Particle Physics

4. Occasionally, high-energy muons collide with electrons and produce two neutrinos according to the reaction $\mu^+ + e^- \rightarrow 2\nu$. What kind of neutrinos are these?

5. One of the mediators of the weak interaction is the Z^0 boson, with mass 93 GeV/c^2. Use this information to find the order of magnitude of the range of the weak interaction.

6. Calculate the range of the force that might be produced by the virtual exchange of a proton.

7. A neutral pion at rest decays into two photons according to

$$\pi^0 \longrightarrow \gamma + \gamma$$

Find the energy, momentum, and frequency of each photon.

8. When a high-energy proton or pion traveling near the speed of light collides with a nucleus, it travels an average distance of 3×10^{-15} m before interacting. From this information, find the order of magnitude of the time interval required for the strong interaction to occur.

9. A free neutron beta decays by creating a proton, an electron and an antineutrino according to the reaction $n \rightarrow p + e^- + \bar{\nu}$. **What If?** Imagine that a free neutron decays by creating a proton and electron according to the reaction

$$n \longrightarrow p + e^-$$

and assume that the neutron is initially at rest in the laboratory. (a) Determine the energy released in this reaction. (b) Determine the speeds of the proton and electron after the reaction. (Energy and momentum are conserved in the reaction.) (c) Is either of these particles moving at a relativistic speed? Explain.

Section 46.4   Classification of Particles

10. Identify the unknown particle on the left side of the following reaction:

$$? + p \longrightarrow n + \mu^+$$

11. Name one possible decay mode (see Table 46.2) for Ω^+, $\overline{K}_S^0$, Λ^0 and $\bar{n}$.

Section 46.5   Conservation Laws

12. Each of the following reactions is forbidden. Determine a conservation law that is violated for each reaction.
    (a) $p + \bar{p} \rightarrow \mu^+ + e^-$
    (b) $\pi^- + p \rightarrow p + \pi^+$
    (c) $p + p \rightarrow p + \pi^+$
    (d) $p + p \rightarrow p + p + n$
    (e) $\gamma + p \rightarrow n + \pi^0$

13. (a) Show that baryon number and charge are conserved in the following reactions of a pion with a proton.

$$\pi^+ + p \longrightarrow K^+ + \Sigma^+ \qquad (1)$$

$$\pi^+ + p \longrightarrow \pi^+ + \Sigma^+ \qquad (2)$$

(b) The first reaction is observed, but the second never occurs. Explain.

14. The first of the following two reactions can occur, but the second cannot. Explain.

$$K_S^0 \longrightarrow \pi^+ + \pi^- \qquad \text{(can occur)}$$

$$\Lambda^0 \longrightarrow \pi^+ + \pi^- \qquad \text{(cannot occur)}$$

15. The following reactions or decays involve one or more neutrinos. In each case, supply the missing neutrino ($\nu_e, \nu_\mu,$ or ν_τ) or antineutrino.
    (a) $\pi^- \rightarrow \mu^- + ?$
    (b) $K^+ \rightarrow \mu^+ + ?$
    (c) $? + p \rightarrow n + e^+$
    (d) $? + n \rightarrow p + e^-$
    (e) $? + n \rightarrow p + \mu^-$
    (f) $\mu^- \rightarrow e^- + ? + ?$

16. A K_S^0 particle at rest decays into a π^+ and a π^-. What will be the speed of each of the pions? The mass of the K_S^0 is 497.7 MeV/c^2, and the mass of each π is 139.6 MeV/c^2.

17. Determine which of the following reactions can occur. For those that cannot occur, determine the conservation law (or laws) violated:
    (a) $p \rightarrow \pi^+ + \pi^0$
    (b) $p + p \rightarrow p + p + \pi^0$
    (c) $p + p \rightarrow p + \pi^+$
    (d) $\pi^+ \rightarrow \mu^+ + \nu_\mu$
    (e) $n \rightarrow p + e^- + \bar{\nu}_e$
    (f) $\pi^+ \rightarrow \mu^+ + n$

18. (a) Show that the proton-decay reaction

$$p \longrightarrow e^+ + \gamma$$

cannot occur, because it violates conservation of baryon number. (b) **What If?** Imagine that this reaction does occur, and that the proton is initially at rest. Determine the energy and momentum of the positron and photon after the reaction. (*Suggestion:* Recall that energy and momentum must be conserved in the reaction.) (c) Determine the speed of the positron after the reaction.

19. Determine the type of neutrino or antineutrino involved in each of the following processes:
    (a) $\pi^+ \rightarrow \pi^0 + e^+ + ?$
    (b) $? + p \rightarrow \mu^- + p + \pi^+$
    (c) $\Lambda^0 \rightarrow p + \mu^- + ?$
    (d) $\tau^+ \rightarrow \mu^+ + ? + ?$

Section 46.6   Strange Particles and Strangeness

20. The neutral meson ρ^0 decays by the strong interaction into two pions: $\rho^0 \rightarrow \pi^+ + \pi^-$, half-life 10^{-23} s. The neutral kaon also decays into two pions: $K_S^0 \rightarrow \pi^+ + \pi^-$, half-life 10^{-10} s. How do you explain the difference in half-lives?

21. Determine whether or not strangeness is conserved in the following decays and reactions.
    (a) $\Lambda^0 \rightarrow p + \pi^-$
    (b) $\pi^- + p \rightarrow \Lambda^0 + K^0$
    (c) $\bar{p} + p \rightarrow \overline{\Lambda}^0 + \Lambda^0$
    (d) $\pi^- + p \rightarrow \pi^- + \Sigma^+$
    (e) $\Xi^- \rightarrow \Lambda^0 + \pi^-$
    (f) $\Xi^0 \rightarrow p + \pi^-$

22. For each of the following forbidden decays, determine which conservation law is violated:
    (a) $\mu^- \rightarrow e^- + \gamma$
    (b) $n \rightarrow p + e^- + \nu_e$
    (c) $\Lambda^0 \rightarrow p + \pi^0$

(d) $p \rightarrow e^+ + \pi^0$
(e) $\Xi^0 \rightarrow n + \pi^0$

23. Which of the following processes are allowed by the strong interaction, the electromagnetic interaction, the weak interaction, or no interaction at all?
    (a) $\pi^- + p \rightarrow 2\eta$
    (b) $K^- + n \rightarrow \Lambda^0 + \pi^-$
    (c) $K^- \rightarrow \pi^- + \pi^0$
    (d) $\Omega^- \rightarrow \Xi^- + \pi^0$
    (e) $\eta \rightarrow 2\gamma$

24. Identify the conserved quantities in the following processes:
    (a) $\Xi^- \rightarrow \Lambda^0 + \mu^- + \overline{\nu}_\mu$
    (b) $K_S^0 \rightarrow 2\pi^0$
    (c) $K^- + p \rightarrow \Sigma^0 + n$
    (d) $\Sigma^0 \rightarrow \Lambda^0 + \gamma$
    (e) $e^+ + e^- \rightarrow \mu^+ + \mu^-$
    (f) $\overline{p} + n \rightarrow \overline{\Lambda}^0 + \Sigma^-$

25. Fill in the missing particle. Assume that (a) occurs via the strong interaction and (b) and (c) involve the weak interaction.
    (a) $K^+ + p \rightarrow ? + p$
    (b) $\Omega^- \rightarrow ? + \pi^-$
    (c) $K^+ \rightarrow ? + \mu^+ + \nu_\mu$

Section 46.7   Making Elementary Particles and Measuring their Properties

26. The particle decay $\Sigma^+ \rightarrow \pi^+ + n$ is observed in a bubble chamber. Figure P46.26 represents the curved tracks of the particles Σ^+ and π^+, and the invisible track of the neutron, in the presence of a uniform magnetic field of 1.15 T directed out of the page. The measured radii of curvature are 1.99 m for the Σ^+ particle and 0.580 m for the π^+ particle. (a) Find the momenta of the Σ^+ and the π^+ particles, in units of MeV/c. (b) The angle between the momenta of the Σ^+ and the π^+ particles at the moment of decay is 64.5°. Find the momentum of the neutron. (c) Calculate the total energy of the π^+ particle, and of the neutron, from their known masses ($m_\pi = 139.6$ MeV/c^2, $m_n = 939.6$ MeV/c^2) and the relativistic energy–momentum relation. What is the total energy of the Σ^+ particle? (d) Calculate the mass and speed of the Σ^+ particle.

27. If a K_S^0 meson at rest decays in 0.900×10^{-10} s, how far will a K_S^0 meson travel if it is moving at $0.960c$?

28. A particle of mass m_1 is fired at a stationary particle of mass m_2, and a reaction takes place in which new particles are created out of the incident kinetic energy. Taken together, the product particles have total mass m_3. The minimum kinetic energy that the bombarding particle must have in order to induce the reaction is called the threshold energy. At this energy, the kinetic energy of the products is a minimum, so that the fraction of the incident kinetic energy that is available to create new particles is a maximum. This occurs when all the product particles have the same velocity, so that the particles have no kinetic energy of motion relative to one another. (a) By using conservation of relativistic energy and momentum, and the relativistic energy-momentum relation, show that the threshold energy is given by

$$K_{min} = \frac{\left[m_3^2 - (m_1 + m_2)^2\right]c^2}{2m_2}$$

Calculate the threshold energy for each of the following reactions:
    (b) $p + p \rightarrow p + p + p + \overline{p}$
       (One of the initial protons is at rest. Antiprotons are produced.)
    (c) $\pi^- + p \rightarrow K^0 + \Lambda^0$
       (The proton is at rest. Strange particles are produced.)
    (d) $p + p \rightarrow p + p + \pi^0$
       (One of the initial protons is at rest. Pions are produced.)
    (e) $p + \overline{p} \rightarrow Z^0$
       (One of the initial particles is at rest. Z^0 particles (mass 91.2 GeV/c^2) are produced.)

Section 46.8   Finding Patterns in the Particles
Section 46.9   Quarks
Section 46.10   Multicolored Quarks
Section 46.11   The Standard Model

> *Note:* Problem 71 in Chapter 39 can be assigned with Section 46.11.

29. (a) Find the number of electrons and the number of each species of quarks in 1 L of water. (b) Make an order-of-magnitude estimate of the number of each kind of fundamental matter particle in your body. State your assumptions and the quantities you take as data.

30. The quark composition of the proton is uud, and that of the neutron is udd. Show that in each case the charge, baryon number, and strangeness of the particle equal, respectively, the sums of these numbers for the quark constituents.

31. **What If?** Imagine that binding energies could be ignored. Find the masses of the u and d quarks from the masses of the proton and neutron.

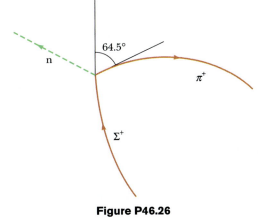

Figure P46.26

32. The quark compositions of the K^0 and Λ^0 particles are $\bar{s}d$ and uds, respectively. Show that the charge, baryon number, and strangeness of these particles equal, respectively, the sums of these numbers for the quark constituents.

33. Analyze each reaction in terms of constituent quarks:
(a) $\pi^- + p \rightarrow K^0 + \Lambda^0$
(b) $\pi^+ + p \rightarrow K^+ + \Sigma^+$
(c) $K^- + p \rightarrow K^+ + K^0 + \Omega^-$
(d) $p + p \rightarrow K^0 + p + \pi^+ + ?$

In the last reaction, identify the mystery particle.

34. The text states that the reaction $\pi^- + p \rightarrow K^0 + \Lambda^0$ occurs with high probability, whereas the reaction $\pi^- + p \rightarrow K^0 + n$ never occurs. Analyze these reactions at the quark level. Show that the first reaction conserves the total number of each type of quark, and the second reaction does not.

35. A Σ^0 particle traveling through matter strikes a proton; then a Σ^+ and a gamma ray emerge, as well as a third particle. Use the quark model of each to determine the identity of the third particle.

36. Identify the particles corresponding to the quark combinations (a) suu, (b) $\bar{u}d$, (c) $\bar{s}d$, and (d) ssd.

37. What is the electrical charge of the baryons with the quark compositions (a) $\bar{u}\,\bar{u}\,\bar{d}$ and (b) $\bar{u}\,\bar{d}\,\bar{d}$? What are these baryons called?

Section 46.12   The Cosmic Connection

Note: Problem 20 in Chapter 39 can be assigned with this section.

38. Review problem. Refer to Section 39.4. Prove that the Doppler shift in wavelength of electromagnetic waves is described by

$$\lambda' = \lambda \sqrt{\frac{1 + v/c}{1 - v/c}}$$

where λ' is the wavelength measured by an observer moving at speed v away from a source radiating waves of wavelength λ.

(a)

(b)

Figure P46.39 (a) Image of the quasar 3C273. (b) Spectrum of the quasar above a comparison spectrum emitted by stationary hydrogen and helium atoms. Both parts of the figure are printed as black-and-white photographic negatives to reveal detail.

Maarten Schmidt/Palomar Observatory/California Institute of Technology

39. A distant quasar is moving away from Earth at such high speed that the blue 434-nm H_γ line of hydrogen is observed at 510 nm, in the green portion of the spectrum (Fig. P46.39). (a) How fast is the quasar receding? You may use the result of Problem 38. (b) Edwin Hubble discovered that all objects outside the local group of galaxies are moving away from us, with speeds proportional to their distances. Hubble's law is expressed as $v = HR$, where Hubble's constant has the approximate value $H = 17 \times 10^{-3}$ m/s·ly. Determine the distance from Earth to this quasar.

40. The various spectral lines observed in the light from a distant quasar have longer wavelengths λ_n' than the wavelengths λ_n measured in light from a stationary source. Here n is an index taking different values for different spectral lines. The fractional change in wavelength toward the red is the same for all spectral lines. That is, the redshift parameter Z defined by

$$Z = \frac{\lambda_n' - \lambda_n}{\lambda_n}$$

is common to all spectral lines for one object. In terms of Z, determine (a) the speed of recession of the quasar and (b) the distance from Earth to this quasar. Use the result of Problem 38 and Hubble's law.

41. Using Hubble's law, find the wavelength of the 590-nm sodium line emitted from galaxies (a) 2.00×10^6 ly away from Earth, (b) 2.00×10^8 ly away, and (c) 2.00×10^9 ly away. You may use the result of Problem 38.

42. Review problem. The cosmic background radiation is blackbody radiation from a source at a temperature of 2.73 K. (a) Use Wien's law to determine the wavelength at which this radiation has its maximum intensity. (b) In what part of the electromagnetic spectrum is the peak of the distribution?

43. Review problem. Use Stefan's law to find the intensity of the cosmic background radiation emitted by the fireball of the Big Bang at a temperature of 2.73 K.

44. It is mostly your roommate's fault. Nosy astronomers have discovered enough junk and clutter in your dorm room to constitute the missing mass required to close the Universe. After observing your floor, closet, bed, and computer files, they extrapolate to slobs in other galaxies and calculate the average density of the observable Universe as $1.20\rho_c$. How many times larger will the Universe become before it begins to collapse? That is, by what factor will the distance between remote galaxies increase in the future?

45. The early Universe was dense with gamma-ray photons of energy $\sim k_{B}T$ and at such a high temperature that protons and antiprotons were created by the process $\gamma \rightarrow p + \bar{p}$ as rapidly as they annihilated each other. As the Universe cooled in adiabatic expansion, its temperature fell below a certain value, and proton pair production became rare. At that time slightly more protons than antiprotons existed, and essentially all of the protons in the Universe today date from that time. (a) Estimate the order of magnitude of the temperature of the Universe when protons condensed out. (b) Estimate the order of magnitude of the temperature of the Universe when electrons condensed out.

46. If the average density of the Universe is small compared to the critical density, the expansion of the Universe described by Hubble's law proceeds with speeds that are nearly constant over time. (a) Prove that in this case the age of the Universe is given by the inverse of Hubble's constant. (b) Calculate $1/H$ and express it in years.

47. Assume that the average density of the Universe is equal to the critical density. (a) Prove that the age of the Universe is given by $2/3H$. (b) Calculate $2/3H$ and express it in years.

48. Hubble's law can be stated in vector form as $\mathbf{v} = H\mathbf{R}$: Outside the local group of galaxies, all objects are moving away from us with velocities proportional to their displacements from us. In this form, it sounds as if our location in the Universe is specially privileged. Prove that Hubble's law would be equally true for an observer elsewhere in the Universe. Proceed as follows. Assume that we are at the origin of coordinates, that one galaxy cluster is at location $\mathbf{R}_1$ and has velocity $\mathbf{v}_1 = H\mathbf{R}_1$ relative to us, and that another galaxy cluster has radius vector $\mathbf{R}_2$ and velocity $\mathbf{v}_2 = H\mathbf{R}_2$. Suppose the speeds are nonrelativistic. Consider the frame of reference of an observer in the first of these galaxy clusters. Show that our velocity relative to her, together with the displacement vector of our galaxy cluster from hers, satisfies Hubble's law. Show that the displacement and velocity of cluster 2 relative to cluster 1 satisfy Hubble's law.

Section 46.13   Problems and Perspectives

49. Classical general relativity views the structure of space–time as deterministic and well defined down to arbitrarily small distances. On the other hand, quantum general relativity forbids distances smaller than the Planck length given by $L = (\hbar G/c^3)^{1/2}$. (a) Calculate the value of the Planck length. The quantum limitation suggests that after the Big Bang, when all the presently observable section of the Universe was contained within a point-like singularity, nothing could be observed until that singularity grew larger than the Planck length. Because the size of the singularity grew at the speed of light, we can infer that no observations were possible during the time interval required for light to travel the Planck length. (b) Calculate this time interval, known as the Planck time T, and compare it with the ultrahot epoch mentioned in the text. (c) Does this suggest we may never know what happened between the time $t = 0$ and the time $t = T$?

Additional Problems

50. Review problem. Supernova Shelton 1987A, located about 170 000 ly from the Earth, is estimated to have emitted a burst of neutrinos carrying energy $\sim 10^{46}$ J (Fig. P46.50).

Anglo-Australian Telescope Board

Figure P46.50 (Problems 50 and 51) The giant star Sanduleak −69° 202 in the "before" picture became Supernova Shelton 1987A in the "after" picture.

Suppose the average neutrino energy was 6 MeV and your body presented cross-sectional area 5 000 cm². To an order of magnitude, how many of these neutrinos passed through you?

51. The most recent naked-eye supernova was Supernova Shelton 1987A (Fig. P46.50). It was 170 000 ly away in the next galaxy to ours, the Large Magellanic Cloud. About 3 h before its optical brightening was noticed, two continuously running neutrino detection experiments simultaneously registered the first neutrinos from an identified source other than the Sun. The Irvine–Michigan–Brookhaven experiment in a salt mine in Ohio registered 8 neutrinos over a 6-s period, and the Kamiokande II experiment in a zinc mine in Japan counted 11 neutrinos in 13 s. (Because the supernova is far south in the sky, these neutrinos entered the detectors from below. They passed through the Earth before they were by chance absorbed by nuclei in the detectors.) The neutrino energies were between about 8 MeV and 40 MeV. If neutrinos have no mass, then neutrinos of all energies should travel together at the speed of light—the data are consistent with this possibility. The arrival times could show scatter simply because neutrinos were created at different moments as the core of the star collapsed into a neutron star. If neutrinos have nonzero mass, then lower-energy neutrinos should move comparatively slowly. The data are consistent with a 10-MeV neutrino requiring at most about 10 s more than a photon would require to travel from the supernova to us. Find the upper limit that this observation sets on the mass of a neutrino. (Other evidence sets an even tighter limit.)

52. Name at least one conservation law that prevents each of the following reactions: (a) $\pi^- + p \rightarrow \Sigma^+ + \pi^0$ (b) $\mu^- \rightarrow \pi^- + \nu_e$ (c) $p \rightarrow \pi^+ + \pi^+ + \pi^-$

53. The energy flux carried by neutrinos from the Sun is estimated to be on the order of 0.4 W/m² at Earth's surface. Estimate the fractional mass loss of the Sun over 10^9 yr due to the emission of neutrinos. (The mass of the Sun is 2×10^{30} kg. The Earth–Sun distance is 1.5×10^{11} m.)

54. Two protons approach each other head-on, each with 70.4 MeV of kinetic energy, and engage in a reaction in which a proton and positive pion emerge at rest. What third particle, obviously uncharged and therefore difficult to detect, must have been created?

55. A rocket engine for space travel using photon drive and matter–antimatter annihilation has been suggested. Suppose the fuel for a short-duration burn consists of N protons and N antiprotons, each with mass m. (a) Assume all of the fuel is annihilated to produce photons. When the photons are ejected from the rocket, what momentum can be imparted to it? (b) **What If?** If half of the protons and antiprotons annihilate each other and the energy released is used to eject the remaining particles, what momentum could be given to the rocket? (c) Which scheme results in the greater change in speed for the rocket?

56. The nuclear force can be attributed to the exchange of an elementary particle between protons and neutrons when they are sufficiently close. Take the range of the nuclear force as approximately 1.4×10^{-15} m. (a) Use the uncertainty principle $\Delta E \Delta t \geq \hbar/2$ to estimate the mass of the elementary particle. Assume that it moves at nearly the speed of light. (b) Using Table 46.2, identify the particle.

57. Determine the kinetic energies of the proton and pion resulting from the decay of a Λ^0 at rest:

$$\Lambda^0 \longrightarrow p + \pi^-$$

58. A gamma-ray photon strikes a stationary electron. Determine the minimum gamma-ray energy to make this reaction occur:

$$\gamma + e^- \longrightarrow e^- + e^- + e^+$$

59. An unstable particle, initially at rest, decays into a proton (rest energy 938.3 MeV) and a negative pion (rest energy 139.6 MeV). A uniform magnetic field of 0.250 T exists perpendicular to the velocities of the created particles. The radius of curvature of each track is found to be 1.33 m. What is the mass of the original unstable particle?

60. A Σ^0 particle at rest decays according to

$$\Sigma^0 \longrightarrow \Lambda^0 + \gamma$$

Find the gamma-ray energy.

61. Two protons approach each other with velocities of equal magnitude in opposite directions. What is the minimum kinetic energy of each of the protons if they are to produce a π^+ meson at rest in the following reaction?

$$p + p \longrightarrow p + n + \pi^+$$

62. A π-meson at rest decays according to $\pi^- \rightarrow \mu^- + \bar{\nu}_\mu$. What is the energy carried off by the neutrino? (Assume that the neutrino has no mass and moves off with the speed of light. Take $m_\pi c^2 = 139.6$ MeV and $m_\mu c^2 = 105.7$ MeV.)

63. **Review problem.** Use the Boltzmann distribution function $e^{-E/k_B T}$ to calculate the temperature at which 1.00% of a population of photons will have energy greater than 1.00 eV. The energy required to excite an atom is on the order of 1 eV. Therefore as the temperature of the Universe fell below the value you calculate, neutral atoms could form from plasma, and the Universe became transparent. The cosmic background radiation represents our vastly red-shifted view of the opaque fireball of the Big Bang as it was at this time and temperature. The fireball surrounds us; we are embers.

64. What processes are described by the Feynman diagrams in Figure P46.64? What is the exchanged particle in each process?

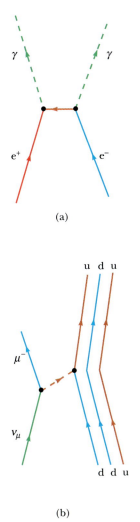

(a)

(b)

Figure P46.64

65. Identify the mediators for the two interactions described in the Feynman diagrams shown in Figure P46.65.

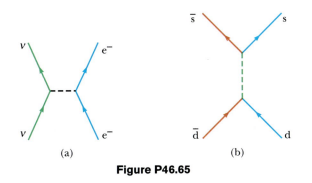

(a)   (b)

Figure P46.65

66. The cosmic rays of highest energy are mostly protons, accelerated by unknown sources. Their spectrum shows a cutoff at an energy on the order of 10^{20} eV. Above that energy, a proton will interact with a photon of cosmic

microwave background radiation to produce mesons, for example according to

$$p + \gamma \longrightarrow p + \pi^0$$

Demonstrate this fact by taking the following steps: (a) Find the minimum photon energy required to produce this reaction in the reference frame where the total momentum of the photon–proton system is zero. The reaction was observed experimentally in the 1950s with photons of a few hundred MeV. (b) Use Wien's displacement law to find the wavelength of a photon at the peak of the blackbody spectrum of the primordial microwave background radiation, with a temperature of 2.73 K. (c) Find the energy of this photon. (d) Consider the reaction in part (a) in a moving reference frame so that the photon is the same as that in part (c). Calculate the energy of the proton in this frame, which represents the Earth reference frame.

Answers to Quick Quizzes

46.1 (a). The right-hand rule for the positive particle tells you that this is the direction that leads to a force directed toward the center of curvature of the path.

46.2 (c), (d). There is a baryon, the neutron, on the left of the reaction, but no baryon on the right. Therefore, baryon number is not conserved. The neutron has spin $\frac{1}{2}$. On the right side of the reaction, the pions each have integral spin, and the combination of two muons must also have integral spin. Therefore, the total spin of the particles on the right-hand side must be integral and angular momentum is not conserved.

46.3 (a). The sum of the proton and pion masses is larger than the mass of the neutron, so energy conservation is violated.

46.4 (b), (e), (f). The pion on the left has integral spin while the three spin-$\frac{1}{2}$ leptons on the right must result in a total spin that is half-integral. Therefore, angular momentum is not conserved. Electron lepton number is zero on the left and -1 on the right. There are no muons on the left, but a muon and its neutrino on the right (both with $L_\mu = +1$). Therefore, muon lepton number is not conserved.

46.5 (b), (e). There is one spin-$\frac{1}{2}$ particle on the left and two on the right, so angular momentum is not conserved. There are no leptons on the left and an electron on the right, so electron lepton number is not conserved.

46.6 (b), (c), (e). Angular momentum is not conserved because there is one spin-$\frac{1}{2}$ particle before the decay and two spin-$\frac{1}{2}$ particles afterward. Electric charge is not conserved because the negative charge on the electron disappears. Electron lepton number is not conserved, because there is an electron with $L_e = 1$ before the decay and two neutrinos, each with $L_e = 1$, afterward.

46.7 The diagram would look this:

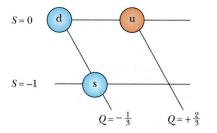

46.8 False. Because the charges on quarks are $+2e/3$ and $-e/3$, the maximum possible charge of a combination of a quark and an antiquark is $\pm e$.

46.9 (a) (i), (ii), (iii). The order of increasing quark mass is c, b, t. (b) (i) = (ii) = (iii). All three baryons are combinations of three spin-$\frac{1}{2}$ quarks. (c) (i) = (iii) > (ii). The total charge of ccc and ttt is $2e$, while the total charge of bbb is $-e$. (d) (i) = (ii) = (iii). All three particles have a baryon number of 1.

"Particles, particles, particles."

© 2003 Sidney Harris

The Meaning of Success

To earn the respect of intelligent people and to win the affection of children;
To appreciate the beauty in nature and all that surrounds us;
To seek out and nurture the best in others;
To give the gift of yourself to others without the slightest thought of return, for it is in giving that we receive;
To have accomplished a task, whether it be saving a lost soul, healing a sick child, writing a book, or risking your life for a friend;
To have celebrated and laughed with great joy and enthusiasm and sung with exultation;
To have hope even in times of despair, for as long as you have hope, you have life;
To love and be loved;
To be understood and to understand;
To know that even one life has breathed easier because you have lived;
This is the meaning of success.

—**Ralph Waldo Emerson**
Modified by Ray Serway, December 1989

Appendix A • Tables

Table A.1

Conversion Factors						

Length

	m	cm	km	in.	ft	mi
1 meter	1	10^2	10^{-3}	39.37	3.281	6.214×10^{-4}
1 centimeter	10^{-2}	1	10^{-5}	0.393 7	3.281×10^{-2}	6.214×10^{-6}
1 kilometer	10^3	10^5	1	3.937×10^4	3.281×10^3	0.621 4
1 inch	2.540×10^{-2}	2.540	2.540×10^{-5}	1	8.333×10^{-2}	1.578×10^{-5}
1 foot	0.304 8	30.48	3.048×10^{-4}	12	1	1.894×10^{-4}
1 mile	1 609	1.609×10^5	1.609	6.336×10^4	5 280	1

Mass

	kg	g	slug	u
1 kilogram	1	10^3	6.852×10^{-2}	6.024×10^{26}
1 gram	10^{-3}	1	6.852×10^{-5}	6.024×10^{23}
1 slug	14.59	1.459×10^4	1	8.789×10^{27}
1 atomic mass unit	1.660×10^{-27}	1.660×10^{-24}	1.137×10^{-28}	1

Note: 1 metric ton = 1 000 kg.

Time

	s	min	h	day	yr
1 second	1	1.667×10^{-2}	2.778×10^{-4}	1.157×10^{-5}	3.169×10^{-8}
1 minute	60	1	1.667×10^{-2}	6.994×10^{-4}	1.901×10^{-6}
1 hour	3 600	60	1	4.167×10^{-2}	1.141×10^{-4}
1 day	8.640×10^4	1 440	24	1	2.738×10^{-5}
1 year	3.156×10^7	5.259×10^5	8.766×10^3	365.2	1

Speed

	m/s	cm/s	ft/s	mi/h
1 meter per second	1	10^2	3.281	2.237
1 centimeter per second	10^{-2}	1	3.281×10^{-2}	2.237×10^{-2}
1 foot per second	0.304 8	30.48	1	0.681 8
1 mile per hour	0.447 0	44.70	1.467	1

Note: 1 mi/min = 60 mi/h = 88 ft/s.

Force

	N	lb
1 newton	1	0.224 8
1 pound	4.448	1

Table A.1

Conversion Factors *continued*

Work, Energy, Heat

	J	ft·lb	eV
1 joule	1	0.737 6	6.242×10^{18}
1 foot-pound	1.356	1	8.464×10^{18}
1 electron volt	1.602×10^{-19}	1.182×10^{-19}	1
1 calorie	4.186	3.087	2.613×10^{19}
1 British thermal unit	1.055×10^{3}	7.779×10^{2}	6.585×10^{21}
1 kilowatt hour	3.600×10^{6}	2.655×10^{6}	2.247×10^{25}

	cal	Btu	kWh
1 joule	0.238 9	9.481×10^{-4}	2.778×10^{-7}
1 foot-pound	0.323 9	1.285×10^{-3}	3.766×10^{-7}
1 electron volt	3.827×10^{-20}	1.519×10^{-22}	4.450×10^{-26}
1 calorie	1	3.968×10^{-3}	1.163×10^{-6}
1 British thermal unit	2.520×10^{2}	1	2.930×10^{-4}
1 kilowatt hour	8.601×10^{5}	3.413×10^{2}	1

Pressure

	Pa	atm	
1 pascal	1	9.869×10^{-6}	
1 atmosphere	1.013×10^{5}	1	
1 centimeter mercury[a]	1.333×10^{3}	1.316×10^{-2}	
1 pound per square inch	6.895×10^{3}	6.805×10^{-2}	
1 pound per square foot	47.88	4.725×10^{-4}	

	cm Hg	lb/in.^{2}	lb/ft^2
1 pascal	7.501×10^{-4}	1.450×10^{-4}	2.089×10^{-2}
1 atmosphere	76	14.70	2.116×10^{3}
1 centimeter mercury[a]	1	0.194 3	27.85
1 pound per square inch	5.171	1	144
1 pound per square foot	3.591×10^{-2}	6.944×10^{-3}	1

[a] At 0°C and at a location where the free-fall acceleration has its "standard" value, 9.806 65 m/s^2.

Table A.2

Symbols, Dimensions, and Units of Physical Quantities

Quantity	Common Symbol	Unit[a]	Dimensions[b]	Unit in Terms of Base SI Units
Acceleration	$\mathbf{a}$	m/s^2	L/T^2	m/s^2
Amount of substance	n	MOLE		mol
Angle	θ, ϕ	radian (rad)	1	
Angular acceleration	α	rad/s^2	T^{-2}	s^{-2}
Angular frequency	ω	rad/s	T^{-1}	s^{-1}
Angular momentum	$\mathbf{L}$	kg·m^2/s	ML^2/T	kg·m^2/s
Angular velocity	ω	rad/s	T^{-1}	s^{-1}
Area	A	m^2	L^2	m^2
Atomic number	Z			

continued

Table A.2

Symbols, Dimensions, and Units of Physical Quantities *continued*

Quantity	Common Symbol	Unit[a]	Dimensions[b]	Unit in Terms of Base SI Units
Capacitance	C	farad (F)	Q^2T^2/ML^2	$A^2 \cdot s^4/kg \cdot m^2$
Charge	q, Q, e	coulomb (C)	Q	$A \cdot s$
Charge density				
Line	λ	C/m	Q/L	$A \cdot s/m$
Surface	σ	C/m^2	Q/L^2	$A \cdot s/m^2$
Volume	ρ	C/m^3	Q/L^3	$A \cdot s/m^3$
Conductivity	σ	$1/\Omega \cdot m$	Q^2T/ML^3	$A^2 \cdot s^3/kg \cdot m^3$
Current	I	AMPERE	Q/T	A
Current density	$\mathbf{J}$	A/m^2	Q/T^2	A/m^2
Density	ρ	kg/m^3	M/L^3	kg/m^3
Dielectric constant	κ			
Length	ℓ, L	METER	L	m
Position	$x, y, z, \mathbf{r}$			
Displacement	$\Delta x, \Delta \mathbf{r}$			
Distance	d, h			
Electric dipole moment	$\mathbf{p}$	$C \cdot m$	QL	$A \cdot s \cdot m$
Electric field	$\mathbf{E}$	V/m	ML/QT^2	$kg \cdot m/A \cdot s^3$
Electric flux	Φ_E	$V \cdot m$	ML^3/QT^2	$kg \cdot m^3/A \cdot s^3$
Electromotive force	$\mathcal{E}$	volt (V)	ML^2/QT^2	$kg \cdot m^2/A \cdot s^3$
Energy	E, U, K	joule (J)	ML^2/T^2	$kg \cdot m^2/s^2$
Entropy	S	J/K	$ML^2/T^2 \cdot K$	$kg \cdot m^2/s^2 \cdot K$
Force	$\mathbf{F}$	newton (N)	ML/T^2	$kg \cdot m/s^2$
Frequency	f	hertz (Hz)	T^{-1}	s^{-1}
Heat	Q	joule (J)	ML^2/T^2	$kg \cdot m^2/s^2$
Inductance	L	henry (H)	ML^2/Q^2	$kg \cdot m^2/A^2 \cdot s^2$
Magnetic dipole moment	$\boldsymbol{\mu}$	$N \cdot m/T$	QL^2/T	$A \cdot m^2$
Magnetic field	$\mathbf{B}$	tesla (T) $(= Wb/m^2)$	M/QT	$kg/A \cdot s^2$
Magnetic flux	Φ_B	weber (Wb)	ML^2/QT	$kg \cdot m^2/A \cdot s^2$
Mass	m, M	KILOGRAM	M	kg
Molar specific heat	C	$J/mol \cdot K$		$kg \cdot m^2/s^2 \cdot mol \cdot K$
Moment of inertia	I	$kg \cdot m^2$	ML^2	$kg \cdot m^2$
Momentum	$\mathbf{p}$	$kg \cdot m/s$	ML/T	$kg \cdot m/s$
Period	T	s	T	s
Permeability of free space	μ_0	$N/A^2 \ (= H/m)$	ML/Q^2T	$kg \cdot m/A^2 \cdot s^2$
Permittivity of free space	ϵ_0	$C^2/N \cdot m^2 \ (= F/m)$	Q^2T^2/ML^3	$A^2 \cdot s^4/kg \cdot m^3$
Potential	V	volt (V) $(= J/C)$	ML^2/QT^2	$kg \cdot m^2/A \cdot s^3$
Power	$\mathcal{P}$	watt (W) $(= J/s)$	ML^2/T^3	$kg \cdot m^2/s^3$
Pressure	P	pascal (Pa) $(= N/m^2)$	M/LT^2	$kg/m \cdot s^2$
Resistance	R	ohm $(\Omega) \ (= V/A)$	ML^2/Q^2T	$kg \cdot m^2/A^2 \cdot s^3$
Specific heat	c	$J/kg \cdot K$	$L^2/T^2 \cdot K$	$m^2/s^2 \cdot K$
Speed	v	m/s	L/T	m/s
Temperature	T	KELVIN	K	K
Time	t	SECOND	T	s
Torque	τ	$N \cdot m$	ML^2/T^2	$kg \cdot m^2/s^2$
Velocity	$\mathbf{v}$	m/s	L/T	m/s
Volume	V	m^3	L^3	m^3
Wavelength	λ	m	L	m
Work	W	joule (J) $(= N \cdot m)$	ML^2/T^2	$kg \cdot m^2/s^2$

[a] The base SI units are given in uppercase letters.

[b] The symbols M, L, T, and Q denote mass, length, time, and charge, respectively.

Table A.3

Table of Atomic Masses[a]						
Atomic Number Z	Element	Symbol	Chemical Atomic Mass (u)	Mass Number (*Indicates Radioactive) A	Atomic Mass (u)	Percent Abundance	Half-Life (If Radioactive) $T_{1/2}$
0	(Neutron)	n		1*	1.008 665		10.4 min
1	Hydrogen	H	1.007 94	1	1.007 825	99.988 5	
	Deuterium	D		2	2.014 102	0.011 5	
	Tritium	T		3*	3.016 049		12.33 yr
2	Helium	He	4.002 602	3	3.016 029	0.000 137	
				4	4.002 603	99.999 863	
				6*	6.018 888		0.81 s
3	Lithium	Li	6.941	6	6.015 122	7.5	
				7	7.016 004	92.5	
				8*	8.022 487		0.84 s
4	Beryllium	Be	9.012 182	7*	7.016 929		53.3 days
				9	9.012 182	100	
				10*	10.013 534		1.5×10^6 yr
5	Boron	B	10.811	10	10.012 937	19.9	
				11	11.009 306	80.1	
				12*	12.014 352		0.020 2 s
6	Carbon	C	12.010 7	10*	10.016 853		19.3 s
				11*	11.011 434		20.4 min
				12	12.000 000	98.93	
				13	13.003 355	1.07	
				14*	14.003 242		5 730 yr
				15*	15.010 599		2.45 s
7	Nitrogen	N	14.006 7	12*	12.018 613		0.011 0 s
				13*	13.005 739		9.96 min
				14	14.003 074	99.632	
				15	15.000 109	0.368	
				16*	16.006 101		7.13 s
				17*	17.008 450		4.17 s
8	Oxygen	O	15.999 4	14*	14.008 595		70.6 s
				15*	15.003 065		122 s
				16	15.994 915	99.757	
				17	16.999 132	0.038	
				18	17.999 160	0.205	
				19*	19.003 579		26.9 s
9	Fluorine	F	18.998 403 2	17*	17.002 095		64.5 s
				18*	18.000 938		109.8 min
				19	18.998 403	100	
				20*	19.999 981		11.0 s
				21*	20.999 949		4.2 s
10	Neon	Ne	20.179 7	18*	18.005 697		1.67 s
				19*	19.001 880		17.2 s
				20	19.992 440	90.48	
				21	20.993 847	0.27	
				22	21.991 385	9.25	
				23*	22.994 467		37.2 s
11	Sodium	Na	22.989 77	21*	20.997 655		22.5 s
				22*	21.994 437		2.61 yr

continued

Table A.3

Table of Atomic Masses[a] *continued*

Atomic Number Z	Element	Symbol	Chemical Atomic Mass (u)	Mass Number (*Indicates Radioactive) A	Atomic Mass (u)	Percent Abundance	Half-Life (If Radioactive) $T_{1/2}$
(11)	Sodium			23	22.989 770	100	
				24*	23.990 963		14.96 h
12	Magnesium	Mg	24.305 0	23*	22.994 125		11.3 s
				24	23.985 042	78.99	
				25	24.985 837	10.00	
				26	25.982 593	11.01	
				27*	26.984 341		9.46 min
13	Aluminum	Al	26.981 538	26*	25.986 892		7.4×10^5 yr
				27	26.981 539	100	
				28*	27.981 910		2.24 min
14	Silicon	Si	28.085 5	28	27.976 926	92.229 7	
				29	28.976 495	4.683 2	
				30	29.973 770	3.087 2	
				31*	30.975 363		2.62 h
				32*	31.974 148		172 yr
15	Phosphorus	P	30.973 761	30*	29.978 314		2.50 min
				31	30.973 762	100	
				32*	31.973 907		14.26 days
				33*	32.971 725		25.3 days
16	Sulfur	S	32.066	32	31.972 071	94.93	
				33	32.971 458	0.76	
				34	33.967 869	4.29	
				35*	34.969 032		87.5 days
				36	35.967 081	0.02	
17	Chlorine	Cl	35.452 7	35	34.968 853	75.78	
				36*	35.968 307		3.0×10^5 yr
				37	36.965 903	24.22	
18	Argon	Ar	39.948	36	35.967 546	0.336 5	
				37*	36.966 776		35.04 days
				38	37.962 732	0.063 2	
				39*	38.964 313		269 yr
				40	39.962 383	99.600 3	
				42*	41.963 046		33 yr
19	Potassium	K	39.098 3	39	38.963 707	93.258 1	
				40*	39.963 999	0.011 7	1.28×10^9 yr
				41	40.961 826	6.730 2	
20	Calcium	Ca	40.078	40	39.962 591	96.941	
				41*	40.962 278		1.0×10^5 yr
				42	41.958 618	0.647	
				43	42.958 767	0.135	
				44	43.955 481	2.086	
				46	45.953 693	0.004	
				48	47.952 534	0.187	
21	Scandium	Sc	44.955 910	41*	40.969 251		0.596 s
				45	44.955 910	100	
22	Titanium	Ti	47.867	44*	43.959 690		49 yr
				46	45.952 630	8.25	

continued

Table A.3

Table of Atomic Masses[a] *continued*

Atomic Number Z	Element	Symbol	Chemical Atomic Mass (u)	Mass Number (*Indicates Radioactive) A	Atomic Mass (u)	Percent Abundance	Half-Life (If Radioactive) $T_{1/2}$
(22)	Titanium			47	46.951 764	7.44	
				48	47.947 947	73.72	
				49	48.947 871	5.41	
				50	49.944 792	5.18	
23	Vanadium	V	50.941 5	48*	47.952 254		15.97 days
				50*	49.947 163	0.250	1.5×10^{17} yr
				51	50.943 964	99.750	
24	Chromium	Cr	51.996 1	48*	47.954 036		21.6 h
				50	49.946 050	4.345	
				52	51.940 512	83.789	
				53	52.940 654	9.501	
				54	53.938 885	2.365	
25	Manganese	Mn	54.938 049	54*	53.940 363		312.1 days
				55	54.938 050	100	
26	Iron	Fe	55.845	54	53.939 615	5.845	
				55*	54.938 298		2.7 yr
				56	55.934 942	91.754	
				57	56.935 399	2.119	
				58	57.933 280	0.282	
				60*	59.934 077		1.5×10^6 yr
27	Cobalt	Co	58.933 200	59	58.933 200	100	
				60*	59.933 822		5.27 yr
28	Nickel	Ni	58.693 4	58	57.935 348	68.076 9	
				59*	58.934 351		7.5×10^4 yr
				60	59.930 790	26.223 1	
				61	60.931 060	1.139 9	
				62	61.928 349	3.634 5	
				63*	62.929 673		100 yr
				64	63.927 970	0.925 6	
29	Copper	Cu	63.546	63	62.929 601	69.17	
				65	64.927 794	30.83	
30	Zinc	Zn	65.39	64	63.929 147	48.63	
				66	65.926 037	27.90	
				67	66.927 131	4.10	
				68	67.924 848	18.75	
				70	69.925 325	0.62	
31	Gallium	Ga	69.723	69	68.925 581	60.108	
				71	70.924 705	39.892	
32	Germanium	Ge	72.61	70	69.924 250	20.84	
				72	71.922 076	27.54	
				73	72.923 459	7.73	
				74	73.921 178	36.28	
				76	75.921 403	7.61	
33	Arsenic	As	74.921 60	75	74.921 596	100	
34	Selenium	Se	78.96	74	73.922 477	0.89	
				76	75.919 214	9.37	
				77	76.919 915	7.63	

continued

Table A.3

Table of Atomic Masses[a] *continued*						
Atomic Number Z	Element	Symbol	Chemical Atomic Mass (u)	Mass Number (*Indicates Radioactive) A	Atomic Mass (u)	Percent Abundance	Half-Life (If Radioactive) $T_{1/2}$
(34)	Selenium			78	77.917 310	23.77	
				79*	78.918 500		$\leq 6.5 \times 10^4$ yr
				80	79.916 522	49.61	
				82*	81.916 700	8.73	1.4×10^{20} yr
35	Bromine	Br	79.904	79	78.918 338	50.69	
				81	80.916 291	49.31	
36	Krypton	Kr	83.80	78	77.920 386	0.35	
				80	79.916 378	2.28	
				81*	80.916 592		2.1×10^5 yr
				82	81.913 485	11.58	
				83	82.914 136	11.49	
				84	83.911 507	57.00	
				85*	84.912 527		10.76 yr
				86	85.910 610	17.30	
37	Rubidium	Rb	85.467 8	85	84.911 789	72.17	
				87*	86.909 184	27.83	4.75×10^{10} yr
38	Strontium	Sr	87.62	84	83.913 425	0.56	
				86	85.909 262	9.86	
				87	86.908 880	7.00	
				88	87.905 614	82.58	
				90*	89.907 738		29.1 yr
39	Yttrium	Y	88.905 85	89	88.905 848	100	
40	Zirconium	Zr	91.224	90	89.904 704	51.45	
				91	90.905 645	11.22	
				92	91.905 040	17.15	
				93*	92.906 476		1.5×10^6 yr
				94	93.906 316	17.38	
				96	95.908 276	2.80	
41	Niobium	Nb	92.906 38	91*	90.906 990		6.8×10^2 yr
				92*	91.907 193		3.5×10^7 yr
				93	92.906 378	100	
				94*	93.907 284		2×10^4 yr
42	Molybdenum	Mo	95.94	92	91.906 810	14.84	
				93*	92.906 812		3.5×10^3 yr
				94	93.905 088	9.25	
				95	94.905 842	15.92	
				96	95.904 679	16.68	
				97	96.906 021	9.55	
				98	97.905 408	24.13	
				100	99.907 477	9.63	
43	Technetium	Tc		97*	96.906 365		2.6×10^6 yr
				98*	97.907 216		4.2×10^6 yr
				99*	98.906 255		2.1×10^5 yr
44	Ruthenium	Ru	101.07	96	95.907 598	5.54	
				98	97.905 287	1.87	
				99	98.905 939	12.76	
				100	99.904 220	12.60	

continued

Table A.3

Table of Atomic Masses[a] *continued*						
Atomic Number Z	Element	Symbol	Chemical Atomic Mass (u)	Mass Number (*Indicates Radioactive) A	Atomic Mass (u)	Percent Abundance	Half-Life (If Radioactive) $T_{1/2}$
(44)	Ruthenium			101	100.905 582	17.06	
				102	101.904 350	31.55	
				104	103.905 430	18.62	
45	Rhodium	Rh	102.905 50	103	102.905 504	100	
46	Palladium	Pd	106.42	102	101.905 608	1.02	
				104	103.904 035	11.14	
				105	104.905 084	22.33	
				106	105.903 483	27.33	
				107*	106.905 128		6.5×10^6 yr
				108	107.903 894	26.46	
				110	109.905 152	11.72	
47	Silver	Ag	107.868 2	107	106.905 093	51.839	
				109	108.904 756	48.161	
48	Cadmium	Cd	112.411	106	105.906 458	1.25	
				108	107.904 183	0.89	
				109*	108.904 986		462 days
				110	109.903 006	12.49	
				111	110.904 182	12.80	
				112	111.902 757	24.13	
				113*	112.904 401	12.22	9.3×10^{15} yr
				114	113.903 358	28.73	
				116	115.904 755	7.49	
49	Indium	In	114.818	113	112.904 061	4.29	
				115*	114.903 878	95.71	4.4×10^{14} yr
50	Tin	Sn	118.710	112	111.904 821	0.97	
				114	113.902 782	0.66	
				115	114.903 346	0.34	
				116	115.901 744	14.54	
				117	116.902 954	7.68	
				118	117.901 606	24.22	
				119	118.903 309	8.59	
				120	119.902 197	32.58	
				121*	120.904 237		55 yr
				122	121.903 440	4.63	
				124	123.905 275	5.79	
51	Antimony	Sb	121.760	121	120.903 818	57.21	
				123	122.904 216	42.79	
				125*	124.905 248		2.7 yr
52	Tellurium	Te	127.60	120	119.904 020	0.09	
				122	121.903 047	2.55	
				123*	122.904 273	0.89	1.3×10^{13} yr
				124	123.902 820	4.74	
				125	124.904 425	7.07	
				126	125.903 306	18.84	
				128*	127.904 461	31.74	$> 8 \times 10^{24}$ yr
				130*	129.906 223	34.08	$\leq 1.25 \times 10^{21}$ yr

continued

Table A.3

Table of Atomic Masses[a] *continued*

Atomic Number Z	Element	Symbol	Chemical Atomic Mass (u)	Mass Number (*Indicates Radioactive) A	Atomic Mass (u)	Percent Abundance	Half-Life (If Radioactive) $T_{1/2}$
53	Iodine	I	126.904 47	127	126.904 468	100	
				129*	128.904 988		1.6×10^7 yr
54	Xenon	Xe	131.29	124	123.905 896	0.09	
				126	125.904 269	0.09	
				128	127.903 530	1.92	
				129	128.904 780	26.44	
				130	129.903 508	4.08	
				131	130.905 082	21.18	
				132	131.904 145	26.89	
				134	133.905 394	10.44	
				136*	135.907 220	8.87	$\geq 2.36 \times 10^{21}$ yr
55	Cesium	Cs	132.905 45	133	132.905 447	100	
				134*	133.906 713		2.1 yr
				135*	134.905 972		2×10^6 yr
				137*	136.907 074		30 yr
56	Barium	Ba	137.327	130	129.906 310	0.106	
				132	131.905 056	0.101	
				133*	132.906 002		10.5 yr
				134	133.904 503	2.417	
				135	134.905 683	6.592	
				136	135.904 570	7.854	
				137	136.905 821	11.232	
				138	137.905 241	71.698	
57	Lanthanum	La	138.905 5	137*	136.906 466		6×10^4 yr
				138*	137.907 107	0.090	1.05×10^{11} yr
				139	138.906 349	99.910	
58	Cerium	Ce	140.116	136	135.907 144	0.185	
				138	137.905 986	0.251	
				140	139.905 434	88.450	
				142*	141.909 240	11.114	$>5 \times 10^{16}$ yr
59	Praseodymium	Pr	140.907 65	141	140.907 648	100	
60	Neodymium	Nd	144.24	142	141.907 719	27.2	
				143	142.909 810	12.2	
				144*	143.910 083	23.8	2.3×10^{15} yr
				145	144.912 569	8.3	
				146	145.913 112	17.2	
				148	147.916 888	5.7	
				150*	149.920 887	5.6	$>1 \times 10^{18}$ yr
61	Promethium	Pm		143*	142.910 928		265 days
				145*	144.912 744		17.7 yr
				146*	145.914 692		5.5 yr
				147*	146.915 134		2.623 yr
62	Samarium	Sm	150.36	144	143.911 995	3.07	
				146*	145.913 037		1.0×10^8 yr
				147*	146.914 893	14.99	1.06×10^{11} yr
				148*	147.914 818	11.24	7×10^{15} yr

continued

Table A.3

Table of Atomic Masses[a] *continued*

Atomic Number Z	Element	Symbol	Chemical Atomic Mass (u)	Mass Number (*Indicates Radioactive) A	Atomic Mass (u)	Percent Abundance	Half-Life (If Radioactive) $T_{1/2}$
(62)	Samarium			149*	148.917 180	13.82	$> 2 \times 10^{15}$ yr
				150	149.917 272	7.38	
				151*	150.919 928		90 yr
				152	151.919 728	26.75	
				154	153.922 205	22.75	
63	Europium	Eu	151.964	151	150.919 846	47.81	
				152*	151.921 740		13.5 yr
				153	152.921 226	52.19	
				154*	153.922 975		8.59 yr
				155*	154.922 889		4.7 yr
64	Gadolinium	Gd	157.25	148*	147.918 110		75 yr
				150*	149.918 656		1.8×10^{6} yr
				152*	151.919 788	0.20	1.1×10^{14} yr
				154	153.920 862	2.18	
				155	154.922 619	14.80	
				156	155.922 120	20.47	
				157	156.923 957	15.65	
				158	157.924 100	24.84	
				160	159.927 051	21.86	
65	Terbium	Tb	158.925 34	159	158.925 343	100	
66	Dysprosium	Dy	162.50	156	155.924 278	0.06	
				158	157.924 405	0.10	
				160	159.925 194	2.34	
				161	160.926 930	18.91	
				162	161.926 795	25.51	
				163	162.928 728	24.90	
				164	163.929 171	28.18	
67	Holmium	Ho	164.930 32	165	164.930 320	100	
				166*	165.932 281		1.2×10^{3} yr
68	Erbium	Er	167.6	162	161.928 775	0.14	
				164	163.929 197	1.61	
				166	165.930 290	33.61	
				167	166.932 045	22.93	
				168	167.932 368	26.78	
				170	169.935 460	14.93	
69	Thulium	Tm	168.934 21	169	168.934 211	100	
				171*	170.936 426		1.92 yr
70	Ytterbium	Yb	173.04	168	167.933 894	0.13	
				170	169.934 759	3.04	
				171	170.936 322	14.28	
				172	171.936 378	21.83	
				173	172.938 207	16.13	
				174	173.938 858	31.83	
				176	175.942 568	12.76	
71	Lutecium	Lu	174.967	173*	172.938 927		1.37 yr
				175	174.940 768	97.41	
				176*	175.942 682	2.59	3.78×10^{10} yr

continued

Table A.3

Table of Atomic Masses[a] *continued*

Atomic Number Z	Element	Symbol	Chemical Atomic Mass (u)	Mass Number (*Indicates Radioactive) A	Atomic Mass (u)	Percent Abundance	Half-Life (If Radioactive) $T_{1/2}$
72	Hafnium	Hf	178.49	174*	173.940 040	0.16	2.0×10^{15} yr
				176	175.941 402	5.26	
				177	176.943 220	18.60	
				178	177.943 698	27.28	
				179	178.945 815	13.62	
				180	179.946 549	35.08	
73	Tantalum	Ta	180.947 9	180*	179.947 466	0.012	8.152 h
				181	180.947 996	99.988	
74	Tungsten (Wolfram)	W	183.84	180	179.946 706	0.12	
				182	181.948 206	26.50	
				183	182.950 224	14.31	
				184*	183.950 933	30.64	$>3 \times 10^{17}$ yr
				186	185.954 362	28.43	
75	Rhenium	Re	186.207	185	184.952 956	37.40	
				187*	186.955 751	62.60	4.4×10^{10} yr
76	Osmium	Os	190.23	184	183.952 491	0.02	
				186*	185.953 838	1.59	2.0×10^{15} yr
				187	186.955 748	1.96	
				188	187.955 836	13.24	
				189	188.958 145	16.15	
				190	189.958 445	26.26	
				192	191.961 479	40.78	
				194*	193.965 179		6.0 yr
77	Iridium	Ir	192.217	191	190.960 591	37.3	
				193	192.962 924	62.7	
78	Platinum	Pt	195.078	190*	189.959 930	0.014	6.5×10^{11} yr
				192	191.961 035	0.782	
				194	193.962 664	32.967	
				195	194.964 774	33.832	
				196	195.964 935	25.242	
				198	197.967 876	7.163	
79	Gold	Au	196.966 55	197	196.966 552	100	
80	Mercury	Hg	200.59	196	195.965 815	0.15	
				198	197.966 752	9.97	
				199	198.968 262	16.87	
				200	199.968 309	23.10	
				201	200.970 285	13.18	
				202	201.970 626	29.86	
				204	203.973 476	6.87	
81	Thallium	Tl	204.383 3	203	202.972 329	29.524	
				204*	203.973 849		3.78 yr
				205	204.974 412	70.476	
		(Ra E″)		206*	205.976 095		4.2 min
		(Ac C″)		207*	206.977 408		4.77 min
		(Th C″)		208*	207.982 005		3.053 min
		(Ra C″)		210*	209.990 066		1.30 min

continued

Table A.3

Table of Atomic Masses[a] *continued*

Atomic Number Z	Element	Symbol	Chemical Atomic Mass (u)	Mass Number (*Indicates Radioactive) A	Atomic Mass (u)	Percent Abundance	Half-Life (If Radioactive) $T_{1/2}$
82	Lead	Pb	207.2	202*	201.972 144		5×10^4 yr
				204*	203.973 029	1.4	$\geq 1.4 \times 10^{17}$ yr
				205*	204.974 467		1.5×10^7 yr
				206	205.974 449	24.1	
				207	206.975 881	22.1	
				208	207.976 636	52.4	
		(Ra D)		210*	209.984 173		22.3 yr
		(Ac B)		211*	210.988 732		36.1 min
		(Th B)		212*	211.991 888		10.64 h
		(Ra B)		214*	213.999 798		26.8 min
83	Bismuth	Bi	208.980 38	207*	206.978 455		32.2 yr
				208*	207.979 727		3.7×10^5 yr
				209	208.980 383	100	
		(Ra E)		210*	209.984 105		5.01 days
		(Th C)		211*	210.987 258		2.14 min
				212*	211.991 272		60.6 min
		(Ra C)		214*	213.998 699		19.9 min
				215*	215.001 832		7.4 min
84	Polonium	Po		209*	208.982 416		102 yr
		(Ra F)		210*	209.982 857		138.38 days
		(Ac C′)		211*	210.986 637		0.52 s
		(Th C′)		212*	211.988 852		0.30 μs
		(Ra C′)		214*	213.995 186		164 μs
		(Ac A)		215*	214.999 415		0.001 8 s
		(Th A)		216*	216.001 905		0.145 s
		(Ra A)		218*	218.008 966		3.10 min
85	Astatine	At		215*	214.998 641		≈ 100 μs
				218*	218.008 682		1.6 s
				219*	219.011 297		0.9 min
86	Radon	Rn					
		(An)		219*	219.009 475		3.96 s
		(Tn)		220*	220.011 384		55.6 s
		(Rn)		222*	222.017 570		3.823 days
87	Francium	Fr					
		(Ac K)		223*	223.019 731		22 min
88	Radium	Ra					
		(Ac X)		223*	223.018 497		11.43 days
		(Th X)		224*	224.020 202		3.66 days
		(Ra)		226*	226.025 403		1 600 yr
		(Ms Th$_1$)		228*	228.031 064		5.75 yr
89	Actinium	Ac		227*	227.027 747		21.77 yr
		(Ms Th$_2$)		228*	228.031 015		6.15 h
90	Thorium	Th	232.038 1				
		(Rd Ac)		227*	227.027 699		18.72 days
		(Rd Th)		228*	228.028 731		1.913 yr
				229*	229.031 755		7 300 yr
		(Io)		230*	230.033 127		75.000 yr

continued

Table A.3

Table of Atomic Masses[a] *continued*							
Atomic Number Z	Element	Symbol	Chemical Atomic Mass (u)	Mass Number (*Indicates Radioactive) A	Atomic Mass (u)	Percent Abundance	Half-Life (If Radioactive) $T_{1/2}$
(90)	Thorium	(UY)		231*	231.036 297		25.52 h
		(Th)		232*	232.038 050	100	1.40×10^{10} yr
		(UX$_1$)		234*	234.043 596		24.1 days
91	Protactinium	Pa	231.035 88	231*	231.035 879		32.760 yr
		(Uz)		234*	234.043 302		6.7 h
92	Uranium	U	238.028 9	232*	232.037 146		69 yr
				233*	233.039 628		1.59×10^5 yr
				234*	234.040 946	0.005 5	2.45×10^5 yr
		(Ac U)		235*	235.043 923	0.720 0	7.04×10^8 yr
				236*	236.045 562		2.34×10^7 yr
		(UI)		238*	238.050 783	99.274 5	4.47×10^9 yr
93	Neptunium	Np		235*	235.044 056		396 days
				236*	236.046 560		1.15×10^5 yr
				237*	237.048 167		2.14×10^6 yr
94	Plutonium	Pu		236*	236.046 048		2.87 yr
				238*	238.049 553		87.7 yr
				239*	239.052 156		2.412×10^4 yr
				240*	240.053 808		6 560 yr
				241*	241.056 845		14.4 yr
				242*	242.058 737		3.73×10^6 yr
				244*	244.064 198		8.1×10^7 yr

[a]   Chemical atomic masses are from T. B. Coplen, "Atomic Weights of the Elements 1999," a technical report to the International Union of Pure and Applied Chemistry, and published in *Pure and Applied Chemistry*, 73(4), 667–683, 2001. Atomic masses of the isotopes are from G. Audi and A. H. Wapstra, "The 1995 Update to the Atomic Mass Evaluation," *Nuclear Physics*, A595, vol. 4, 409–480, December 25, 1995. Percent abundance values are from K. J. R. Rosman and P. D. P. Taylor, "Isotopic Compositions of the Elements 1999", a technical report to the International Union of Pure and Applied Chemistry, and published in *Pure and Applied Chemistry*, 70(1), 217–236, 1998.

Appendix B • Mathematics Review

These appendices in mathematics are intended as a brief review of operations and methods. Early in this course, you should be totally familiar with basic algebraic techniques, analytic geometry, and trigonometry. The appendices on differential and integral calculus are more detailed and are intended for those students who have difficulty applying calculus concepts to physical situations.

B.1 Scientific Notation

Many quantities that scientists deal with often have very large or very small values. For example, the speed of light is about 300 000 000 m/s, and the ink required to make the dot over an i in this textbook has a mass of about 0.000 000 001 kg. Obviously, it is very cumbersome to read, write, and keep track of numbers such as these. We avoid this problem by using a method dealing with powers of the number 10:

$$10^0 = 1$$

$$10^1 = 10$$

$$10^2 = 10 \times 10 = 100$$

$$10^3 = 10 \times 10 \times 10 = 1000$$

$$10^4 = 10 \times 10 \times 10 \times 10 = 10\ 000$$

$$10^5 = 10 \times 10 \times 10 \times 10 \times 10 = 100\ 000$$

and so on. The number of zeros corresponds to the power to which 10 is raised, called the **exponent** of 10. For example, the speed of light, 300 000 000 m/s, can be expressed as 3×10^8 m/s.

In this method, some representative numbers smaller than unity are

$$10^{-1} = \frac{1}{10} = 0.1$$

$$10^{-2} = \frac{1}{10 \times 10} = 0.01$$

$$10^{-3} = \frac{1}{10 \times 10 \times 10} = 0.001$$

$$10^{-4} = \frac{1}{10 \times 10 \times 10 \times 10} = 0.000\ 1$$

$$10^{-5} = \frac{1}{10 \times 10 \times 10 \times 10 \times 10} = 0.000\ 01$$

In these cases, the number of places the decimal point is to the left of the digit 1 equals the value of the (negative) exponent. Numbers expressed as some power of 10 multiplied by another number between 1 and 10 are said to be in **scientific notation.** For example, the scientific notation for 5 943 000 000 is 5.943×10^9 and that for 0.000 083 2 is 8.32×10^{-5}.

When numbers expressed in scientific notation are being multiplied, the following general rule is very useful:

$$10^n \times 10^m = 10^{n+m} \tag{B.1}$$

where n and m can be *any* numbers (not necessarily integers). For example, $10^2 \times 10^5 = 10^7$. The rule also applies if one of the exponents is negative: $10^3 \times 10^{-8} = 10^{-5}$.

When dividing numbers expressed in scientific notation, note that

$$\frac{10^n}{10^m} = 10^n \times 10^{-m} = 10^{n-m} \qquad\text{(B.2)}$$

Exercises

With help from the above rules, verify the answers to the following:

1. $86\,400 = 8.64 \times 10^4$
2. $9\,816\,762.5 = 9.816\,762\,5 \times 10^6$
3. $0.000\,000\,039\,8 = 3.98 \times 10^{-8}$
4. $(4 \times 10^8)(9 \times 10^9) = 3.6 \times 10^{18}$
5. $(3 \times 10^7)(6 \times 10^{-12}) = 1.8 \times 10^{-4}$
6. $\dfrac{75 \times 10^{-11}}{5 \times 10^{-3}} = 1.5 \times 10^{-7}$
7. $\dfrac{(3 \times 10^6)(8 \times 10^{-2})}{(2 \times 10^{17})(6 \times 10^5)} = 2 \times 10^{-18}$

B.2 Algebra

Some Basic Rules

When algebraic operations are performed, the laws of arithmetic apply. Symbols such as x, y, and z are usually used to represent quantities that are not specified, what are called the **unknowns.**

First, consider the equation

$$8x = 32$$

If we wish to solve for x, we can divide (or multiply) each side of the equation by the same factor without destroying the equality. In this case, if we divide both sides by 8, we have

$$\frac{8x}{8} = \frac{32}{8}$$

$$x = 4$$

Next consider the equation

$$x + 2 = 8$$

In this type of expression, we can add or subtract the same quantity from each side. If we subtract 2 from each side, we obtain

$$x + 2 - 2 = 8 - 2$$

$$x = 6$$

In general, if $x + a = b$, then $x = b - a$.

Now consider the equation

$$\frac{x}{5} = 9$$

If we multiply each side by 5, we are left with x on the left by itself and 45 on the right:

$$\left(\frac{x}{5}\right)(5) = 9 \times 5$$

$$x = 45$$

In all cases, *whatever operation is performed on the left side of the equality must also be performed on the right side.*

The following rules for multiplying, dividing, adding, and subtracting fractions should be recalled, where a, b, and c are three numbers:

	Rule	Example
Multiplying	$\left(\dfrac{a}{b}\right)\left(\dfrac{c}{d}\right) = \dfrac{ac}{bd}$	$\left(\dfrac{2}{3}\right)\left(\dfrac{4}{5}\right) = \dfrac{8}{15}$
Dividing	$\dfrac{(a/b)}{(c/d)} = \dfrac{ad}{bc}$	$\dfrac{2/3}{4/5} = \dfrac{(2)(5)}{(4)(3)} = \dfrac{10}{12}$
Adding	$\dfrac{a}{b} \pm \dfrac{c}{d} = \dfrac{ad \pm bc}{bd}$	$\dfrac{2}{3} - \dfrac{4}{5} = \dfrac{(2)(5) - (4)(3)}{(3)(5)} = -\dfrac{2}{15}$

Exercises

In the following exercises, solve for x:

Answers

1. $a = \dfrac{1}{1 + x}$ $x = \dfrac{1 - a}{a}$

2. $3x - 5 = 13$ $x = 6$

3. $ax - 5 = bx + 2$ $x = \dfrac{7}{a - b}$

4. $\dfrac{5}{2x + 6} = \dfrac{3}{4x + 8}$ $x = -\dfrac{11}{7}$

Powers

When powers of a given quantity x are multiplied, the following rule applies:

$$x^n x^m = x^{n+m} \tag{B.3}$$

For example, $x^2 x^4 = x^{2+4} = x^6$.

When dividing the powers of a given quantity, the rule is

$$\frac{x^n}{x^m} = x^{n-m} \tag{B.4}$$

For example, $x^8/x^2 = x^{8-2} = x^6$.

A power that is a fraction, such as $\frac{1}{3}$, corresponds to a root as follows:

$$x^{1/n} = \sqrt[n]{x} \tag{B.5}$$

For example, $4^{1/3} = \sqrt[3]{4} = 1.5874$. (A scientific calculator is useful for such calculations.)

Finally, any quantity x^n raised to the mth power is

$$(x^n)^m = x^{nm} \tag{B.6}$$

Table B.1 summarizes the rules of exponents.

Table B.1

Rules of Exponents

$x^0 = 1$
$x^1 = x$
$x^n x^m = x^{n+m}$
$x^n/x^m = x^{n-m}$
$x^{1/n} = \sqrt[n]{x}$
$(x^n)^m = x^{nm}$

Exercises

Verify the following:

1. $3^2 \times 3^3 = 243$

2. $x^5 x^{-8} = x^{-3}$

3. $x^{10}/x^{-5} = x^{15}$

4. $5^{1/3} = 1.709\ 975$ (Use your calculator.)

5. $60^{1/4} = 2.783\ 158$ (Use your calculator.)

6. $(x^4)^3 = x^{12}$

Factoring

Some useful formulas for factoring an equation are

$$ax + ay + az = a(x + y + z) \qquad \text{common factor}$$

$$a^2 + 2ab + b^2 = (a + b)^2 \qquad \text{perfect square}$$

$$a^2 - b^2 = (a + b)(a - b) \qquad \text{differences of squares}$$

Quadratic Equations

The general form of a quadratic equation is

$$ax^2 + bx + c = 0 \qquad \text{(B.7)}$$

where x is the unknown quantity and a, b, and c are numerical factors referred to as **coefficients** of the equation. This equation has two roots, given by

$$x = \frac{-b \pm \sqrt{b^2 - 4ac}}{2a} \qquad \text{(B.8)}$$

If $b^2 \geq 4ac$, the roots are real.

Example 1

The equation $x^2 + 5x + 4 = 0$ has the following roots corresponding to the two signs of the square-root term:

$$x = \frac{-5 \pm \sqrt{5^2 - (4)(1)(4)}}{2(1)} = \frac{-5 \pm \sqrt{9}}{2} = \frac{-5 \pm 3}{2}$$

$$x_+ = \frac{-5 + 3}{2} = -1 \qquad x_- = \frac{-5 - 3}{2} = -4$$

where x_+ refers to the root corresponding to the positive sign and x_- refers to the root corresponding to the negative sign.

Exercises

Solve the following quadratic equations:

Answers

1. $x^2 + 2x - 3 = 0$ $x_+ = 1$ $x_- = -3$
2. $2x^2 - 5x + 2 = 0$ $x_+ = 2$ $x_- = \frac{1}{2}$
3. $2x^2 - 4x - 9 = 0$ $x_+ = 1 + \sqrt{22}/2$ $x_- = 1 - \sqrt{22}/2$

Linear Equations

A linear equation has the general form

$$y = mx + b \qquad \text{(B.9)}$$

where m and b are constants. This equation is referred to as being linear because the graph of y versus x is a straight line, as shown in Figure B.1. The constant b, called the **y-intercept,** represents the value of y at which the straight line intersects the y axis. The constant m is equal to the **slope** of the straight line. If any two points on the straight line are specified by the coordinates (x_1, y_1) and (x_2, y_2), as in Figure B.1, then

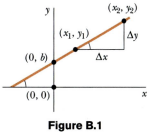

Figure B.1

the slope of the straight line can be expressed as

$$\text{Slope} = \frac{y_2 - y_1}{x_2 - x_1} = \frac{\Delta y}{\Delta x} \tag{B.10}$$

Note that m and b can have either positive or negative values. If $m > 0$, the straight line has a *positive* slope, as in Figure B.1. If $m < 0$, the straight line has a *negative* slope. In Figure B.1, both m and b are positive. Three other possible situations are shown in Figure B.2.

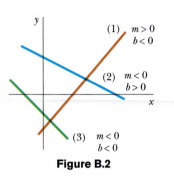

Figure B.2

Exercises

1. Draw graphs of the following straight lines:
   (a) $y = 5x + 3$ (b) $y = -2x + 4$ (c) $y = -3x - 6$

2. Find the slopes of the straight lines described in Exercise 1.

Answers   (a) 5   (b) -2 (c) -3

3. Find the slopes of the straight lines that pass through the following sets of points:
   (a) $(0, -4)$ and $(4, 2)$ (b) $(0, 0)$ and $(2, -5)$ (c) $(-5, 2)$ and $(4, -2)$

Answers   (a) $3/2$ (b) $-5/2$ (c) $-4/9$

Solving Simultaneous Linear Equations

Consider the equation $3x + 5y = 15$, which has two unknowns, x and y. Such an equation does not have a unique solution. For example, note that $(x = 0, y = 3)$, $(x = 5, y = 0)$, and $(x = 2, y = 9/5)$ are all solutions to this equation.

If a problem has two unknowns, a unique solution is possible only if we have *two* equations. In general, if a problem has n unknowns, its solution requires n equations. In order to solve two simultaneous equations involving two unknowns, x and y, we solve one of the equations for x in terms of y and substitute this expression into the other equation.

Example 2

Solve the following two simultaneous equations:

(1) $5x + y = -8$

(2) $2x - 2y = 4$

Solution From Equation (2), $x = y + 2$. Substitution of this into Equation (1) gives

$$5(y + 2) + y = -8$$

$$6y = -18$$

$$y = \boxed{-3}$$

$$x = y + 2 = \boxed{-1}$$

Alternate Solution Multiply each term in Equation (1) by the factor 2 and add the result to Equation (2):

$$10x + 2y = -16$$

$$\underline{2x - 2y = 4}$$

$$12x = -12$$

$$x = \boxed{-1}$$

$$y = x - 2 = \boxed{-3}$$

Two linear equations containing two unknowns can also be solved by a graphical method. If the straight lines corresponding to the two equations are plotted in a conventional coordinate system, the intersection of the two lines represents the solution. For example, consider the two equations

$$x - y = 2$$

$$x - 2y = -1$$

These are plotted in Figure B.3. The intersection of the two lines has the coordinates $x = 5$, $y = 3$. This represents the solution to the equations. You should check this solution by the analytical technique discussed above.

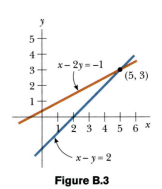

Figure B.3

Exercises

Solve the following pairs of simultaneous equations involving two unknowns:

		Answers
1.	$x + y = 8$ $x - y = 2$	$x = 5, y = 3$
2.	$98 - T = 10a$ $T - 49 = 5a$	$T = 65, a = 3.3$
3.	$6x + 2y = 6$ $8x - 4y = 28$	$x = 2, y = -3$

Logarithms

Suppose that a quantity x is expressed as a power of some quantity a:

$$x = a^y \tag{B.11}$$

The number a is called the **base** number. The **logarithm** of x with respect to the base a is equal to the exponent to which the base must be raised in order to satisfy the expression $x = a^y$:

$$y = \log_a x \tag{B.12}$$

Conversely, the **antilogarithm** of y is the number x:

$$x = \text{antilog}_a y \tag{B.13}$$

In practice, the two bases most often used are base 10, called the *common* logarithm base, and base $e = 2.718\,282$, called Euler's constant or the *natural* logarithm base. When common logarithms are used,

$$y = \log_{10} x \qquad (\text{or } x = 10^y) \tag{B.14}$$

When natural logarithms are used,

$$y = \ln x \qquad (\text{or } x = e^y) \tag{B.15}$$

For example, $\log_{10} 52 = 1.716$, so that $\text{antilog}_{10} 1.716 = 10^{1.716} = 52$. Likewise, $\ln 52 = 3.951$, so $\text{antiln } 3.951 = e^{3.951} = 52$.

In general, note that you can convert between base 10 and base e with the equality

$$\ln x = (2.302\,585) \log_{10} x \tag{B.16}$$

Finally, some useful properties of logarithms are

$$\left. \begin{aligned} \log(ab) &= \log a + \log b \\ \log(a/b) &= \log a - \log b \\ \log(a^n) &= n \log a \end{aligned} \right\} \text{any base}$$

$$\ln e = 1$$
$$\ln e^a = a$$
$$\ln\left(\frac{1}{a}\right) = -\ln a$$

B.3   Geometry

The **distance** d between two points having coordinates (x_1, y_1) and (x_2, y_2) is

$$d = \sqrt{(x_2 - x_1)^2 + (y_2 - y_1)^2} \tag{B.17}$$

Radian measure: The arc length s of a circular arc (Fig. B.4) is proportional to the radius r for a fixed value of θ (in radians):

$$s = r\theta$$
$$\theta = \frac{s}{r} \tag{B.18}$$

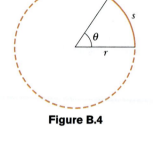

Figure B.4

Table B.2 gives the areas and volumes for several geometric shapes used throughout this text:

Table B.2

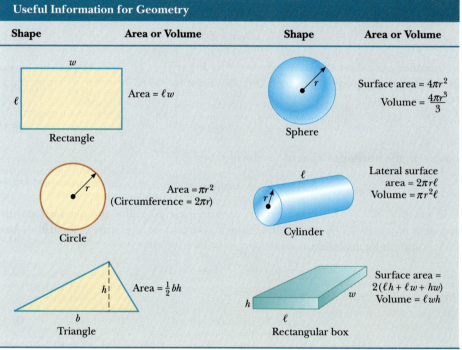

Useful Information for Geometry			
Shape	**Area or Volume**	**Shape**	**Area or Volume**

Rectangle — Area = ℓw

Sphere — Surface area = $4\pi r^2$, Volume = $\frac{4\pi r^3}{3}$

Circle — Area = πr^2 (Circumference = $2\pi r$)

Cylinder — Lateral surface area = $2\pi r \ell$, Volume = $\pi r^2 \ell$

Triangle — Area = $\frac{1}{2}bh$

Rectangular box — Surface area = $2(\ell h + \ell w + hw)$, Volume = $\ell w h$

The equation of a **straight line** (Fig. B.5) is

$$y = mx + b \tag{B.19}$$

where b is the y intercept and m is the slope of the line.

The equation of a **circle** of radius R centered at the origin is

$$x^2 + y^2 = R^2 \tag{B.20}$$

The equation of an **ellipse** having the origin at its center (Fig. B.6) is

$$\frac{x^2}{a^2} + \frac{y^2}{b^2} = 1 \tag{B.21}$$

Figure B.5

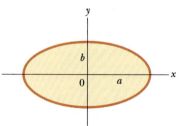

Figure B.6

where a is the length of the semimajor axis (the longer one) and b is the length of the semiminor axis (the shorter one).

The equation of a **parabola** the vertex of which is at $y = b$ (Fig. B.7) is

$$y = ax^2 + b \tag{B.22}$$

The equation of a **rectangular hyperbola** (Fig. B.8) is

$$xy = \text{constant} \tag{B.23}$$

Figure B.7

B.4 Trigonometry

That portion of mathematics based on the special properties of the right triangle is called trigonometry. By definition, a right triangle is one containing a 90° angle. Consider the right triangle shown in Figure B.9, where side a is opposite the angle θ, side b is adjacent to the angle θ, and side c is the hypotenuse of the triangle. The three basic trigonometric functions defined by such a triangle are the sine (sin), cosine (cos), and tangent (tan) functions. In terms of the angle θ, these functions are defined by

$$\sin\theta \equiv \frac{\text{side opposite }\theta}{\text{hypotenuse}} = \frac{a}{c} \tag{B.24}$$

$$\cos\theta \equiv \frac{\text{side adjacent to }\theta}{\text{hypotenuse}} = \frac{b}{c} \tag{B.25}$$

$$\tan\theta \equiv \frac{\text{side opposite }\theta}{\text{side adjacent to }\theta} = \frac{a}{b} \tag{B.26}$$

Figure B.8

The Pythagorean theorem provides the following relationship among the sides of a right triangle:

$$c^2 = a^2 + b^2 \tag{B.27}$$

From the above definitions and the Pythagorean theorem, it follows that

$$\sin^2\theta + \cos^2\theta = 1$$

$$\tan\theta = \frac{\sin\theta}{\cos\theta}$$

The cosecant, secant, and cotangent functions are defined by

$$\csc\theta \equiv \frac{1}{\sin\theta} \qquad \sec\theta \equiv \frac{1}{\cos\theta} \qquad \cot\theta \equiv \frac{1}{\tan\theta}$$

The relationships below follow directly from the right triangle shown in Figure B.9:

$$\sin\theta = \cos(90° - \theta)$$

$$\cos\theta = \sin(90° - \theta)$$

$$\cot\theta = \tan(90° - \theta)$$

Some properties of trigonometric functions are

$$\sin(-\theta) = -\sin\theta$$

$$\cos(-\theta) = \cos\theta$$

$$\tan(-\theta) = -\tan\theta$$

a = opposite side
b = adjacent side
c = hypotenuse

Figure B.9

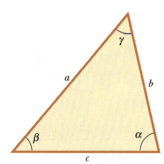

Figure B.10

The following relationships apply to *any* triangle, as shown in Figure B.10:

$$\alpha + \beta + \gamma = 180°$$

Law of cosines
$$\begin{aligned} a^2 &= b^2 + c^2 - 2bc \cos \alpha \\ b^2 &= a^2 + c^2 - 2ac \cos \beta \\ c^2 &= a^2 + b^2 - 2ab \cos \gamma \end{aligned}$$

Law of sines
$$\frac{a}{\sin \alpha} = \frac{b}{\sin \beta} = \frac{c}{\sin \gamma}$$

Table B.3 lists a number of useful trigonometric identities.

Table B.3

Some Trigonometric Identities

$$\sin^2 \theta + \cos^2 \theta = 1 \qquad\qquad\qquad \csc^2 \theta = 1 + \cot^2 \theta$$

$$\sec^2 \theta = 1 + \tan^2 \theta \qquad\qquad\qquad \sin^2 \frac{\theta}{2} = \tfrac{1}{2}(1 - \cos \theta)$$

$$\sin 2\theta = 2 \sin \theta \cos \theta \qquad\qquad\qquad \cos^2 \frac{\theta}{2} = \tfrac{1}{2}(1 + \cos \theta)$$

$$\cos 2\theta = \cos^2 \theta - \sin^2 \theta \qquad\qquad\qquad 1 - \cos \theta = 2 \sin^2 \frac{\theta}{2}$$

$$\tan 2\theta = \frac{2 \tan \theta}{1 - \tan^2 \theta} \qquad\qquad\qquad \tan \frac{\theta}{2} = \sqrt{\frac{1 - \cos \theta}{1 + \cos \theta}}$$

$$\sin(A \pm B) = \sin A \cos B \pm \cos A \sin B$$
$$\cos(A \pm B) = \cos A \cos B \mp \sin A \sin B$$
$$\sin A \pm \sin B = 2 \sin[\tfrac{1}{2}(A \pm B)]\cos[\tfrac{1}{2}(A \mp B)]$$
$$\cos A + \cos B = 2 \cos[\tfrac{1}{2}(A + B)]\cos[\tfrac{1}{2}(A - B)]$$
$$\cos A - \cos B = 2 \sin[\tfrac{1}{2}(A + B)]\sin[\tfrac{1}{2}(B - A)]$$

Example 3

Consider the right triangle in Figure B.11, in which $a = 2$, $b = 5$, and c is unknown. From the Pythagorean theorem, we have

$$c^2 = a^2 + b^2 = 2^2 + 5^2 = 4 + 25 = 29$$

$$c = \sqrt{29} = \boxed{5.39}$$

To find the angle θ, note that

$$\tan \theta = \frac{a}{b} = \frac{2}{5} = 0.400$$

From a table of functions or from a calculator, we have

$$\theta = \tan^{-1}(0.400) = \boxed{21.8°}$$

where $\tan^{-1}(0.400)$ is the notation for "angle whose tangent is 0.400," sometimes written as arctan (0.400).

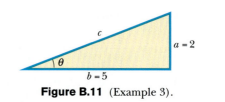

Figure B.11 (Example 3).

Exercises

Figure B.12

1. In Figure B.12, identify (a) the side opposite θ (b) the side adjacent to ϕ. Then find (c) $\cos \theta$ (d) $\sin \phi$ (e) $\tan \phi$.

Answers   (a) 3   (b) 3   (c) $\tfrac{4}{5}$ (d) $\tfrac{4}{5}$ (e) $\tfrac{4}{3}$

2. In a certain right triangle, the two sides that are perpendicular to each other are 5 m and 7 m long. What is the length of the third side?

Answer   8.60 m

3. A right triangle has a hypotenuse of length 3 m, and one of its angles is 30°. What is the length of (a) the side opposite the 30° angle (b) the side adjacent to the 30° angle?

Answers (a) 1.5 m (b) 2.60 m

B.5 Series Expansions

$$(a + b)^n = a^n + \frac{n}{1!} a^{n-1}b + \frac{n(n-1)}{2!} a^{n-2}b^2 + \cdots$$

$$(1 + x)^n = 1 + nx + \frac{n(n-1)}{2!} x^2 + \cdots$$

$$e^x = 1 + x + \frac{x^2}{2!} + \frac{x^3}{3!} + \cdots$$

$$\ln(1 \pm x) = \pm x - \tfrac{1}{2}x^2 \pm \tfrac{1}{3}x^3 - \cdots$$

$$\left.\begin{array}{l} \sin x = x - \dfrac{x^3}{3!} + \dfrac{x^5}{5!} - \cdots \\[2mm] \cos x = 1 - \dfrac{x^2}{2!} + \dfrac{x^4}{4!} - \cdots \\[2mm] \tan x = x + \dfrac{x^3}{3} + \dfrac{2x^5}{15} + \cdots \qquad |x| < \pi/2 \end{array}\right\} \quad x \text{ in radians}$$

For $x \ll 1$, the following approximations can be used[1]:

$$(1 + x)^n \approx 1 + nx \qquad \sin x \approx x$$
$$e^x \approx 1 + x \qquad \cos x \approx 1$$
$$\ln(1 \pm x) \approx \pm x \qquad \tan x \approx x$$

B.6 Differential Calculus

In various branches of science, it is sometimes necessary to use the basic tools of calculus, invented by Newton, to describe physical phenomena. The use of calculus is fundamental in the treatment of various problems in Newtonian mechanics, electricity, and magnetism. In this section, we simply state some basic properties and "rules of thumb" that should be a useful review to the student.

First, a **function** must be specified that relates one variable to another (such as a coordinate as a function of time). Suppose one of the variables is called y (the dependent variable), the other x (the independent variable). We might have a function relationship such as

$$y(x) = ax^3 + bx^2 + cx + d$$

If a, b, c, and d are specified constants, then y can be calculated for any value of x. We usually deal with continuous functions, that is, those for which y varies "smoothly" with x.

The **derivative** of y with respect to x is defined as the limit, as Δx approaches zero, of the slopes of chords drawn between two points on the y versus x curve. Mathematically, we write this definition as

$$\frac{dy}{dx} = \lim_{\Delta x \to 0} \frac{\Delta y}{\Delta x} = \lim_{\Delta x \to 0} \frac{y(x + \Delta x) - y(x)}{\Delta x} \qquad \text{(B.28)}$$

where Δy and Δx are defined as $\Delta x = x_2 - x_1$ and $\Delta y = y_2 - y_1$ (Fig. B.13). It is important to note that dy/dx *does not* mean dy divided by dx, but is simply a notation of the limiting process of the derivative as defined by Equation B.28.

Figure B.13

---

[1] The approximations for the functions sin x, cos x, and tan x are for $x \leq 0.1$ rad.

A useful expression to remember when $y(x) = ax^n$, where a is a *constant* and n is *any* positive or negative number (integer or fraction), is

$$\frac{dy}{dx} = nax^{n-1} \qquad \text{(B.29)}$$

If $y(x)$ is a polynomial or algebraic function of x, we apply Equation B.29 to *each* term in the polynomial and take $d[\text{constant}]/dx = 0$. In Examples 4 through 7, we evaluate the derivatives of several functions.

Special Properties of the Derivative

A. Derivative of the product of two functions If a function $f(x)$ is given by the product of two functions, say, $g(x)$ and $h(x)$, then the derivative of $f(x)$ is defined as

$$\frac{d}{dx} f(x) = \frac{d}{dx} [g(x) h(x)] = g \frac{dh}{dx} + h \frac{dg}{dx} \qquad \text{(B.30)}$$

B. Derivative of the sum of two functions If a function $f(x)$ is equal to the sum of two functions, then the derivative of the sum is equal to the sum of the derivatives:

$$\frac{d}{dx} f(x) = \frac{d}{dx} [g(x) + h(x)] = \frac{dg}{dx} + \frac{dh}{dx} \qquad \text{(B.31)}$$

C. Chain rule of differential calculus If $y = f(x)$ and $x = g(z)$, then dy/dz can be written as the product of two derivatives:

$$\frac{dy}{dz} = \frac{dy}{dx} \frac{dx}{dz} \qquad \text{(B.32)}$$

D. The second derivative The second derivative of y with respect to x is defined as the derivative of the function dy/dx (the derivative of the derivative). It is usually written

$$\frac{d^2y}{dx^2} = \frac{d}{dx} \left(\frac{dy}{dx} \right) \qquad \text{(B.33)}$$

Example 4

Suppose $y(x)$ (that is, y as a function of x) is given by

$$y(x) = ax^3 + bx + c$$

where a and b are constants. Then it follows that

$$y(x + \Delta x) = a(x + \Delta x)^3 + b(x + \Delta x) + c$$
$$y(x + \Delta x) = a(x^3 + 3x^2 \Delta x + 3x \Delta x^2 + \Delta x^3) + b(x + \Delta x) + c$$

so

$$\Delta y = y(x + \Delta x) - y(x) = a(3x^2 \Delta x + 3x \Delta x^2 + \Delta x^3) + b\Delta x$$

Substituting this into Equation B.28 gives

$$\frac{dy}{dx} = \lim_{\Delta x \to 0} \frac{\Delta y}{\Delta x} = \lim_{\Delta x \to 0} [3ax^2 + 3x\Delta x + \Delta x^2] + b$$

$$\frac{dy}{dx} = 3ax^2 + b$$

Example 5

Find the derivative of

$$y(x) = 8x^5 + 4x^3 + 2x + 7$$

Solution Applying Equation B.29 to each term independently, and remembering that $d/dx (\text{constant}) = 0$, we have

$$\frac{dy}{dx} = 8(5)x^4 + 4(3)x^2 + 2(1)x^0 + 0$$

$$\frac{dy}{dx} = 40x^4 + 12x^2 + 2$$

Example 6

Find the derivative of $y(x) = x^3/(x+1)^2$ with respect to x.

Solution We can rewrite this function as $y(x) = x^3(x+1)^{-2}$ and apply Equation B.30:

$$\frac{dy}{dx} = (x+1)^{-2}\frac{d}{dx}(x^3) + x^3\frac{d}{dx}(x+1)^{-2}$$

$$= (x+1)^{-2}3x^2 + x^3(-2)(x+1)^{-3}$$

$$\frac{dy}{dx} = \frac{3x^2}{(x+1)^2} - \frac{2x^3}{(x+1)^3}$$

Example 7

A useful formula that follows from Equation B.30 is the derivative of the quotient of two functions. Show that

$$\frac{d}{dx}\left[\frac{g(x)}{h(x)}\right] = \frac{h\dfrac{dg}{dx} - g\dfrac{dh}{dx}}{h^2}$$

Solution We can write the quotient as gh^{-1} and then apply Equations B.29 and B.30:

$$\frac{d}{dx}\left(\frac{g}{h}\right) = \frac{d}{dx}(gh^{-1}) = g\frac{d}{dx}(h^{-1}) + h^{-1}\frac{d}{dx}(g)$$

$$= -gh^{-2}\frac{dh}{dx} + h^{-1}\frac{dg}{dx}$$

$$= \frac{h\dfrac{dg}{dx} - g\dfrac{dh}{dx}}{h^2}$$

Some of the more commonly used derivatives of functions are listed in Table B.4.

B.7 Integral Calculus

We think of integration as the inverse of differentiation. As an example, consider the expression

$$f(x) = \frac{dy}{dx} = 3ax^2 + b \tag{B.34}$$

which was the result of differentiating the function

$$y(x) = ax^3 + bx + c$$

in Example 4. We can write Equation B.34 as $dy = f(x)\,dx = (3ax^2 + b)\,dx$ and obtain $y(x)$ by "summing" over all values of x. Mathematically, we write this inverse operation

$$y(x) = \int f(x)\,dx$$

For the function $f(x)$ given by Equation B.34, we have

$$y(x) = \int (3ax^2 + b)\,dx = ax^3 + bx + c$$

where c is a constant of the integration. This type of integral is called an *indefinite integral* because its value depends on the choice of c.

A general **indefinite integral** $I(x)$ is defined as

$$I(x) = \int f(x)\,dx \tag{B.35}$$

where $f(x)$ is called the *integrand* and $f(x) = dI(x)/dx$.

For a *general continuous* function $f(x)$, the integral can be described as the area under the curve bounded by $f(x)$ and the x axis, between two specified values of x, say, x_1 and x_2, as in Figure B.14.

The area of the blue element is approximately $f(x_i)\,\Delta x_i$. If we sum all these area elements from x_1 and x_2 and take the limit of this sum as $\Delta x_i \rightarrow 0$, we obtain the *true*

Table B.4

Derivative for Several Functions

$$\frac{d}{dx}(a) = 0$$

$$\frac{d}{dx}(ax^n) = nax^{n-1}$$

$$\frac{d}{dx}(e^{ax}) = ae^{ax}$$

$$\frac{d}{dx}(\sin ax) = a\cos ax$$

$$\frac{d}{dx}(\cos ax) = -a\sin ax$$

$$\frac{d}{dx}(\tan ax) = a\sec^2 ax$$

$$\frac{d}{dx}(\cot ax) = -a\csc^2 dx$$

$$\frac{d}{dx}(\sec x) = \tan x\sec x$$

$$\frac{d}{dx}(\csc x) = -\cot x\csc x$$

$$\frac{d}{dx}(\ln ax) = \frac{1}{x}$$

Note: The symbols a and n represent constants.

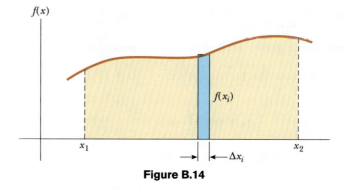

Figure B.14

area under the curve bounded by $f(x)$ and x, between the limits x_1 and x_2:

$$\text{Area} = \lim_{\Delta x \to 0} \sum_i f(x_i)\, \Delta x_i = \int_{x_1}^{x_2} f(x)\, dx \qquad\qquad (\text{B.36})$$

Integrals of the type defined by Equation B.36 are called **definite integrals.**

One common integral that arises in practical situations has the form

$$\int x^n\, dx = \frac{x^{n+1}}{n+1} + c \qquad (n \neq -1) \qquad\qquad (\text{B.37})$$

This result is obvious, being that differentiation of the right-hand side with respect to x gives $f(x) = x^n$ directly. If the limits of the integration are known, this integral becomes a *definite integral* and is written

$$\int_{x_1}^{x_2} x^n\, dx = \frac{x^{n+1}}{n+1}\Big|_{x_1}^{x_2} = \frac{x_2^{n+1} - x_1^{n+1}}{n+1} \qquad (n \neq -1) \qquad\qquad (\text{B.38})$$

Examples

1. $\displaystyle\int_0^a x^2\, dx = \frac{x^3}{3}\bigg]_0^a = \frac{a^3}{3}$

2. $\displaystyle\int_0^b x^{3/2}\, dx = \frac{x^{5/2}}{5/2}\bigg]_0^b = \tfrac{2}{5}b^{5/2}$

3. $\displaystyle\int_3^5 x\, dx = \frac{x^2}{2}\bigg]_3^5 = \frac{5^2 - 3^2}{2} = 8$

Partial Integration

Sometimes it is useful to apply the method of *partial integration* (also called "integrating by parts") to evaluate certain integrals. The method uses the property that

$$\int u\, dv = uv - \int v\, du \qquad\qquad (\text{B.39})$$

where u and v are *carefully* chosen so as to reduce a complex integral to a simpler one. In many cases, several reductions have to be made. Consider the function

$$I(x) = \int x^2 e^x\, dx$$

This can be evaluated by integrating by parts twice. First, if we choose $u = x^2$, $v = e^x$, we obtain

$$\int x^2 e^x\, dx = \int x^2\, d(e^x) = x^2 e^x - 2\int e^x x\, dx + c_1$$

Now, in the second term, choose $u = x$, $v = e^x$, which gives

$$\int x^2 e^x \, dx = x^2 e^x - 2x e^x + 2 \int e^x \, dx + c_1$$

or

$$\int x^2 e^x \, dx = x^2 e^x - 2x e^x + 2 e^x + c_2$$

The Perfect Differential

Another useful method to remember is the use of the *perfect differential*, in which we look for a change of variable such that the differential of the function is the differential of the independent variable appearing in the integrand. For example, consider the integral

$$I(x) = \int \cos^2 x \sin x \, dx$$

This becomes easy to evaluate if we rewrite the differential as $d(\cos x) = -\sin x \, dx$. The integral then becomes

$$\int \cos^2 x \sin x \, dx = - \int \cos^2 x \, d(\cos x)$$

If we now change variables, letting $y = \cos x$, we obtain

$$\int \cos^2 x \sin x \, dx = - \int y^2 dy = -\frac{y^3}{3} + c = -\frac{\cos^3 x}{3} + c$$

Table B.5 lists some useful indefinite integrals. Table B.6 gives Gauss's probability integral and other definite integrals. A more complete list can be found in various handbooks, such as *The Handbook of Chemistry and Physics*, CRC Press.

Table B.5

Some Indefinite Integrals (An arbitrary constant should be added to each of these integrals.)	
$\displaystyle\int x^n dx = \frac{x^{n+1}}{n+1}$ (provided $n \neq -1$)	$\displaystyle\int \frac{dx}{\sqrt{a^2 - x^2}} = \sin^{-1}\frac{x}{a} = -\cos^{-1}\frac{x}{a}$ $(a^2 - x^2 > 0)$
$\displaystyle\int \frac{dx}{x} = \int x^{-1} dx = \ln x$	$\displaystyle\int \frac{dx}{\sqrt{x^2 \pm a^2}} = \ln(x + \sqrt{x^2 \pm a^2})$
$\displaystyle\int \frac{dx}{a + bx} = \frac{1}{b}\ln(a + bx)$	$\displaystyle\int \frac{x \, dx}{\sqrt{a^2 - x^2}} = -\sqrt{a^2 - x^2}$
$\displaystyle\int \frac{xdx}{a + bx} = \frac{x}{b} - \frac{a}{b^2}\ln(a + bx)$	$\displaystyle\int \frac{x \, dx}{\sqrt{x^2 \pm a^2}} = \sqrt{x^2 \pm a^2}$
$\displaystyle\int \frac{dx}{x(x + a)} = -\frac{1}{a}\ln\frac{x + a}{x}$	$\displaystyle\int \sqrt{a^2 - x^2} \, dx = \frac{1}{2}\left(x\sqrt{a^2 - x^2} + a^2 \sin^{-1}\frac{x}{a}\right)$
$\displaystyle\int \frac{dx}{(a + bx)^2} = -\frac{1}{b(a + bx)}$	$\displaystyle\int x\sqrt{a^2 - x^2} \, dx = -\frac{1}{3}(a^2 - x^2)^{3/2}$
$\displaystyle\int \frac{dx}{a^2 + x^2} = \frac{1}{a}\tan^{-1}\frac{x}{a}$	$\displaystyle\int \sqrt{x^2 \pm a^2} \, dx = \frac{1}{2}[x\sqrt{x^2 \pm a^2} \pm a^2 \ln(x + \sqrt{x^2 \pm a^2})]$
$\displaystyle\int \frac{dx}{a^2 - x^2} = \frac{1}{2a}\ln\frac{a + x}{a - x}$ $(a^2 - x^2 > 0)$	$\displaystyle\int x(\sqrt{x^2 \pm a^2}) \, dx = \frac{1}{3}(x^2 \pm a^2)^{3/2}$
$\displaystyle\int \frac{dx}{x^2 - a^2} = \frac{1}{2a}\ln\frac{x - a}{x + a}$ $(x^2 - a^2 > 0)$	$\displaystyle\int e^{ax} \, dx = \frac{1}{a}e^{ax}$
$\displaystyle\int \frac{x \, dx}{a^2 \pm x^2} = \pm\frac{1}{2}\ln(a^2 \pm x^2)$	$\displaystyle\int \ln ax \, dx = (x \ln ax) - x$

continued

Table B.5

Some Indefinite Integrals (An arbitrary constant should be added to each of these integrals.) *continued*

$$\int xe^{ax}\, dx = \frac{e^{ax}}{a^2}(ax - 1)$$

$$\int \frac{dx}{a + be^{cx}} = \frac{x}{a} - \frac{1}{ac}\ln(a + be^{cx})$$

$$\int \sin ax\, dx = -\frac{1}{a}\cos ax$$

$$\int \cos ax\, dx = \frac{1}{a}\sin ax$$

$$\int \tan ax\, dx = -\frac{1}{a}\ln(\cos ax) = \frac{1}{a}\ln(\sec ax)$$

$$\int \cot ax\, dx = \frac{1}{a}\ln(\sin ax)$$

$$\int \sec ax\, dx = \frac{1}{a}\ln(\sec ax + \tan ax) = \frac{1}{a}\ln\left[\tan\left(\frac{ax}{2} + \frac{\pi}{4}\right)\right]$$

$$\int \csc ax\, dx = \frac{1}{a}\ln(\csc ax - \cot ax) = \frac{1}{a}\ln\left(\tan\frac{ax}{2}\right)$$

$$\int \sin^2 ax\, dx = \frac{x}{2} - \frac{\sin 2ax}{4a}$$

$$\int \cos^2 ax\, dx = \frac{x}{2} + \frac{\sin 2ax}{4a}$$

$$\int \frac{dx}{\sin^2 ax} = -\frac{1}{a}\cot ax$$

$$\int \frac{dx}{\cos^2 ax} = \frac{1}{a}\tan ax$$

$$\int \tan^2 ax\, dx = \frac{1}{a}(\tan ax) - x$$

$$\int \cot^2 ax\, dx = -\frac{1}{a}(\cot ax) - x$$

$$\int \sin^{-1} ax\, dx = x(\sin^{-1} ax) + \frac{\sqrt{1 - a^2 x^2}}{a}$$

$$\int \cos^{-1} ax\, dx = x(\cos^{-1} ax) - \frac{\sqrt{1 - a^2 x^2}}{a}$$

$$\int \frac{dx}{(x^2 + a^2)^{3/2}} = \frac{x}{a^2\sqrt{x^2 + a^2}}$$

$$\int \frac{x\, dx}{(x^2 + a^2)^{3/2}} = -\frac{1}{\sqrt{x^2 + a^2}}$$

Table B.6

Gauss's Probability Integral and Other Definite Integrals

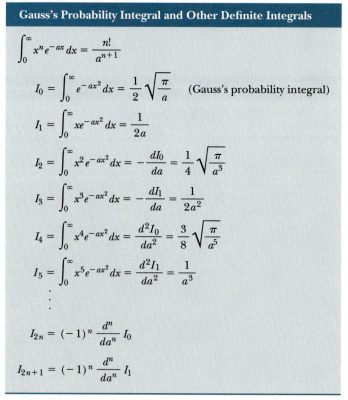

$$\int_0^\infty x^n e^{-ax}\, dx = \frac{n!}{a^{n+1}}$$

$$I_0 = \int_0^\infty e^{-ax^2}\, dx = \frac{1}{2}\sqrt{\frac{\pi}{a}} \quad \text{(Gauss's probability integral)}$$

$$I_1 = \int_0^\infty xe^{-ax^2}\, dx = \frac{1}{2a}$$

$$I_2 = \int_0^\infty x^2 e^{-ax^2}\, dx = -\frac{dI_0}{da} = \frac{1}{4}\sqrt{\frac{\pi}{a^3}}$$

$$I_3 = \int_0^\infty x^3 e^{-ax^2}\, dx = -\frac{dI_1}{da} = \frac{1}{2a^2}$$

$$I_4 = \int_0^\infty x^4 e^{-ax^2}\, dx = \frac{d^2 I_0}{da^2} = \frac{3}{8}\sqrt{\frac{\pi}{a^5}}$$

$$I_5 = \int_0^\infty x^5 e^{-ax^2}\, dx = \frac{d^2 I_1}{da^2} = \frac{1}{a^3}$$

$$\vdots$$

$$I_{2n} = (-1)^n \frac{d^n}{da^n} I_0$$

$$I_{2n+1} = (-1)^n \frac{d^n}{da^n} I_1$$

B.8   Propagation of Uncertainty

In laboratory experiments, a common activity is to take measurements that act as raw data. These measurements are of several types—length, time interval, temperature, voltage, etc.—and are taken by a variety of instruments. Regardless of the measurement and the quality of the instrumentation, **there is always uncertainty associated with a physical measurement.** This uncertainty is a combination of that associated with the instrument and that related to the system being measured. An example of the former is the inability to exactly determine the position of a length measurement between the lines on a meter stick. An example of uncertainty related to the system being measured is the variation of temperature within a sample of water so that a single temperature for the sample is difficult to determine.

Uncertainties can be expressed in two ways. **Absolute uncertainty** refers to an uncertainty expressed in the same units as the measurement. Thus, a length might be expressed as (5.5 ± 0.1) cm, as was the length of the computer disk label in Section 1.7. The uncertainty of ± 0.1 cm by itself is not descriptive enough for some purposes, however. This is a large uncertainty if the measurement is 1.0 cm, but it is a small uncertainty if the measurement is 100 m. To give a more descriptive account of the uncertainty, **fractional uncertainty** or **percent uncertainty** is used. In this type of description, the uncertainty is divided by the actual measurement. Thus, the length of the computer disk label could be expressed as

$$\ell = 5.5 \text{ cm} \pm \frac{0.1 \text{ cm}}{5.5 \text{ cm}} = 5.5 \text{ cm} \pm 0.018 \qquad \text{(fractional uncertainty)}$$

or as

$$\ell = 5.5 \text{ cm} \pm 1.8\% \qquad \text{(percent uncertainty)}$$

When combining measurements in a calculation, the uncertainty in the final result is larger than the uncertainty in the individual measurements. This is called **propagation of uncertainty** and is one of the challenges of experimental physics. As a calculation becomes more complicated, there is increased propagation of uncertainty and the uncertainty in the value of the final result can grow to be quite large.

There are simple rules that can provide a reasonable estimate of the uncertainty in a calculated result:

Multiplication and division: When measurements with uncertainties are multiplied or divided, add the *percent uncertainties* to obtain the percent uncertainty in the result.

Example: The Area of a Rectangular Plate

$$A = \ell w = (5.5 \text{ cm} \pm 1.8\%) \times (6.4 \text{ cm} \pm 1.6\%) = 35 \text{ cm}^2 \pm 3.4\%$$
$$= (35 \pm 1) \text{ cm}^2$$

Addition and subtraction: When measurements with uncertainties are added or subtracted, add the *absolute uncertainties* to obtain the absolute uncertainty in the result.

Example: A Change in Temperature

$$\Delta T = T_2 - T_1 = (99.2 \pm 1.5)°\text{C} - (27.6 \pm 1.5)°\text{C} = (71.6 \pm 3.0)°\text{C}$$
$$= 71.6°\text{C} \pm 4.2\%$$

Powers: If a measurement is taken to a power, the percent uncertainty is multiplied by that power to obtain the percent uncertainty in the result.

Example: The Volume of a Sphere

$$V = \tfrac{4}{3}\pi r^3 = \tfrac{4}{3}\pi(6.20 \text{ cm} \pm 2.0\%)^3 = 998 \text{ cm}^3 \pm 6.0\%$$
$$= (998 \pm 60) \text{ cm}^3$$

Notice that uncertainties in a calculation always add. As a result, an experiment involving a subtraction should be avoided if possible. This is especially true if the measurements being subtracted are close together. The result of such a calculation is a small difference in the measurements and uncertainties that add together. It is possible that the uncertainty in the result could be larger than the result itself!

Appendix C • Periodic Table of the Elements

Group I	Group II	Transition elements							
H 1 1.007 9 $1s$									
Li 3 6.941 $2s^1$	**Be** 4 9.0122 $2s^2$								
Na 11 22.990 $3s^1$	**Mg** 12 24.305 $3s^2$								
K 19 39.098 $4s^1$	**Ca** 20 40.078 $4s^2$	**Sc** 21 44.956 $3d^14s^2$	**Ti** 22 47.867 $3d^24s^2$	**V** 23 50.942 $3d^34s^2$	**Cr** 24 51.996 $3d^54s^1$	**Mn** 25 54.938 $3d^54s^2$	**Fe** 26 55.845 $3d^64s^2$	**Co** 27 58.933 $3d^74s^2$
Rb 37 85.468 $5s^1$	**Sr** 38 87.62 $5s^2$	**Y** 39 88.906 $4d^15s^2$	**Zr** 40 91.224 $4d^25s^2$	**Nb** 41 92.906 $4d^45s^1$	**Mo** 42 95.94 $4d^55s^1$	**Tc** 43 (98) $4d^55s^2$	**Ru** 44 101.07 $4d^75s^1$	**Rh** 45 102.91 $4d^85s^1$
Cs 55 132.91 $6s^1$	**Ba** 56 137.33 $6s^2$	57-71*	**Hf** 72 178.49 $5d^26s^2$	**Ta** 73 180.95 $5d^36s^2$	**W** 74 183.84 $5d^46s^2$	**Re** 75 186.21 $5d^56s^2$	**Os** 76 190.23 $5d^66s^2$	**Ir** 77 192.2 $5d^76s^2$
Fr 87 (223) $7s^1$	**Ra** 88 (226) $7s^2$	89-103**	**Rf** 104 (261) $6d^27s^2$	**Db** 105 (262) $6d^37s^2$	**Sg** 106 (266)	**Bh** 107 (264)	**Hs** 108 (269)	**Mt** 109 (268)

Symbol — **Ca** 20 — Atomic number
Atomic mass † — 40.078
$4s^2$ — Electron configuration

*Lanthanide series

La 57 138.91 $5d^16s^2$	**Ce** 58 140.12 $5d^14f^16s^2$	**Pr** 59 140.91 $4f^36s^2$	**Nd** 60 144.24 $4f^46s^2$	**Pm** 61 (145) $4f^56s^2$	**Sm** 62 150.36 $4f^66s^2$

**Actinide series

Ac 89 (227) $6d^17s^2$	**Th** 90 232.04 $6d^27s^2$	**Pa** 91 231.04 $5f^26d^17s^2$	**U** 92 238.03 $5f^36d^17s^2$	**Np** 93 (237) $5f^46d^17s^2$	**Pu** 94 (244) $5f^66d^07s^2$

▫ Atomic mass values given are averaged over isotopes in the percentages in which they exist in nature.
† For an unstable element, mass number of the most stable known isotope is given in parentheses.
†† Elements 110, 111, 112, and 114 have not yet been named.
††† For a description of the atomic data, visit *physics.nist.gov/atomic*

			Group III	Group IV	Group V	Group VI	Group VII	Group 0
							H 1 1.007 9 $1s^1$	**He** 2 4.002 6 $1s^2$
			B 5 10.811 $2p^1$	**C** 6 12.011 $2p^2$	**N** 7 14.007 $2p^3$	**O** 8 15.999 $2p^4$	**F** 9 18.998 $2p^5$	**Ne** 10 20.180 $2p^6$
			Al 13 26.982 $3p^1$	**Si** 14 28.086 $3p^2$	**P** 15 30.974 $3p^3$	**S** 16 32.066 $3p^4$	**Cl** 17 35.453 $3p^5$	**Ar** 18 39.948 $3p^6$
Ni 28 58.693 $3d^8 4s^2$	**Cu** 29 63.546 $3d^{10}4s^1$	**Zn** 30 65.39 $3d^{10}4s^2$	**Ga** 31 69.723 $4p^1$	**Ge** 32 72.61 $4p^2$	**As** 33 74.922 $4p^3$	**Se** 34 78.96 $4p^4$	**Br** 35 79.904 $4p^5$	**Kr** 36 83.80 $4p^6$
Pd 46 106.42 $4d^{10}$	**Ag** 47 107.87 $4d^{10}5s^1$	**Cd** 48 112.41 $4d^{10}5s^2$	**In** 49 114.82 $5p^1$	**Sn** 50 118.71 $5p^2$	**Sb** 51 121.76 $5p^3$	**Te** 52 127.60 $5p^4$	**I** 53 126.90 $5p^5$	**Xe** 54 131.29 $5p^6$
Pt 78 195.08 $5d^9 6s^1$	**Au** 79 196.97 $5d^{10}6s^1$	**Hg** 80 200.59 $5d^{10}6s^2$	**Tl** 81 204.38 $6p^1$	**Pb** 82 207.2 $6p^2$	**Bi** 83 208.98 $6p^3$	**Po** 84 (209) $6p^4$	**At** 85 (210) $6p^5$	**Rn** 86 (222) $6p^6$
110†† (271)	111†† (272)	112†† (285)		114†† (289)				

Eu 63 151.96 $4f^7 6s^2$	**Gd** 64 157.25 $5d^1 4f^7 6s^2$	**Tb** 65 158.93 $5d^1 4f^8 6s^2$	**Dy** 66 162.50 $4f^{10}6s^2$	**Ho** 67 164.93 $4f^{11}6s^2$	**Er** 68 167.26 $4f^{12}6s^2$	**Tm** 69 168.93 $4f^{13}6s^2$	**Yb** 70 173.04 $4f^{14}6s^2$	**Lu** 71 174.97 $5d^1 4f^{14}6s^2$
Am 95 (243) $5f^7 6d^0 7s^2$	**Cm** 96 (247) $5f^7 6d^1 7s^2$	**Bk** 97 (247) $5f^8 6d^1 7s^2$	**Cf** 98 (251) $5f^{10}6d^0 7s^2$	**Es** 99 (252) $5f^{11}6d^0 7s^2$	**Fm** 100 (257) $5f^{12}6d^0 7s^2$	**Md** 101 (258) $5f^{13}6d^0 7s^2$	**No** 102 (259) $6d^0 7s^2$	**Lr** 103 (262) $6d^1 7s^2$

Appendix D • SI Units

Table D.1

SI Units		
	SI Base Unit	
Base Quantity	**Name**	**Symbol**
Length	Meter	m
Mass	Kilogram	kg
Time	Second	s
Electric current	Ampere	A
Temperature	Kelvin	K
Amount of substance	Mole	mol
Luminous intensity	Candela	cd

Table D.2

Some Derived SI Units				
Quantity	**Name**	**Symbol**	**Expression in Terms of Base Units**	**Expression in Terms of Other SI Units**
Plane angle	radian	rad	m/m	
Frequency	hertz	Hz	s^{-1}	
Force	newton	N	$kg \cdot m/s^2$	J/m
Pressure	pascal	Pa	$kg/m \cdot s^2$	N/m^2
Energy; work	joule	J	$kg \cdot m^2/s^2$	$N \cdot m$
Power	watt	W	$kg \cdot m^2/s^3$	J/s
Electric charge	coulomb	C	$A \cdot s$	
Electric potential	volt	V	$kg \cdot m^2/A \cdot s^3$	W/A
Capacitance	farad	F	$A^2 \cdot s^4/kg \cdot m^2$	C/V
Electric resistance	ohm	Ω	$kg \cdot m^2/A^2 \cdot s^3$	V/A
Magnetic flux	weber	Wb	$kg \cdot m^2/A \cdot s^2$	$V \cdot s$
Magnetic field	tesla	T	$kg/A \cdot s^2$	
Inductance	henry	H	$kg \cdot m^2/A^2 \cdot s^2$	$T \cdot m^2/A$

Appendix E • Nobel Prizes

All Nobel Prizes in physics are listed (and marked with a P), as well as relevant Nobel Prizes in Chemistry (C). The key dates for some of the scientific work are supplied; they often antedate the prize considerably.

1901 (P) *Wilhelm Roentgen* for discovering x-rays (1895).

1902 (P) *Hendrik A. Lorentz* for predicting the Zeeman effect and *Pieter Zeeman* for discovering the Zeeman effect, the splitting of spectral lines in magnetic fields.

1903 (P) *Antoine-Henri Becquerel* for discovering radioactivity (1896) and *Pierre* and *Marie Curie* for studying radioactivity.

1904 (P) *Lord Rayleigh* for studying the density of gases and discovering argon.

(C) *William Ramsay* for discovering the inert gas elements helium, neon, xenon, and krypton, and placing them in the periodic table.

1905 (P) *Philipp Lenard* for studying cathode rays, electrons (1898–1899).

1906 (P) *J. J. Thomson* for studying electrical discharge through gases and discovering the electron (1897).

1907 (P) *Albert A. Michelson* for inventing optical instruments and measuring the speed of light (1880s).

1908 (P) *Gabriel Lippmann* for making the first color photographic plate, using interference methods (1891).

(C) *Ernest Rutherford* for discovering that atoms can be broken apart by alpha rays and for studying radioactivity.

1909 (P) *Guglielmo Marconi* and *Carl Ferdinand Braun* for developing wireless telegraphy.

1910 (P) *Johannes D. van der Waals* for studying the equation of state for gases and liquids (1881).

1911 (P) *Wilhelm Wien* for discovering Wien's law giving the peak of a blackbody spectrum (1893).

(C) *Marie Curie* for discovering radium and polonium (1898) and isolating radium.

1912 (P) *Nils Dalén* for inventing automatic gas regulators for lighthouses.

1913 (P) *Heike Kamerlingh Onnes* for the discovery of superconductivity and liquefying helium (1908).

1914 (P) *Max T. F. von Laue* for studying x-rays from their diffraction by crystals, showing that x-rays are electromagnetic waves (1912).

(C) *Theodore W. Richards* for determining the atomic weights of sixty elements, indicating the existence of isotopes.

1915 (P) *William Henry Bragg* and *William Lawrence Bragg*, his son, for studying the diffraction of x-rays in crystals.

1917 (P) *Charles Barkla* for studying atoms by x-ray scattering (1906).

1918 (P) *Max Planck* for discovering energy quanta (1900).

1919 (P) *Johannes Stark*, for discovering the Stark effect, the splitting of spectral lines in electric fields (1913).

1920 (P) *Charles-Édouard Guillaume* for discovering invar, a nickel–steel alloy with low coefficient of expansion.

(C) *Walther Nernst* for studying heat changes in chemical reactions and formulating the third law of thermodynamics (1918).

1921 (P) *Albert Einstein* for explaining the photoelectric effect and for his services to theoretical physics (1905).

(C) *Frederick Soddy* for studying the chemistry of radioactive substances and discovering isotopes (1912).

1922 (P) *Niels Bohr* for his model of the atom and its radiation (1913).

(C) *Francis W. Aston* for using the mass spectrograph to study atomic weights, thus discovering 212 of the 287 naturally occurring isotopes.

1923 (P) *Robert A. Millikan* for measuring the charge on an electron (1911) and for studying the photoelectric effect experimentally (1914).

1924 (P) *Karl M. G. Siegbahn* for his work in x-ray spectroscopy.

1925 (P) *James Franck* and *Gustav Hertz* for discovering the Franck–Hertz effect in electron–atom collisions.

1926 (P) *Jean-Baptiste Perrin* for studying Brownian motion to validate the discontinuous structure of matter and measure the size of atoms.

1927 (P) *Arthur Holly Compton* for discovering the Compton effect on x-rays, their change in wavelength when they collide with matter (1922), and *Charles T. R. Wilson* for inventing the cloud chamber, used to study charged particles (1906).

1928 (P) *Owen W. Richardson* for studying the thermionic effect and electrons emitted by hot metals (1911).

1929 (P) *Louis Victor de Broglie* for discovering the wave nature of electrons (1923).

1930 (P) *Chandrasekhara Venkata Raman* for studying Raman scattering, the scattering of light by atoms and molecules with a change in wavelength (1928).

1932 (P) *Werner Heisenberg* for creating quantum mechanics (1925).

1933 (P) *Erwin Schrödinger* and *Paul A. M. Dirac* for developing wave mechanics (1925) and relativistic quantum mechanics (1927).

(C) *Harold Urey* for discovering heavy hydrogen, deuterium (1931).

1935 (P) *James Chadwick* for discovering the neutron (1932).

(C) *Irène* and *Frédéric Joliot-Curie* for synthesizing new radioactive elements.

1936 (P) *Carl D. Anderson* for discovering the positron in particular and antimatter in general (1932) and *Victor F. Hess* for discovering cosmic rays.

(C) *Peter J. W. Debye* for studying dipole moments and diffraction of x-rays and electrons in gases.

1937 (P) *Clinton Davisson* and *George Thomson* for discovering the diffraction of electrons by crystals, confirming de Broglie's hypothesis (1927).

1938 (P) *Enrico Fermi* for producing the transuranic radioactive elements by neutron irradiation (1934–1937).

1939 (P) *Ernest O. Lawrence* for inventing the cyclotron.

1943 (P) *Otto Stern* for developing molecular-beam studies (1923) and using them to discover the magnetic moment of the proton (1933).

1944 (P) *Isidor I. Rabi* for discovering nuclear magnetic resonance in atomic and molecular beams.

(C) *Otto Hahn* for discovering nuclear fission (1938).

1945 (P) *Wolfgang Pauli* for discovering the exclusion principle (1924).

1946 (P) *Percy W. Bridgman* for studying physics at high pressures.

1947 (P) *Edward V. Appleton* for studying the ionosphere.

1948 (P) *Patrick M. S. Blackett* for studying nuclear physics with cloud-chamber photographs of cosmic-ray interactions.

1949 (P) *Hideki Yukawa* for predicting the existence of mesons (1935).

1950 (P) *Cecil F. Powell* for developing the method of studying cosmic rays with photographic emulsions and discovering new mesons.

1951 (P) *John D. Cockcroft* and *Ernest T. S. Walton* for transmuting nuclei in an accelerator (1932).

(C) *Edwin M. McMillan* for producing neptunium (1940) and *Glenn T. Seaborg* for producing plutonium (1941) and further transuranic elements.

1952 (P) *Felix Bloch* and *Edward Mills Purcell* for discovering nuclear magnetic resonance in liquids and gases (1946).

1953 (P) *Frits Zernike* for inventing the phase-contrast microscope, which uses interference to provide high contrast.

1954 (P) *Max Born* for interpreting the wave function as a probability (1926) and other quantum-mechanical discoveries and *Walther Bothe* for developing the coincidence method to study subatomic particles (1930–1931), producing, in particular, the particle interpreted by Chadwick as the neutron.

1955 (P) *Willis E. Lamb, Jr.*, for discovering the Lamb shift in the hydrogen spectrum (1947) and *Polykarp Kusch* for determining the magnetic moment of the electron (1947).

1956 (P) *John Bardeen, Walter H. Brattain*, and *William Shockley* for inventing the transistor (1956).

1957 (P) *T.-D. Lee* and *C.-N. Yang* for predicting that parity is not conserved in beta decay (1956).

1958 (P) *Pavel A. Čerenkov* for discovering Čerenkov radiation (1935) and *Ilya M. Frank* and *Igor Tamm* for interpreting it (1937).

1959 (P) *Emilio G. Segrè* and *Owen Chamberlain* for discovering the antiproton (1955).

1960 (P) *Donald A. Glaser* for inventing the bubble chamber to study elementary particles (1952).

(C) *Willard Libby* for developing radiocarbon dating (1947).

1961 (P) *Robert Hofstadter* for discovering internal structure in protons and neutrons and *Rudolf L. Mössbauer* for discovering the Mössbauer effect of recoilless gamma-ray emission (1957).

1962 (P) *Lev Davidovich Landau* for studying liquid helium and other condensed matter theoretically.

1963 (P) *Eugene P. Wigner* for applying symmetry principles to elementary-particle theory and *Maria Goeppert Mayer* and *J. Hans D. Jensen* for studying the shell model of nuclei (1947).

1964 (P) *Charles H. Townes, Nikolai G. Basov*, and *Alexandr M. Prokhorov* for developing masers (1951–1952) and lasers.

1965 (P) *Sin-itiro Tomonaga, Julian S. Schwinger*, and *Richard P. Feynman* for developing quantum electrodynamics (1948).

1966 (P) *Alfred Kastler* for his optical methods of studying atomic energy levels.

1967 (P) *Hans Albrecht Bethe* for discovering the routes of energy production in stars (1939).

1968 (P) *Luis W. Alvarez* for discovering resonance states of elementary particles.

1969 (P) *Murray Gell-Mann* for classifying elementary particles (1963).

1970 (P) *Hannes Alfvén* for developing magnetohydrodynamic theory and *Louis Eugène Félix Néel* for discovering antiferromagnetism and ferrimagnetism (1930s).

1971 (P) *Dennis Gabor* for developing holography (1947).

(C) *Gerhard Herzberg* for studying the structure of molecules spectroscopically.

1972 (P) *John Bardeen, Leon N. Cooper*, and *John Robert Schrieffer* for explaining superconductivity (1957).

1973 (P) *Leo Esaki* for discovering tunneling in semiconductors, *Ivar Giaever* for discovering tunneling in superconductors, and *Brian D. Josephson* for predicting the Josephson effect, which involves tunneling of paired electrons (1958–1962).

1974 (P) *Anthony Hewish* for discovering pulsars and *Martin Ryle* for developing radio interferometry.

1975 (P) *Aage N. Bohr, Ben R. Mottelson*, and *James Rainwater* for discovering why some nuclei take asymmetric shapes.

1976 (P) *Burton Richter* and *Samuel C. C. Ting* for discovering the J/psi particle, the first charmed particle (1974).

1977 (P) *John H. Van Vleck, Nevill F. Mott*, and *Philip W. Anderson* for studying solids quantum-mechanically.

(C) *Ilya Prigogine* for extending thermodynamics to show how life could arise in the face of the second law.

1978 (P) *Arno A. Penzias* and *Robert W. Wilson* for discovering the cosmic background radiation (1965) and *Pyotr Kapitsa* for his studies of liquid helium.

1979 (P) *Sheldon L. Glashow, Abdus Salam*, and *Steven Weinberg* for developing the theory that unified the weak and electromagnetic forces (1958–1971).

1980   (P) *Val Fitch* and *James W. Cronin* for discovering CP (charge-parity) violation (1964), which possibly explains the cosmological dominance of matter over antimatter.

1981   (P) *Nicolaas Bloembergen* and *Arthur L. Schawlow* for developing laser spectroscopy and *Kai M. Siegbahn* for developing high-resolution electron spectroscopy (1958).

1982   (P) *Kenneth G. Wilson* for developing a method of constructing theories of phase transitions to analyze critical phenomena.

1983   (P) *William A. Fowler* for theoretical studies of astrophysical nucleosynthesis and *Subramanyan Chandrasekhar* for studying physical processes of importance to stellar structure and evolution, including the prediction of white dwarf stars (1930).

1984   (P) *Carlo Rubbia* for discovering the W and Z particles, verifying the electroweak unification, and *Simon van der Meer*, for developing the method of stochastic cooling of the CERN beam that allowed the discovery (1982–1983).

1985   (P) *Klaus von Klitzing* for the quantized Hall effect, relating to conductivity in the presence of a magnetic field (1980).

1986   (P) *Ernst Ruska* for inventing the electron microscope (1931), and *Gerd Binnig* and *Heinrich Rohrer* for inventing the scanning-tunneling electron microscope (1981).

1987   (P) *J. Georg Bednorz* and *Karl Alex Müller* for the discovery of high-temperature superconductivity (1986).

1988   (P) *Leon M. Lederman, Melvin Schwartz,* and *Jack Steinberger* for a collaborative experiment that led to the development of a new tool for studying the weak nuclear force, which affects the radioactive decay of atoms.

1989   (P) *Norman Ramsay* for various techniques in atomic physics; and *Hans Dehmelt* and *Wolfgang Paul* for the development of techniques for trapping single-charge particles.

1990   (P) *Jerome Friedman, Henry Kendall* and *Richard Taylor* for experiments important to the development of the quark model.

1991   (P) *Pierre-Gilles de Gennes* for discovering that methods developed for studying order phenomena in simple systems can be generalized to more complex forms of matter, in particular to liquid crystals and polymers.

1992   (P) *George Charpak* for developing detectors that trace the paths of evanescent subatomic particles produced in particle accelerators.

1993   (P) *Russell Hulse* and *Joseph Taylor* for discovering evidence of gravitational waves.

1994   (P) *Bertram N. Brockhouse* and *Clifford G. Shull* for pioneering work in neutron scattering.

1995   (P) *Martin L. Perl* and *Frederick Reines* for discovering the tau particle and the neutrino, respectively.

1996   (P) *David M. Lee, Douglas C. Osheroff,* and *Robert C. Richardson* for developing a superfluid using helium-3.

1997   (P) *Steven Chu, Claude Cohen-Tannoudji,* and *William D. Phillips* for developing methods to cool and trap atoms with laser light.

1998   (P) *Robert B. Laughlin, Horst L. Störmer,* and *Daniel C. Tsui* for discovering a new form of quantum fluid with fractionally charged excitations.

1999   (P) *Gerardus 'T Hooft* and *Martinus J. G. Veltman* for studies in the quantum structure of electroweak interactions in physics.

2000   (P) *Zhores I. Alferov* and *Herbert Kroemer* for developing semiconductor heterostructures used in high-speed electronics and optoelectronics and *Jack St. Clair Kilby* for participating in the invention of the integrated circuit.

2001   (P) *Eric A. Cornell, Wolfgang Ketterle,* and *Carl E. Wieman* for the achievement of Bose–Einstein condensation in dilute gases of alkali atoms.

2002   (P) *Raymond Davis Jr.* and *Masatoshi Koshiba* for the detection of cosmic neutrinos and *Riccardo Giacconi* for contributions to astrophysics that led to the discovery of cosmic x-ray sources.

Answers to Odd-Numbered Problems

CHAPTER 1

1. 0.141 nm

3. $2.15 \times 10^4 \text{ kg/m}^3$

5. $4\pi\rho(r_2{}^3 - r_1{}^3)/3$

7. (a) $4.00 \text{ u} = 6.64 \times 10^{-24} \text{ g}$ (b) $55.9 \text{ u} = 9.28 \times 10^{-23} \text{ g}$ (c) $207 \text{ u} = 3.44 \times 10^{-22} \text{ g}$

9. $8.72 \times 10^{11} \text{ atom/s}$

11. (a) 72.6 kg (b) 7.82×10^{26} atoms

13. No.

15. (b) only

17. The units of G are $\text{m}^3/\text{kg} \cdot \text{s}^2$

19. 9.19 nm/s

21. $1.39 \times 10^3 \text{ m}^2$

23. (a) 0.071 4 gal/s (b) $2.70 \times 10^{-4} \text{ m}^3/\text{s}$ (c) 1.03 h

25. $11.4 \times 10^3 \text{ kg/m}^3$

27. 667 lb/s

29. (a) 190 yr (b) 2.32×10^4 times

31. $151 \mu\text{m}$

33. $1.00 \times 10^{10} \text{ lb}$

35. (a) 2.07 mm (b) 8.62×10^{13} times as large

37. 5.0 m

39. 2.86 cm

41. $\sim 10^6$ balls

43. $\sim 10^7$

45. $\sim 10^2 \text{ kg}; \sim 10^3 \text{ kg}$

47. $\sim 10^2$ tuners

49. (a) $(346 \pm 13) \text{ m}^2$ (b) $(66.0 \pm 1.3) \text{ m}$

51. $(1.61 \pm 0.17) \times 10^3 \text{ kg/m}^3$

53. 31 556 926.0 s

55. 5.2 m^3, 3%

57. $2.57 \times 10^{-10} \text{ m}$

59. $0.579 t \text{ ft}^3/\text{s} + 1.19 \times 10^{-9} t^2 \text{ ft}^3/\text{s}^2$

61. 3.41 m

63. 0.449%

65. (a) 0.529 cm/s (b) 11.5 cm/s

67. $1 \times 10^{10} \text{ gal/yr}$

69. $\sim 10^{11}$ stars

71. (a) $3.16 \times 10^7 \text{ s/yr}$ (b) $6.05 \times 10^{10} \text{ yr}$

CHAPTER 2

1. (a) 2.30 m/s (b) 16.1 m/s (c) 11.5 m/s

3. (a) 5 m/s (b) 1.2 m/s (c) -2.5 m/s (d) -3.3 m/s (e) 0

5. (a) 3.75 m/s (b) 0

7. (a) -2.4 m/s (b) -3.8 m/s (c) 4.0 s

9. (a) 5.0 m/s (b) -2.5 m/s (c) 0 (d) 5.0 m/s

11. $1.34 \times 10^4 \text{ m/s}^2$

13. (a) 52.4 ft/s, 55.0 ft/s, 55.5 ft/s, 57.4 ft/s (b) 0.598 ft/s^2

15. (a) 2.00 m (b) -3.00 m/s (c) -2.00 m/s^2

17. (a) 1.3 m/s^2 (b) 2.0 m/s^2 at 3 s (c) at $t = 6$ s and for $t > 10$ s (d) -1.5 m/s^2 at 8 s

19. $2.74 \times 10^5 \text{ m/s}^2$, which is $2.79 \times 10^4 \text{ g}$

21. -16.0 cm/s^2

23. (a) 4.53 s (b) 14.1 m/s

25. (a) 2.56 m (b) -3.00 m/s

27. (a) 20.0 s (b) no

29. 3.10 m/s

31. (a) -202 m/s^2 (b) 198 m

33. (a) $4.98 \times 10^{-9} \text{ s}$ (b) $1.20 \times 10^{15} \text{ m/s}^2$

35. (a) v_c/t_m (c) $v_c t_0/2$ (d) $v_c t_0$ (e) yes, no

37. (a) 3.00 m/s (b) 6.00 s (c) -0.300 m/s^2 (d) 2.05 m/s

39. 31 s

41. \$99.3/h

43. (a) 10.0 m/s up (b) 4.68 m/s down

45. (a) 2.17 s (b) -21.2 m/s (c) 2.23 s

47. (a) 29.4 m/s (b) 44.1 m

49. (a) 7.82 m (b) 0.782 s

51. 7.96 s

53. (a) $a_x(t) = a_{xi} + Jt$, $v_x(t) = v_{xi} + a_{xi}t + (1/2)Jt^2$, $x(t) = x_i + v_{xi}t + (1/2)a_{xi}t^2 + (1/6)Jt^3$

55. (a) $a = -(10.0 \times 10^7 \text{ m/s}^3) t + 3.00 \times 10^5 \text{ m/s}^2$; $x = -(1.67 \times 10^7 \text{ m/s}^3) t^3 + (1.50 \times 10^5 \text{ m/s}^2) t^2$ (b) $3.00 \times 10^{-3} \text{ s}$ (c) 450 m/s (d) 0.900 m

59. (a) Acela steadily cruises out of the city center at 45 mi/h. In less than a minute it smoothly speeds up to 150 mi/h; then its speed is nudged up to 170 mi/h. Next it smoothly slows to a very low speed, which it maintains as it rolls into a railroad yard. When it stops, it immediately begins backing up and smoothly speeds up to 50 mi/h in reverse, all in less than seven minutes after it started. (b) $2.2 \text{ mi/h/s} = 0.98 \text{ m/s}^2$ (c) 6.7 mi

61. 48.0 mm

63. (a) 15.0 s (b) 30.0 m/s (c) 225 m

65. (a) 5.43 m/s^2 and 3.83 m/s^2 (b) 10.9 m/s and 11.5 m/s (c) Maggie by 2.62 m

67. $\sim 10^3 \text{ m/s}^2$

69. (a) 3.00 s   (b) -15.3 m/s   (c) 31.4 m/s down and 34.8 m/s down

71. (c) $v_{\text{boy}}^2/h, 0$ (d) $v_{\text{boy}}, 0$

73. (a) 5.46 s   (b) 73.0 m
(c) $v_{\text{Stan}} = 22.6$ m/s, $v_{\text{Kathy}} = 26.7$ m/s

75. $0.577v$

CHAPTER 3

1. $(-2.75, -4.76)$ m

3. (a) 2.24 m   (b) 2.24 m at 26.6°

5. $y = 1.15$; $r = 2.31$

7. 70.0 m

9. 310 km at 57° S of W

11. (a) 10.0 m   (b) 15.7 m   (c) 0

13. (a) $\sim 10^5$ m   vertically   upward   (b) $\sim 10^3$ m   vertically upward

15. (a) 5.2 m at 60°   (b) 3.0 m at 330°   (c) 3.0 m at 150°
(d) 5.2 m at 300°

17. approximately 420 ft at $-3°$

19. 47.2 units at 122°

21. (a) $(-11.1\hat{\mathbf{i}} + 6.40\hat{\mathbf{j}})$ m   (b) $(1.65\hat{\mathbf{i}} + 2.86\hat{\mathbf{j}})$ cm
(c) $(-18.0\hat{\mathbf{i}} - 12.6\hat{\mathbf{j}})$ in.

23. (a) 5.00 blocks at 53.1° N of E   (b) 13.0 blocks

25. 358 m at 2.00° S of E

27. (a)

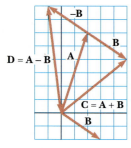

(b) $\mathbf{C} = 5.00\hat{\mathbf{i}} + 4.00\hat{\mathbf{j}}$ or 6.40 at 38.7°; $\mathbf{D} = -1.00\hat{\mathbf{i}} + 8.00\hat{\mathbf{j}}$ or 8.06 at 97.2°

29. 196 cm at 345°

31. (a) $2.00\hat{\mathbf{i}} - 6.00\hat{\mathbf{j}}$ (b) $4.00\hat{\mathbf{i}} + 2.00\hat{\mathbf{j}}$ (c) 6.32
(d) 4.47   (e) 288°; 26.6°

33. 9.48 m at 166°

35. (a) 185 N at 77.8° from the $+x$ axis
(b) $(-39.3\hat{\mathbf{i}} - 181\hat{\mathbf{j}})$ N

37. $\mathbf{A} + \mathbf{B} = (2.60\hat{\mathbf{i}} + 4.50\hat{\mathbf{j}})$ m

39. $|\mathbf{B}| = 7.81$, $\theta_x = 59.2°$, $\theta_y = 39.8°$, $\theta_z = 67.4°$

41. (a) $8.00\hat{\mathbf{i}} + 12.0\hat{\mathbf{j}} - 4.00\hat{\mathbf{k}}$ (b) $2.00\hat{\mathbf{i}} + 3.00\hat{\mathbf{j}} - 1.00\hat{\mathbf{k}}$
(c) $-24.0\hat{\mathbf{i}} - 36.0\hat{\mathbf{j}} + 12.0\hat{\mathbf{k}}$

43. (a) 5.92 m is the magnitude of $(5.00\hat{\mathbf{i}} - 1.00\hat{\mathbf{j}} - 3.00\hat{\mathbf{k}})$ m
(b) 19.0 m is the magnitude of $(4.00\hat{\mathbf{i}} - 11.0\hat{\mathbf{j}} - 15.0\hat{\mathbf{k}})$ m

45. 157 km

47. (a) $-3.00\hat{\mathbf{i}} + 2.00\hat{\mathbf{j}}$ (b) 3.61 at 146°
(c) $3.00\hat{\mathbf{i}} - 6.00\hat{\mathbf{j}}$

49. (a) $49.5\hat{\mathbf{i}} + 27.1\hat{\mathbf{j}}$ (b) 56.4 units at 28.7°

51. 1.15°

53. (a) 2.00, 1.00, 3.00   (b) 3.74   (c) $\theta_x = 57.7°$, $\theta_y = 74.5°$, $\theta_z = 36.7°$

55. 2.29 km

57. (a) 11.2 m   (b) 12.9 m at 36.4°

59. 240 m at 237°

61. 390 mi/h at 7.37° north of east

63. (a) zero   (b) zero

65. 106°

CHAPTER 4

1. (a) 4.87 km at 209° from east   (b) 23.3 m/s
(c) 13.5 m/s at 209°

3. 2.50 m/s

5. (a) $(2.00\hat{\mathbf{i}} + 3.00\hat{\mathbf{j}})$ m/s^2
(b) $(3.00t + t^2)\hat{\mathbf{i}}$ m $+ (1.50t^2 - 2.00t)\hat{\mathbf{j}}$ m

7. (a) $(0.800\hat{\mathbf{i}} - 0.300\hat{\mathbf{j}})$ m/s^2 (b) 339°
(c) $(360\hat{\mathbf{i}} - 72.7\hat{\mathbf{j}})$ m, $-15.2°$

9. (a) $x = 0.010\ 0$ m, $y = 2.41 \times 10^{-4}$ m
(b) $\mathbf{v} = (1.84 \times 10^7\hat{\mathbf{i}} + 8.78 \times 10^5\hat{\mathbf{j}})$ m/s
(c) $v = 1.85 \times 10^7$ m/s
(d) $\theta = 2.73°$

11. (a) $3.34\hat{\mathbf{i}}$ m/s   (b) $-50.9°$

13. (a) 20.0°   (b) 3.05 s

15. 53.1°

17. (a) 22.6 m   (b) 52.3 m   (c) 1.18 s

19. (a) The ball clears by 0.889 m while
(b) descending

21. (a) 18.1 m/s   (b) 1.13 m   (c) 2.79 m

23. 9.91 m/s

25. (a) 30.3 m/s   (b) 2.09 s

27. 377 m/s^2

29. 10.5 m/s, 219 m/s^2 inward

31. (a) 6.00 rev/s   (b) 1.52 km/s^2
(c) 1.28 km/s^2

33. 1.48 m/s^2 inward and 29.9° backward

35. (a) 13.0 m/s^2 (b) 5.70 m/s   (c) 7.50 m/s^2

37. $\theta = \tan^{-1}(1/4\pi) = 4.55°$

39. (a) 57.7 km/h at 60.0° west of vertical
(b) 28.9 km/h downward

41. 2.02×10^3 s; 21.0% longer

43. $t_{\text{Alan}} = \dfrac{2L/c}{1 - v^2/c^2}$, $t_{\text{Beth}} = \dfrac{2L/c}{\sqrt{1 - v^2/c^2}}$.
Beth returns first.

45. 15.3 m

47. (a) 101 m/s   (b) 32 700 ft   (c) 20.6 s
(d) 180 m/s

49. 54.4 m/s^2

51. (a) 41.7 m/s   (b) 3.81 s   (c) $(34.1\hat{\mathbf{i}} - 13.4\hat{\mathbf{j}})$ m/s;
36.7 m/s

53. (a) 25.0 m/s^2; 9.80 m/s^2

(b)

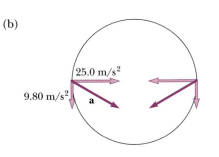

(c) 26.8 m/s^2 inward at 21.4° below the horizontal

55. (a) 26.6°   (b) 0.949

57. (a) 0.600 m   (b) 0.402 m   (c) 1.87 m/s^2 toward center
(d) 9.80 m/s^2 down

59. (a) 6.80 km   (b) 3.00 km vertically above the impact
point   (c) 66.2°

61. (a) 46.5 m/s   (b) −77.6°   (c) 6.34 s

63. (a) 20.0 m/s, 5.00 s   (b) $(16.0\hat{\mathbf{i}} − 27.1\hat{\mathbf{j}})$ m/s   (c) 6.53 s
(d) $24.5\hat{\mathbf{i}}$ m

65. (a) 22.9 m/s   (b) 360 m from the base of the cliff
(c) $\mathbf{v} = (114\hat{\mathbf{i}} − 44.3\hat{\mathbf{j}})$ m/s

67. (a) 43.2 m   (b) $(9.66\hat{\mathbf{i}} − 25.5\hat{\mathbf{j}})$ m/s

69. (a) 4.00 km/h   (b) 4.00 km/h

71. Safe distances are less than 270 m or greater than
3.48×10^3 m from the western shore.

CHAPTER 5

1. (a) 1/3   (b) 0.750 m/s^2

3. $(6.00\hat{\mathbf{i}} + 15.0\hat{\mathbf{j}})$ N; 16.2 N

5. (a) $(2.50\hat{\mathbf{i}} + 5.00\hat{\mathbf{j}})$ N   (b) 5.59 N

7. (a) 3.64×10^{-18} N   (b) 8.93×10^{-30} N is 408 billion
times smaller

9. 2.38 kN

11. (a) 5.00 m/s^2 at 36.9°   (b) 6.08 m/s^2 at 25.3°

13. (a) $\sim 10^{-22}$ m/s^2 (b) $\sim 10^{-23}$ m

15. (a) 15.0 lb up   (b) 5.00 lb up   (c) 0

17. 613 N

21. (a) 49.0 N   (b) 98.0 N   (c) 24.5 N

23. 8.66 N east

25. 3.73 m

27. A is in compression 3.83 kN and B is in tension 3.37 kN

29. 950 N

31. (a) $F_x > 19.6$ N   (b) $F_x \le − 78.4$ N

(c)

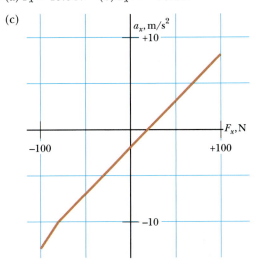

33. (a) 706 N   (b) 814 N   (c) 706 N   (d) 648 N

35. (a) 0.404   (b) 45.8 lb

37. (a) 256 m   (b) 42.7 m

39. (a) 1.10 s   (b) 0.875 s

41. (a) 1.78 m/s^2 (b) 0.368   (c) 9.37 N   (d) 2.67 m/s

43. 37.8 N

45. (a)

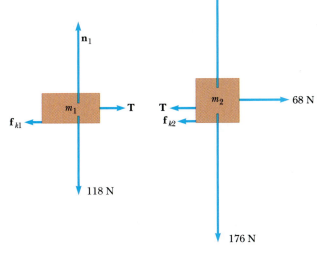

(b) 27.2 N, 1.29 m/s^2

47. $\mu_k = (3/5)\tan\theta$

49. (a) 8.05 N   (b) 53.2 N   (c) 42.0 N

51. (a)

(b) 0.408 m/s² (c) 83.3 N

53. (a) $F_A = mg(\sin \theta - \mu_s \cos \theta)$
(b) $F_B = mg(\sin \theta - \mu_s \cos \theta)/(\cos \theta + \mu_s \sin \theta)$
(c) A's job is easier
(d) B's job is easier

55. (a) $Mg/2$, $Mg/2$, $Mg/2$, $3Mg/2$, Mg (b) $Mg/2$

57. (b)

θ	0	15°	30°	45°	60°
$P(N)$	40.0	46.4	60.1	94.3	260

59. (a) 19.3° (b) 4.21 N

61. $(M + m_1 + m_2)(m_2 g/m_1)$

63. (a) $m_2 g \left[\dfrac{m_1 M}{m_1 M + m_2(m_1 + M)} \right]$

(b) $\dfrac{m_2 g(M + m_1)}{m_1 M + m_2(m_1 + M)}$

(c) $\dfrac{m_1 m_2 g}{m_1 M + m_2(m_1 + M)}$

(d) $\dfrac{M m_2 g}{m_1 M + m_2(m_1 + M)}$

65. (c) 3.56 N

67. (a) $T = f/(2 \sin \theta)$ (b) 410 N

69. (a) 30.7° (b) 0.843 N

71. 0.060 0 m

73. (a) $T_1 = \dfrac{2mg}{\sin \theta_1}$

$T_2 = \dfrac{mg}{\sin \theta_2} = \dfrac{mg}{\sin[\tan^{-1}(\frac{1}{2}\tan \theta_1)]}$

$T_3 = 2mg/\tan \theta_1$

(b) $\theta_2 = \tan^{-1}\left(\dfrac{\tan \theta_1}{2} \right)$

CHAPTER 6

1. Any speed up to 8.08 m/s

3. (a) 8.32×10^{-8} N toward the nucleus
(b) 9.13×10^{22} m/s² inward

5. (a) static friction (b) 0.085 0

7. $v \le 14.3$ m/s

9. (a) 68.6 N toward the center of the circle and 784 N up
(b) 0.857 m/s²

11. (a) 108 N (b) 56.2 N

13. (a) 4.81 m/s (b) 700 N up

15. No. The jungle lord needs a vine of tensile strength 1.38 kN.

17. 3.13 m/s

19. (a) 2.49×10^4 N up (b) 12.1 m/s

21. (a) 3.60 m/s² (b) zero (c) An observer in the car (a noninertial frame) claims an 18.0-N force toward the left and an 18.0-N force toward the right. An inertial observer (outside the car) claims only an 18.0-N force toward the right.

23. (a) 17.0° (b) 5.12 N

25. (a) 491 N (b) 50.1 kg (c) 2.00 m/s²

27. (a) $v = [2(a - \mu_k g)\ell]^{1/2}$; (b) $v' = (2\mu_k g\ell/v)$, where $v = [2(a - \mu_k g)\ell]^{1/2}$

29. 93.8 N

31. 0.092 7°

33. (a) 32.7 s⁻¹ (b) 9.80 m/s² down (c) 4.90 m/s² down

35. 3.01 N up

37. (a) 1.47 N·s/m (b) 2.04×10^{-3} s (c) 2.94×10^{-2} N

39. (a) 0.034 7 s⁻¹ (b) 2.50 m/s (c) $a = -cv$

41. (a) $x = k^{-1} \ln(1 + kv_0 t)$ (b) $v = v_0 e^{-kx}$

43. $\sim 10^1$ N

45. (a) 13.7 m/s down

(b)

t (s)	x (m)	v (m/s)
0	0	0
0.2	0	−1.96
0.4	−0.392	−3.88
1.0	−3.77	−8.71
2.0	−14.4	−12.56
4.0	−41.0	−13.67

47. (a) 49.5 m/s down and 4.95 m/s down

(b)

t (s)	y (m)	v (m/s)
0	1 000	0
1	995	−9.7
2	980	−18.6
10	674	−47.7
10.1	671	−16.7
12	659	−4.95
145	0	−4.95

49. (a) 2.33×10^{-4} kg/m (b) 57.0 m/s (c) 44.9 m/s. The second trajectory is higher and shorter than the first. In both cases, the ball attains maximum height when it has covered 56% of its horizontal range, and attains minimum speed a little later. The impact speeds are also similar, 30 m/s and 29 m/s.

51. (a) 11.5 kN   (b) 14.1 m/s = 50.9 km/h

53. (a) 0.016 2 kg/m   (b) $\frac{1}{2}D\rho A$ (c) 0.778   (d) 1.5%
(e) For stacked coffee filters falling in air at terminal speed, the graph of resistive force as a function of squared speed demonstrates that the force is proportional to the speed squared, within the experimental uncertainty, estimated as ± 2%. This proportionality agrees with that predicted by the theoretical equation $R = \frac{1}{2}D\rho Av^2$. The value of the constant slope of the graph implies that the drag coefficient for coffee filters is $D = 0.78 \pm 2\%$.

55. $g(\cos \phi \tan \theta - \sin \phi)$

57. (b) 732 N down at the equator and 735 N down at the poles

59. (a) 967 lb   (b) − 647 lb (pilot must be strapped in)   (c) Speed and radius of path can be adjusted so that $v^2 = gR$.

61. (a) 1.58 m/s²   (b) 455 N   (c) 329 N   (d) 397 N upward and 9.15° inward

63. (a) 5.19 m/s   (b) T = 555 N

T cos 28.0°

T sin 28.0°

490 N

65. (b) 2.54 s; 23.6 rev/min

67. (a) $v_{\text{min}} = \sqrt{\dfrac{Rg(\tan \theta - \mu_s)}{1 + \mu_s \tan \theta}}$, $v_{\text{max}} = \sqrt{\dfrac{Rg(\tan \theta + \mu_s)}{1 - \mu_s \tan \theta}}$
(b) $\mu_s = \tan\theta$ (c) 8.57 m/s ≤ v ≤ 16.6 m/s

69. (a) 0.013 2 m/s   (b) 1.03 m/s   (c) 6.87 m/s

71. 12.8 N

73. $\Sigma\mathbf{F} = -km\mathbf{v}$

CHAPTER 7

1. (a) 31.9 J   (b) 0   (c) 0   (d) 31.9 J

3. − 4.70 kJ

5. 28.9

7. (a) 16.0 J   (b) 36.9°

9. (a) 11.3°   (b) 156°   (c) 82.3°

11. (a) 24.0 J   (b) − 3.00 J   (c) 21.0 J

13. (a) 7.50 J   (b) 15.0 J   (c) 7.50 J   (d) 30.0 J

15. (a) 0.938 cm   (b) 1.25 J

17. (a) 0.768 m   (b) 1.68 × 10⁵ J

19. 12.0 J

21. (a) 0.020 4 m   (b) 720 N/m

23. kg/s²

25. (a) 33.8 J   (b) 135 J

27. 878 kN up

29. (a) 4.56 kJ   (b) 6.34 kN   (c) 422 km/s²   (d) 6.34 kN

31. (a) 650 J   (b) 588 J   (c) 0   (d) 0   (e) 62.0 J
(f) 1.76 m/s

33. (a) − 168 J   (b) 184 J   (c) 500 J   (d) 148 J
(e) 5.65 m/s

35. 2.04 m

37. 875 W

39. (a) 20.6 kJ   (b) 686 W

41. $46.2

43. (a) 423 mi/gal   (b) 776 mi/gal

45. (a) 0.013 5 gal   (b) 73.8   (c) 8.08 kW

47. 2.92 m/s

49. (a) $(2 + 24t^2 + 72t^4)$ J   (b) $12t$ m/s²; $48t$ N
(c) $(48t + 288t^3)$ W   (d) 1 250 J

51. $k_1 x_{\text{max}}^2/2 + k_2 x_{\text{max}}^3/3$

53. (a) $\sqrt{2W/m}$ (b) W/d

55. (b) 240 W

57. (a) 1.38 × 10⁴ J   (b) 3.02 × 10⁴ W

59. (a) $\mathscr{P} = 2Mgv_T$ (b) $\mathscr{P} = 24Mgv_T$

61. (a) 4.12 m   (b) 3.35 m

63. 1.68 m/s

65. − 1.37 × 10⁻²¹ J

67. 0.799 J

69. (b) For a block of weight w pushed over a rough horizontal surface at constant velocity, $b = \mu_k$. For a load pulled vertically upward at constant velocity, $b = 1$.

CHAPTER 8

1. (a) 259 kJ, 0, − 259 kJ   (b) 0, − 259 kJ, − 259 kJ

3. 22.0 kW

5. (a) $v = (3gR)^{1/2}$ (b) 0.098 0 N down

7. (a) 1.47 m/s   (b) 1.35 m/s

9. (a) 2.29 m/s   (b) 1.98 m/s

11. 10.2 m

13. (a) 4.43 m/s   (b) 5.00 m

15. 5.49 m/s

17. (a) 18.5 km, 51.0 km   (b) 10.0 MJ

19. (a) 25.8 m   (b) 27.1 m/s²

21. (a) − 196 J   (b) − 196 J   (c) − 196 J. The force is conservative.

23. (a) 125 J   (b) 50.0 J   (c) 66.7 J   (d) Nonconservative. The results differ.

25. (a) − 9.00 J; no; the force is conservative.   (b) 3.39 m/s
(c) 9.00 J

27. 26.5 m/s

29. 6.92 m/s

31. 3.74 m/s

33. (a) -160 J   (b) 73.5 J   (c) 28.8 N   (d) 0.679

35. (a) 1.40 m/s   (b) 4.60 cm after release   (c) 1.79 m/s

37. (a) 0.381 m   (b) 0.143 m   (c) 0.371 m

39. (a) $a_x = -\mu_k gx/L$ (b) $v = (\mu_k gL)^{1/2}$

41. (a) 40.0 J   (b) -40.0 J   (c) 62.5 J

43. (A/r^2) away from the other particle

45. (a) $+$ at Ⓑ, $-$ at Ⓓ, 0 at Ⓐ, Ⓒ, and Ⓔ   (b) Ⓒ stable; Ⓐ and Ⓔ unstable

    (c)

47. (b)

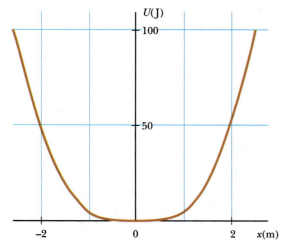

    Equilibrium at $x = 0$

    (c) 0.823 m/s

49. $\sim 10^3$ W peak or $\sim 10^2$ W sustainable

51. 48.2°

53. (a) 0.225 J   (b) $\Delta E_{\text{mech}} = -0.363$ J   (c) No; the normal force changes in a complicated way.

55. (a) 23.6 cm   (b) 5.90 m/s² up the incline; no.   (c) Gravitational potential energy is transformed into kinetic energy plus elastic potential energy and then entirely into elastic potential energy.

57. 0.328

59. 1.24 m/s

61. (a) 0.400 m   (b) 4.10 m/s   (c) The block stays on the track.

63. $(h/5)(4\sin^2\theta + 1)$

65. (a) 6.15 m/s   (b) 9.87 m/s

67. (a) 11.1 m/s   (b) 19.6 m/s²   upward   (c) 2.23×10^3 N upward   (d) 1.01×10^3 J   (e) 5.14 m/s   (f) 1.35 m   (g) 1.39 s

69. (b) 1.44 m   (c) 0.400 m   (d) No. A very strong wind pulls the string out horizontally. The largest possible equilibrium height is equal to L.

73. (a) $2.5R$

CHAPTER 9

1. (a) $(9.00\hat{\mathbf{i}} - 12.0\hat{\mathbf{j}})$ kg·m/s   (b) 15.0 kg·m/s at 307°

3. $\sim 10^{-23}$ m/s

5. (b) $p = \sqrt{2mK}$

7. (a) 13.5 N·s   (b) 9.00 kN   (c) 18.0 kN

9. 260 N normal to the wall

11. (a) $(9.05\hat{\mathbf{i}} + 6.12\hat{\mathbf{j}})$ N·s   (b) $(377\hat{\mathbf{i}} + 255\hat{\mathbf{j}})$ N

13. 15.0 N in the direction of the initial velocity of the exiting water stream

15. 65.2 m/s

17. 301 m/s

19. (a) 2.50 m/s   (b) 37.5 kJ   (c) Each process is the time-reversal of the other. The same momentum conservation equation describes both.

21. (a) $v_{gx} = 1.15$ m/s   (b) $v_{px} = -0.346$ m/s

23. (a) 0.284   (b) 115 fJ and 45.4 fJ

25. 91.2 m/s

27. (a) 2.24 m/s to the right   (b) No. Coupling order makes no difference.

29. $v_{\text{orange}} = 3.99$ m/s, $v_{\text{yellow}} = 3.01$ m/s

31. $v_{\text{green}} = 7.07$ m/s, $v_{\text{blue}} = 5.89$ m/s

33. 2.50 m/s at $-60.0°$

35. $(3.00\hat{\mathbf{i}} - 1.20\hat{\mathbf{j}})$ m/s

37. (a) $(-9.33\hat{\mathbf{i}} - 8.33\hat{\mathbf{j}})$ Mm/s   (b) 439 fJ

39. 0.006 73 nm from the oxygen nucleus along the bisector of the angle

41. $\mathbf{r}_{\text{CM}} = (11.7\hat{\mathbf{i}} + 13.3\hat{\mathbf{j}})$ cm

43. (a) 15.9 g   (b) 0.153 m

45. (a) $(1.40\hat{\mathbf{i}} + 2.40\hat{\mathbf{j}})$ m/s   (b) $(7.00\hat{\mathbf{i}} + 12.0\hat{\mathbf{j}})$ kg·m/s

47. 0.700 m

49. (a) 39.0 MN   (b) 3.20 m/s² up

51. (a) 442 metric tons   (b) 19.2 metric tons

53. 4.41 kg

55. (a) $1.33\hat{\mathbf{i}}$ m/s   (b) $-235\hat{\mathbf{i}}$ N   (c) 0.680 s   (d) $-160\hat{\mathbf{i}}$ N·s and $+160\hat{\mathbf{i}}$ N·s   (e) 1.81 m   (f) 0.454 m   (g) -427 J   (h) $+107$ J   (i) Equal friction forces act through different distances on person and cart, to do different amounts of work on them. The total work on both together, -320 J, becomes $+320$ J of extra internal energy in this perfectly inelastic collision.

57. 240 s

59. (a) 0; inelastic (b) $(-0.250\hat{\mathbf{i}} + 0.750\hat{\mathbf{j}} - 2.00\hat{\mathbf{k}})$ m/s; perfectly inelastic (c) either $a = -6.74$ with $\mathbf{v} = -0.419\hat{\mathbf{k}}$ m/s or $a = 2.74$ with $\mathbf{v} = -3.58\hat{\mathbf{k}}$ m/s

61. (a) $v_i = v(m + \rho V)/m$ (b) The cart slows with constant acceleration and eventually comes to rest.

63. (a) $m/M = 0.403$ (b) No changes; no difference.

65. (a) 6.29 m/s (b) 6.16 m/s

67. (a) 100 m/s (b) 374 J

69. (a) $(20.0\hat{\mathbf{i}} + 7.00\hat{\mathbf{j}})$ m/s (b) $4.00\hat{\mathbf{i}}$ m/s^2 (c) $4.00\hat{\mathbf{i}}$ m/s^2 (d) $(50.0\hat{\mathbf{i}} + 35.0\hat{\mathbf{j}})$ m (e) 600 J (f) 674 J (g) 674 J

71. $(3Mgx/L)\hat{\mathbf{j}}$

73. $\dfrac{m_1(R + \ell/2)}{(m_1 + m_2)}$

CHAPTER 10

1. (a) 5.00 rad, 10.0 rad/s, 4.00 rad/s^2 (b) 53.0 rad, 22.0 rad/s, 4.00 rad/s^2

3. (a) 4.00 rad/s^2 (b) 18.0 rad

5. (a) 5.24 s (b) 27.4 rad

7. 50.0 rev

9. (a) 7.27×10^{-5} rad/s (b) 2.57×10^4 s = 428 min

11. $\sim 10^7$ rev

13. (a) 8.00 rad/s (b) 8.00 m/s, $a_r = -64.0$ m/s^2, $a_t = 4.00$ m/s^2 (c) 9.00 rad

15. (a) 25.0 rad/s (b) 39.8 rad/s^2 (c) 0.628 s

17. (a) 126 rad/s (b) 3.77 m/s (c) 1.26 km/s^2 (d) 20.1 m

19. (a) $\omega(2h^3/g)^{1/2}$ (b) 0.011 6 m (c) Yes; the deflection is only 0.02% of the original height.

21. (a) 143 kg·m^2 (b) 2.57 kJ

23. $11mL^2/12$

25. 5.80 kg·m^2; the height makes no difference

29. $(23/48)MR^2\omega^2$

31. -3.55 N·m

33. 8.02×10^3 N

35. (a) 24.0 N·m (b) 0.035 6 rad/s^2 (c) 1.07 m/s^2

37. (a) 0.309 m/s^2 (b) 7.67 N and 9.22 N

39. 21.5 N

41. 24.5 km

43. (a) 1.59 m/s (b) 53.1 rad/s

45. (a) 11.4 N, 7.57 m/s^2, 9.53 m/s down (b) 9.53 m/s

49. (a) $2(Rg/3)^{1/2}$ (b) $4(Rg/3)^{1/2}$ (c) $(Rg)^{1/2}$

51. (a) 500 J (b) 250 J (c) 750 J

53. (a) $\frac{2}{3} g \sin\theta$ for the disk, larger than $\frac{1}{2} g \sin\theta$ for the hoop (b) $\frac{1}{3} \tan\theta$

55. 1.21×10^{-4} kg·m^2; height is unnecessary

57. $\frac{1}{3}\ell$

59. (a) 4.00 J (b) 1.60 s (c) yes

61. (a) $(3g/L)^{1/2}$ (b) $3g/2L$ (c) $-\frac{3}{2}g\hat{\mathbf{i}} - \frac{3}{4}g\hat{\mathbf{j}}$ (d) $-\frac{3}{2}Mg\hat{\mathbf{i}} + \frac{1}{4}Mg\hat{\mathbf{j}}$

63. -0.322 rad/s^2

65. (b) $2gM(\sin\theta - \mu\cos\theta)(m + 2M)^{-1}$

67. (a) $\sim -10^{-22}$ s^{-2} (b) $\sim -10^{16}$ N·m (c) $\sim 10^{13}$ m

71. (a) 118 N and 156 N (b) 1.17 kg·m^2

73. (a) $\alpha = -0.176$ rad/s^2 (b) 1.29 rev (c) 9.26 rev

75. (a) 61.2 J (b) 50.8 J

79. (a) $2.70R$ (b) $F_x = -20mg/7$, $F_y = -mg$

81. $\sim 10^1$ m

83. (a) $(3gh/4)^{1/2}$ (b) $(3gh/4)^{1/2}$

85. (c) $(8Fd/3M)^{1/2}$

87. $\mathbf{F}_1$ to right, $\mathbf{F}_2$ no rolling, $\mathbf{F}_3$ and $\mathbf{F}_4$ to left

CHAPTER 11

1. $-7.00\hat{\mathbf{i}} + 16.0\hat{\mathbf{j}} - 10.0\hat{\mathbf{k}}$

3. (a) $-17.0\hat{\mathbf{k}}$ (b) 70.6°

5. 0.343 N·m horizontally north

7. 45.0°

9. $F_3 = F_1 + F_2$; no

11. $(17.5\hat{\mathbf{k}})$ kg·m^2/s

13. $(60.0\hat{\mathbf{k}})$ kg·m^2/s

15. $mvR[\cos(vt/R) + 1]\hat{\mathbf{k}}$

17. (a) zero (b) $[-mv_i^3 \sin^2\theta \cos\theta/2g]\hat{\mathbf{k}}$ (c) $[-2mv_i^3 \sin^2\theta \cos\theta/g]\hat{\mathbf{k}}$ (d) The downward gravitational force exerts a torque in the $-z$ direction.

19. $-m\ell gt \cos\theta\hat{\mathbf{k}}$

23. (a) 0.360 kg·m^2/s (b) 0.540 kg·m^2/s

25. (a) 0.433 kg·m^2/s (b) 1.73 kg·m^2/s

27. (a) 1.57×10^8 kg·m^2/s (b) 6.26×10^3 s = 1.74 h

29. 7.14 rev/min

31. (a) 9.20 rad/s (b) 9.20 rad/s

33. (a) 0.360 rad/s counterclockwise (b) 99.9 J

35. (a) $mv\ell$ down (b) $M/(M + m)$

37. (a) $\omega = 2mv_i d/(M + 2m)R^2$ (b) No; some mechanical energy changes into internal energy

39. $\sim 10^{-13}$ rad/s

41. 5.45×10^{22} N·m

43. 7.50×10^{-11} s

45. (a) $7md^2/3$ (b) $mgd\,\hat{\mathbf{k}}$ (c) $3g/7d$ counterclockwise (d) $2g/7$ upward (e) mgd (f) $\sqrt{6g/7d}$ (g) $m\sqrt{14gd^3/3}$ (h) $\sqrt{2gd/21}$

47. 0.910 km/s

49. (a) $v_i r_i/r$ (b) $T = (mv_i^2 r_i^2)r^{-3}$ (c) $\frac{1}{2}mv_i^2(r_i^2/r^2 - 1)$ (d) 4.50 m/s, 10.1 N, 0.450 J

51. (a) 3 750 kg·m^2/s (b) 1.88 kJ (c) 3 750 kg·m^2/s (d) 10.0 m/s (e) 7.50 kJ (f) 5.62 kJ

53. An increase of 0.550 s

55. $4[ga(\sqrt{2} - 1)/3]^{1/2}$

CHAPTER 12

1. 10.0 N up; 6.00 N·m counterclockwise

3. $[(m_1 + m_b)d + (m_1\ell/2)]/m_2$

5. (3.85 cm, 6.85 cm)

7. $(-1.50$ m, -1.50 m)

9. 177 kg

11. 8.33%

13. (a) $f_s = 268$ N, $n = 1\,300$ N   (b) 0.324

15. (a) 1.04 kN at 60.0°   (b) $(370\hat{i} + 900\hat{j})$ N

17. 2.94 kN on each rear wheel and 4.41 kN on each front wheel

19. (a) 29.9 N   (b) 22.2 N

21. (a) 1.73 rad/s²   (b) 1.56 rad/s
    (c) $(-4.72\hat{i} + 6.62\hat{j})$ kN   (d) $38.9\hat{j}$ kN

23. 2.82 m

25. 88.2 N and 58.8 N

27. 4.90 mm

29. 10×10^{10} N/m²

31. 23.8 μm

33. (a) 3.14×10^4 N   (b) 6.28×10^4 N

35. 1.65×10^8 N/m²

37. 0.860 mm

39. $n_A = 5.98 \times 10^5$ N, $n_B = 4.80 \times 10^5$ N

41. 9.00 ft

43. (a)

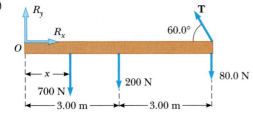

(b) $T = 343$ N; $R_x = 171$ N to the right, $R_y = 683$ N up
(c) 5.13 m

45. (a) $T = F_g(L + d)/\sin\theta\,(2L + d)$
    (b) $R_x = F_g(L + d)\cot\theta/(2L + d)$; $R_y = F_gL/(2L + d)$

47. $\mathbf{F}_A = (-6.47 \times 10^5\hat{i} + 1.27 \times 10^5\hat{j})$ N,
    $\mathbf{F}_B = 6.47 \times 10^5\hat{i}$ N

49. 5.08 kN; $R_x = 4.77$ kN, $R_y = 8.26$ kN

51. $T = 2.71$ kN, $R_x = 2.65$ kN

53. (a) 20.1 cm to the left of the front edge; $\mu_k = 0.571$
    (b) 0.501 m

55. (a) Assume $\mu_s < \cot\theta$.
    $M = (m/2)(2\mu_s\sin\theta - \cos\theta)(\cos\theta - \mu_s\sin\theta)^{-1}$
    (b) $R = (m + M)g(1 + \mu_s^2)^{1/2}$;
    $F = g[M^2 + \mu_s^2(m + M)^2]^{1/2}$

57. (a) 133 N   (b) $n_A = 429$ N and $n_B = 257$ N
    (c) $R_x = 133$ N and $R_y = -257$ N

59. 66.7 N

63. 1.09 m

65. (a) 4 500 N   (b) 4.50×10^6 N/m²   (c) The board will break.

67. 5.73 rad/s

69. $n_A = 11.0$ kN, $n_E = 3.67$ kN; $F_{AB} = F_{DE} = 7.35$ kN   compression; $F_{AC} = F_{CE} = 6.37$ kN   compression; $F_{BC} = F_{CD} = 4.24$ kN tension; $F_{BD} = 8.49$ kN compression

71. (a) $P_y = (F_g/L)(d - ah/g)$ (b) 0.306 m
    (c) $(-306\hat{i} + 5.53\hat{j})$ N

73. Decrease h, increase d

CHAPTER 13

1. $\sim10^{-7}$ N toward you

3. (a) 2.50×10^{-5} N toward the 500-kg object
   (b) between the objects and 0.245 m from the 500-kg object

5. $(-100\hat{i} + 59.3\hat{j})$ pN

7. 7.41×10^{-10} N

9. 0.613 m/s² toward the Earth

11. (a) 3.46×10^8 m   (b) 3.34×10^{-3} m/s² toward the Earth

13. 1.26×10^{32} kg

15. 1.90×10^{27} kg

17. 35.2 AU

19. 8.92×10^7 m

21. After 393 yr, Mercury would be farther from the Sun than Pluto

23. $\mathbf{g} = (Gm/\ell^2)(\frac{1}{2} + \sqrt{2})$ toward the opposite corner

25. $\mathbf{g} = 2MGr(r^2 + a^2)^{-3/2}$ toward the center of mass

27. 4.17×10^{10} J

29. (a) 1.84×10^9 kg/m³   (b) 3.27×10^6 m/s²
    (c) -2.08×10^{13} J

31. (a) -1.67×10^{-14} J   (b) at the center

33. 1.66×10^4 m/s

37. (a) 5.30×10^3 s   (b) 7.79 km/s   (c) 6.43×10^9 J

39. 469 MJ

41. 15.6 km/s

43. (b) 1.00×10^7 m   (c) 1.00×10^4 m/s

45. (a) 0.980   (b) 127 yr   (c) -2.13×10^{17} J

49. (b) $2[Gm^3(1/2r - 1/R)]^{1/2}$

51. (b) 1.10×10^{32} kg

53. (a) -7.04×10^4 J   (b) -1.57×10^5 J   (c) 13.2 m/s

55. 7.79×10^{14} kg

57. $\omega = 0.057\,2$ rad/s or 1 rev in 110 s

59. $v_{esc} = (8\pi G\rho/3)^{1/2}R$

61. (a) $m_2(2G/d)^{1/2}(m_1 + m_2)^{-1/2}$ and
    $m_1(2G/d)^{1/2}(m_1 + m_2)^{-1/2}$;
    relative speed $(2G/d)^{1/2}(m_1 + m_2)^{1/2}$
    (b) 1.07×10^{32} J and 2.67×10^{31} J

63. (a) 8.50×10^8 J   (b) 2.71×10^9 J

65. (a) 200 Myr   (b) $\sim10^{41}$ kg; $\sim10^{11}$ stars

67. $(GM_E/4R_E)^{1/2}$

71.

t (s)	x (m)	y (m)	v_x (m/s)	v_y (m/s)
0	0	12 740 000	5 000	0
10	50 000	12 740 000	4 999.9	−24.6
20	99 999	12 739 754	4 999.7	−49.1
30	149 996	12 739 263	4 999.4	−73.7...

The object does not hit the Earth; its minimum radius is $1.33R_E$ as shown in the diagram below. Its period is 1.09×10^4 s. A circular orbit would require speed 5.60 km/s.

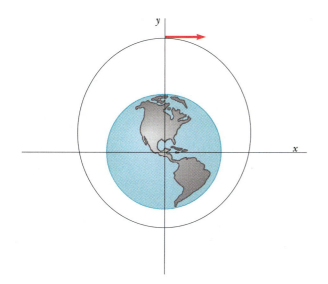

CHAPTER 14

1. 0.111 kg

3. 6.24 MPa

5. 5.27×10^{18} kg

7. 1.62 m

9. 7.74×10^{-3} m^2

11. 271 kN horizontally backward

13. $P_0 + \frac{1}{2}\rho d\sqrt{g^2 + a^2}$

15. 0.722 mm

17. 10.5 m; no; some alcohol and water evaporate

19. 98.6 kPa

21. (a) 1.57 Pa, 1.55×10^{-2} atm, 11.8 mm Hg   (b) The fluid level in the tap should rise.   (c) Blockage of flow of the cerebrospinal fluid

23. 0.258 N

25. (a) 9.80 N   (b) 6.17 N

27. (a) $1.017\ 9 \times 10^3$ N down, $1.029\ 7 \times 10^3$ N up   (b) 86.2 N

29. (a) 7.00 cm   (b) 2.80 kg

33. 1 430 m^3

35. 1 250 kg/m^3 and 500 kg/m^3

37. 1.28×10^4 m^2

39. (a) 17.7 m/s   (b) 1.73 mm

41. 31.6 m/s

43. 0.247 cm

45. (a) 1 atm + 15.0 MPa   (b) 2.95 m/s   (c) 4.34 kPa

47. 2.51×10^{-3} m^3/s

49. 103 m/s

51. (a) 4.43 m/s
   (b) The siphon can be no higher than 10.3 m.

53. 12.6 m/s

55. 1.91 m

59. 0.604 m

63. 17.3 N and 31.7 N

65. 90.04%

67. 758 Pa

69. 4.43 m/s

71. (a) 1.25 cm   (b) 13.8 m/s

73. (a) 3.307 g   (b) 3.271 g   (c) 3.48×10^{-4} N

75. (c) 1.70 m^2

CHAPTER 15

1. (a) The motion repeats precisely.   (b) 1.82 s
   (c) No, the force is not in the form of Hooke's law.

3. (a) 1.50 Hz, 0.667 s   (b) 4.00 m   (c) π rad
   (d) 2.83 m

5. (b) 18.8 cm/s, 0.333 s   (c) 178 cm/s^2, 0.500 s
   (d) 12.0 cm

7. (a) 2.40 s   (b) 0.417 Hz   (c) 2.62 rad/s

9. 40.9 N/m

11. (a) 40.0 cm/s, 160 cm/s^2 (b) 32.0 cm/s, -96.0 cm/s^2
   (c) 0.232 s

13. 0.628 m/s

15. (a) 0.542 kg   (b) 1.81 s   (c) 1.20 m/s^2

17. 2.23 m/s

19. (a) 28.0 mJ   (b) 1.02 m/s   (c) 12.2 mJ   (d) 15.8 mJ

21. (a) E increases by a factor of 4.   (b) $v_{\max}$ is doubled.
   (c) $a_{\max}$ is doubled.   (d) Period is unchanged.

23. 2.60 cm and -2.60 cm

25. (b) 0.628 s

27. (a) 35.7 m   (b) 29.1 s

29. $\sim 10^0$ s

31. Assuming simple harmonic motion, (a) 0.820 m/s
   (b) 2.57 rad/s^2 (c) 0.641 N. More precisely,
   (a) 0.817 m/s   (b) 2.54 rad/s^2 (c) 0.634 N

35. 0.944 kg·m^2

39. (a) 5.00×10^{-7} kg·m^2 (b) 3.16×10^{-4} N·m/rad

41. 1.00×10^{-3} s^{-1}

43. (a) 7.00 Hz   (b) 2.00%   (c) 10.6 s

45. (a) 1.00 s   (b) 5.09 cm

47. 318 N

49. 1.74 Hz

51. (a) $2Mg$; $Mg(1 + y/L)$
   (b) $T = (4\pi/3)(2L/g)^{1/2}$; 2.68 s

53. 6.62 cm

55. 9.19×10^{13} Hz

57. (a)

(b) $\dfrac{dT}{dt} = \dfrac{\pi(dM/dt)}{2\rho a^2 g^{1/2}[L_i + (dM/dt)\,t/2\rho a^2]^{1/2}}$

(c) $T = 2\pi g^{-1/2}\,[L_i + (dM/dt)\,t/2\rho a^2]^{1/2}$

59. $f = (2\pi L)^{-1}\,(gL + kh^2/M)^{1/2}$

61. (b) 1.23 Hz

63. (a) 3.00 s   (b) 14.3 J   (c) 25.5°

65. If the cyclist goes over them at one certain speed, the washboard bumps can excite a resonance vibration of the bike, so large in amplitude as to make the rider lose control. $\sim 10^1$ m

73. For $\theta_{max} = 5.00°$ there is precise agreement.
For $\theta_{max} = 100°$ there are large differences, and the period is 23% greater than small-angle period.

75. (b) after 42.1 min

CHAPTER 16

1. $y = 6\,[(x - 4.5t)^2 + 3]^{-1}$

3. (a) left   (b) 5.00 m/s

5. 184 km

7. 0.319 m

9. 2.00 cm, 2.98 m, 0.576 Hz, 1.72 m/s

11. (a) 3.77 m/s   (b) 118 m/s^2

13. (a) 0.250 m   (b) 40.0 rad/s   (c) 0.300 rad/m
(d) 20.9 m   (e) 133 m/s   (f) $+x$

15. (a) $y = (8.00\text{ cm})\sin(7.85x + 6\pi t)$
(b) $y = (8.00\text{ cm})\sin(7.85x + 6\pi t - 0.785)$

17. (a) -1.51 m/s, 0   (b) 16.0 m, 0.500 s, 32.0 m/s

19. (a) 0.500 Hz, 3.14 rad/s   (b) 3.14 rad/m
(c) $(0.100\text{ m})\sin(3.14x/\text{m} - 3.14t/\text{s})$
(d) $(0.100\text{ m})\sin(-3.14t/\text{s})$
(e) $(0.100\text{ m})\sin(4.71\text{ rad} - 3.14t/\text{s})$ (f) 0.314 m/s

21. 80.0 N

23. 520 m/s

25. 1.64 m/s^2

27. 13.5 N

29. 185 m/s

31. 0.329 s

33. (a) s and kg·m/s^2 (b) time interval (period) and force (tension)

37. 55.1 Hz

39. (a) 62.5 m/s   (b) 7.85 m   (c) 7.96 Hz   (d) 21.1 W

41. $\sqrt{2}\mathcal{P}_0$

43. (a) $A = 40$ (b) $A = 7.00$, $B = 0$, $C = 3.00$. One can take the dot product of the given equation with each one of $\hat{\mathbf{i}}, \hat{\mathbf{j}}$, and $\hat{\mathbf{k}}$.   (c) $A = 0$, $B = 7.00$ mm, $C = 3.00/\text{m}$, $D = 4.00/\text{s}$, $E = 2.00$. Consider the average value of both sides of the given equation to find A. Then consider the maximum value of both sides to find B. One can evaluate the partial derivative of both sides of the given equation with respect to x and separately with respect to t to obtain equations yielding C and D upon chosen substitutions for x and t. Then substitute $x = 0$ and $t = 0$ to obtain E.

47. ~ 1 min

49. (a) $(3.33\hat{\mathbf{i}})$ m/s   (b) -5.48 cm   (c) 0.667 m, 5.00 Hz
(d) 11.0 m/s

51. 0.456 m/s

53. (a) 39.2 N   (b) 0.892 m   (c) 83.6 m/s

55. (a) 179 m/s   (b) 17.7 kW

57. 0.084 3 rad

61. (a) $(0.707)2(L/g)^{1/2}$ (b) $L/4$

63. 3.86×10^{-4}

65. (b) 31.6 m/s

67. (a) $\dfrac{\mu\omega^3}{2k} A_0^2 e^{-2bx}$ (b) $\dfrac{\mu\omega^3}{2k} A_0^2$ (c) e^{-2bx}

69. (a) $\mu_0 + (\mu_L - \mu_0)x/L$

CHAPTER 17

1. 5.56 km

3. 7.82 m

5. (a) 826 m   (b) 1.47 s

7. 5.67 mm

9. 1.50 mm to 75.0 μm

11. (a) 2.00 μm, 40.0 cm, 54.6 m/s   (b) -0.433 μm
(c) 1.72 mm/s

13. $\Delta P = (0.200\text{ N/m}^2)\sin(62.8\,x/\text{m} - 2.16 \times 10^4\,t/\text{s})$

15. 5.81 m

19. 66.0 dB

21. (a) 3.75 W/m^2 (b) 0.600 W/m^2

23. (a) 2.34 m and 0.390 m   (b) 0.161 N/m^2 for both notes
(c) 4.25×10^{-7} m and 7.09×10^{-8} m   (d) The wavelengths and displacement amplitudes would be larger by a factor of 1.09. The answer to (b) is unchanged.

25. (a) 1.32×10^{-4} W/m^2 (b) 81.2 dB

27. (a) 0.691 m   (b) 691 km

29. 65.6 dB

31. (a) 65.0 dB   (b) 67.8 dB   (c) 69.6 dB

33. (a) 30.0 m   (b) 9.49×10^5 m

35. (a) 332 J   (b) 46.4 dB

37. (a) 338 Hz   (b) 483 Hz

39. 26.4 m/s

41. 19.3 m

43. (a) 0.364 m   (b) 0.398 m   (c) 941 Hz   (d) 938 Hz

45. 2.82×10^8 m/s

47. (a) 56.3 s   (b) 56.6 km farther along

49. 22.3° left of center

51. $f \sim 300$ Hz, $\lambda \sim 10^0$ m, duration $\sim 10^{-1}$ s

55. 6.01 km

57. (a) 55.8 m/s   (b) 2 500 Hz

59. 1 204.2 Hz

61. 1.60

63. 2.34 m

65. (a) 0.948°   (b) 4.40°

67. 1.34×10^4 N

69. (b) 531 Hz

71. (a) 6.45   (b) 0

CHAPTER 18

1. (a) −1.65 cm   (b) −6.02 cm   (c) 1.15 cm

3. (a) $+x, -x$ (b) 0.750 s   (c) 1.00 m

5. (a) 9.24 m   (b) 600 Hz

7. (a) zero   (b) 0.300 m

9. (a) 2   (b) 9.28 m and 1.99 m

11. (a) 156°   (b) 0.058 4 cm

13. 15.7 m, 31.8 Hz, 500 m/s

15. At 0.089 1 m, 0.303 m, 0.518 m, 0.732 m, 0.947 m, 1.16 m from one speaker

17. (a) 4.24 cm   (b) 6.00 cm   (c) 6.00 cm
(d) 0.500 cm, 1.50 cm, 2.50 cm

19. 0.786 Hz, 1.57 Hz, 2.36 Hz, 3.14 Hz

21. (a) 350 Hz   (b) 400 kg

23. 1.27 cm

25. (a) reduced by 1/2   (b) reduced by $1/\sqrt{2}$
(c) increased by $\sqrt{2}$

27. (a) 163 N   (b) 660 Hz

29. $\dfrac{Mg}{4Lf^2 \tan \theta}$

31. (a) 3 loops   (b) 16.7 Hz   (c) 1 loop

33. (a) 3.66 m/s   (b) 0.200 Hz

35. 9.00 kHz

37. (a) 0.357 m   (b) 0.715 m

39. 57.6 Hz

41. $n(206$ Hz$)$ for $n = 1$ to 9 and $n(84.5$ Hz$)$ for $n = 2$ to 23

43. 50.0 Hz, 1.70 m

45. (a) 350 m/s   (b) 1.14 m

47. (a) 162 Hz   (b) 1.06 m

49. (a) 1.59 kHz   (b) odd-numbered harmonics
(c) 1.11 kHz

51. 5.64 beats/s

53. (a) 1.99 beats/s   (b) 3.38 m/s

55. The second harmonic of E is close to the third harmonic of A, and the fourth harmonic of C$^{\#}$ is close to the fifth harmonic of A.

57. (a) 34.8 m/s   (b) 0.977 m

59. 3.85 m/s away from the station or 3.77 m/s toward the station

61. 21.5 m

63. (a) 59.9 Hz   (b) 20.0 cm

65. (a) 1/2   (b) $[n/(n + 1)]^2 T$ (c) 9/16

67. $y_1 + y_2 = 11.2 \sin(2.00x - 10.0t + 63.4°)$

69. (a) 78.9 N   (b) 211 Hz

CHAPTER 19

1. (a) −274°C   (b) 1.27 atm   (c) 1.74 atm

3. (a) −320°F   (b) 77.3 K

5. (a) 810°F   (b) 450 K

7. (a) 1 337 K, 2 993 K   (b) 1 596°C = 1 596 K

9. 3.27 cm

11. 55.0°C

13. (a) 0.176 mm   (b) 8.78 μm   (c) 0.093 0 cm^3

15. (a) −179°C (attainable)
(b) −376°C (below 0 K, unattainable)

17. 0.548 gal

19. (a) 99.8 mL
(b) about 6% of the volume change of the acetone

21. (a) 99.4 cm^3 (b) 0.943 cm

23. 1.14°C

25. 5 336 images

27. (a) 400 kPa   (b) 449 kPa

29. 1.50×10^{29} molecules

31. 1.61 MPa = 15.9 atm

33. 472 K

35. (a) 41.6 mol   (b) 1.20 kg, nearly in agreement with the tabulated density

37. (a) 1.17 g   (b) 11.5 mN   (c) 1.01 kN
(d) The molecules must be moving very fast.

39. 4.39 kg

41. 3.55 L

43. $m_1 - m_2 = \dfrac{P_0 V M}{R}\left(\dfrac{1}{T_1} - \dfrac{1}{T_2}\right)$

45. (a) 94.97 cm   (b) 95.03 cm

47. 3.55 cm

49. It falls by 0.094 3 Hz

51. (a) Expansion makes density drop. (b) $5 \times 10^{-5}/°C$

53. (a) $h = nRT/(mg + P_0A)$ (b) 0.661 m

55. We assume that $\alpha \Delta T$ is much less than 1.

57. (a) 0.340% (b) 0.480%

59. 0.750

61. 2.74 m

63. (b) 1.33 kg/m^3

67. No. Steel is not strong enough.

69. (a) $L_f = L_i e^{\alpha \Delta T}$ (b) $2.00 \times 10^{-4}\%$; 59.4%

71. (a) 6.17×10^{-3} kg/m (b) 632 N (c) 580 N; 192 Hz

73. 4.54 m

CHAPTER 20

1. $(10.0 + 0.117)°C$

3. 0.234 kJ/kg$\cdot$°C

5. 1.78×10^4 kg

7. 29.6°C

9. (a) 0.435 cal/g$\cdot$°C (b) beryllium

11. 23.6°C

13. 50.7 ks

15. 1.22×10^5 J

17. 0.294 g

19. 0.414 kg

21. (a) 0°C (b) 114 g

23. -1.18 MJ

25. -466 J

27. (a) $-4P_iV_i$ (b) It is proportional to the square of the volume, according to $T = (P_i/nRV_i)V^2$

29. $Q = -720$ J

31.

	Q	W	ΔE_{int}
BC	$-$	0	$-$
CA	$-$	$+$	$-$
AB	$+$	$-$	$+$

33. 3.60 kJ

35. (a) 7.50 kJ (b) 900 K

37. -3.10 kJ; 37.6 kJ

39. (a) $0.041\ 0$ m^3 (b) $+5.48$ kJ (c) -5.48 kJ

41. 2.22×10^{-2} W/m$\cdot$°C

43. 51.2°C

45. 67.9°C

47. 3.77×10^{26} J/s

49. 3.49×10^3 K

51. 277 K $= 4°C$

53. 2.27 km

55. (a) 16.8 L (b) 0.351 L/s

57. $c = \mathcal{P}/\rho R \Delta T$

59. -1.87 kJ

61. 5.87×10^4 °C

63. 5.31 h

65. 1.44 kg

67. 38.6 m^3/d

71. 9.32 kW

73. (a) 3.16×10^{22} W (b) 5.78×10^3 K, 0.327% less than 5 800 K (c) 3.17×10^{22} W, 0.408% larger

CHAPTER 21

1. 0.943 N; 1.57 Pa

3. 3.65×10^4 N

5. 3.32 mol

7. (a) 3.54×10^{23} atoms (b) 6.07×10^{-21} J (c) 1.35 km/s

9. (a) 8.76×10^{-21} J for both (b) 1.62 km/s for helium and 514 m/s for argon

13. (a) 3.46 kJ (b) 2.45 kJ (c) -1.01 kJ

15. (a) 209 J (b) zero (c) 317 K

17. 1.18 atm

19. Between 10^{-2} and 10^{-3} °C

21. (a) 316 K (b) 200 J

23. (a) $C = \dfrac{n_1 C_1 + n_2 C_2}{n_1 + n_2}$ (b) $C = \dfrac{\sum\limits_{i=1}^{m} n_i C_i}{\sum\limits_{i=1}^{m} n_i}$

25. (a) 1.39 atm (b) 366 K, 253 K (c) 0, -4.66 kJ, -4.66 kJ

27. 227 K

29. (a)

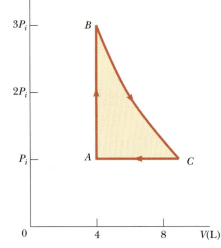

(b) 8.77 L (c) 900 K (d) 300 K (e) -336 J

31. (a) 28.0 kJ (b) 46.1 kJ (c) Isothermal process, $P_f = 10.0$ atm; adiabatic process, $P_f = 25.1$ atm

33. (a) 9.95 cal/K, 13.9 cal/K (b) 13.9 cal/K, 17.9 cal/K

35. 2.33×10^{-21} J

37. (a) 6.80 m/s (b) 7.41 m/s (c) 7.00 m/s

41. (a) 2.37×10^4 K (b) 1.06×10^3 K

43. (b) 0.278

45. (a) 3.21×10^{12} molecules (b) 779 km
(c) 6.42×10^{-4} s^{-1}

49. (a) 9.36×10^{-8} m (b) 9.36×10^{-8} atm (c) 302 atm

51. (a) 100 kPa, 66.5 L, 400 K, 5.82 kJ, 7.48 kJ, -1.66 kJ

(b) 133 kPa, 49.9 L, 400 K, 5.82 kJ, 5.82 kJ, 0

(c) 120 kPa, 41.6 L, 300 K, 0, -910 J, $+910$ J

(d) 120 kPa, 43.3 L, 312 K, 722 J, 0, $+722$ J

55. 510 K and 290 K

57. 0.623

59. (a) Pressure increases as volume decreases
(d) 0.500 atm^{-1}, 0.300 atm^{-1}

61. (a) 0.514 m^3 (b) 2.06 m^3 (c) 2.38×10^3 K
(d) -480 kJ (e) 2.28 MJ

63. 1.09×10^{-3}; 2.69×10^{-2}; 0.529; 1.00; 0.199; 1.01×10^{-41}; 1.25×10^{-1082}

67. (a) 0.203 mol (b) $T_B = T_C = 900$ K, $V_C = 15.0$ L

(c, d)	P, atm	V, L	T, K	E_{int}, kJ
A	1.00	5.00	300	0.760
B	3.00	5.00	900	2.28
C	1.00	15.0	900	2.28
A	1.00	5.00	300	0.760

(e) Lock the piston in place and put the cylinder into an oven at 900 K. Keep the gas in the oven while gradually letting the gas expand to lift a load on the piston as far as it can. Move the cylinder from the oven back to the 300-K room and let the gas cool and contract.

(f, g)	Q, kJ	W, kJ	ΔE_{int}, kJ
AB	1.52	0	1.52
BC	1.67	-1.67	0
CD	-2.53	$+1.01$	-1.52
ABCA	0.656	-0.656	0

69. 1.60×10^4 K

CHAPTER 22

1. (a) 6.94% (b) 335 J

3. (a) 10.7 kJ (b) 0.533 s

5. (a) 29.4 L/h (b) 185 hp (c) 527 N·m
(d) 1.91×10^5 W

7. (a) 24.0 J (b) 144 J

9. (a) 2.93 (b) coefficient of performance for a refrigerator
(c) $300 is twice as large as $150

11. (a) 67.2% (b) 58.8 kW

13. (a) 741 J (b) 459 J

15. (a) 4.20 W (b) 31.2 g

17. (a) 564 K (b) 212 kW (c) 47.5%

19. (b) $1 - T_c/T_h$ (c) $(T_c + T_h)/2$ (d) $(T_hT_c)^{1/2}$

21. (a) 214 J, 64.3 J
(b) -35.7 J, -35.7 J. The net effect is the transport of energy by heat from the cold to the hot reservoir without expenditure of external work. (c) 333 J, 233 J

(d) 83.3 J, 83.3 J, 0. The net effect is converting energy, taken in by heat, entirely into energy output by work in a cyclic process.
(e) -0.111 J/K. The entropy of the Universe has decreased.

23. 9.00

27. 72.2 J

29. 1.86

31. (a) 244 kPa (b) 192 J

33. 146 kW, 70.8 kW

35. -610 J/K

37. 195 J/K

39. 236 J/K

41. 1.02 kJ/K

43. $\sim 10^0$ W/K from metabolism; much more if you are using high-power electric appliances or an automobile, or if your taxes are paying for a war.

45. 5.76 J/K; temperature is constant if the gas is ideal

47. 18.4 J/K

49. (a) 1 (b) 6

51. (a)

Result	Number of Ways to Draw
All R	1
2 R, 1 G	3
1R, 2 G	3
All G	1

(b)

Result	Number of Ways to Draw
All R	1
4R, 1G	5
3R, 2G	10
2R, 3G	10
1R, 4G	5
All G	1

53. (a) 5.00 kW (b) 763 W

55. (a) 0.476 J/K (b) 417 J (c) $W_{net} = T_1 \Delta S_U = 167$ J

57. (a) $2nRT_i \ln 2$ (b) 0.273

59. 5.97×10^4 kg/s

61. (a) 3.19 cal/K (b) 98.19°F, 2.59 cal/K

63. (a) 8.48 kW (b) 1.52 kW (c) 1.09×10^4 J/K
(d) COP drops by 20.0%

65. (a) $10.5nRT_i$ (b) $8.50nRT_i$ (c) 0.190 (d) 0.833

67. (a) $nC_P \ln 3$
(b) Both ask for the change in entropy between the same two states of the same system. Entropy is a function of state. The change in entropy does not depend on path, but only on original and final states.

71. (a) 20.0°C (c) $\Delta S = +4.88$ J/K (d) Yes

CHAPTER 23

1. (a) $+160$ zC, 1.01 u (b) $+160$ zC, 23.0 u
(c) -160 zC, 35.5 u (d) $+320$ zC, 40.1 u
(e) -480 zC, 14.0 u (f) $+640$ zC, 14.0 u
(g) $+1.12$ aC, 14.0 u (h) -160 zC, 18.0 u

3. The force is $\sim 10^{26}$ N.

5. (a) 1.59 nN away from the other
   (b) 1.24×10^{36} times larger
   (c) 8.61×10^{-11} C/kg

7. 0.872 N at 330°

9. (a) 2.16×10^{-5} N toward the other   (b) 8.99×10^{-7} N away from the other

11. (a) 82.2 nN   (b) 2.19 Mm/s

13. (a) 55.8 pN/C down   (b) 102 nN/C up

15. 1.82 m to the left of the negative charge

17. $-9Q$ and $+27Q$

19. (a) $(-0.599\hat{\mathbf{i}} - 2.70\hat{\mathbf{j}})$ kN/C   (b) $(-3.00\hat{\mathbf{i}} - 13.5\hat{\mathbf{j}})$ μN

21. (a) $5.91 k_e q/a^2$ at 58.8°   (b) $5.91 k_e q^2/a^2$ at 58.8°

23. (a) $[k_e Qx/(R^2 + x^2)^{3/2}]\hat{\mathbf{i}}$ (b) As long as the charge is symmetrically placed, the number of charges does not matter. A continuous ring corresponds to n becoming larger without limit.

25. 1.59×10^6 N/C toward the rod

27. (a) $6.64\hat{\mathbf{i}}$ MN/C   (b) $24.1\hat{\mathbf{i}}$ MN/C   (c) $6.40\hat{\mathbf{i}}$ MN/C
    (d) $0.664\hat{\mathbf{i}}$ MN/C, taking the axis of the ring as the x axis

31. (a) 93.6 MN/C; the near-field approximation is 104 MN/C, about 11% high   (b) 0.516 MN/C; the point-charge approximation is 0.519 MN/C, about 0.6% high

33. $-21.6\hat{\mathbf{i}}$ MN/C

37. (a) 86.4 pC for each
    (b) 324 pC, 459 pC, 459 pC, 432 pC
    (c) 57.6 pC, 106 pC, 154 pC, 96.0 pC

39.

41. (a)

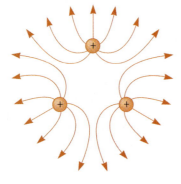

The field is zero at the center of the triangle.
(b) $1.73 k_e q\hat{\mathbf{j}}/a^2$

43. (a) 61.3 Gm/s²   (b) 19.5 μs   (c) 11.7 m   (d) 1.20 fJ

45. K/ed in the direction of motion

47. (a) 111 ns   (b) 5.68 mm   (c) $(450\hat{\mathbf{i}} + 102\hat{\mathbf{j}})$ km/s

49. (a) 36.9°, 53.1°   (b) 167 ns, 221 ns

51. (a) 21.8 μm   (b) 2.43 cm

53. (a) mv^2/qR (b) $mv^2/2^{1/2}qR$ oriented at 135° to the x axis

55. (a) 10.9 nC   (b) 5.44 mN

57. 40.9 N at 263°

59. $Q = 2L \sqrt{\dfrac{k(L - L_i)}{k_e}}$

63. $-707\hat{\mathbf{j}}$ mN

65. (a) $\theta_1 = \theta_2$

67. (a) 0.307 s   (b) Yes. Ignoring gravity makes a difference of 2.28%.

69. (a) $\mathbf{F} = 1.90(k_e q^2/s^2)(\hat{\mathbf{i}} + \hat{\mathbf{j}} + \hat{\mathbf{k}})$ (b) $\mathbf{F} = 3.29(k_e q^2/s^2)$ in the direction away from the diagonally opposite vertex

CHAPTER 24

1. (a) 858 N·m²/C   (b) 0   (c) 657 N·m²/C

3. 4.14 MN/C

5. (a) aA (b) bA (c) 0

7. 1.87 kN·m²/C

9. (a) -6.89 MN·m²/C   (b) The number of lines entering exceeds the number leaving by 2.91 times or more.

11. $-Q/\epsilon_0$ for S_1; 0 for S_2; $-2Q/\epsilon_0$ for S_3; 0 for S_4

13. $E_0 \pi r^2$

15. (a) $+Q/2\epsilon_0$ (b) $-Q/2\epsilon_0$

17. -18.8 kN·m²/C

19. 0 if $R \le d$; $\dfrac{2\lambda}{\epsilon_0} \sqrt{R^2 - d^2}$ if $R > d$

21. (a) 3.20 MN·m²/C   (b) 19.2 MN·m²/C   (c) The answer to (a) could change, but the answer to (b) would stay the same.

23. 2.33×10^{21} N/C

25. -2.48 μC/m²

27. 5.94×10^5 m/s

29. $\mathbf{E} = \rho r/2\epsilon_0$ away from the axis

31. (a) 0   (b) 7.19 MN/C away from the center

33. (a) ~1 mN   (b) ~100 nC   (c) ~10 kN/C
    (d) ~10 kN·m²/C

35. (a) 51.4 kN/C outward   (b) 646 N·m²/C

37. 508 kN/C up

39. (a) 0   (b) 5 400 N/C outward   (c) 540 N/C outward

41. $\mathbf{E} = Q/2\epsilon_0 A$ vertically upward in each case if $Q > 0$

43. (a) $+708$ nC/m² and -708 nC/m²   (b) $+177$ nC and -177 nC

45. 2.00 N

47. (a) $-\lambda, +3\lambda$ (b) $3\lambda/2\pi\epsilon_0 r$ radially outward

49. (a) 80.0 nC/m² on each face   (b) $9.04\hat{\mathbf{k}}$ kN/C
    (c) $-9.04\hat{\mathbf{k}}$ kN/C

51. $\mathbf{E} = 0$ inside the sphere and within the material of the shell. $\mathbf{E} = k_e Q/r^2$ radially inward between the sphere and the shell. $\mathbf{E} = 2k_e Q/r^2$ radially outward outside the shell. Charge $-Q$ resides on the outer surface of the sphere.

$+Q$ is on the inner surface of the shell. $+2Q$ is on the outer surface of the shell.

53. (b) $Q/2\epsilon_0$ (c) Q/ϵ_0

55. (a) $+2Q$ (b) radially outward (c) $2k_eQ/r^2$ (d) 0 (e) 0
(f) $3Q$ (g) $3k_eQ/r^2$ radially outward (h) $3Qr^3/a^3$
(i) $3k_eQr/a^3$ radially outward (j) $-3Q$ (k) $+2Q$
(l) See below.

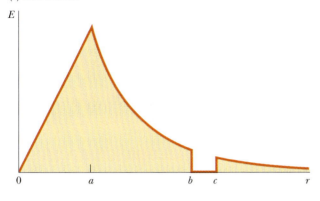

57. (a) $\rho r/3\epsilon_0$; $Q/4\pi\epsilon_0 r^2$; 0; $Q/4\pi\epsilon_0 r^2$, all radially outward
(b) $-Q/4\pi b^2$ and $+Q/4\pi c^2$

59. $\theta = \tan^{-1}[qQ/(2\pi\epsilon_0 dmv^2)]$

61. For $r < a$, $\mathbf{E} = \lambda/2\pi\epsilon_0 r$ radially outward. For $a < r < b$,
$\mathbf{E} = [\lambda+\rho\pi(r^2 - a^2)]/2\pi\epsilon_0 r$ radially outward. For $r > b$,
$\mathbf{E} = [\lambda+\rho\pi(b^2 - a^2)]/2\pi\epsilon_0 r$ radially outward.

63. (a) σ/ϵ_0 away from both plates (b) 0 (c) σ/ϵ_0 away from both plates

65. $\sigma/2\epsilon_0$ radially outward

69. $\mathbf{E} = a/2\epsilon_0$ radially outward

73. (b) $\mathbf{g} = GM_E r/R_E^3$ radially inward

CHAPTER 25

1. 1.35 MJ

3. (a) 152 km/s (b) 6.49 Mm/s

5. (a) $-600\ \mu$J (b) -50.0 V

7. 38.9 V; the origin

9. $+260$ V

11. (a) $2QE/k$ (b) QE/k (c) $2\pi\sqrt{m/k}$ (d) $2(QE - \mu_k mg)/k$

13. (a) 0.400 m/s (b) the same

15. (a) 1.44×10^{-7} V (b) -7.19×10^{-8} V
(c) -1.44×10^{-7} V, $+7.19 \times 10^{-8}$ V

17. (a) 6.00 m (b) $-2.00\ \mu$C

19. -11.0 MV

21. 8.95 J

25. (a) no point at a finite distance from the charges
(b) $2k_eq/a$

27. (a) $v_1 = \sqrt{\dfrac{2m_2 k_e q_1 q_2}{m_1(m_1 + m_2)}\left(\dfrac{1}{r_1 + r_2} - \dfrac{1}{d}\right)}$

$v_2 = \sqrt{\dfrac{2m_1 k_e q_1 q_2}{m_2(m_1 + m_2)}\left(\dfrac{1}{r_1 + r_2} - \dfrac{1}{d}\right)}$

(b) faster than calculated in (a)

29. $5k_eq^2/9d$

31. 0.720 m, 1.44 m, 2.88 m. No. The radii of the equipotentials are inversely proportional to the potential.

33. 7.26 Mm/s

35. $\left[\left(1 + \sqrt{\dfrac{1}{8}}\right)\dfrac{k_e q^2}{mL}\right]^{1/2}$

37. (a) 10.0 V, -11.0 V, -32.0 V
(b) 7.00 N/C in the $+x$ direction

39. $\mathbf{E} = (-5 + 6xy)\hat{\mathbf{i}} + (3x^2 - 2z^2)\hat{\mathbf{j}} - 4yz\hat{\mathbf{k}}$; 7.07 N/C

41. $E_y = \dfrac{k_e Q}{y\sqrt{\ell^2 + y^2}}$

43. (a) C/m^2 (b) $k_e\alpha[L - d\ln(1 + L/d)]$

45. -1.51 MV

47. $k_e\lambda(\pi + 2\ln 3)$

49. (a) 0, 1.67 MV (b) 5.84 MN/C away, 1.17 MV
(c) 11.9 MN/C away, 1.67 MV

51. (a) 450 kV (b) 7.51 μC

53. 253 MeV

55. (a) -27.2 eV (b) -6.80 eV (c) 0

59. $k_e Q^2/2R$

63. $V_2 - V_1 = (-\lambda/2\pi\epsilon_0)\ln(r_2/r_1)$

69. (b) $E_r = (2k_ep\cos\theta)/r^3$; $E_\theta = (k_ep\sin\theta)/r^3$; yes; no
(c) $V = k_epy(x^2 + y^2)^{-3/2}$;
$\mathbf{E} = 3k_epxy(x^2 + y^2)^{-5/2}\hat{\mathbf{i}} + k_ep(2y^2 - x^2)(x^2 + y^2)^{-5/2}\hat{\mathbf{j}}$

71. $V = \pi k_e C\left[R\sqrt{x^2 + R^2} + x^2\ln\left(\dfrac{x}{R + \sqrt{x^2 + R^2}}\right)\right]$

73. (a) 8 876 V (b) 112 V

CHAPTER 26

1. (a) 48.0 μC (b) 6.00 μC

3. (a) 1.33 μC/m^2 (b) 13.3 pF

5. (a) 5.00 μC on the larger and 2.00 μC on the smaller sphere (b) 89.9 kV

7. (a) 11.1 kV/m toward the negative plate.
(b) 98.3 nC/m^2 (c) 3.74 pF (d) 74.7 pC

9. 4.42 μm

11. (a) 2.68 nF (b) 3.02 kV

13. (a) 15.6 pF (b) 256 kV

15. 708 μF

17. (a) 3.53 μF (b) 6.35 V and 2.65 V (c) 31.8 μC on each

19. 6.00 pF and 3.00 pF

21. (a) 5.96 μF (b) 89.5 μC on 20 μF, 63.2 μC on 6 μF, 26.3 μC on 15 μF and on 3 μF

23. 120 μC; 80.0 μC and 40.0 μC

25. 10

27. 6.04 μF

29. 12.9 μF

31. (a) 216 μJ (b) 54.0 μJ

33. (a) Circuit diagram:

Stored energy = 0.150 J
(b) Potential difference = 268 V
Circuit diagram:

35. (a) 1.50 μC (b) 1.83 kV

39. 9.79 kg

43. (a) 81.3 pF (b) 2.40 kV

45. 1.04 m

47. (a) 369 pC (b) 118 pF, 3.12 V (c) -45.5 nJ

49. 22.5 V

51. (b) -8.78×10^6 N/C$\cdot$m; -5.53×10^{-2} $\hat{\mathbf{i}}$ N

55. (a) 11.2 pF (b) 134 pC (c) 16.7 pF (d) 66.9 pC

57. (a) $-2Q/3$ on upper plate, $-Q/3$ on lower plate
(b) $2Qd/3\epsilon_0 A$

59. 0.188 m^2

61. (a) $C = \dfrac{\epsilon_0 A}{d}\left(\dfrac{\kappa_1}{2} + \dfrac{\kappa_2 \kappa_3}{\kappa_2 + \kappa_3}\right)$ (b) 1.76 pF

63. (b) $1/C$ approaches $\dfrac{1}{4\pi\epsilon_0 a} + \dfrac{1}{4\pi\epsilon_0 b}$

65. (a) $Q_0^2\, d(\ell - x)/(2\ell^3\epsilon_0)$ (b) $Q_0^2\, d/(2\ell^3\epsilon_0)$ to the right
(c) $Q_0^2/(2\ell^4\epsilon_0)$ (d) $Q_0^2/(2\ell^4\epsilon_0)$

67. 4.29 μF

69. (a) The additional energy comes from work done by the electric field in the wires as it forces more charge onto the already-charged plates. (b) $Q/Q_0 = \kappa$

71. 750 μC on C_1 and 250 μC on C_2

73. 19.0 kV

75. $\tfrac{4}{3}C$

CHAPTER 27

1. 7.50×10^{15} electrons

3. (a) 0.632 $I_0\tau$ (b) 0.999 95 $I_0\tau$ (c) $I_0\tau$

5. $q\omega/2\pi$

7. 0.265 C

9. (a) 2.55 A/m^2 (b) 5.31×10^{10} m^{-3} (c) 1.20×10^{10} s

11. 0.130 mm/s

13. 500 mA

15. 6.43 A

17. (a) 1.82 m (b) 280 μm

19. (a) $\sim 10^{18}$ Ω (b) $\sim 10^{-7}$ Ω (c) ~ 100 aA, ~ 1 GA

21. $R/9$

23. $6.00 \times 10^{-15}/\Omega\cdot$m

25. 0.181 V/m

27. 21.2 nm

29. $1.44 \times 10^{3}{}^\circ$C

31. (a) 31.5 n$\Omega\cdot$m (b) 6.35 MA/m^2 (c) 49.9 mA
(d) 659 μm/s (e) 0.400 V

33. 0.125

35. 67.6°C

37. 7.50 W

39. 28.9 Ω

41. 36.1%

43. (a) 5.97 V/m (b) 74.6 W (c) 66.1 W

45. 0.833 W

47. $0.232

49. 26.9 cents/d

51. (a) 184 W (b) 461°C

53. $\sim$\$1

55. (a) $Q/4C$ (b) $Q/4$ and $3Q/4$ (c) $Q^2/32C$ and $3Q^2/32C$ (d) $3Q^2/8C$

59. Experimental resistivity = 1.47 $\mu\Omega\cdot$m $\pm$ 4%, in agreement with 1.50 $\mu\Omega\cdot$m

61. (a) $(8.00\hat{\mathbf{i}})$ V/m (b) 0.637 Ω (c) 6.28 A
(d) $(200\hat{\mathbf{i}})$ MA/m^2

63. 2 020°C

65. (a) 667 A (b) 50.0 km

67.

Material	$\alpha' = \alpha/(1 - 20\alpha)$
Silver	$4.1 \times 10^{-3}/$°C
Copper	$4.2 \times 10^{-3}/$°C
Gold	$3.6 \times 10^{-3}/$°C
Aluminum	$4.2 \times 10^{-3}/$°C
Tungsten	$4.9 \times 10^{-3}/$°C
Iron	$5.6 \times 10^{-3}/$°C
Platinum	$4.25 \times 10^{-3}/$°C
Lead	$4.2 \times 10^{-3}/$°C
Nichrome	$0.4 \times 10^{-3}/$°C
Carbon	$-0.5 \times 10^{-3}/$°C
Germanium	$-24 \times 10^{-3}/$°C
Silicon	$-30 \times 10^{-3}/$°C

69. No. The fuses should pass no more than 3.87 A.

73. (b) 1.79 PΩ

75. (a) $\dfrac{\epsilon_0 \ell}{2d}(\ell + 2x + \kappa\ell - 2\kappa x)$

(b) $\dfrac{\epsilon_0 \ell v\, \Delta V(\kappa - 1)}{d}$ clockwise

CHAPTER 28

1. (a) 6.73 Ω   (b) 1.97 Ω

3. (a) 4.59 Ω   (b) 8.16%

5. 12.0 Ω

7. Circuit diagram:

power 73.8 W

9. (a) 227 mA   (b) 5.68 V

11. (a) 75.0 V   (b) 25.0 W, 6.25 W, and 6.25 W; 37.5 W

13. 1.00 kΩ

15. 14.2 W to 2 Ω, 28.4 W to 4 Ω, 1.33 W to 3 Ω, 4.00 W to 1 Ω

17. (a) $\Delta t_p = 2\,\Delta t/3$ (b) $\Delta t_s = 3\,\Delta t$

19. (a) $\Delta V_4 > \Delta V_3 > \Delta V_1 > \Delta V_2$
    (b) $\Delta V_1 = \mathcal{E}/3$, $\Delta V_2 = 2\mathcal{E}/9$, $\Delta V_3 = 4\mathcal{E}/9$, $\Delta V_4 = 2\mathcal{E}/3$
    (c) $I_1 > I_4 > I_2 = I_3$ (d) $I_1 = I$, $I_2 = I_3 = I/3$, $I_4 = 2I/3$
    (e) I_4 increases while I_1, I_2, and I_3 decrease
    (f) $I_1 = 3I/4$, $I_2 = I_3 = 0$, $I_4 = 3I/4$

21. 846 mA down in the 8-Ω resistor; 462 mA down in the middle branch; 1.31 A up in the right-hand branch

23. (a) − 222 J and 1.88 kJ   (b) 687 J, 128 J, 25.6 J, 616 J, 205 J
    (c) 1.66 kJ of chemical energy is transformed into internal energy

25. 50.0 mA from a to e

27. starter 171 A; battery 0.283 A

29. (a) 909 mA   (b) − 1.82 V = $V_b − V_a$

31. (a) 5.00 s   (b) 150 μC   (c) 4.06 μA

33. $U_0/4$

37. (a) 6.00 V   (b) 8.29 μs

39. (a) 12.0 s   (b) $I(t) = (3.00\ \mu A)e^{-t/12.0\text{ s}}$;
    $q(t) = (36.0\ \mu C)(1 − e^{-t/12.0\text{ s}})$

41. 0.302 Ω

43. 16.6 kΩ

45.

47. 145 Ω, 0.756 mA

49. (a) 12.5 A, 6.25 A, 8.33 A   (b) No; together they would require 27.1 A.

51. (a) ~10^{-14} A   (b) $V_h/2 + (\sim 10^{-10}$ V) and $V_h/2 − (\sim 10^{-10}$ V), where V_h is the potential of the live wire, ~10^2 V

53. (a) either 3.84 Ω or 0.375 Ω   (b) impossible

55. (a) $\mathcal{E}^2/3R$ (b) $3\mathcal{E}^2/R$ (c) in the parallel connection

57. (a) $R \to \infty$ (b) $R \to 0$ (c) $R = r$

59. 6.00 Ω; 3.00 Ω

61. (a) 4.40 Ω   (b) 32.0 W, 9.60 W, 70.4 W   (c) 48.0 W

63. (a) $R \le 1\,050$ Ω   (b) $R \ge 10.0$ Ω

65. (a) 9.93 μC   (b) 33.7 nA   (c) 334 nW   (d) 337 nW

67. (a) 40.0 W   (b) 80.0 V, 40.0 V, 40.0 V

69. (a) 0.991   (b) 0.648   (c) Insulation should be added to the ceiling.

71. (a) 0 in 3 kΩ and 333 μA in 12 kΩ and 15 kΩ   (b) 50.0 μC
    (c) $(278\ \mu A)\,e^{-t/180\text{ ms}}$ (d) 290 ms

73. (a) $\ln(\mathcal{E}/\Delta V) = (0.011\,8)t + 0.088\,2$ (b) 84.7 s, 8.47 μF

75. $q_1 = (240\ \mu C)(1 − e^{-1\,000t/6})$; $q_2 = (360\ \mu C)(1 − e^{-1\,000t/6})$

CHAPTER 29

1. (a) up   (b) out of the plane of the paper   (c) no deflection   (d) into the plane of the paper

3. negative z direction

5. $(− 20.9\hat{j})$ mT

7. 48.9° or 131°

9. 2.34 aN

11. 0.245 T east

13. (a) 4.73 N   (b) 5.46 N   (c) 4.73 N

15. 1.07 m/s

17. $2\pi rIB \sin \theta$ up

19. 2.98 μN west

21. 18.4 mA·m²

23. 9.98 N·m clockwise as seen looking down from above

27. (a) 118 μN·m   (b) − 118 μJ ≤ U ≤ 118 μJ

29. (a) 49.6 aN south   (b) 1.29 km

31. 115 keV

33. $r_\alpha = r_d = \sqrt{2}r_p$

35. 4.98×10^8 rad/s

37. 7.88 pT

39. $m = 2.99$ u, either $^3_1H^+$ or $^3_2He^+$

41. (a) 8.28 cm   (b) 8.23 cm; ratio is independent of both ΔV and B

43. (a) 4.31×10^7 rad/s   (b) 51.7 Mm/s

45. (a) 7.66×10^7 rad/s   (b) 26.8 Mm/s   (c) 3.76 MeV
    (d) 3.13×10^3 rev   (e) 257 μs

47. 70.1 mT

49. 1.28×10^{29} m^{-3}, 1.52

51. 43.3 μT

53. (a) The electric current experiences a magnetic force.

55. (a) $− 8.00 \times 10^{-21}$ kg·m/s   (b) 8.90°

57. (a) $(3.52\hat{i} − 1.60\hat{j})$ aN   (b) 24.4°

59. $(2\pi/d)(2m_e\,\Delta V/e)^{1/2}$

61. 0.588 T

63. 0.713 A counterclockwise as seen from above

65. 438 kHz

67. 3.70×10^{-24} N·m

69. (a) 0.501 m   (b) 45.0°

71. (a) 1.33 m/s   (b) Positive ions moving toward you in magnetic field to the right feel upward magnetic force, and migrate upward in the blood vessel. Negative ions moving toward you feel downward magnetic force and accumulate at the bottom of this section of vessel. Thus both species can participate in the generation of the emf.

CHAPTER 30

1. 12.5 T

3. (a) 28.3 μT into the paper   (b) 24.7 μT into the paper

5. $\dfrac{\mu_0 I}{4\pi x}$ into the paper

7. 26.2 μT into the paper

9. (a) $2I_1$ out of the page   (b) $6I_1$ into the page

11. (a) along the line $(y = -0.420$ m, $z = 0)$
   (b) $(-34.7\hat{\mathbf{j}})$ mN   (c) $(17.3\hat{\mathbf{j}})$ kN/C

13. (a) $4.5\,\dfrac{\mu_0 I}{\pi L}$ (b) stronger

15. $(-13.0\hat{\mathbf{j}})$ μT

17. $(-27.0\hat{\mathbf{i}})$ μN

19. (a) 12.0 cm to the left of wire 1   (b) 2.40 A, downward

21. 20.0 μT toward the bottom of the page

23. 200 μT toward the top of the page; 133 μT toward the bottom of the page

25. (a) 6.34 mN/m inward   (b) greater

27. (a) 0   (b) $\dfrac{\mu_0 I}{2\pi R}$ tangent to the wall in a

counterclockwise sense   (c) $\dfrac{\mu_0 I^2}{(2\pi R)^2}$ inward

29. (a) $\frac{1}{3}\mu_0 b r_1^2$ (b) $\dfrac{\mu_0 b R^3}{3 r_2}$

31. 31.8 mA

33. 226 μN away from the center of the loop, 0

35. (a) 3.13 mWb   (b) 0

37. (a) 11.3 GV·m/s   (b) 0.100 A

39. (a) 9.27×10^{-24} A·m^2 (b) down

41. 0.191 T

43. 2.62 MA/m

45. (b) 6.45×10^4 K·A/T·m

47. (a) 8.63×10^{45} electrons   (b) 4.01×10^{20} kg

49. $\dfrac{\mu_0 I}{2\pi w}\ln\left(1 + \dfrac{w}{b}\right)\hat{\mathbf{k}}$

51. 12 layers, 120 m

53. 143 pT away along the axis

59. (a) 2.46 N up   (b) 107 m/s^2 up

61. (a) 274 μT   (b) $(-274\hat{\mathbf{j}})$ μT   (c) $(1.15\hat{\mathbf{i}})$ mN
   (d) $(0.384\hat{\mathbf{i}})$ m/s^2 (e) acceleration is constant
   (f) $(0.999\hat{\mathbf{i}})$ m/s

63. 81.7 A

65. $\dfrac{\mu_0 I_1 I_2 L}{\pi R}$ to the right

69. $\dfrac{\mu_0 I}{4\pi}\,(1 - e^{-2\pi})$ out of the plane of the paper

71. $\frac{1}{3}\rho\mu_0\omega R^2$

73. (a) $\dfrac{\mu_0 I(2r^2 - a^2)}{\pi r(4r^2 - a^2)}$ to the left   (b) $\dfrac{\mu_0 I(2r^2 + a^2)}{\pi r(4r^2 + a^2)}$ toward the top of the page

CHAPTER 31

1. 500 mV

3. 9.82 mV

5. 160 A

7. (a) 1.60 A counterclockwise   (b) 20.1 μT   (c) up

9. (a) $(\mu_0 IL/2\pi)\ln(1 + w/h)$ (b) -4.80 μV; current is counterclockwise

11. 283 μA upward

13. $(68.2$ mV$)\,e^{-1.6t}$, tending to produce counterclockwise current

15. 272 m

17. $(0.422$ V$)\cos\omega t$

19. (a) eastward   (b) 458 μV

21. (a) 3.00 N to the right   (b) 6.00 W

23. 360 T

25. (a) 233 Hz   (b) 1.98 mV

27. 2.83 mV

29. (a) $F = N^2 B^2 w^2 v/R$ to the left   (b) 0   (c) $F = N^2 B^2 w^2 v/R$ to the left

31. 145 μA

33. 1.80 mN/C upward and to the left, perpendicular to r_1

35. (a) 7.54 kV   (b) The plane of the coil is parallel to **B**.

37. $(28.6$ mV$)\sin(4\pi t)$

39. (a) 110 V   (b) 8.53 W   (c) 1.22 kW

41. (a) $(8.00$ mWb$)\cos(377t)$ (b) $(3.02$ V$)\sin(377t)$
   (c) $(3.02$ A$)\sin(377t)$ (d) $(9.10$ W$)\sin^2(377t)$
   (e) $(24.1$ mN·m$)\sin^2(377t)$

43. (b) Larger R makes current smaller, so the loop must travel faster to maintain equality of magnetic force and weight. (c) The magnetic force is proportional to the product of field and current, while the current is itself proportional to field. If B becomes two times smaller, the speed must become four times larger to compensate.

45. $(-2.87\hat{\mathbf{j}} + 5.75\hat{\mathbf{k}})$ Gm/s^2

47. (a) Doubling N doubles the amplitude. (b) Doubling ω doubles the amplitude and halves the period. (c) Doubling ω and halving N leaves the amplitude the same and cuts the period in half.

49. 62.3 mA down through 6.00 Ω, 860 mA down through 5.00 Ω, 923 mA up through 3.00 Ω

51. $\sim 10^{-4}$ V, by reversing a 20-turn coil of diameter 3 cm in 0.1 s in a field of 10^{-3} T

53. (a) 254 km/s (b) 215 V

55. 1.20 μC

57. (a) 0.900 A (b) 0.108 N (c) b (d) no

59. (a) $a\pi r^2$ (b) $-b\pi r^2$ (c) $-b\pi r^2/R$ (d) $b^2\pi^2 r^4/R$

61. (a) 36.0 V (b) 600 mWb/s (c) 35.9 V (d) 4.32 N·m

65. 6.00 A

67. (a) $(1.19$ V$) \cos(120\pi t)$ (b) 88.5 mW

71. $(-87.1$ mV$) \cos(200\pi t + \phi)$

CHAPTER 32

1. 19.5 mV

3. 100 V

5. $(18.8$ V$) \cos(377t)$

7. -0.421 A/s

9. (a) 188 μT (b) 33.3 nT·m^2 (c) 0.375 mH (d) B and Φ_B are proportional to current; L is independent of current

11. 0.750 m

13. $\mathcal{E}_0/k^2 L$

15. (a) 0.139 s (b) 0.461 s

17. (a) 2.00 ms (b) 0.176 A (c) 1.50 A (d) 3.22 ms

19. (a) 0.800 (b) 0

21. (a) 6.67 A/s (b) 0.332 A/s

23. $(500$ mA$)(1 - e^{-10t/s})$, 1.50 A $- (0.25$ A$) e^{-10t/s}$

25. 0 for $t < 0$; $(10$ A$)(1 - e^{-10\,000t})$ for $0 < t < 200$ μs; $(63.9$ A$) e^{-10\,000t}$ for $t > 200$ μs

27. (a) 5.66 ms (b) 1.22 A (c) 58.1 ms

29. 0.064 8 J

31. 2.44 μJ

33. 44.2 nJ/m^3 for the **E**-field and 995 μJ/m^3 for the **B**-field

35. (a) 0.500 J (b) 17.0 W (c) 11.0 W

37. 2.27 mT

39. 1.73 mH

41. 80.0 mH

43. (a) 18.0 mH (b) 34.3 mH (c) -9.00 mV

45. $(L_1 L_2 - M^2)/(L_1 + L_2 - 2M)$

47. 20.0 V

49. 608 pF

51. (a) 135 Hz (b) 119 μC (c) -114 mA

53. (a) 6.03 J (b) 0.529 J (c) 6.56 J

55. (a) 4.47 krad/s (b) 4.36 krad/s (c) 2.53%

57. $L = 199$ mH; $C = 127$ nF

59. (b) $\mu_0 J_s^2/2$ away from the other sheet (c) $\mu_0 J_s$ and zero (d) $\mu_0 J_s^2/2$

61. (a) -20.0 mV (b) $-(10.0$ MV/s$^2) t^2$ (c) 63.2 μs

63. $(Q/2N)(3L/C)^{1/2}$

65. (a) $L \approx (\pi/2) N^2 \mu_0 R$ (b) ~ 100 nH (c) ~ 1 ns

71. (a) 72.0 V; b

(b)

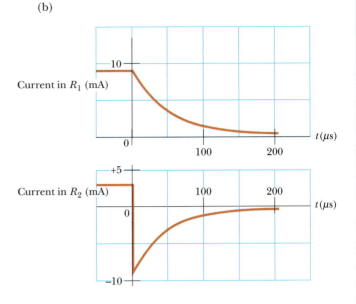

(c) 75.2 μs

73. 300 Ω

75. (a) It creates a magnetic field. (b) The long narrow rectangular area between the conductors encloses all of the magnetic flux.

77. (a) 62.5 GJ (b) 2 000 N

79. (a) 2.93 mT up (b) 3.42 Pa (c) clockwise as seen from above (d) up (e) 1.30 mN

CHAPTER 33

1. $\Delta v(t) = (283$ V$) \sin(628t)$

3. 2.95 A, 70.7 V

5. 14.6 Hz

7. 3.38 W

9. (a) 42.4 mH (b) 942 rad/s

11. 5.60 A

13. 0.450 Wb

15. (a) 141 mA (b) 235 mA

17. 100 mA

19. (a) 194 V (b) current leads by 49.9°

21. (a) 78.5 Ω (b) 1.59 kΩ (c) 1.52 kΩ (d) 138 mA (e) $-84.3°$

23. (a) 17.4° (b) voltage leads the current

25. 1.88 V

27.

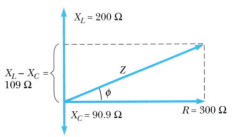

29. (a) either 123 nF or 124 nF   (b) 51.5 kV

31. 8.00 W

33. (a) 16.0 Ω (b) $-12.0\ \Omega$

35. $\sqrt{\dfrac{800\ \rho \mathscr{P} d}{\pi (\Delta V)^2}}$

37. 1.82 pF

39. (a) 633 fF   (b) 8.46 mm   (c) 25.1 Ω

41. 242 mJ

43. 0.591 and 0.987; the circuit in Problem 23

45. 687 V

47. 87.5 Ω

49. (a) 29.0 kW   (b) 5.80×10^{-3} (c) If the generator were limited to 4 500 V, no more than 17.5 kW could be delivered to the load, never 5 000 kW.

51. (b) 0; 1   (c) $f_h = (10.88RC)^{-1}$

53. (a) 613 μF   (b) 0.756

55. (a) 580 μH and 54.6 μF   (b) 1   (c) 894 Hz   (d) ΔV_{out} leads ΔV_{in} by 60.0° at 200 Hz. ΔV_{out} and ΔV_{in} are in phase at 894 Hz. ΔV_{out} lags ΔV_{in} by 60.0° at 4 000 Hz. (e) 1.56 W, 6.25 W, 1.56 W   (f) 0.408

57. 56.7 W

59. 99.6 mH

61. (a) 225 mA   (b) 450 mA

63. (a) 1.25 A   (b) Current lags voltage by 46.7°.

65. (a) 200 mA; voltage leads by 36.8°   (b) 40.0 V; $\phi = 0°$
(c) 20.0 V; $\phi = -90.0°$ (d) 50.0 V; $\phi = +90.0°$

67. (b) 31.6

71.

f (Hz)	X_L (Ω)	X_C (Ω)	Z (Ω)
300	283	12 600	12 300
600	565	6 280	5 720
800	754	4 710	3 960
1 000	942	3 770	2 830
1 500	1 410	2 510	1 100
2 000	1 880	1 880	40.0
3 000	2 830	1 260	1 570
4 000	3 770	942	2 830
6 000	5 650	628	5 020
10 000	9 420	377	9 040

(b)

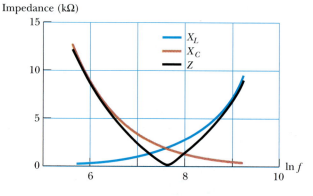

Impedance (kΩ)

73. (a) 1.84 kHz

(b)

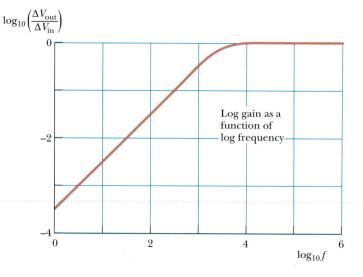

$\log_{10}\left(\dfrac{\Delta V_{\text{out}}}{\Delta V_{\text{in}}}\right)$

Log gain as a function of log frequency

$\log_{10} f$

CHAPTER 34

1. (a) $(3.15\hat{\mathbf{j}})$ kN/C   (b) $(525\hat{\mathbf{k}})$ nT   (c) $(-483\hat{\mathbf{j}})$ aN

3. 2.25×10^8 m/s

5. (a) 6.00 MHz   (b) $(-73.3\hat{\mathbf{k}})$ nT
(c) $\mathbf{B} = [(-73.3\hat{\mathbf{k}})\ \text{nT}]\cos(0.126x - 3.77 \times 10^7 t)$

7. (a) 0.333 μT   (b) 0.628 μm   (c) 477 THz

9. 75.0 MHz

11. 3.33 μJ/m^3

13. 307 μW/m^2

15. 3.33×10^3 m^2

17. (a) 332 kW/m^2 radially inward   (b) 1.88 kV/m and 222 μT

19. (a) $\mathbf{E} \cdot \mathbf{B} = 0$ (b) $(11.5\hat{\mathbf{i}} - 28.6\hat{\mathbf{j}})$ W/m^2

21. 29.5 nT

23. (a) 2.33 mT   (b) 650 MW/m^2 (c) 510 W

25. (a) 540 V/m   (b) 2.58 μJ/m^3 (c) 773 W/m^2
(d) 77.3% of the intensity in Example 34.5

27. 83.3 nPa

29. (a) 1.90 kN/C   (b) 50.0 pJ   (c) 1.67×10^{-19} kg·m/s

31. (a) 11.3 kJ   (b) 1.13×10^{-4} kg·m/s

33. (a) 134 m   (b) 46.9 m

35. (a) away along the perpendicular bisector of the line segment joining the antennas   (b) along the extensions of the line segment joining the antennas

37. (a) $\mathbf{E} = \frac{1}{2}\mu_0 c J_{\text{max}}[\cos(kx - \omega t)]\hat{\mathbf{j}}$

(b) $\mathbf{S} = \frac{1}{4}\mu_0 c J_{\text{max}}^2[\cos^2(kx - \omega t)]\hat{\mathbf{i}}$

(c) $I = \dfrac{\mu_0 c J_{\text{max}}^2}{8}$ (d) 3.48 A/m

39. 545 THz

41. (a) 6.00 pm   (b) 7.50 cm

43. 60.0 km

45. 1.00 Mm = 621 mi; not very practical

47. (a) 3.77×10^{26} W   (b) 1.01 kV/m and 3.35 μT

49. (a) $2\pi^2 r^2 f B_{max} \cos \theta$, where θ is the angle between the magnetic field and the normal to the loop   (b) The loop should be in the vertical plane containing the line of sight to the transmitter.

51. (a) 6.67×10^{-16} T   (b) 5.31×10^{-17} W/m²
(c) 1.67×10^{-14} W   (d) 5.56×10^{-23} N

53. 95.1 mV/m

55. (a) B_{max} = 583 nT, k = 419 rad/m, ω = 126 Grad/s; **B** vibrates in xz plane   (b) $\mathbf{S}_{av} = (40.6\hat{\mathbf{i}})$ W/m²
(c) 271 nPa   (d) $(406\hat{\mathbf{i}})$ nm/s²

57. (a) 22.6 h   (b) 30.6 s

59. (a) 8.32×10^7 W/m²   (b) 1.05 kW

61. (a) 1.50 cm   (b) 25.0 μJ   (c) 7.37 mJ/m³
(d) 40.8 kV/m, 136 μT   (e) 83.3 μN

63. 637 nPa

65. $\epsilon_0 E^2 A/2m$

67. (a) 16.1 cm   (b) 0.163 m²   (c) 470 W/m²   (d) 76.8 W
(e) 595 N/C   (f) 1.98 μT   (g) The cats are nonmagnetic and carry no macroscopic charge or current. Oscillating charges within molecules make them emit infrared radiation.   (h) 119 W

69. 4.77 Gm

CHAPTER 35

1. 299.5 Mm/s

3. 114 rad/s

5. (c) 0.055 7 °

9. 23.3°

11. 15.4°; 2.56 m

13. 19.5° above the horizon

15. (a) 1.52   (b) 417 nm   (c) 474 THz   (d) 198 Mm/s

17. 158 Mm/s

19. 30.0° and 19.5° at entry; 19.5° and 30.0° at exit

21. 3.88 mm

23. 30.4° and 22.3°

25. $\sim 10^{-11}$ s; between 10^3 and 10^4 wavelengths

29. 0.171°

31. 86.8°

33. 27.9°

35. 4.61°

37. (a) 33.4°   (b) 53.4°   (c) There is no critical angle.

39. 1.000 08

41. 1.08 cm < d < 1.17 cm

43. Skylight incident from above travels down the plastic. If the index of refraction of the plastic is greater than 1.41, the rays close in direction to the vertical are totally reflected from the side walls of the slab and from both facets at the bottom of the plastic, where it is not immersed in gasoline. This light returns up inside the plastic and makes it look bright. Where the plastic is immersed in gasoline, total internal reflection is frustrated and the downward-propagating light passes from the plastic out into the gasoline. Little light is reflected up, and the gauge looks dark.

45. Scattered light leaving the photograph in all forward horizontal directions in air is gathered by refraction into a fan in the water of half-angle 48.6°. At larger angles you see things on the other side of the globe, reflected by total internal reflection at the back surface of the cylinder.

47. 77.5°

49. 2.27 m

51. (a) 0.172 mm/s   (b) 0.345 mm/s   (c) northward at 50.0° below the horizontal   (d) northward at 50.0° below the horizontal

53. 62.2%

55. 82 reflections

57. (b) 68.5%

59. 27.5°

61. (a) It always happens.   (b) 30.3°   (c) It cannot happen.

63. 2.36 cm

67. 1.93

69. (a) $n = [1 + (4t/d)^2]^{1/2}$ (b) 2.10 cm   (c) violet

71. (a) 1.20   (b) 3.40 ns

CHAPTER 36

1. $\sim 10^{-9}$ s younger

3. 35.0 in.

5. 10.0 ft, 30.0 ft, 40.0 ft

7. (a) 13.3 cm, -0.333, real and inverted   (b) 20.0 cm, -1.00, real and inverted   (c) no image is formed

9. (a) -12.0 cm; 0.400   (b) -15.0 cm; 0.250   (c) upright

11. (a) q = 45.0 cm; $M = -0.500$ (b) $q = -60.0$ cm; M = 3.00   (c) Image (a) is real, inverted, and diminished. Image (b) is virtual, upright, and enlarged. The ray diagrams are like Figures 36.15a and 36.15b, respectively.

13. At 0.708 cm in front of the reflecting surface. Image is virtual, upright, and diminished.

15. 7.90 mm

17. (a) a concave mirror with radius of curvature 2.08 m
(b) 1.25 m from the object

19. (a) 25.6 m   (b) 0.058 7 rad   (c) 2.51 m   (d) 0.023 9 rad
(e) 62.8 m from your eyes

21. 38.2 cm below the top surface of the ice

23. 8.57 cm

25. (a) 45.0 cm   (b) -90.0 cm   (c) -6.00 cm

27. 1.50 cm/s

29. (a) 16.4 cm   (b) 16.4 cm

31. (a) 650 cm from the lens on the opposite side from the object; real, inverted, enlarged   (b) 600 cm from the lens on the same side as the object; virtual, upright, enlarged

33. 2.84 cm

37. (a) -12.3 cm, to the left of the lens   (b) 0.615
(c)

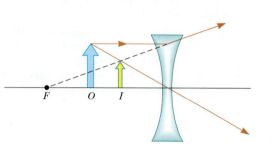

39. (a) 7.10 cm   (b) 0.074 0 mm   (c) 23.3 MW/m^2

41. (a) $p = \dfrac{d}{2} \pm \sqrt{\dfrac{d^2}{4} - fd}$ (b) Both images are real and inverted. One is enlarged, the other diminished.

43. 1.24 cm

45. 21.3 cm

47. -4.00 diopters, a diverging lens

49. -3.70 diopters

51. -575

53. (a) -800 (b) image is inverted

55. (a) virtual   (b) infinity   (c) 15.0 cm, -5.00 cm

57. -40.0 cm

59. (a) 23.1 cm   (b) 0.147 cm

61. (a) 67.5 cm   (b) The lenses can be displaced in two ways. The first lens can be displaced 1.28 cm farther away from the object, and the second lens 17.7 cm toward the object. Alternatively, the first lens can be displaced 0.927 cm toward the object and the second lens 4.44 cm toward the object.

63. $q = 5.71$ cm; real

65. 0.107 m to the right of the vertex of the hemispherical face

67. 8.00 cm
Ray diagram:

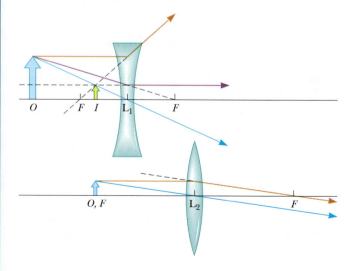

69. 1.50 m in front of the mirror; 1.40 cm (inverted)

71. (a) 30.0 cm and 120 cm   (b) 24.0 cm   (c) real, inverted, diminished with $M = -0.250$

73. -75.0

75. (a) 44.6 diopters   (b) 3.03 diopters

77. (a) 20.0 cm to the right of the second lens, -6.00
(b) inverted   (c) 6.67 cm to the right of the second lens, -2.00, inverted

CHAPTER 37

1. 1.58 cm

3. (a) 55.7 m   (b) 124 m

5. 1.54 mm

7. (a) 2.62 mm   (b) 2.62 mm

9. 11.3 m

11. (a) 10.0 m   (b) 516 m   (c) Only the runway center-line is a maximum for the interference patterns for both frequencies. If the frequencies were related by a ratio of small integers k/ℓ, the plane could by mistake fly along the kth side maximum of one signal where it coincides with the ℓth side maximum of the other.

13. (a) 13.2 rad   (b) 6.28 rad   (c) 0.012 7 degree
(d) 0.059 7 degree

15. (a) 1.93 μm   (b) 3.00λ (c) maximum

17. 48.0 μm

19. (a) 7.95 rad   (b) 0.453

21. (a) and (b) 19.7 kN/C at 35.0°
(c) 9.36 kN/C at 169°

23. $10.0 \sin(100\pi t + 0.927)$

25. $26.2 \sin(\omega t + 36.6°)$

27. $\pi/2$

29. $360°/N$

31. (a) green   (b) violet

33. 0.500 cm

35. no reflection maxima in the visible spectrum

37. 290 nm

39. 4.35 μm

41. 39.6 μm

43. $1 + N\lambda/2L$

45. 1.25 m

47. (a) $\sim 10^{-3}$ degree
(b) $\sim 10^{11}$ Hz, microwave

49. 20.0×10^{-6} °C^{-1}

51. 3.58°

53. 1.62 km

55. 421 nm

59. (a) $2(4h^2 + d^2)^{1/2} - 2d$
(b) $(4h^2 + d^2)^{1/2} - d$

61. $y' = (n - 1)tL/d$

63. (a) 70.6 m   (b) 136 m

65. 1.73 cm

67. (a) 4.86 cm from the top   (b) 78.9 nm and 128 nm
(c) 2.63×10^{-6} rad

69. 0.505 mm

CHAPTER 38

1. 4.22 mm

3. 0.230 mm

5. three maxima, at 0° and near 46° on both sides

7. 51.8 μm wide and 949 μm high

9. 0.016 2

11. 1.00 mrad

13. 3.09 m

15. violet; between 186 m and 271 m

17. 13.1 m

19. Neither. It can resolve objects no closer than several centimeters apart.

21. 0.244 rad = 14.0°

23. 7.35°

25. 5.91° in first order, 13.2° in second order, 26.5° in third order

27. (a) 478.7 nm, 647.6 nm, and 696.6 nm   (b) 20.51°, 28.30°, and 30.66°

29. (a) 12 000, 24 000, 36 000   (b) 11.1 pm

31. (a) 2 800 grooves   (b) 4.72 μm

33. (a) 5 orders   (b) 10 orders in the short-wavelength region

35. 93.4 pm

37. 14.4°

39. 5.51 m, 2.76 m, 1.84 m

41. (a) 54.7°   (b) 63.4°   (c) 71.6°

43. 1.11

45. 60.5°

47. 36.9° above the horizon

49. (a) 6   (b) 7.50°

51. 632.8 nm

53. (a) 25.6°   (b) 19.0°

55. 545 nm

57. (a) 3.53×10^3 grooves/cm
(c) Eleven maxima

59. 4.58 μm $< d <$ 5.23 μm

61. 15.4

63. (a) 41.8°   (b) 0.593   (c) 0.262 m

67. (b) 3.77 nm/cm

69. (b) 15.3 μm

71. $\phi = 1.391\,557\,4$ after seventeen steps or fewer

73. $a = 99.5\ \mu$m $\pm 1\%$

CHAPTER 39

5. $0.866c$

7. (a) 64.9/min   (b) 10.6/min

9. 1.54 ns

11. $0.800c$

13. (a) 39.2 μs   (b) accurate to one digit

15. (a) 20.0 m   (b) 19.0 m   (c) $0.312c$

17. (a) 21.0 yr   (b) 14.7 ly   (c) 10.5 ly   (d) 35.7 yr

19. (c) 2.00 kHz   (d) $\pm 0.075\,0$ m/s ≈ 0.2 mi/h

21. $0.220c = 6.59 \times 10^7$ m/s

23. (a) 17.4 m   (b) 3.30°

25. (a) 2.50×10^8 m/s   (b) 4.97 m   (c) -1.33×10^{-8} s

27. $0.960c$

29. (a) 2.73×10^{-24} kg·m/s   (b) 1.58×10^{-22} kg·m/s
(c) 5.64×10^{-22} kg·m/s

31. 4.50×10^{-14}

33. $0.285c$

35. (a) 5.37×10^{-11} J   (b) 1.33×10^{-9} J

37. 1.63×10^3 MeV/c

39. (a) 938 MeV   (b) 3.00 GeV   (c) 2.07 GeV

41. 8.84×10^{-28} kg and 2.51×10^{-28} kg

45. (a) 3.91×10^4 (b) $u = 0.999\,999\,999\,7c$ (c) 7.67 cm

47. 4.08 MeV and 29.6 MeV

49. $\sim 10^{-15}$

51. 0.842 kg

53. 4.19×10^9 kg/s

55. (a) 26.6 Mm   (b) 3.87 km/s   (c) -8.34×10^{-11}
(d) 5.29×10^{-10} (e) $+4.46 \times 10^{-10}$

57. (a) a few hundred seconds   (b) $\sim 10^8$ km

59. (a) $0.800c$ (b) 7.50 ks   (c) 1.44 Tm, $-0.385c$ (d) 4.88 ks

61. 0.712%

63. (a) $0.946c$ (b) 0.160 ly   (c) 0.114 yr   (d) 7.50×10^{22} J

65. (a) 76.0 min   (b) 52.1 min

67. yes, with 18.8 m to spare

69. (b) For u small compared to c, the relativistic expression agrees with the classical expression. As u approaches c, the acceleration approaches zero, so that the object can never reach or surpass the speed of light.
(c) Perform $\int (1 - u^2/c^2)^{-3/2} \, du = (qE/m) \int dt$ to obtain $u = qEct(m^2c^2 + q^2E^2t^2)^{-1/2}$ and then $\int dx = \int qEct(m^2c^2 + q^2E^2t^2)^{-1/2} \, dt$ to obtain $x = (c/qE)[(m^2c^2 + q^2E^2t^2)^{1/2} - mc]$

75. 1.82×10^{-3} eV

CHAPTER 40

1. 5.18×10^3 K

3. 1.30×10^{15}/s

5. (a) 5.75×10^3 K   (b) 504 nm

7. (a) 2.57 eV   (b) 12.8 μeV   (c) 191 neV   (d) 484 nm (visible); 9.68 cm and 6.52 m (radio waves)

9. 2.27×10^{30} photons/s

11. 7.73×10^3 K

13. (a) 296 nm, 1.01 PHz   (b) 2.71 V

15. (a) only lithium   (b) 0.808 eV

17. (a) 1.90 eV   (b) 0.216 V

19. 8.41 pC

21. 1.78 eV, 9.47×10^{-28} kg·m/s

23. 70.0°

25. (a) 43.0°   (b) 602 keV, 3.21×10^{-22} kg·m/s   (c) 278 keV, 3.21×10^{-22} kg·m/s

27. (a) 0.101 nm   (b) 81.1°

29. 0.004 86 nm

31. By this definition, ionizing radiation is the ultraviolet light, x-rays, and γ rays with wavelength shorter than 124 nm; that is, with frequency higher than 2.42×10^{15} Hz.

33. 397 fm

35. (a) 0.709 nm   (b) 414 nm

37. (a) ~100 MeV or more   (b) No. The kinetic energy of the electron is much larger than the typical binding energy, so the electron would immediately escape.

39. (b) No. $\lambda^{-2} + \lambda_C^{-2}$ cannot be equal to λ^{-2}.

41. (a) 14.9 keV or, neglecting relativistic correction, 15.1 keV   (b) 124 keV

43. $v_{\text{phase}} = u/2$

45. (a) 993 nm   (b) 4.96 mm   (c) If its detection forms part of an interference pattern, the neutron must have passed through both slits. If we test to see which slit a particular neutron passes through, it will not form part of the interference pattern.

47. 2.27 pA

49. Within 1.16 mm for the electron, 5.28×10^{-32} m for the bullet

51. The electron energy must be $\sim 100mc^2$ or larger. The proton energy can be as small as $1.001mc^2$, which is within the range well described classically.

53. (b) 519 am

55. (a)

(b) 6.4×10^{-34} J·s $\pm 8\%$ (c) 1.4 eV

57. (a) 191 MeV   (b) 9.20 MeV

67. 2.81×10^{-8}

69. (a) The light is unpolarized. It contains both horizontal and vertical electric field oscillations.   (b) The interference pattern appears, but with diminished overall intensity.   (c) The results are the same in each case.   (d) The interference pattern appears and disappears as the polarizer turns, with alternately increasing and decreasing contrast between the bright and dark fringes. The intensity on the screen is precisely zero at the center of a dark fringe four times in each revolution, when the filter axis has turned by 45°, 135°, 225°, and 315° from the vertical.   (e) Looking at the overall light energy arriving at the screen, we see a low-contrast interference pattern. After we sort out the individual photon runs into those for trial 1, those for trial 2, and those for trial 3, we have the original results replicated: The runs for trials 1 and 2 form the two blue graphs in Figure 40.24, and the runs for trial 3 build up the red graph.

CHAPTER 41

1. (a) 126 pm   (b) 5.27×10^{-24} kg·m/s   (c) 95.5 eV

3. (a) 0.434 nm   (b) 6.00 eV

5.   (a) n

```
4 ———————— 603 eV



3 ———————— 339 eV


2 ———————— 151 eV

1 ———————— 37.7 eV
```
ENERGY

(b) 2.20 nm, 2.75 nm, 4.12 nm, 4.71 nm, 6.60 nm, 11.0 nm

7. 0.793 nm

9. 6.17 MeV, 202 fm, a gamma ray

11. 0.513 MeV, 2.05 MeV, 4.62 MeV; yes

19. at $L/4$ and at $3L/4$

21. (a) 0.196   (b) 0.609   (c) The classical probability, 1/3, is very different.

23. (a) $E = \hbar^2/mL^2$ (b) Requiring $\int_{-L}^{L} A^2(1 - x^2/L^2)^2 \, dx = 1$ gives $A = (15/16L)^{1/2}$ (c) $47/81 = 0.580$

25. (a)

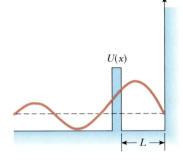

(b) $2L$

27. (a) 0.010 3   (b) 0.990

29. by 0.959 nm, to 1.91 nm

31. 1.35

35. 600 nm

37. (b) The acceleration is equal to a negative constant times the excursion from equilibrium. The frequency is

$$\frac{1}{2\pi} \sqrt{\frac{k}{\mu}}.$$

39. $\sim 10^{-10^{30}}$

41.

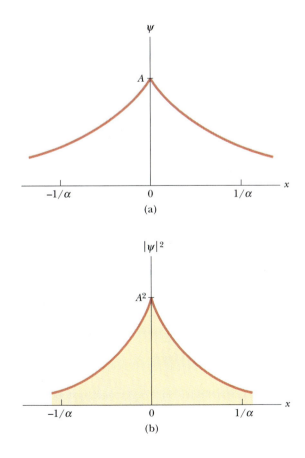

(a)

(b)

(c) The wave function is continuous. It shows localization by approaching zero as $x \rightarrow \pm\infty$. It is everywhere finite and can be normalized.   (d) $A = \sqrt{\alpha}$ (e) 0.632

43. 0.029 4

45. (a) 434 THz   (b) 691 nm   (c) 165 peV or more

49. (a)

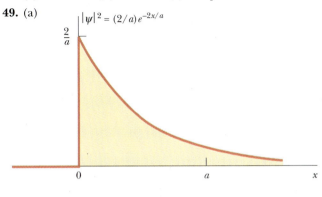

(b) 0   (c) 0.865

51. (a) $-7k_e e^2/3d$ (b) $h^2/36m_e d^2$ (c) 49.9 pm
(d) 280 pm is 5.62 times larger.

53. (a) $A = (2/17L)^{1/2}$ (b) $|A|^2 + |B|^2 = 1/a$

55. 2.25

CHAPTER 42

1. (a) 5   (b) no; no

3. (b) 0.846 ns

5. (a) 2.19 Mm/s   (b) 13.6 eV   (c) -27.2 eV

7. (a) 0.212 nm   (b) 9.95×10^{-25} kg·m/s
(c) 2.11×10^{-34} kg·m^2/s   (d) 3.40 eV   (e) -6.80 eV
(f) -3.40 eV

9. (a) 3.03 eV   (b) 411 nm   (c) 732 THz

11. 4.42×10^4 m/s

13. (a) $E_n = -54.4$ eV/n^2 for $n = 1, 2, 3, \ldots$

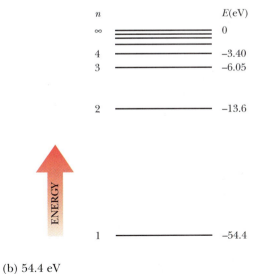

(b) 54.4 eV

15. (a) 2.89×10^{34} kg·m^2/s   (b) 2.74×10^{68}
(c) 7.30×10^{-69}

17. (a) $\Delta p \geq \hbar/2r$ (b) Choosing $p \approx \hbar/r$,
$E = K + U = \hbar^2/2m_e r^2 - k_e e^2/r$ (c) $r - \hbar^2/m_e k_e c^2 - a_0$
and $E = -13.6$ eV, in agreement with the Bohr theory.

19. (b) 0.497

21. It does, with $E = -k_e e^2/2a_0$

23. (a)

n	ℓ	m_ℓ	m_s
3	2	2	1/2
3	2	2	−1/2
3	2	1	1/2
3	2	1	−1/2
3	2	0	1/2
3	2	0	−1/2
3	2	−1	1/2
3	2	−1	−1/2
3	2	−2	1/2
3	2	−2	−1/2

(b)

n	ℓ	m_ℓ	m_s
3	1	1	1/2
3	1	1	−1/2
3	1	0	1/2
3	1	0	−1/2
3	1	−1	1/2
3	1	−1	−1/2

25. $\ell = 4$

27. (a) 2   (b) 8   (c) 18   (d) 32   (e) 50

29. (a) 3.99×10^{17} kg/m^3 (b) 81.7 am   (c) 1.77 Tm/s
(d) $5.91 \times 10^3 c$

31. $n = 3$; $\ell = 2$; $m_\ell = -2, -1, 0, 1,$ or 2; $s = 1$; $m_s = -1, 0,$
or 1, for a total of 15 states

33. The $4s$ subshell is filled first. We would expect $[\text{Ar}]3d^4 4s^2$
to have lower energy, but $[\text{Ar}]3d^5 4s^1$ has more unpaired
spins and lower energy according to Hund's rule. It is the
ground-state configuration of chromium.

35. Aluminum

37. (a) $1s, 2s, 2p, 3s, 3p, 4s, 3d, 4p, 5s, 4d, 5p, 6s, 4f, 5d, 6p, 7s$
(b) Element 15 should have valence $+5$ or -3, and it
does. Element 47 should have valence -1, but it has valence $+1$. Element 86 should be inert, and it is.

39. 18.4 T

41. 124 V

43. 0.072 5 nm

45. iron

47. 28.2 THz, 10.6 μm, infrared

49. 3.49×10^{16} photons

51. (a) 217 nm   (b) 93.1 nm

53. (a) 609 μeV   (b) 6.9 μeV   (c) 147 GHz, 2.04 mm

55. $r_n = (0.106 \text{ nm})n^2$, $E_n = -6.80 \text{ eV}/n^2$, for $n = 1, 2, 3, \ldots$

57. The classical frequency is $4\pi^2 m_e k_e^2 e^4/h^3 n^3$.

59. (a) 1.57×10^{14} m$^{-3/2}$ (b) 2.47×10^{28} m^{-3}
(c) 8.69×10^8 m^{-1}

61. (a) 4.20 mm   (b) 1.05×10^{19} photons
(c) 8.82×10^{16}/mm^3

63. $3\hbar^2/4mL^2$

67. 5.39 keV

69. 0.125

71. 9.79 GHz

73. (a) $\sim -10^6$ m/s^2 (b) ~ 1 m

CHAPTER 43

1. (a) 921 pN toward the other ion   (b) -2.88 eV

3. (a) $(2A/B)^{1/6}$ (b) $B^2/4A$ (c) 74.2 pm, 4.46 eV

5. ~ 10 K

7. 5.63 Trad/s

11. 2.72×10^{-47} kg$\cdot$m^2

13. 0.358 eV

15. $(18.4 \ \mu\text{eV}) J(J + 1)$, where $J = 0, 1, 2, 3, \ldots$

17. 558

19. (a) 4.60×10^{-48} kg$\cdot$m^2 (b) 1.32×10^{14} Hz
(c) 0.074 1 nm

21. 6.25×10^9

23. -7.84 eV

27. An average atom contributes 0.981 electron to the conduction band.

29. (a) 2.54×10^{28} electrons/m^3 (b) 3.15 eV

31. 5.28 eV

35. (a) 1.10   (b) 1.47×10^{-25}, much smaller

37. All of the Balmer lines are absorbed, except for the red
line at 656 nm, which is transmitted.

39. 1.24 eV or less; yes

41. $-0.021 9$ eV, 2.81 nm

43. 4.4 V

45. (a) 2.98 mA   (b) 67.1 Ω (c) 8.39 Ω

47. 203 A to produce a magnetic field in the direction of the
original field

49. (a) 6.15×10^{13} Hz   (b) 1.59×10^{-46} kg$\cdot$m^2
(c) 4.79 μm or 4.96 μm

51. 7

55. (a) 0.350 nm   (b) -7.02 eV   (c) $-1.20\hat{\mathbf{i}}$ nN

57. (a) r_0 (b) B (c) $(a/\pi)[B/2\mu]^{1/2}$
(d) $B - (ha/\pi)[B/8\mu]^{1/2}$

59. (b) No. The fourth term is larger than the sum of the first
three.

CHAPTER 44

1. $\sim 10^{28}$; $\sim 10^{28}$; $\sim 10^{28}$

3. (a) 27.6 N   (b) 4.17×10^{27} m/s^2 away from the nucleus
(c) 1.73 MeV

5. (a) 455 fm   (b) 6.04 Mm/s

7. (a) 1.90 fm   (b) 7.44 fm

9. 16.0 km

11. Z magic: He, O, Ca, Ni, Sn, Pb. N magic: isotopes of H, He,
N, O, Cl, K, Ca, V, Cr, Sr, Y, Zr, Xe, Ba, La, Ce, Pr, Nd, Pb,
Bi, Po.

13. (a) $^{36}_{16}S$, $^{6}_{2}He$ (b) $^{6}_{2}He$ is unstable. Two neutrons must be removed to make it the stable $^{4}_{2}He$.

15. They agree with the figure.

17. (a) $^{139}_{55}Cs$ (b) $^{139}_{57}La$ (c) $^{139}_{55}Cs$

19. greater for $^{15}_{7}N$ by 3.54 MeV

21. 200 MeV

23. (a) 491 MeV   (b) 179%, -53.0%, -24.6%, -1.37%

25. 1.16 ks

27. (a) $1.55 \times 10^{-5}/s$, 12.4 h   (b) 2.39×10^{13} atoms
   (c) 1.88 mCi

29. 9.47×10^{9} nuclei

31. 2.66 d

33. 4.27 MeV

35. 9.96×10^{3} yr

37. $^{3}_{1}H$ atom $\rightarrow$ $^{3}_{2}He$ atom $+ \bar{\nu}$; 18.6 keV

39. (a) $e^{-} + p \rightarrow n + \nu$ (b) $^{15}_{8}O$ atom $\rightarrow$ $^{15}_{7}N$ atom $+ \nu$
   (c) 2.75 MeV

41.

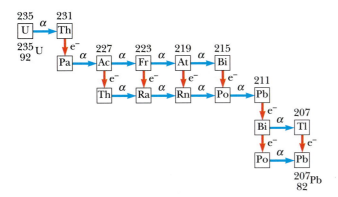

43. (a) X: 3.66 days, Y: 10.6 h   (b) ^{224}Ra, ^{212}Pb (c) three

45. -2.64 MeV

47. (a) $^{197}_{79}Au + ^{1}_{0}n \rightarrow ^{198}_{80}Hg + ^{0}_{-1}e + \bar{\nu}$ (b) 7.89 MeV

49. 8.005 3 u; 10.013 5 u

51. 165 MeV

53. (a) 29.2 MHz   (b) 42.6 MHz   (c) 2.13 kHz

55. (a) 5.70 MeV   (b) 3.27 MeV, exothermic

57. 1.88 MeV

59. (a) cannot occur   (b) can occur   (c) K_e between 0 and 156 keV

61. (a) conservation of energy   (b) electric potential energy of the nucleus   (c) 1.20 MeV

63. (a) $^{141}_{59}Pr$ (b) 2.32 MeV   (c) 97.2%

65. (a) 61.8 Bq/L   (b) 40.3 d

67. 5.94 Gyr

69. 2.20 μeV

73. 0.400%

75. 2.64 min

CHAPTER 45

1. 0.387 g

3. $^{144}_{54}Xe$, $^{143}_{54}Xe$, $^{142}_{54}Xe$

5. $^{1}_{0}n + ^{232}Th \rightarrow ^{233}Th \rightarrow ^{233}Pa + e^{-} + \bar{\nu}$,
   $^{233}Pa \rightarrow ^{233}U + e^{-} + \bar{\nu}$

7. (a) 8.68 MeV   (b) The proton and the boron nucleus have positive charges. The colliding particles must have enough kinetic energy to approach very closely in spite of their electric repulsion.

9. 5.80 Mm

11. about 3 000 yr

13. 6.25×10^{19} Bq

15. (a) $2.30 \times 10^{-14} Z_1 Z_2$ J   (b) 0.144 MeV

17. (a) 2.22 Mm/s   (b) $\sim 10^{-7}$ s

19. (a) 1.7×10^{7} J   (b) 7.3 kg

21. 1.30×10^{25} ^{6}Li; 1.61×10^{26} ^{7}Li

23. 1.66×10^{3} yr

25. (a) 2.5 mrem per x-ray
   (b) 5 rem/yr is 38 times 0.13 rem/yr

27. 2.09×10^{6} s

29. 1.14 rad

31. (a) 3.12×10^{7} (b) 3.12×10^{10} electrons

33. (a) about 8 min   (b) 27.6 min   (c) 30 min $\pm$ 30%

35. $\sim 10^{3}$ Bq

37. (b) R/λ

39. (a) 8×10^{4} eV   (b) 4.62 MeV and 13.9 MeV
   (c) 1.03×10^{7} kWh

41. (a) 177 MeV   (b) $K_{Br} = 112$ MeV, $K_{La} = 65.4$ MeV
   (c) $v_{Br} = 15.8$ Mm/s, $v_{La} = 9.20$ Mm/s

43. 482 Ci, less than the fission inventory by on the order of a hundred million times

45. 2.56×10^{4} kg

47. (a) 2.65 GJ   (b) The fusion energy is 78.0 times larger.

49. (a) 4.91×10^{8} kg/h $= 4.91 \times 10^{5}$ m^3/h   (b) 0.141 kg/h

51. (a) 15.5 cm   (b) 51.7 MeV   (c) The number of decays per second is the decay rate R, and the energy released in each decay is Q. Then the energy released per unit time interval is $\mathcal{P} = QR$.   (d) 227 kJ/yr   (e) 3.18 J/yr

53. 14.0 MeV or, ignoring relativistic correction, 14.1 MeV

55. (a) 2.24×10^{7} kWh   (b) 17.6 MeV   (c) 2.34×10^{8} kWh
   (d) 9.36 kWh   (e) Coal is cheap at this moment in human history. We hope that safety and waste-disposal problems can be solved so that nuclear energy can be affordable before scarcity drives up the price of fossil fuels.

57. (b) 26.7 MeV

59. (a) 5×10^{7} K   (b) 1.94 MeV,   1.20 MeV,   1.02 MeV, 7.55 MeV,   7.30 MeV,   1.73 MeV,   1.02 MeV,   4.97 MeV, 26.7 MeV   (c) Most of the neutrinos leave the star directly after their creation, without interacting with any other particles.

61. (b) 1.00 μs   (c) 2.83 km/s   (d) 14.6 μs   (e) yes, 107 kilotons of TNT

CHAPTER 46

1. 453 ZHz; 662 am

3. 118 MeV

5. $\sim 10^{-18}$ m

7. 67.5 MeV, 67.5 MeV/c, 16.3 ZHz

9. (a) 0.782 MeV (b) $v_e = 0.919\ c$, $v_p = 380$ km/s (c) The electron is relativistic; the proton is not.

11. $\Omega^+ \to \overline{\Lambda}^0 + K^+$ $\overline{K}_S^0 \to \pi^+ + \pi^-$ $\overline{\Lambda}^0 \to \overline{p} + \pi^+$ $\overline{n} \to \overline{p} + e^+ + \nu_e$

13. (b) The second violates strangeness conservation.

15. (a) $\overline{\nu}_\mu$ (b) ν_μ (c) $\overline{\nu}_e$ (d) ν_e (e) ν_μ (f) $\overline{\nu}_e + \nu_\mu$

17. (a), (c), and (f) violate baryon number conservation. (b), (d), and (e) can occur. (f) violates muon–lepton number conservation.

19. (a) ν_e (b) ν_μ (c) $\overline{\nu}_\mu$ (d) $\nu_\mu + \overline{\nu}_\tau$

21. (b) and (c) conserve strangeness. (a), (d), (e), and (f) violate strangeness conservation.

23. (a) not allowed; violates conservation of baryon number (b) strong interaction (c) weak interaction (d) weak interaction (e) electromagnetic interaction

25. (a) K^+ (b) Ξ^0 (c) π^0

27. 9.26 cm

29. (a) 3.34×10^{26} e$^-$, 9.36×10^{26} u, 8.70×10^{26} d (b) $\sim 10^{28}$ e$^-$, $\sim 10^{29}$ u, $\sim 10^{29}$ d. You have zero strangeness, charm, truth, and beauty.

31. $m_u = 312$ MeV/c^2; $m_d = 314$ MeV/c^2

33. (a) The reaction $\overline{u}d + uud \to \overline{s}d + uds$ has a total of 1 u, 2 d, and 0 s quarks originally and finally. (b) The reaction $\overline{d}u + uud \to \overline{s}d + uus$ has a net of 3 u, 0 d, and 0 s quarks before and after. (c) $\overline{u}s + uud \to \overline{s}u + \overline{s}d + sss$ shows conservation at 1 u, 1 d, and 1 s quark. (d) The process $uud + uud \to \overline{s}d + uud + \overline{d}u + uds$ nets 4 u, 2 d, 0 s quarks initially and finally; the mystery particle is a Λ^0 or a Σ^0.

35. a neutron, udd

37. (a) $-e$, antiproton (b) 0, antineutron

39. (a) $0.160c$ (b) 2.82×10^9 ly

41. (a) 590.07 nm (b) 597 nm (c) 661 nm

43. 3.15 μW/m^2

45. (a) $\sim 10^{13}$ K (b) $\sim 10^{10}$ K

47. (b) 11.8 Gyr

49. (a) 1.61×10^{-35} m (b) 5.38×10^{-44} s (c) yes

51. 19 eV/c^2

53. one part in 50 000 000

55. (a) $2Nmc$ (b) $3^{1/2}Nmc$ (c) method (a)

57. 5.35 MeV and 32.3 MeV

59. 1 116 MeV/c^2

61. 70.4 MeV

63. 2.52×10^3 K

65. (a) Z^0 boson (b) gluon or photon

Credits

Photographs

This page constitutes an extension of the copyright page. We have made every effort to trace the ownership of all copyrighted material and to secure permission from copyright holders. In the event of any question arising as to the use of any material, we will be pleased to make the necessary corrections in future printings. Thanks are due to the following authors, publishers, and agents for permission to use the material indicated.

Chapter 1. **xxiv:** Courtesy of NASA **2:** elektraVision/Index Stock Imagery **6:** top left, Courtesy of National Institute of Standards and Technology, U.S. Dept. of Commerce; top right, Courtesy of National Institute of Standards and Technology, U.S. Dept. of Commerce **13:** Phil Boorman/Getty Images **19:** Sylvain Grandadam/Photo Researchers, Inc.

Chapter 2. **23:** George Lepp/Getty Images **26:** Ken White/Allsport/Getty Images **46:** North Wind Picture Archive **52:** bottom left, Courtesy U.S. Air Force; bottom right, Photri, Inc. **53:** George Semple **55:** Courtesy Amtrak Nec Media Relations

Chapter 3. **58:** Mark Wagner/Getty Images

Chapter 4. **77:** © Arndt/Premium Stock/PictureQuest **79:** Mark C. Burnett/Photo Researchers, Inc. **84:** The Telegraph Colour Library/Getty Images **86:** Mike Powell/Allsport/Getty Images **88:** Central Scientific Company **103:** bottom left, McKinney/Getty Images **104:** top left, Jed Jacobsohn/Allsport/Getty Images; top right, Bill Lee/Dembinsky Photo Associates; bottom left, Sam Sargent/Liaison International; bottom right, Courtesy of NASA **106:** Courtesy of NASA

Chapter 5. **111:** © Steve Raymer/CORBIS **114:** Giraudon/Art Resource **119:** Courtesy of NASA **120:** John Gillmoure/corbisstockmarket.com **122:** © John Elk III/Stock, Boston Inc./PictureQuest **139:** Roger Violet, Mill Valley, CA, University Science Books, 1992

Chapter 6. **150:** © Paul Hardy/CORBIS **151:** Mike Powell/Allsport/Getty Images **152:** © Tom Carroll/Index Stock Imagery/PictureQuest **157:** Robin Smith/Getty Images **165:** Jump Run Productions/Getty Images **166:** Charles D. Winters **174:** Frank Cezus/Getty Images **178:** Color Box/Getty Images

Chapter 7. **181:** Billy Hustace/Getty Images **183:** all, Charles D. Winters **184:** Gerard Vandystandt/Photo Researchers, Inc. **198:** all except top right, George Semple **198:** top right, Digital Vision/Getty Images **213:** Ron Chapple/Getty Images

Chapter 8. **217:** Harold E. Edgerton/Courtesy of Palm Press, Inc. **242:** Gamma **250:** Engraving from Scientific American, July 1888

Chapter 9. **251:** Mark Cooper/corbisstockmarket.com **257:** Courtesy of Saab **259:** top, Harold & Esther Edgerton Foundation 2002, courtesy of Palm Press, Inc; bottom, © Tim Wright/CORBIS **263:** No credit available **264:** Courtesy of Central Scientific Company **276:** Richard Megna/Fundamental Photographs **278:** Courtesy of NASA **279:** Courtesy of NASA **281:** Bill Stormont/corbisstockmarket.com **286:** Eye Ubiquitous/CORBIS

Chapter 10. **292:** Courtesy Tourism Malaysia **299:** George Semple **316:** Henry Leap and Jim Lehman **323:** Bruce Ayers/Getty Images **327:** John Lawrence/Getty Images **329:** Jerry Wachter/Photo Researchers, Inc.

Chapter 11. **336:** Otto Gruele/Getty Images **346:** both, © 1998 David Madison **354:** Gerard Lacz/NHPA

Chapter 12. **362:** John W. Jewett, Jr. **366:** Charles D. Winters

Chapter 13. **389:** University of Arizona/JPL/NASA **394:** Courtesy of NASA **410:** Courtesy H. Ford, et al., & NASA **417:** Courtesy of NASA

Chapter 14. **420:** Austin MacRae **422:** Earl Young/Getty Images **424:** right, David Frazier **427:** © Hulton-Deutsch Collection/CORBIS **430:** Geraldine Prentice/Getty Images **431:** Andy Sachs/Getty Images **431:** Werner Wolff/stockphoto.com **432:** George Semple **434:** Bettmann/CORBIS **435:** Courtesy of Central Scientific Company **438:** Galen Rowell/Peter Arnold, Inc. **439:** bottom left, Pamela Zilly/Getty Images **439:** top right, Henry Leap and Jim Lehman **439:** bottom right, Henry Leap and Jim Lehman **444:** George Semple **445:** Stan Osolinski/Dembinsky Photo Associates **447:** The Granger Collection **448:** Courtesy of Jeanne Maier

Chapter 15. **450:** Don Bonsey/Getty Images **452:** both, www.comstock.com **465:** top left, Courtesy of Ford Motor Company; bottom left, © Link/Visuals Unlimited **474:** both, UPI-Bettmann/CORBIS **478:** Telegraph Colour Library/Getty Images **479:** George Semple

Chapter 16. **486:** Kathy Ferguson Johnson/PhotoEdit/PictureQuest **509:** Gregg Adams/Getty Images

Chapter 17. **512:** Getty Images **518:** Courtesy Kenneth Burger Museum Archives/Kent State University **524:** Courtesy of the Educational Development Center, Newton, MA **527:** © 1973 Kim Vandiver and Harold E. Edgerton/Courtesy of Palm Press, Inc. **528:** www.comstock.com

Bohr Library. **1110:** David Parker/Science Photo Library/Photo Researchers, Inc. **1113:** Courtesy of Henry Leap and Jim Lehman **1114:** left, Dennis O'Clair/Getty Images; right, Hank Morgan/Photo Researchers, Inc. **1117:** Courtesy of U.S. Air Force, Langley Air Force Base **1120:** Roy Atkeson Photo Library **1122:** Courtesy Edwin Lo

Chapter 36. **1126:** Don Hammond/CORBIS **1128:** George Semple **1130:** Courtesy of Henry Leap/Jim Lehman **1131:** Ken Kay/Fundamental Photographs **1133:** top, Henry Leap/Jim Lehman **1133:** Courtesy of Thompson Consumer Electronics **1135:** David Rogers **1136:** NASA **1138:** © 1990 Paul Silverman/Fundamental Photographs **1144:** Henry Leap/Jim Lehman **1149:** Henry Leap/Jim Lehman **1155:** From Lennart Nilsson, in collaboration with Jan Lindberg, *Behold Man: A Photographic Journey of Discovery Inside the Body*, Boston, Little, Brown & Co. 1974. **1160:** George Semple **1161:** © Tony Freeman/PhotoEdit **1163:** © Tony Freeman/PhotoEdit **1164:** Orion® Sky View Pro™ 6 EQ Reflector, Orion Telescopes & Binoculars. Copyright © 2003 OrionTelescopes & Binoculars. All rights reserved. "Orion" is the trademark and service mark of Optronic Technologies, Inc. 800-676-1343, www.telescope.com **1166:** © 2003 by Sidney Harris. All rights reserved. **1166:** Richard Megna/Fundamental Photographs **1167:** George Semple **1167:** M.C. Escher/Cordon Art-Baarn-Holland. All Rights Reserved. **1169:** THE FAR SIDE © 1985 FARWORKS, INC. Used by permission. All rights reserved. **1174:** © Michael Levin/Opti-Gone Associates

Chapter 37. **1176:** RO-MA/Index Stock Imagery **1177:** © Kolar, Richard/Animals Animals/Earth Scenes **1178:** Courtesy of Sabina Zigman/Benjamin Cardozo High School, and by permission of *The Physics Teacher*, vol. 37, January 1999, p. 55. **1179:** Photograph from M. Cagnet, M. Francon, and J.C. Thierr, *Atlas of Optical Phenomena*, Berlin, Springer-Verlag, 1962. **1179:** Richard Megna/Fundamental Photographs **1184:** Photograph from M. Cagnet, M. Francon, and J.C. Thierr, *Atlas of Optical Phenomena*, Berlin, Springer-Verlag, 1962. **1191:** left, Dr. Jeremy Burgess/Science Photo Library/Photo Researchers, Inc. **1191:** right, Peter Aprahamian/Science Photo Library/Photo Researchers, Inc. **1191:** Courtesy Bausch and Lomb Optical Company **1192:** From Physical Science Study Committee, College Physics, Lexington, MA, Heath, 1968. **1193:** Kristen Brochmann/Fundamental Photographs **1194:** Richard Megna/Fundamental Photographs **1196:** LIGO Hanford Observatory **1201:** Courtesy of Central Scientific Supply

Chapter 38. **1205:** Denis Scott/CORBIS **1207:** right, P.M. Rinard, *Am. J. Phys.* 44:70, 1976 **1207:** From M. Cagnet, M. Francon, and J.C. Thierr, *Atlas of Optical Phenomena*, Berlin, Springer-Verlag, 1962, plate 18. **1212:** From M. Cagnet, M. Francon, and J.C. Thierr, *Atlas of Optical Phenomena*, Berlin, Springer-Verlag, 1962, plate 18. **1213:** Photograph courtesy of Central Scientific Company **1215:** From M. Cagnet, M. Francon, and J.C. Thierr, *Atlas of Optical Phenomena*, Berlin, Springer-Verlag, 1962, plate 16. **1218:** left, U.S. Naval Observatory/James W. Christy; right, Photo courtesy of Gemini Observatory **1220:** Courtesy Silicon Light Machines **1221:** Kristen Brochmann/Fundamental Photographs **1223:** Photo by Ronald R. Erickson; hologram by Nicklaus

Philips **1224:** bottom left, Used with permission of Eastman Kodak Company; bottom right, I. Andersson, Oxford Molecular Biophysics Laboratory/Science Photo Library/Photo Researchers, Inc. **1227:** Henry Leap/Jim Lehman **1230:** left, Henry Leap/Jim Lehman **1230:** bottom: a) Sepp Seitz 1981; b) Peter Aprahamian/Science Photo Library/Photo Researchers, Inc. **1231:** Photograph by Gary Friedman. Copyright © 2003, Los Angeles Times. Reprinted with permission. **1235:** SuperStock **1238:** top, Susan Allen Sigma/Getty Images **1238:** bottom, left: Riccardo Giovanni and Martha Haynes, Cornell University **1238:** center, right: Diane Schiumo 1988, Fundamental Photographs **1241:** © 2003 by Sidney Harris. All rights reserved.

Chapter 39. **1242:** Fermilab Photo, Courtesy of Fermi National Accelerator Laboratory **1244:** Emily Serway **1251:** AIP Niels Bohr Library **1275:** Courtesy NASA **1279:** Trent Steffler/David R. Frazier Photolibrary **1281:** Photo courtesy of Garmin Ltd. **1283:** © 2003 by Sidney Harris. All rights reserved.

Chapter 40. **1284:** Steve Cole/Getty Images **1286:** CORBIS **1287:** John Chumack/Photo Researchers **1288:** © Bettmann/CORBIS **1290:** Photodisc/Getty Images **1297:** Courtesy of AIP Niels Bohr Library **1301:** Courtesy of AIP Niels Bohr Library **1303:** © David Parker/Photo Researchers, Inc. **1304:** © Eye of Science/Science Source/Photo Researchers **1308:** a), b), c) From E.R. Huggins, *Physics 1*, New York, W.A. Benjamin, 1968. d) From C. Jonsson, *Zietschrift fur Physik* 161:454, 1961; used with permission **1310:** Courtesy of the University of Hamburg **1314:** Alexandra Héder **1314:** Alexandra Héder

Chapter 41. **1321:** Courtesy of IBM Research, Almaden Research Center. Unauthorized use prohibited. **1331:** AIP Emilio Segré Visual Archives **1340:** Courtesy of IBM Research, Almaden Research Center. Unauthorized use prohibited.

Chapter 42. **1351:** NASA/GRIN **1353:** (K.W. Whitten, R.E. Davis, M.L. Peck and G.G. Stanley, *General Chemistry*, 7th ed., Belmont, CA, Brooks/Cole, 2004). **1354:** Dembinsky Photo Assoc. **1355:** Stock Montage, Inc. **1356:** Photo courtesy of AIP Niels Bohr Library, Margarethe Bohr Collection **1371:** Photo courtesy of AIP Niels Bohr Library, Margarethe Bohr Collection **1375:** CERN, courtesy of AIP Emilio Segré Visual Archives **1387:** Philippe Plailly/SPL/Photo Researchers **1389:** Courtesy of Mark Helfer/National Institute Standards and Technology

Chapter 43. **1398:** © Bruce Dale/Getty Images **1403:** Douglas Struthers/Getty Images **1413:** Charles D. Winters **1414:** Charles D. Winters **1428:** Courtesy of Texas Instruments, Inc. **1429:** Brand X Pictures/Getty Images

Chapter 44. **1440:** © Paul Hanny/Gamma **1451:** Courtesy of Louise Barker/AIP Niels Bohr Library **1452:** Getty Images **1459:** © Michael Dalton/Fundamental Photographs **1469:** SBHA/Getty Images **1472:** © Richard Megna/Fundamental Photographs

Chapter 45. **1479:** Tony Freeman/Index Stock Imagery **1484:** Courtesy of Chicago Historical Society **1484:** National

Accelerator Laboratory **1491:** b) Courtesy of Princeton Plasma Physics Laboratory **1491:** c) Courtesy of Princeton University **1492:** Courtesy of ITER **1494:** Courtesy of University of California Lawrence Livermore National Laboratory and the U.S. Department of Energy **1494:** Courtesy of Los Alamos National Laboratory **1498:** SPL/Photo Researchers **1498:** Dan McCoy/Rainbow **1499:** © Hank Morgan/Photo Researchers **1500:** Courtesy of Brookhaven National Laboratory/RHIC-STAR **1502:** AP/Wide World Photos **1502:** Martin Dohrn/Science Photo Library/Photo Researchers **1503:** Council for Agricultural Science and Technology **1508:** Courtesy of NASA Ames

Chapter 46. **1511:** Courtesy of Brookhaven National Laboratory/RHIC-STAR **1513:** Courtesy of AIP Emilio Segré Visual Archives **1514:** Courtesy Lawrence Berkeley Laboratory, University of California **1515:** National Institutes of Health **1516:** UPI/Bettman-CORBIS **1517:** © Shelley Gazin/CORBIS **1521:** Courtesy of KRR (Institute for Cosmic Ray Research), University of Tokyo **1523:** Courtesy Lawrence Berkeley Laboratory, University of California, Photographic Services **1527:** Courtesy of Michael R. Dressler **1528:** Courtesy of Brookhaven National Laboratory **1535:** Courtesy of CERN **1536:** Courtesy of Fermi National Accelerator Laboratory **1537:** AT&T Bell Laboratories **1548:** Maarten Schmidt/Palomar Observatory/California Institute of Technology **1548:** Maarten Schmidt/Palomar Observatory/California Institute of Technology Photograph **1549:** Anglo-Australian Telescope Board **1549:** Anglo-Australian Telescope Board

Tables and Illustrations

This page constitutes an extension of the copyright page. We have made every effort to trace the ownership of all copyrighted material and to secure permission from copyright holders. In the event of any question arising as to the use of any material, we will be pleased to make the necessary corrections in future printings. Thanks are due to the following authors, publishers, and agents for permission to use the material indicated.

Chapter 2. **56:** By permission of John Hart and Creators Syndicate, Inc.

Chapter 13. **415:** By permission of John Hart and Creators Syndicate, Inc.

Chapter 14. **449:** Calvin & Hobbes © 1992 Watterson. Reprinted with permission of Universal Press Syndicate. All rights reserved.

Chapter 21. **666:** By permission of John Hart and Creators Syndicate, Inc.

Chapter 30. **966:** Calvin and Hobbes © Watterson. Reprinted with permission of Universal Press Syndicate. All rights reserved.

Chapter 36. **1166:** © 2003 Sidney Harris **1169:** The Far Side® by Gary Larson © 1985 FarWorks, Inc. All Rights Reserved. Used with permission.

Chapter 38. **1241:** © 2003 by Sidney Harris

Chapter 39. **1283:** © 2003 by Sidney Harris

Chapter 40. **1320:** ROSE IS ROSE reprinted by permission of United Features Syndicate, Inc.

Chapter 41. **1340:** Based on a drawing from P.K. Hansma, V.B. Elings, O. Marti, and C. Bracker, *Science* 242:209, 1988. © 1988 by AAAS. **1342:** from C.W. Sherwin, *Introduction to Quantum Mechanics*, New York, Holt, Reinhart and Winston, 1959. Used with permission.

Chapter 46. **1552:** © 2003 by Sidney Harris

Index

Locator notes: **boldface** indicates a definition; *italics* indicates a figure; *t* indicates a table

Some Physical Constants[a]

Quantity	Symbol	Value[b]
Atomic mass unit	u	$1.660\ 538\ 73\ (13) \times 10^{-27}$ kg $931.494\ 013\ (37)$ MeV/c^2
Avogadro's number	N_A	$6.022\ 141\ 99\ (47) \times 10^{23}$ particles/mol
Bohr magneton	$\mu_B = \dfrac{e\hbar}{2m_e}$	$9.274\ 008\ 99\ (37) \times 10^{-24}$ J/T
Bohr radius	$a_0 = \dfrac{\hbar^2}{m_e e^2 k_e}$	$5.291\ 772\ 083\ (19) \times 10^{-11}$ m
Boltzmann's constant	$k_B = \dfrac{R}{N_A}$	$1.380\ 650\ 3\ (24) \times 10^{-23}$ J/K
Compton wavelength	$\lambda_C = \dfrac{h}{m_e c}$	$2.426\ 310\ 215\ (18) \times 10^{-12}$ m
Coulomb constant	$k_e = \dfrac{1}{4\pi\epsilon_0}$	$8.987\ 551\ 788\ \ldots \times 10^9$ N·m²/C² (exact)
Deuteron mass	m_d	$3.343\ 583\ 09\ (26) \times 10^{-27}$ kg $2.013\ 553\ 212\ 71\ (35)$ u
Electron mass	m_e	$9.109\ 381\ 88\ (72) \times 10^{-31}$ kg $5.485\ 799\ 110\ (12) \times 10^{-4}$ u $0.510\ 998\ 902\ (21)$ MeV/c^2
Electron volt	eV	$1.602\ 176\ 462\ (63) \times 10^{-19}$ J
Elementary charge	e	$1.602\ 176\ 462\ (63) \times 10^{-19}$ C
Gas constant	R	$8.314\ 472\ (15)$ J/K·mol
Gravitational constant	G	$6.673\ (10) \times 10^{-11}$ N·m²/kg²
Josephson frequency– voltage ratio	$\dfrac{2e}{h}$	$4.835\ 978\ 98\ (19) \times 10^{14}$ Hz/V
Magnetic flux quantum	$\Phi_0 = \dfrac{h}{2e}$	$2.067\ 833\ 636\ (81) \times 10^{-15}$ T·m²
Neutron mass	m_n	$1.674\ 927\ 16\ (13) \times 10^{-27}$ kg $1.008\ 664\ 915\ 78\ (55)$ u $939.565\ 330\ (38)$ MeV/c^2
Nuclear magneton	$\mu_n = \dfrac{e\hbar}{2m_p}$	$5.050\ 783\ 17\ (20) \times 10^{-27}$ J/T
Permeability of free space	μ_0	$4\pi \times 10^{-7}$ T·m/A (exact)
Permittivity of free space	$\epsilon_0 = \dfrac{1}{\mu_0 c^2}$	$8.854\ 187\ 817\ldots \times 10^{-12}$ C²/N·m² (exact)
Planck's constant	h	$6.626\ 068\ 76\ (52) \times 10^{-34}$ J·s
	$\hbar = \dfrac{h}{2\pi}$	$1.054\ 571\ 596\ (82) \times 10^{-34}$ J·s
Proton mass	m_p	$1.672\ 621\ 58\ (13) \times 10^{-27}$ kg $1.007\ 276\ 466\ 88\ (13)$ u $938.271\ 998\ (38)$ MeV/c^2
Rydberg constant	R_H	$1.097\ 373\ 156\ 854\ 9\ (83) \times 10^7$ m⁻¹
Speed of light in vacuum	c	$2.997\ 924\ 58 \times 10^8$ m/s (exact)

[a] These constants are the values recommended in 1998 by CODATA, based on a least-squares adjustment of data from different measurements. For a more complete list, see P. J. Mohr and B. N. Taylor, *Rev. Mod. Phys.* 72:351, 2000.

[b] The numbers in parentheses for the values above represent the uncertainties of the last two digits.

Solar System Data

Body	Mass (kg)	Mean Radius (m)	Period (s)	Distance from the Sun (m)
Mercury	3.18×10^{23}	2.43×10^{6}	7.60×10^{6}	5.79×10^{10}
Venus	4.88×10^{24}	6.06×10^{6}	1.94×10^{7}	1.08×10^{11}
Earth	5.98×10^{24}	6.37×10^{6}	3.156×10^{7}	1.496×10^{11}
Mars	6.42×10^{23}	3.37×10^{6}	5.94×10^{7}	2.28×10^{11}
Jupiter	1.90×10^{27}	6.99×10^{7}	3.74×10^{8}	7.78×10^{11}
Saturn	5.68×10^{26}	5.85×10^{7}	9.35×10^{8}	1.43×10^{12}
Uranus	8.68×10^{25}	2.33×10^{7}	2.64×10^{9}	2.87×10^{12}
Neptune	1.03×10^{26}	2.21×10^{7}	5.22×10^{9}	4.50×10^{12}
Pluto	$\approx 1.4 \times 10^{22}$	$\approx 1.5 \times 10^{6}$	7.82×10^{9}	5.91×10^{12}
Moon	7.36×10^{22}	1.74×10^{6}	—	—
Sun	1.991×10^{30}	6.96×10^{8}	—	—

Physical Data Often Used[a]

Average Earth–Moon distance	3.84×10^{8} m
Average Earth–Sun distance	1.496×10^{11} m
Average radius of the Earth	6.37×10^{6} m
Density of air (20°C and 1 atm)	1.20 kg/m^3
Density of water (20°C and 1 atm)	1.00×10^{3} kg/m^3
Free-fall acceleration	9.80 m/s^2
Mass of the Earth	5.98×10^{24} kg
Mass of the Moon	7.36×10^{22} kg
Mass of the Sun	1.99×10^{30} kg
Standard atmospheric pressure	1.013×10^{5} Pa

[a] These are the values of the constants as used in the text.

Some Prefixes for Powers of Ten

Power	Prefix	Abbreviation	Power	Prefix	Abbreviation
10^{-24}	yocto	y	10^{1}	deka	da
10^{-21}	zepto	z	10^{2}	hecto	h
10^{-18}	atto	a	10^{3}	kilo	k
10^{-15}	femto	f	10^{6}	mega	M
10^{-12}	pico	p	10^{9}	giga	G
10^{-9}	nano	n	10^{12}	tera	T
10^{-6}	micro	μ	10^{15}	peta	P
10^{-3}	milli	m	10^{18}	exa	E
10^{-2}	centi	c	10^{21}	zetta	Z
10^{-1}	deci	d	10^{24}	yotta	Y

Standard Abbreviations and Symbols for Units

Symbol	Unit	Symbol	Unit
A	ampere	K	kelvin
u	atomic mass unit	kg	kilogram
atm	atmosphere	kmol	kilomole
Btu	British thermal unit	L	liter
C	coulomb	lb	pound
°C	degree Celsius	ly	lightyear
cal	calorie	m	meter
d	day	min	minute
eV	electron volt	mol	mole
°F	degree Fahrenheit	N	newton
F	farad	Pa	pascal
ft	foot	rad	radian
G	gauss	rev	revolution
g	gram	s	second
H	henry	T	tesla
h	hour	V	volt
hp	horsepower	W	watt
Hz	hertz	Wb	weber
in.	inch	yr	year
J	joule	Ω	ohm

Mathematical Symbols Used in the Text and Their Meaning

Symbol	Meaning
$=$	is equal to
$\equiv$	is defined as
$\neq$	is not equal to
$\propto$	is proportional to
$\sim$	is on the order of
$>$	is greater than
$<$	is less than
$\gg (\ll)$	is much greater (less) than
$\approx$	is approximately equal to
Δx	the change in x
$\sum\limits_{i=1}^{N} x_i$	the sum of all quantities x_i from $i = 1$ to $i = N$
$\lvert x \rvert$	the magnitude of x (always a nonnegative quantity)
$\Delta x \to 0$	Δx approaches zero
$\dfrac{dx}{dt}$	the derivative of x with respect to t
$\dfrac{\partial x}{\partial t}$	the partial derivative of x with respect to t
$\displaystyle\int$	integral

Conversions[a]

Length

1 in. = 2.54 cm (exact)

1 m = 39.37 in. = 3.281 ft

1 ft = 0.304 8 m

12 in. = 1 ft

3 ft = 1 yd

1 yd = 0.914 4 m

1 km = 0.621 mi

1 mi = 1.609 km

1 mi = 5 280 ft

$1\ \mu m = 10^{-6}\ m = 10^{3}\ nm$

$1\ \text{lightyear} = 9.461 \times 10^{15}\ m$

Area

$1\ m^2 = 10^4\ cm^2 = 10.76\ ft^2$

$1\ ft^2 = 0.092\ 9\ m^2 = 144\ in.^2$

$1\ in.^2 = 6.452\ cm^2$

Volume

$1\ m^3 = 10^6\ cm^3 = 6.102 \times 10^4\ in.^3$

$1\ ft^3 = 1\ 728\ in.^3 = 2.83 \times 10^{-2}\ m^3$

$1\ L = 1\ 000\ cm^3 = 1.057\ 6\ qt = 0.035\ 3\ ft^3$

$1\ ft^3 = 7.481\ gal = 28.32\ L = 2.832 \times 10^{-2}\ m^3$

$1\ gal = 3.786\ L = 231\ in.^3$

Mass

1 000 kg = 1 t (metric ton)

1 slug = 14.59 kg

$1\ u = 1.66 \times 10^{-27}\ kg = 931.5\ MeV/c^2$

Some Approximations Useful for Estimation Problems

1 m ≈ 1 yd

1 kg ≈ 2 lb

$1\ N \approx \frac{1}{4}\ lb$

$1\ L \approx \frac{1}{4}\ gal$

Force

1 N = 0.224 8 lb

1 lb = 4.448 N

Velocity

1 mi/h = 1.47 ft/s = 0.447 m/s = 1.61 km/h

1 m/s = 100 cm/s = 3.281 ft/s

1 mi/min = 60 mi/h = 88 ft/s

Acceleration

$1\ m/s^2 = 3.28\ ft/s^2 = 100\ cm/s^2$

$1\ ft/s^2 = 0.304\ 8\ m/s^2 = 30.48\ cm/s^2$

Pressure

$1\ bar = 10^5\ N/m^2 = 14.50\ lb/in.^2$

1 atm = 760 mm Hg = 76.0 cm Hg

$1\ atm = 14.7\ lb/in.^2 = 1.013 \times 10^5\ N/m^2$

$1\ Pa = 1\ N/m^2 = 1.45 \times 10^{-4}\ lb/in.^2$

Time

$1\ yr = 365\ days = 3.16 \times 10^7\ s$

$1\ day = 24\ h = 1.44 \times 10^3\ min = 8.64 \times 10^4\ s$

Energy

1 J = 0.738 ft · lb

1 cal = 4.186 J

$1\ Btu = 252\ cal = 1.054 \times 10^3\ J$

$1\ eV = 1.6 \times 10^{-19}\ J$

$1\ kWh = 3.60 \times 10^6\ J$

Power

1 hp = 550 ft · lb/s = 0.746 kW

1 W = 1 J/s = 0.738 ft · lb/s

1 Btu/h = 0.293 W

1 m/s ≈ 2 mi/h

$1\ yr \approx \pi \times 10^7\ s$

60 mi/h ≈ 100 ft/s

$1\ km \approx \frac{1}{2}\ mi$

[a] See Table A.1 of Appendix A for a more complete list.

The Greek Alphabet

Name	Cap	Low	Name	Cap	Low	Name	Cap	Low
Alpha	A	α	Iota	I	ι	Rho	P	ρ
Beta	B	β	Kappa	K	κ	Sigma	Σ	σ
Gamma	Γ	γ	Lambda	Λ	λ	Tau	T	τ
Delta	Δ	δ	Mu	M	μ	Upsilon	Y	υ
Epsilon	E	ϵ	Nu	N	ν	Phi	Φ	ϕ
Zeta	Z	ζ	Xi	Ξ	ξ	Chi	X	χ
Eta	H	η	Omicron	O	o	Psi	Ψ	ψ
Theta	Θ	θ	Pi	Π	π	Omega	Ω	ω